Coral Reefs of the World

Volume 23

Coral Reefs of the World is a series presenting the status of knowledge of the world's coral reefs authored by leading scientists. The volumes are organized according to political or regional oceanographic boundaries. Emphasis is put on providing authoritative overviews of biology and geology, explaining the origins and peculiarities of coral reefs in each region. The information is so organized that it is up to date and can be used as a general reference and entry-point for further study. The series will cover all recent and many of the fossil coral reefs of the world.

Prospective authors and/or editors should consult the Series Editors B.M. Riegl and R.E. Dodge for more details. Any comments or suggestions for future volumes are welcomed:

Dr. Bernhard M. Riegl/Dr. Richard E. DodgeNova Southeastern UniversityDania Beach, FL 33004USA

e-mail: rieglb@nova.edu and dodge@nova.edu

Loke Ming Chou • Danwei Huang
Editors

Coral Reefs of Singapore's Urbanised Sea

Springer

Editors
Loke Ming Chou
Department of Biological Sciences
National University of Singapore
Singapore, Singapore

Danwei Huang
Lee Kong Chian Natural History Museum
National University of Singapore
Singapore, Singapore

ISSN 2213-719X ISSN 2213-7203 (electronic)
Coral Reefs of the World
ISBN 978-981-95-0902-7 ISBN 978-981-95-0903-4 (eBook)
https://doi.org/10.1007/978-981-95-0903-4

Cover caption: A coral reef community in Singapore (left; photo by Reef Ecology Study Team), an intertidal community close to its hyper-urbanised coastline (top-right; photo by Mei Lin Neo), and coral transplants on a seawall (bottom-right; photo by Reef Ecology Study Team).

This Springer imprint is published by the registered company Springer Nature Singapore Pte Ltd.
The registered company address is: 152 Beach Road, #21-01/04 Gateway East, Singapore 189721, Singapore

Preface

Intense coastal urbanisation since the 1960s resulted in the loss of more than 65% of Singapore's coral reefs. Coastal development projects prior to the 2000s lacked sediment-management measures and transformed a once clear sea with more than 10-metre underwater visibility into a highly turbid one with less than 2-metre visibility. Sedimentation continues to affect the existing reefs by confining growth to shallow reef slopes, and degrading substrate stability and suitability to support larval recruitment. The reefs have also been subjected to increasing coral bleaching events. Despite these challenges, the reefs have persisted and continue to maintain their coral-dominated integrity, partly owing to the ecological history and local adaptation over the Holocene. The annual mass spawning is generally consistent, and larvae have been modelled to circulate between source and sink reefs within the country's reef system. Enhanced management of the marine environment since the mid-2000s includes measures to contain sediment spread from development projects and relocating corals away from affected sites. Reef restoration efforts have been initiated to enhance resilience of reef habitats against further impacts. Scientific capacity has increased with many more researchers investigating Singapore's coral reef system to understand how it is responding to changing environmental conditions caused by urbanisation and climate change. This will strengthen the scientific basis for effective management responses and demonstrate how reef integrity can be conserved under changing environmental conditions driven by urbanisation and climate change impacts.

In realising this book, the editors had the enthusiastic support of 43 contributors, all of whom have deep knowledge of Singapore's coral reefs through their roles as researcher, manager or community stakeholder. They all share a common vision of strengthening reef resilience against coastal urbanisation and climate change pressures. We thank them for advancing the science that can be applied to more effective management of this natural heritage. The editors wish to thank Nandhini Natchatra, Production Editor, and her team at Springer for seeing through the book's entire publication process.

Singapore, Singapore

Loke Ming Chou
Danwei Huang

Contents

1 **Coral Reefs of Singapore** . . . 1
Loke Ming Chou and Danwei Huang

2 **Reefs in Singapore's Murky Waters** . . . 7
Ai Chin Lee and Seng Keat Ooi

3 **Impacts of Sedimentation on Coral Communities** . . . 35
Stacia Loong, Min Choo, Yong Kit Samuel Chan, and Danwei Huang

4 **Energetic Needs of Corals Living in Singapore's Extreme Reef Environment** . . . 53
Jani Thuaibah Isa Tanzil, Wei Long Ow Yong, Yan Xiang Ow, Wai Leong, Chin Soon Lionel Ng, Jonas Chee Haw Khaw, and Karenne Tun

5 **Coral Recruitment Potential on Singapore's Urban Reefs** . . . 67
Kareen Vicentuan, Jani Thuaibah Isa Tanzil, Iris Baula, Pei Ying Liew, and Ai Chin Lee

6 **Diversity and Assemblages of Symbiodiniaceae Dinoflagellates in Singapore's Coral Reefs: A Case Study on Giant Clams (*Tridacna squamosa* and *T. crocea*)** . . . 87
H. R. Samuel Lee, Teresa S. Tay, Yuchen Sun, Maxine A. D. Mowe, Jing Hui Ong, Benjamin J. Wainwright, Li Keat Lee, and Mei Lin Neo

7 **Microbiomes Associated with Benthic Marine Macroorganisms in Singapore** . . . 107
Lik Tong Tan, Beverly Pi Lee Goh, Deepak Kumar Gupta, Nursheena Parveen Katermeran, and Lindsey Kane Deignan

8 **Coral Reef Fishes Thriving in the Urban Reef System of Singapore** . . . 137
Daisuke Taira and Zeehan Jaafar

9 **Long-Term Trends of Coral Biodiversity in Singapore** . . . 155
Denise A. McIntyre, Aubane L. M. Carré de Malberg, Myra Y. Y. Khoo, James Philip D. V. Cabatbat, Z. K. Medric Tan, J. J. Marc Chang, and Danwei Huang

10 **Conservation Status of Corals in Singapore** . . . 173
Chin Soon Lionel Ng, Yong Kit Samuel Chan, Rosa Celia Poquita-Du, Karenne Tun, and Danwei Huang

11 **Biological Connectivity and Conservation of Singapore's Urbanised Seascape** . . . 185
Jing Hui Ong, Benjamin J. Wainwright, and Danwei Huang

12 **Community-Based Conservation of Coral Reefs in Singapore** . . . 197
Tai Chong Toh, Hui Hui Tham, Rui Xuan Veralyn Toh, Shu Qin Sam, Rachel Yuen Yee Mark, and Chin Soon Lionel Ng

13 **Integrated Urban Coastal Management in Singapore: Balancing Sustainable Use with Conservation** . . . 209
Karenne Tun and Hor Yee Collin Tong

About the Contributors

Iris Baula is a Research Associate at the National University of Singapore (Tropical Marine Science Institute) who has been taking underwater photos of permanent quadrats in various reefs in Singapore since 2009 to ascertain factors that contribute to the extent by which the reefs are changing. Iris received her education and training at the University of the City of Manila and University of the Philippines (Marine Science Institute), mainly on the dinoflagellate cyst aspect of harmful algal blooms before being given the opportunity in NUS to work on coral reef ecology assessments.

James Philip D. V. Cabatbat is currently an undergraduate student studying Life Sciences at the National University of Singapore. He is also undertaking a minor in Aquatic Ecology. In the course of his studies, he has been exploring aquatic ecology with a specific focus on the marine realm. He has volunteered actively at the Reef Ecology Laboratory and Lee Kong Chian Natural History Museum, primarily working with its coral collections in addition to participating in fieldwork.

Aubane L. M. Carré De Malberg is a Master of Research graduate from the University College of London. She has worked at the Natural History Museum of London (Department of Herpetology) and presented her work at the World Congress of Herpetology. She has been engaged with coral reef research at the Reef Ecology Laboratory, National University of Singapore, and contributed actively to the archival of the SH Chuang collection at the Lee Kong Chian Natural History Museum.

Yong Kit Samuel Chan is currently based at the Australian Institute of Marine Science in Townsville, Australia, with the ReefCloud team. Before that, he earned his PhD from the National University of Singapore in 2023. His current research focuses on developing tools and indicators in ReefCloud to help coral reef monitoring be more efficient and effective. His research interests revolve around studying reefs using interdisciplinary approaches and understanding their variation through spatial and temporal scales.

J. J. Marc Chang is a Research Fellow at the Lee Kong Chian Natural History Museum, National University of Singapore. He specialises in DNA sequencing methodologies and applies these techniques to characterise marine biodiversity through DNA (meta)barcoding as well as environmental DNA. More recently, he has focused on building up DNA sequence databases to improve the effectiveness of DNA-based identification resources for tropical marine fauna.

Min Choo is an Environmental Consultant at DHI Singapore, specializing in Environmental Monitoring and Management Plans (EMMPs) and Environmental Impact Assessments (EIAs). Prior to her current role, she had carried out coral reef research based on patterns of sediment accumulation on reef corals and palaeoecological coral assemblage records.

Loke Ming Chou (Editor) is Emeritus Professor at the National University of Singapore (Department of Biological Sciences). He has been actively engaged with coral reef research for over four decades. He is a Fellow of the Singapore National Academy of Science and an Honorary Fellow of the Singapore Institute of Biology. He received the "ASEAN Biodiversity Hero Award" from the ASEAN Centre for Biodiversity in 2022 for his contribution to coral reef science.

Lindsey Kane Deignan is a marine biologist and research consultant at Nanyang Technological University's Singapore Centre for Environmental Life Sciences Engineering (SCELSE). She earned her PhD in Marine Biology from the University of North Carolina Wilmington in 2017. Her research focuses on coral reef microbiology, studying the microbiomes of coral and algae to understand reef resilience in Singapore. She has extensive field and lab experience, including work as an "aquanaut" at the Aquarius Reef Base in Florida. Passionate about marine science, she enjoys water sports, diving and exploring new environments while contributing to marine conservation and research.

Beverly Pi Lee Goh is a marine ecotoxicologist with an interest in studying the effects of environmental change from anthropogenic sources on marine organisms and ecosystems. She is a Senior Lecturer in the Natural Sciences and Science Education Department of the National Institute of Education, Nanyang Technological University, and a Fellow of the Singapore Institute of Biology. Her recent research efforts focus on facilitating citizen science through using marine organisms to monitor marine environmental health. The biomarker techniques developed form her core research that aims to identify signals from nature to serve as important early warning signs of environmental stress.

Deepak Kumar Gupta is an Officer involved in bridging innovation and regulatory compliance in healthcare at the Indian Council of Medical Research (ICMR), India. With expertise in pharmaceutics, medical devices and cancer research, he has contributed to multiple scientific publications. Previously, he was a Research Associate at the Bioinformatics Institute (BII), under the Agency for Science, Technology and Research (A*STAR) in Singapore, and a Research Assistant at Nanyang Technological University (NTU), Singapore. He holds a PhD from NTU's National Institute of Education, Singapore. His research focuses on medical devices, drug formulation, medicinal chemistry and bioactive natural products.

Danwei Huang (Editor) is the Deputy Head of the Lee Kong Chian Natural History Museum and Associate Professor at the Department of Biological Sciences, National University of Singapore. He is also a Research Affiliate of the Tropical Marine Science Institute and Centre for Nature-based Climate Solutions. His research integrates the fields of ecology, evolutionary biology and conservation, using the coral reef as a model system to address questions on biodiversity. His team aims to understand the origins, maintenance and loss of marine biodiversity at the local, regional and global scales to support coral reef restoration and integrated coastal management.

Zeehan Jaafar is a marine biologist with research interests in the evolution and ecology of marine fishes, and conservation of their habitats. She has led and participated in many international and regional expeditions to enhance understanding of the diversity and distribution patterns of fish fauna. She is the author of several books focusing on fishes and the marine environments such as *Fishes Out of Water: Biology and Ecology of Mudskippers* and *Endangered Forested Wetlands of Sundaland—Ecology, Connectivity, Conservation.* A keen diver, she enjoys sharing her passion for Singapore's marine areas through public engagement and education.

Nursheena Parveen Katermeran holds a Master in Environmental Science from Nanyang Technological University, Singapore, and is passionate about marine ecosystems. Her research includes studying bioindicators of ocean warming and acidification, as well as bacterial communities in marine sponges, contributing to a deeper understanding of marine biodiversity. With a strong interest in marine conservation and ecosystem dynamics, she is also committed to promoting environmental awareness and sustainable practices.

Jonas Chee Haw Khaw is a Research Assistant with the Tropical Marine Science Institute at the National University of Singapore. He has been actively involved with coral reef restoration research and engaged with citizen scientist training to build capacity for coral reef monitoring efforts in Singapore over the past three years.

Myra Y. Y. Khoo is an undergraduate student majoring in Life Sciences (specialising in Ecology, Evolution, and Biodiversity) with a minor in Aquatic Ecology at the National University of Singapore. She has contributed to cataloguing the SH Chuang collection of corals at the Lee Kong Chian Natural History Museum. Deeply fascinated by the interconnectedness of marine ecosystems, her particular research interests lie in coral reef macroecology and restoration.

Ai Chin Lee is a marine biologist by training and a Research Associate at the Tropical Marine Science Institute, National University of Singapore. She has been actively involved in marine research in Singapore for the past 25 years. Her current research focus revolves around the movement and behaviour of suspended particles around coral reefs and the implications of suspended sediment dynamics on coral reef health and resilience.

H. R. Samuel Lee is a marine ecologist with an interest in mariculture and the adaptive potential of populations. He earned his BSc in Life Sciences with First Class Honours from the National University of Singapore, where he received the Holttum Prize. His research includes invertebrate larval biology, marine microplastics and the dynamics of the giant clam holobiont (specifically with Symbiodiniaceae). Currently, he investigates nutrient dynamics and physiological responses of tropical seagrasses to environmental stressors at the Tropical Marine Science Institute. He seeks to understand how organisms thrive amid Anthropocene challenges and advance efforts managing threatened marine species and habitats.

Li Keat Lee is a marine biologist specialising in molecular ecology, coral reef resilience and symbiotic interactions in marine invertebrates. He earned his PhD from Universiti Malaya, focusing on the genetics and population structure of endangered giant clams. His research includes harmful algal blooms, coral-giant clam holobionts (particularly Symbiodiniaceae) and coral resilience. Currently, he is a postdoctoral researcher at the Tropical Marine Science Institute, National University of Singapore, where he studies the ecological resilience of coral reefs in Singapore. When not diving into data or reefs, he enjoys pour-over coffee, reading books and binge-watching movies.

Wai Leong is a Research Associate with the Tropical Marine Science Institute at the National University of Singapore. Her research interests include coral reef ecology, life history trade-offs, and her current research focuses on enhancing thermal tolerance in reef corals for restoration.

Pei Ying Liew is a Laboratory Technologist at the National University of Singapore's Tropical Marine Science Institute. She graduated from Murdoch University in Western Australia, where she gained hands-on experience working with diverse flora and fauna across distinct ecosystems. Passionate and dedicated about marine conservation, she focuses on monitoring the health of Singapore's southern coral reefs. Her work encompasses in situ reef assessments, water quality analysis, and the bi-annual survey of coral spawning on tiles. Through her work, she contributes to the ongoing preservation and sustainability of these vital marine ecosystems.

Stacia Loong is a Research Associate at the Reef Ecology Laboratory, National University of Singapore. She earned her master's and bachelor's degrees from NUS, focusing on coral reef ecology, conservation management and aquaculture impacts. She has investigated spatial and temporal variations of different reef taxa in Singapore's urban reefs and conducted SCUBA-based benthic assessments. Her research interests lie in integrating ecological monitoring, stakeholder engagement, and modeling to understand reef community dynamics and enhance reef resilience.

Rachel Yuen Yee Mark is a Manager at the coastal and marine branch of National Parks Board, Singapore. She has an interest in marine conservation, coral reef restoration and reef fish ecology. Her current research focuses on upscaling coral propagation and the use of novel methods for coral restoration.

Denise A. Mcintyre is a graduate of the National University of Singapore, with a major in Life Sciences with a specialisation in Ecology, Evolution and Biodiversity, and a minor in Aquatic Ecology. Her research examines long-term trends in coral biodiversity in Singapore. Prior to her undergraduate studies, she worked as a SCUBA diving instructor and managed volunteer marine research projects. She is particularly interested in coral reef restoration, with a focus on community-based conservation in coastal areas.

Maxine A. D. Mowe is a Biology Lecturer at the National University of Singapore in the Department of Biological Sciences and has more than 14 years of research experience in aquatic biology specifically in phytoplankton and zooplankton ecology and has been a lecturer since 2019. She has worked on cyanobacterial ecology (*Microcystis* and *Raphidiopsis*), phytoplankton and zooplankton interactions, microplastic impacts and eutrophication and algal bloom mitigation using aquatic plants in Singapore's reservoirs and coastal environment. She teaches various biology and ecology courses for undergraduate students and graduate courses for the MSc in Biodiversity Conservation and Sustainability at the National University of Singapore.

Mei Lin Neo is a Senior Research Fellow at the Tropical Marine Science Institute, National University of Singapore. Her research interests broadly cover marine invertebrate ecology and conservation. In particular, she has been actively engaged with giant clam research, conservation and management in Southeast Asia region. She is also passionate about science communication, environmental advocacy and motherhood in science. She is the first Singaporean recipient of the Pew Charitable Trust's Pew Fellows Program in Marine Conservation in 2021.

Chin Soon Lionel Ng is a Research Fellow with the Tropical Marine Science Institute at the National University of Singapore. He has broad interests in coral reef ecology, habitat restoration and environmental management. His research focuses on prioritising species and developing innovative strategies to accelerate the recovery of structure and function in degraded coral reefs and urbanised marine environments.

Jing Hui Ong is a Research Fellow with the Tropical Marine Science Institute at the National University of Singapore. She has a keen interest in marine connectivity, especially in how it contributes to population resilience and maintains biodiversity. Her research currently focuses on elucidating patterns of genomic connectivity across the seascape of Singapore and translating them into actionable conservation solutions such as the design of marine protected areas.

Seng Keat Ooi is a Senior Lecturer at the Department of Civil and Environmental Engineering and Head of the Ecological Monitoring, Informatics and Dynamics group at Tropical Marine Science Institute, National University of Singapore. His research interests are in the issues associated with tropical environmental flows, particularly inputs and interactions between the biota and the environment and how to model them. The implications of these interests lie in the domain of the potential impact in light of the changing climate and rapidly changing built environment to water resources issues especially flows, water quality and ecology in the tropics.

Yan Xiang Ow is a Senior Research Fellow with the Tropical Marine Science Institute at the National University of Singapore. She is also one of the Resident Scientists based at the St. John's Island National Marine Laboratory, funded by the National Research Foundation, Singapore. She has been involved in corals and seagrass research for over 15 years. Her current research interests explore climate change impacts on seagrass ecology and the feasibility of nature-based solutions in urbanised coastal zones.

Wei Long Ow Yong is a Research Assistant with the Tropical Marine Science Institute at the National University of Singapore. He actively pursues hard coral research and husbandry at the St. John's Island National Marine Laboratory. His research interests involve the propagation of hard corals and restoring degraded reef areas through active transplantation.

Rosa Celia Poquita-Du is a Postdoctoral Researcher at the Senckenberg Biodiversity and Climate Research Center in Frankfurt am Main, Germany. She earned her PhD (Ecology, Evolution and Biodiversity) from the National University of Singapore in 2019. Her research explores the complexity of obligate symbiotic systems, such as corals and lichens, with a focus on their acclimatisation and adaptation mechanisms in response to environmental changes.

Shu Qin Sam is an Instructor with the NUS College at the National University of Singapore. She has a broad interest in coral reef ecology, coral reef restoration and science communication. Her current research focuses on enhancing biodiversity on artificial structures and applying interdisciplinary approaches to marine conservation with coastal communities.

Yuchen Sun is a Research Assistant at St. John's Island National Marine Laboratory, specialising in marine ecology and mariculture. A 2022 Marine Biology graduate from the University of Melbourne, Sun has researched marine habitats as sinks for macro-litter and the captive breeding of the tiger cowrie. Currently, Sun leads a seagrass project exploring eutrophic thresholds and their effects on seagrass species, enhancing understanding of nutrient impacts on these ecosystems. An avid diver and macrophotographer, Sun captures the delicate balance of marine life, showcasing its intricate beauty through underwater photography.

Daisuke Taira is a Postdoctoral Research Fellow at the Department of Biological Sciences, National University of Singapore. His research focuses on reef fish diversity, coral reef restoration, urban marine ecology and nature-based climate change adaptation. He is particularly interested in understanding the ecological impacts of coastal urbanisation on fish diversity and their interactions with urban seascapes. Daisuke also explores how to incorporate ecological considerations into waterfront urban developments to promote sustainable and resilient coastal ecosystems.

Lik Tong Tan is a Principal Lecturer in Natural Sciences and Science Education at the National Institute of Education, Nanyang Technological University, Singapore. He holds a PhD in Pharmaceutical Sciences from Oregon State University and has extensive expertise in marine natural products, chemical ecology and microbial diversity. Dr Tan has been recognised among the top 2% scientists by Stanford University in Medicinal and Biomolecular Chemistry and has received multiple teaching and research accolades. A Fellow of the Singapore Institute of Biology, he actively mentors students and contributes to marine science research and education initiatives in Singapore.

Z. K. Medric Tan is a Life Sciences and Philosophy Undergraduate at the National University of Singapore. He is interested in marine biodiversity and coral reef research and is an intern at the Lee Kong Chian Natural History Museum and Reef Ecology Laboratory.

Jani Thuaibah Isa Tanzil is the Facility Director at the St. John's Island National Marine Laboratory and a Senior Research Fellow at the Tropical Marine Science Institute, National University of Singapore. Her research on marine ecosystem dynamics and coral resilience aims to develop better strategies for environmental assessments, resource management and ecosystem recovery. She leads several projects including reef restoration and resilience projects to rejuvenate Singapore's degraded and light-limited reefs, as well as the Marine Environment Sensing Network project which aims to expand Singapore's marine monitoring network for research and education and to facilitate collaborations for our shared seas.

Teresa S. Tay is a Research Assistant at the Tropical Marine Science Institute, National University of Singapore. Her research focuses primarily on marine invertebrate ecology, with an emphasis on larval culture. She is particularly interested in understanding species' life cycle and their interactions with the environment. Over the years, she has worked on a diverse range of marine organisms, including mussels, sea urchins, cowrie snails and giant clams, and is always keen to expand her expertise by culturing species from different taxonomic groups.

Hui Hui Tham is a Senior Manager (Philanthropy) at The Majurity Trust. In both her professional and personal capacity, she engages with social and environmental community groups to design strategies, measure outcomes and foster partnerships. She graduated from the National University of Singapore in 2020 with a Bachelor of Environmental Studies (Environmental Biology) and was part of the University Town College Programme at the College of Alice and Peter Tan. As a student, she conducted research on local marine groups in Singapore and studied the effects of ocean acidification on sediment dissolution.

Tai Chong Toh is a Senior Lecturer at the National University of Singapore. Trained as a marine biologist, his research focuses on urban marine ecosystems and coral reef restoration. He co-founded two non-profits, Our Singapore Reefs and Our Blue Spaces, to advance marine conservation. As an educator, he promotes environmental education as an executive committee member of the Singapore National Academy of Sciences and the Singapore Institute of Biology. He also advises Ocean Geographic, Rumah Foundation and a public school in Singapore. In 2019, he received the "ASEAN Youth Eco-Champion Award" for his contribution to ocean sustainability.

Rui Xuan Veralyn Toh is a student at the Methodist Girls' School. She has an interest in community-based work and finds joy in working with the elderly and children. She was exposed to marine environmental issues since young and has developed an interest in marine biodiversity and conservation.

Collin Tong is the Deputy Director at the National Biodiversity Centre, National Parks Board (NParks). He manages the Sisters' Islands Marine Park, the Marine Park Education and Outreach Centre on St. John's Island, the Turtle Hatchery on Small Sister's Island and various outreach and education programmes to promote better understanding and interaction with marine biodiversity in Singapore. His passion for the marine environment started since young, and he is committed to help in the conservation and enhancement efforts of marine habitats for the enjoyment of future generations.

Karenne Tun is the Group Director of the National Biodiversity Centre and the Director of the Coastal and Marine Branch, National Parks Board (NParks). She earned her PhD from the National University of Singapore in 2012. Her current work at NParks covers issues related to the management and conservation of Singapore's natural areas across the terrestrial and coastal and marine environments, and the biodiversity they support. She currently represents Singapore as one of 25 members of Group of Experts for the United Nations Regular Process for the 3rd Cycle tasked to prepare the 3rd World Ocean Assessment (WOA3), which will be published by the end of 2025.

Kareen Vicentuan is a marine biologist and a Research Associate at the Tropical Marine Science Institute, National University of Singapore. She earned her Master's degree from the Marine Science Institute at the University of the Philippines Diliman, where she focused on coral culture, transplantation and the conservation of giant clams. Her research has contributed to reef restoration projects and the restocking of threatened marine species in the Philippines. Her current work in Singapore involves coral reef monitoring and conservation, studying how coral ecosystems respond to environmental changes and human impact.

Benjamin J. Wainwright is research is primarily motivated by the conservation of biodiversity and understanding the mechanisms that create and maintain it. He has a particular interest in the marine regions of Southeast Asia, especially the Coral Triangle.

1 Coral Reefs of Singapore

Loke Ming Chou and Danwei Huang

Abstract

Extensive coastal development since the 1960s led to the loss of 65% of reef habitat area in Singapore. The remaining reefs continue to face challenges from sedimentation and frequent incidents of elevated sea surface temperature. Persisting through these harsh environmental conditions, these habitats provide an invaluable opportunity to study their responses. Active management of marine ecosystems such as coral reefs was non-existent before the mid-2000s, after which proper institutional arrangements were established. This helped to consolidate support for coral reef research, which in turn led to more effective evidence-based management interventions. As Singapore looks towards balancing conservation and development, its coral reefs provide an important role functioning as a living laboratory. This series of reviews and syntheses of published work, long-term datasets and unpublished data is expected to provide insights into the responses and potential trajectories of urbanised reefs.

Keywords

Habitat loss · Coastal urbanisation impact · Climate change · Reef recovery and persistence · Reef management · Ecosystem restoration · Living laboratory · Research and management capacity

1.1 Introduction

Singapore's coral reefs have been exposed to strong anthropogenic pressure since the 1960s and repeated incidents of elevated sea surface temperature, which was first recorded in 1998 (Chou 2000). Land reclamation expanded Singapore's size by more than 25% (Chou 2023) but eliminated 65% of the reef habitat area by the early 1990s (Hilton and Manning 1995). Heavy sedimentation resulting from the reclamation reduced light penetration to the detriment of corals at deeper depths (Guest et al. 2016; Chow et al. 2019). The reefs now cover 835 ha of the intertidal zone and 102 ha of the subtidal zone (Nguyen et al. 2022; Tan et al. 2023) (Fig. 1.1). However, human-engineered structures of the urban seascape like seawalls have provided novel habitats for colonisation by corals and other flora and fauna, with diverse communities establishing over time (Ng et al. 2012; Kikuzawa et al. 2020; Lee et al. 2023). Indeed, it has been noted recently that whilst natural reef decline has slowed since the 1990s, colonisation of seawalls led to a net increase of coral habitats from 2012 (Tan et al. 2025).

No government agency was responsible for marine biodiversity conservation before the mid-2000s, and large-scale land reclamation from the 1960s to mid-2000s proceeded without any significant environmental impact assessment (Hesp 1995). The establishment in 2003 of the National Biodiversity Centre under the National Parks Board (NParks) identified for the first time an agency responsible for marine conservation (Chou, 2023). Effective management measures implemented since the mid-2000s have potentially helped

L. M. Chou (✉)
Department of Biological Sciences, National University of Singapore, Singapore, Singapore

Tropical Marine Science Institute, National University of Singapore, Singapore, Singapore
e-mail: tmsclm@nus.edu.sg

D. Huang
Department of Biological Sciences, National University of Singapore, Singapore, Singapore

Tropical Marine Science Institute, National University of Singapore, Singapore, Singapore

Lee Kong Chian Natural History Museum, National University of Singapore, Singapore, Singapore

Centre for Nature-based Climate Solutions, National University of Singapore, Singapore, Singapore

L. M. Chou, D. Huang (eds.), *Coral Reefs of Singapore's Urbanised Sea*, Coral Reefs of the World 23,
https://doi.org/10.1007/978-981-95-0903-4_1

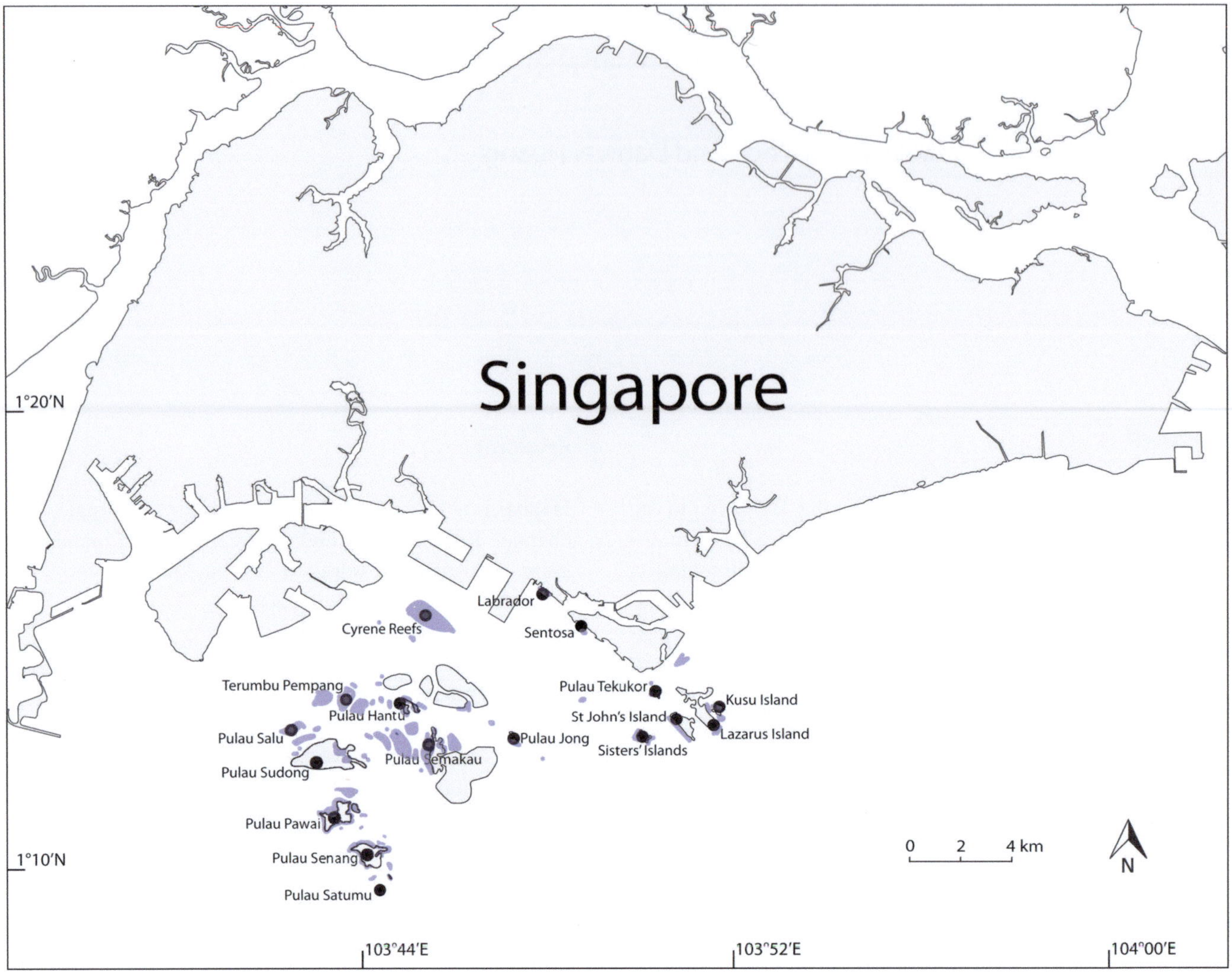

Fig. 1.1 Map of Singapore showing the coral reefs (in blue) fringing its southern shores and islands

alleviate the impacts from new coastal urbanisation projects and facilitate reef recovery (Chng et al. 2022; Chan et al. 2024).

Despite enduring decades of urbanisation impacts and the increasing frequency of marine heatwave pulses without adequate conservation protection, the reefs have managed to retain biodiversity richness and ecological integrity (Wong et al. 2018; Chou et al. 2019; Ip et al. 2023). Renewed hope for the persistence of these biodiverse habitats is now brighter with the stronger management and protection regime (Jaafar et al. 2018), but it remains critically important to understand the processes that influence reef response to impacts and to identify and apply appropriate management interventions for sustaining reef health (Chou and Huang 2020). These reefs play the crucial role of a living laboratory, and the present situation with increased research and management capacities together with raised public awareness and active community participation is appropriate to fully optimise this learning opportunity (Ng et al. 2023; Ying et al. 2025).

Over the last three decades, studies on Singapore's reefs have made significant contributions to coral reef science. They include the discovery of bioactive compounds in reef organisms (Goh et al. 1995; Fung et al. 1997; Goh and Chou 1998; Ding et al. 1999); revelation that synchronous mass coral spawning occurs near the equator (Guest et al. 2002; Guest et al. 2008a; Guest et al. 2008b); and characterisation of morphological plasticity in corals (Todd et al. 2001; Todd et al. 2002a; Todd et al. 2002b; Todd et al. 2004). Detailed studies on corals have also established that coral larval replenishment is maintained by strong connectivity amongst reefs in Singapore (Tay et al. 2012, 2015); adaptation could account for differential species response to bleaching (Guest et al. 2012); and exposure to thermal and sediment stress can drive changes in coral holobiont gene expression (Poquita-Du et al. 2019, 2020b; Bollati et al. 2022), photo-physiology

(Poquita-Du et al. 2019, 2020a), and endosymbiont community composition (Jain et al. 2020).

Research is also helping to provide the scientific basis for improved management. For example, molecular characterisation of corals has provided recommendations on the number of colonies required to maximise genetic diversity in coral transplantation efforts (Afiq-Rosli et al. 2019). Genomic-wide studies have led to the recognition and conservation prioritisation of reef sites which are important larval sources or are highly self-recruiting (Afiq-Rosli et al. 2021; Feldman et al. 2022).

The following chapters of this book contribute further to the science and management of Singapore's coral reef ecosystem by providing deeper insights into the biodiversity of reefs and their responses to the changing environment. They are based on syntheses of published information as well as analyses of long-term datasets and unpublished data.

1.2 Science

Singapore's coral reef habitats remained relatively unexploited throughout the early years following its founding in 1819. Early naturalists focused on reef species richness, cataloguing and depositing numerous specimens in British and European museums until the establishment of the Raffles Library and Museum in 1874, which then became a repository of local specimens. The specimen collection, officially known as the Zoological Reference Collection, is now secured and housed by the Lee Kong Chian Natural History Museum (previously the Raffles Museum of Biodiversity Research). Research expanded beyond documenting what species can be found on coral reefs to which biological processes are crucial for retaining the habitat's existence in a changing environment, and what considerations are important for reef restoration (Loong et al. 2025). This shift became apparent in the 1970s soon after large-scale land reclamation started with one of the earliest documentations on reef ecological patterns provided by Chuang (1977). Since then, research expanded to cover various aspects of reef response to sedimentation impacts and global warming, as well as reef conservation and restoration.

A wide range of coral reef science topics have been investigated, and some of the latest findings contained in this book are as follows:

Chapter 2: Hydrodynamic patterns influence sediment transport and distribution over the reefs. Elevated suspended sediment is sustained by its high horizontal mobility driven by tidal flow and resuspension from island wakes. Coastal development in local waters is the main source of sediment.

Chapter 3: High sedimentation induces coral compositional shifts by affecting species and growth forms differently. Some species and growth forms (particularly laminar, encrusting and massive) accumulate sediment and appear to be tolerant of high sediment loads. Clearly, Singapore's corals live in an extreme reef environment characterised by high sedimentation rate, turbidity and nutrient level, and low light.

Chapter 4: Corals in the shallow reef zone (2–3 m) have positive energy balances, whilst those at the deeper zone (6–7 m) have been sufficient for basic metabolism but not growth or reproduction. Consequently, Singapore's coral growth zone is compressed to depths generally less than 7 m where growth rates and diversity remain high.

Chapter 5: A spatiotemporal analysis of coral recruits over 15 years showed that the composition of adult coral population, other benthic species and environmental conditions on the reef can influence coral recruitment rate. The study also revealed a higher mean recruitment rate of coral spat than what had been reported previously. However, high post-recruitment mortality is observed and attributed to suboptimal environmental conditions including ocean warming.

Chapter 6: The role of Symbiodiniaceae endosymbionts in coral reef resilience is well established, and a first documentation of Symbiodiniaceae diversity in Singapore's two native giant clams, *Tridacna crocea* and *T. squamosa*, showed that they primarily hosted *Durusdinium* and *Cladocopium*, similar to other reef species including hard corals.

Chapter 7: Marine microbiomes associated with reef benthic macroorganisms are a promising source of novel bioactive compounds with medicinal and industrial applications. They are also important for maintaining ecosystem health, and it should be possible to exploit microbes from resilient hard corals to increase the resilience of susceptible corals.

Chapter 8: Singapore's coral reefs continue to support high fish diversity, although current reef fish assemblage patterns are very much shaped by decades of subsistence fishing, coastal development and climate change impacts.

Chapter 9: Long-term trends of coral biodiversity at two reefs from the 1950s to the 2020s revealed a temporal increase in phylogenetic and species richness accompanied by a decline in functional richness that levelled off in the 2010s. This suggests that corals in Singapore's reefs have become more functionally similar, especially for sites in close proximity to coastal urbanisation.

Chapter 10: Reef survey data, museum records and the literature reviews supported the conservation status assessments of over half (142) of the 276 stony coral species recorded in Singapore. One species, *Seriatopora hystrix*, was assessed to be Nationally Extinct, whilst another, *Stylophora pistillata*, was categorised as Critically Endangered. Another seven and 20 species were assessed to be Endangered and Vulnerable, respectively.

Chapter 11: Genetic connectivity patterns across Singapore's reef habitat network are influenced by hydrodynamic forcing, especially for species with a planktonic life stage. Genetic connectivity considerations in conservation planning should contribute towards enhancing resilience efforts.

1.3 Management

Marine biodiversity conservation received almost no attention before the establishment of the National Biodiversity Centre in 2003. Reef habitat loss and degradation were widespread over the preceding decades of large-scale reclamation. Concerns raised by nature enthusiasts were accompanied by volunteer-driven projects to raise awareness. Over 150 sport divers participated in surveys to assess reef health in 1988 (Chou 1991), marking the involvement of citizen science. Another large community project translocated corals from reef sites designated for reclamation to 'safer' locations (Chou and Tun 1997).

Shortly after the establishment of the National Biodiversity Centre at the turn of the century, effective management efforts were implemented to protect marine biodiversity. These included impact assessments, real-time monitoring of development projects to contain sediment spread, mitigation to curb biodiversity declines and habitat restoration to replenish coral abundance (Chou et al. 2019). A major government-led effort was the establishment of the nation's first marine park, Sisters' Islands Marine Park, in 2014 (Jaafar et al. 2018; Ip et al. 2019). The 'Friends of Marine Park' was formed to encourage stakeholders to take ownership of plans to promote responsible use of the Marine Park. Such distinct partnership arrangements with engaged communities set the stage for the implementation of more effective conservation projects, such as the deployment of purpose-built, high-relief artificial reef structures (Chou et al. 2023).

Community efforts were no longer marginalised but taken into the conservation mainstream, including the establishment of the 'Singapore Blue Plan 2018' (Jaafar et al. 2018). This marine conservation proposal was developed by the stakeholder community and presented to the Singapore Government, leveraging capacity built over the previous decade through broad-scale research activities such as the five-year Comprehensive Marine Biodiversity Survey involving numerous scientists and over 500 volunteers (Tan et al. 2015, 2016).

As highlighted in Chap. 12, the present collaborative arrangement has encouraged wider community involvement and increased management efficiency. Partnerships between government, academia and the wider community increase connectivity between different stakeholder networks and facilitate consolidation of resources for quicker and more effective response to crisis situations such as oil spills and marine heatwaves.

Last but not least, Chap. 13 traces the development of marine conservation and management, which effectively began in the mid-2000s with the formation of more clearly defined institutional arrangements. NParks, the agency responsible for biodiversity conservation and management, acknowledges the significance of a science-based approach and the importance of involving the communities. Conservation and management policies have been strengthened further by the adoption of the Integrated Urban Coastal Management (IUCM) framework that seeks a better balance between conservation and development.

1.4 Conclusion

Singapore's coral reef habitats serve as an important living laboratory (Fig. 1.1). Despite substantial habitat loss, they have persisted through six decades of heavy sedimentation and four global-scale marine heatwaves. Understanding the processes that help urbanised coral reefs persist will be crucial for improving management effectiveness.

Research revealing responses at molecular, cellular, organismal and community levels can provide fundamental understanding of how reefs respond to decades of chronic urbanisation impacts and episodic marine heatwave events. It also lays out the environmental and palaeoecological contexts of long-term persistence of marginal reefs. Knowing what drives the persistence of coral-dominated reefs amidst urbanisation and climate change impacts will contribute to formulation of effective conservation strategies. Singapore's reef conservation efforts demonstrate the significance and exemplifies the benefits of protecting coral reef habitats in tropical coastal cities.

References

Afiq-Rosli L, Huang D, Toh TC, Taira D, Ng CSL, Song T, Chou LM (2019) Maximising genetic diversity during coral transplantation from a highly impacted source reef. Conserv Genet 20:629–637. https://doi.org/10.1007/s10592-019-01164-6

Afiq-Rosli L, Wainwright BJ, Gajanur AR, Lee AC, Ooi SK, Chou LM, Huang D (2021) Barriers and corridors of gene flow in an urbanized tropical reef system. Evol Appl 14:2502–2515. https://doi.org/10.1111/eva.13276

Bollati E, Rosenberg Y, Simon-Blecher N, Tamir R, Levy O, Huang D (2022) Untangling the molecular basis of coral response to sedimentation. Mol Ecol 31:884–901. https://doi.org/10.1111/mec.16263

Chan YKS, Ng CSL, Tun KPP, Chou LM, Huang D (2024) Decadal decline of functional diversity despite increasing taxonomic and phylogenetic diversity of coral reefs under chronic urbanisation stress. Ecol Indic 164:112143. https://doi.org/10.1016/j.ecolind.2024.112143
Chng LC, Chou LM, Huang D (2022) Environmental performance indicators for the urban coastal environment of Singapore. Reg Stud Mar Sci 49:102101. https://doi.org/10.1016/j.rsma.2021.102101
Chou LM (1991) A reef conservation project involving sport divers in Singapore. In: Magoon OT, Converse H, Tippie V, Tobin LT, Clark D (eds) Coastal Zone '91: Proceedings of the Seventh Symposium on Coastal and Ocean Management, 8–12 July 1991. American Society of Engineers, New York, pp 1990–1994
Chou LM (2000) Southeast Asian reefs – status update: Cambodia, Indonesia, Malaysia, Philippines, Singapore, Thailand and Vietnam. In: Wilkinson C (ed) Status of Coral Reefs of the World: 2000. Australian Institute of Marine Science, Townsville, pp 117–129
Chou LM (2023) Sustaining marine biodiversity in Singapore's heavily urbanized coast. In: Ting DS-K, O'Brien PG (eds) Progress in sustainable development: sustainable engineering practices. Elsevier, Amsterdam, pp 265–281
Chou LM, Huang D (2020) Towards coral reef resilience. J Mar Sci Eng 8:563. https://doi.org/10.3390/jmse8080563
Chou LM, Tun KPP (1997) Coral transplantation as a reef conservation tool – the Singapore experience. In: Pacific Congress on Marine Science and Technology, PACON International, Hong Kong, 6–8 August 1997, Hawaii, pp 528–537
Chou LM, Huang D, Tan KS, Toh TC, Goh BPL, Tun K (2019) Singapore. In: Sheppard CRC (ed) World Seas: an environmental evaluation. Volume II: The Indian Ocean to the Pacific. Academic, London, pp 539–558. https://doi.org/10.1016/b978-0-08-100853-9.00031-2
Chou LM, Srirangam S, Kiong J, Tun K (2023) High-relief artificial reefs at Sisters' Islands Marine Park. In: Chan L, Ng D, Lim LJ (eds) Handbook on habitat restoration: general principles and case studies in Singapore. National Parks Board, Singapore, pp 188–192
Chow GSE, Chan YKS, Jain SS, Huang D (2019) Light limitation selects for depth generalists in urbanised reef coral communities. Mar Environ Res 147:101–112. https://doi.org/10.1016/j.marenvres.2019.04.010
Chuang SH (1977) Ecology of Singapore and Malayan coral reefs – preliminary classification. In: Proceedings of the third international coral reef symposium, vol 1, pp 55–61
Ding JL, Fung FMY, Ng GWS, Chou LM (1999) Novel bioactivities from a coral, *Galaxea fascicularis*: DNase-like activity and apoptotic activity against a multiple-drug-resistant leukemia cell line. Mar Biotechnol 1:328–336. https://doi.org/10.1007/pl00011782
Feldman B, Afiq-Rosli L, Simon-Blecher N, Bollati E, Wainwright BJ, Bongaerts P, Huang D, Levy O (2022) Distinct lineages and population genomic structure of the coral Pachyseris speciosa in the small equatorial reef system of Singapore. Coral Reefs 41:575–585. https://doi.org/10.1007/s00338-021-02160-4
Fung FMY, Tachibana S, Chou LM, Ding JL (1997) Cytotoxic and anticancer agents in mucus of *Galaxea fascicularis*: purification and characterization. J Mar Biotechnol 5:50–57
Goh NKC, Chou LM (1998) Bioactivity screening of Singapore gorgonians: antimicrobial activity, toxicity to *Artemia salina* and efficacy against *Plasmodium berghei*. Zool Sci 15:805–812. https://doi.org/10.2108/zsj.15.805
Goh NKC, Sim TS, Chou LM (1995) Bioactivity of gorgonians (subclass Octocorallia) in Singapore: preliminary studies. Mar Res 4:33–46
Guest JR, Chou LM, Baird AH, Goh BPL (2002) Multispecific, synchronous coral spawning in Singapore. Coral Reefs 21:422–423. https://doi.org/10.1007/s00338-002-0263-4
Guest JR, Baird AH, Clifton KE, Heyward AJ (2008a) From molecules to moonbeams: Spawning synchrony in coral reef organisms. Invertebr Reprod Dev 51:145–149. https://doi.org/10.1080/07924259.2008.9652264
Guest JR, Edwards AJ, Heyward AJ (2008b) Multi-species spawning of corals in North-western Philippines. Coral Reefs 27:83–83. https://doi.org/10.1007/s00338-007-0325-8
Guest JR, Baird AH, Maynard JA, Muttaqin E, Edwards AJ, Campbell SJ, Yewdall K, Affendi YA, Chou LM (2012) Contrasting patterns of coral bleaching susceptibility in 2010 suggest an adaptive response to thermal stress. PLoS One 7:e33353. https://doi.org/10.1371/journal.pone.0033353
Guest JR, Tun K, Low J, Vergés A, Marzinelli EM, Campbell AH, Bauman AG, Feary DA, Chou LM, Steinberg PD (2016) 27 years of benthic and coral community dynamics on turbid, highly urbanised reefs off Singapore. Sci Rep 6:36260. https://doi.org/10.1038/srep36260
Hesp PA (1995) The environmental impact assessment process in Singapore with particular respect to coastal environments and the role of NGOs. J Coast Conserv 1:135–144. https://doi.org/10.1007/bf02905122
Hilton MJ, Manning SS (1995) Conversion of coastal habitats in Singapore: indications of unsustainable development. Environ Conserv 22:307–322. https://doi.org/10.1017/s0376892900034883
Ip YCA, Tay YC, Gan SX, Ang HP, Tun K, Chou LM, Huang D, Meier R (2019) From marine park to future genomic observatory? Enhancing marine biodiversity assessments using a biocode approach. Biodivers Data J 7:e46833. https://doi.org/10.3897/bdj.7.e46833
Ip YCA, Chang JJM, Oh RM, Quek ZBR, Chan YKS, Bauman AG, Huang D (2023) Seq' and ARMS shall find: DNA (meta)barcoding of Autonomous Reef Monitoring Structures across the tree of life uncovers hidden cryptobiome of tropical urban coral reefs. Mol Ecol 32:6223–6242. https://doi.org/10.1111/mec.16568
Jaafar Z, Huang D, Tanzil JTI, Ow YX, Yap NWL (2018) The Singapore Blue Plan 2018. Singapore Institute of Biology, Singapore
Jain SS, Afiq-Rosli L, Feldman B, Levy O, Phua JW, Wainwright BJ, Huang D (2020) Homogenization of endosymbiont communities hosted by equatorial corals during the 2016 mass bleaching event. Microorganisms 8:1370. https://doi.org/10.3390/microorganisms8091370
Kikuzawa YP, Ng CSL, Toh TC, Sam SQ, Lee Y-L, Loo PL, Chua YZ, Tan KS, Chou LM (2020) Diversity of subtidal benthic and hard coral communities on sloping and vertical seawalls in Singapore. Mar Biodivers 50:95. https://doi.org/10.1007/s12526-020-01118-z
Lee YL, Hui DS-E, Loo PL, Toh TC, Kikuzawa YP, Tan KS (2023) Intertidal seawall communities in Singapore: are they all one and the same? Ecol Eng 187:106860. https://doi.org/10.1016/j.ecoleng.2022.106860
Ng CSL, Chen D, Chou LM (2012) Hard coral assemblages on seawalls in Singapore. In: Tan KS (ed) Contributions to marine science. National University of Singapore, Singapore, pp 75–79
Ng CSL, Toh TC, Toh KB, Sam SQ, Kikuzawa YP, Chou LM, Huang D (2023) Enhancing reef restoration by assessing stakeholder knowledge, attitudes, and preferences. Restor Ecol 31:e13854. https://doi.org/10.1111/rec.13854
Nguyen NTH, Friess DA, Todd PA, Mazor T, Lovelock CE, Lowe R, Gilmour J, Chou LM, Bhatia N, Jaafar Z, Tun K, Yaakub SM, Huang D (2022) Maximising resilience to sea-level rise in urban coastal ecosystems through systematic conservation planning. Landsc Urban Plan 221:104374. https://doi.org/10.1016/j.landurbplan.2022.104374
Poquita-Du RC, Huang D, Chou LM, Mrinalini TPA (2019) Short term exposure to heat and sediment triggers changes in coral gene expression and photo-physiological performance. Front Mar Sci 6:121. https://doi.org/10.3389/fmars.2019.00121
Poquita-Du RC, Goh YL, Huang D, Chou LM, Todd PA (2020a) Gene expression and photophysiological changes in *Pocillopora acuta*

coral holobiont following heat stress and recovery. Microorganisms 8:1227. https://doi.org/10.3390/microorganisms8081227

Poquita-Du RC, Huang D, Chou LM, Todd PA (2020b) The contribution of stress-tolerant endosymbiotic dinoflagellate *Durusdinium* to *Pocillopora acuta* survival in a highly urbanized reef system. Coral Reefs 39:745–755. https://doi.org/10.1007/s00338-020-01902-0

Tan KS, Koh KS, Goh L (2015) Taking stock of Singapore's marine life: the Comprehensive Marine Biodiversity Survey Johor Straits International Workshop 2012. Raffles Bull Zool Suppl 31:1–6

Tan KS, Koh KS, Ng JY, Goh L (2016) The Comprehensive Marine Biodiversity Survey Singapore Strait International Workshop 2013. Raffles Bull Zool Suppl 34:1–7

Tan YHJ, Tham JKQ, Paul A, Rana U, Ang HP, Nguyen NTH, Yee ATK, Leong BPI, Drummond S, Tun KPP (2023) Remote sensing mapping of the regeneration of coastal natural habitats in Singapore: implications for marine conservation in tropical cities. Singap J Trop Geogr 44:130–148. https://doi.org/10.1111/sjtg.12454

Tan YHJ, Lai S, Tun KPP (2025) "Accidental" artificial reefs–lessons from Singapore for coral and seagrass restoration along urban shorelines. Bull Mar Sci 101:197–209. https://doi.org/10.5343/bms.2023.0134

Tay YC, Todd PA, Rosshaug PS, Chou LM (2012) Simulating the transport of broadcast coral larvae among the Southern Islands of Singapore. Aquat Biol 15:283–297. https://doi.org/10.3354/ab00433

Tay YC, Noreen AME, Suharsono CLM, Todd PA (2015) Genetic connectivity of the broadcast spawning reef coral *Platygyra sinensis* on impacted reefs, and the description of new microsatellite markers. Coral Reefs 34:301–311. https://doi.org/10.1007/s00338-014-1206-6

Todd PA, Sanderson PG, Chou LM (2001) Morphological variation in the polyps of the scleractinian coral *Favia speciosa* (Dana) around Singapore. Hydrobiologia 444:227–235. https://doi.org/10.1023/a:1017570100029

Todd PA, Sidle RC, Chou LM (2002a) Plastic corals from Singapore: 1. Coral Reefs 21:391–392. https://doi.org/10.1007/s00338-002-0265-2

Todd PA, Sidle RC, Chou LM (2002b) Plastic corals from Singapore: 2. Coral Reefs 21:407–408. https://doi.org/10.1007/s00338-002-0266-1

Todd PA, Ladle RJ, Lewin-Koh NJI, Chou LM (2004) Genotype × environment interactions in transplanted clones of the massive corals *Favia speciosa* and *Diploastrea heliopora*. Mar Ecol Prog Ser 271:167–182. https://doi.org/10.3354/meps271167

Wong JSY, Chan YKS, Ng CSL, Tun KPP, Darling ES, Huang D (2018) Comparing patterns of taxonomic, functional and phylogenetic diversity in reef coral communities. Coral Reefs 37:737–750. https://doi.org/10.1007/s00338-018-1698-6

Ying LS, Ledet J, Griffin K, Loke LH, Bhatia N, Todd PA (2025) Public perception of coastal ecoengineering interventions in Singapore. Bull Mar Sci 101:419–436. https://doi.org/10.5343/bms.2023.0087

Reefs in Singapore's Murky Waters

2

Ai Chin Lee and Seng Keat Ooi

Abstract

Coral reefs in Singapore have persisted in a sediment-rich environment due to ongoing coastal development. Discerning sediment regimes and sediment movement around the reefs is imperative in understanding how corals respond to elevated sediment load. Through a dedicated coral reef monitoring initiative conducted by the Tropical Marine Science Institute (TMSI), suspended sediment regimes are described for fringing reefs at three islands: Pulau Hantu, Kusu Island and Pulau Satumu. The results indicate that horizontal sediment flux may be influenced by the spring-neap tidal cycles of the diurnal tide. High mobility of sediments in reef environments in Singapore was observed, underscoring the importance of orientation of the reef to main flow direction on local resuspension of suspended sediments due to island wakes. Over the longer time scale, it was apparent that increases in suspended sediment were related to local coastal activity rather than transport from external sources. Reef environments in Singapore appear to be both sources and sinks of suspended sediments, depending on the different hydrodynamic conditions, which may account for the elevated suspended sediment observed around reef environments in Singapore.

Keywords

Hydrodynamic modelling · Monsoon · Reef compression · Reef resilience · Sediment flux · Tidal flow · Water quality

A. C. Lee (✉) · S. K. Ooi
Tropical Marine Science Institute, National University of Singapore, Singapore, Singapore
e-mail: tmsleeac@nus.edu.sg

2.1 Introduction

The Republic of Singapore is a small island-nation, faced by the challenges of economic growth in a land-scarce environment, which has grown into a highly developed country with a maritime port that is one of the busiest ports in the world. Due to economic growth and expansion, many types of coastal development and activity have been witnessed in Singapore. The most notable form of coastal development here is land reclamation. Land scarcity in Singapore was dealt with by creating new land, where the shallower (and sometimes, not so shallow) shoreline was reclaimed with sand, soil and other fill material. Numerous land reclamation development projects have been carried out in the past century and continue to this day. These projects serve to provide much-needed space for industrial, housing and infrastructure purposes (Koh and Lin 2006). The earliest land reclamation dates to 1822, in colonial Singapore, which transformed the south bank of the Singapore River from wet marshland to a bustling waterfront commercial district (Lim 2017). More recent reclamation projects include the building of the Pasir Panjang Container Terminal and Tuas Mega Port. As one of the world's busiest maritime ports, Singapore needed bigger and better equipped container terminals to handle the large volume of vessels arriving at port. As a result of the many land reclamation projects over the years, Singapore's land area has grown from ~581 km^2 in 1960 (Yong et al. 1991) to ~735 km^2 in 2023 (Singapore Department of Statistics 2024). Besides land reclamation, other coastal modifications have also taken place over the years in Singapore. Much of the natural coastline in Singapore has been modified, whether by some form of shoreline armouring or the creation of artificial lagoons and beaches. Most of the small islands in Singapore are bound by hard seawalls that are constructed of large granite blocks, which are designed to protect the shore front from coastal erosion and promote land stabilisation. An estimated 63% of Singapore's coastline consists of man-made armoured shorelines (Lai et al. 2015). Another coastal activ-

L. M. Chou, D. Huang (eds.), *Coral Reefs of Singapore's Urbanised Sea*, Coral Reefs of the World 23,
https://doi.org/10.1007/978-981-95-0903-4_2

ity that is prevalent in Singapore is the maintenance dredging of the port and shipping channels. Maintenance dredging is deployed to keep the port and channels deep and free from any underwater obstructions that could pose a danger to vessels and usually takes at least 6 months to complete (e.g. MPA 2023, 2024). Dredging works are usually carried out by clamshell or grab dredgers that dredge the seabed, with the spoils subsequently removed from the vicinity.

A consequence of the ongoing coastal modifications and shipping-related activities in Singapore is the introduction of additional suspended particles into the water, increasing the sediment load in the environment. The increase in sediment load will also likely result in greater volume of particles settling out from the water column, eventually causing the smothering of benthic habitats and the biota therein (Fabricius and Wolanski 2000; Fabricius et al. 2003; Fabricius 2005). Amid all this coastal activity and modifications are the coral reefs of Singapore, now persisting in a sediment-rich environment. Over the past decades, these reefs have been subjected to deteriorating water clarity and numerous bleaching events and occasional boat/physical damage. Widespread losses to Singapore's reef communities have been documented, where at least 60% of Singapore's initial coral reefs have been lost (Chou 2006; Lai et al. 2015; Tay et al. 2018). The widespread losses of coral reef communities Singapore recorded post-1990 were most noticeable for communities at the lower extent of the reef slope (Chou and Tun 2012). Historical reports of coral reefs have indicated that coral reefs extended up to 20 m below chart datum (Chuang 1977). However, in present-day Singapore, coral reefs do not extend beyond 10 m below chart datum (Dikou and van Woesik 2006; Erftemeijer et al. 2012). The loss is evidenced by the change in the dominant component of the reef community, where these lower reef slope communities transitioned from a mixed biotic-abiotic community to a purely abiotic community dominated by fine sediments and unconsolidated substrata (Chou and Tun 2012). The silt-dominated conditions have persisted to the present day, suggesting that these are in a stable state. In a review by Zweifler et al. (2021), Singapore reefs were classified as transitional reefs where a sustained or stepped increase in turbidity occurred over a relatively short time scale (<100 years).

Loss of water clarity, a very likely cause of the truncation of the reefs in Singapore, is driven by increased suspended particles in the water column. These are not exclusively in the form of abiotic sediment particles but also consist of biotic components such as detritus and plankton. Increase in suspended sediment load levels raises water turbidity (Kirk 1985), which in turn results in poorer water clarity. Poor water clarity attenuates light available to biota at greater depths (Te 1997; Rogers 1990), and severely limits the depth distribution of coral reefs to shallow waters (Cooper et al. 2007; Morgan et al. 2020). Light attenuation by suspended particles greatly reduces the ability of organisms to photosynthesise and can have significant negative effects on the energy budget of photosynthetic organisms. This has many downstream impacts, including negatively affecting the growth and normal physiological functions of organisms. Deterioration in water clarity in Singapore has previously been documented (e.g. Low and Chou 1994). Average water visibility of >10 m vertical depth was noted in the past; however, in the present day, average visibility is often not more than 2 m (Chou and Chia 1991).

The dynamics of suspended particles are an important driver of ecosystem health in sensitive shallow water coastal environments such as coral reefs (Fabricius et al. 2003; Fabricius 2005). In these ecosystems, increases in suspended particles can greatly affect the healthy functioning of the coral reef system. The persistence of suspended sediments is dependent on the sediment availability, sediment characteristics (shape, size, density, and minerology), settling velocity, flow characteristics (maximum and residual tidal flows, tidal cycles), waves and water density and viscosity (Dyer 1986). Sources of sediments in marine environments are diverse, ranging from terrestrial-based inputs to deep-water processes, such as upwelling, to sediment transported from outside of its immediate vicinity through a combination of wind, waves and currents. Differences in transport distances of sediment sources are directly related to the settling velocity of the sediments but are also significantly influenced by hydrodynamics. In particular, the resuspension of settled sediments reintroduces sediment particles into the water column, making them available for further advection. Resuspension can occur through various mechanisms, both physical and biological. Sediment resuspension through wave- and tide-induced bed stress is an important aspect of sediment dynamics in many major reef systems (Hoitink and Hoekstra 2003; Storlazzi et al. 2004; Wolanski et al. 2005; Presto et al. 2006). A unique feature of Singapore's coastline, be it on the main island or on the smaller offshore islands, is the high percentage of the shoreline that is armoured with artificial seawalls. Armouring of shorelines with hard granite structures may play a significant role in the resuspension of sediments in the habitats in the vicinity of these seawalls. Energy dissipation characteristics around these seawalls would be greatly altered from a natural shoreline. The seawalls are constructed with a slope inclination of between 20° and 40° (Lee et al. 2009). These steep slopes encourage the formation of plunging waves, and increased energy reflection will occur when a wave breaks on the steep seawall (USACE 1984). Increased energy reflection will likely change the water flow characteristics in the vicinity, influencing the dynamics of sediment particles over the reefs.

The differences in hydrodynamic flow patterns between monsoon seasons may also affect the sediment dynamics in the Singapore Strait region. The predominant direction of residual currents in the Singapore Straits is east to west (Sin et al. 2016). However, this direction changes with the different monsoons, with weaker westward flows during the Southwest monsoon and stronger westward flows during the Northeast monsoon (Sin et al. 2016). The variation in monsoonal-driven flow patterns could result in transport of sediments from different sources. This in turn will influence seasonality patterns in suspended sediment concentrations in the Singapore Straits. Alternatively, differences in residual currents, which peak during the monsoon periods (van Maren and Gerritsen 2012), may enhance resuspension during one monsoon and not the other.

In this chapter, the effects of hydrodynamics on suspended sediment concentrations (SSC) and long-term trends observed in SSC on reefs in Singapore are presented. Through a dedicated reef monitoring initiative, the Tropical Marine Science Institute initiated monitoring of SSC at three reef sites in Singapore, at the fringing reefs of Pulau Hantu, Kusu Island and Pulau Satumu (Fig. 2.1). Dedicated surveys at the three reefs examined the variation in SSC over short and decadal time scales. Short time-scale variations in SSC were examined over a single spring-neap tidal cycle and an annual (12-month) cycle, while the long-term variation in SSC was recorded from September 2011 to May 2024. The field observation data were evaluated against flow data obtained from numerical 2D shallow-water flow models.

2.2 Seasonal and Tidal Effects on Suspended Sediment Dynamics

The hydrodynamics, in terms of flow velocity and water level changes, around the reef can have large effects on the sediment dynamics in the reef environment. Different flow velocity regimes and the ebbing and flooding of tides can change resuspension and the settling rate of particles, subsequently influencing sediment transport. For the Singapore domain, flow characteristics in terms of tidal flows can potentially exert a large influence on suspended sediment dynamics. Tidal asymmetry, defined as the differences in the magnitude and duration of current velocities between ebb and flood tides (Dronkers 1986), can play a significant role in sediment transport in the marine environment. The differences in the magnitude of velocities between spring and neap tidal cycles can also affect sediment dynamics in the environment.

To capture the effect of the tidal and seasonal water flows on sediment dynamics on the reef, SSC were examined against hydrodynamic data that were generated from calibrated numerical 2D shallow-water flow models. The Singapore region is characterised as a complex tidal region (Kurniawan et al. 2011), driven by interactions of tidally generated flows, seasonally varying monsoon contributions and local winds (Hasan et al. 2011; Kurniawan et al. 2011). Predominant currents in Singapore are east to west (Sin et al. 2016), with a mixed semi-diurnal tidal frequency. Two models previously developed using the Delft3D Modelling suite were used to generate current velocities for the three reefs of

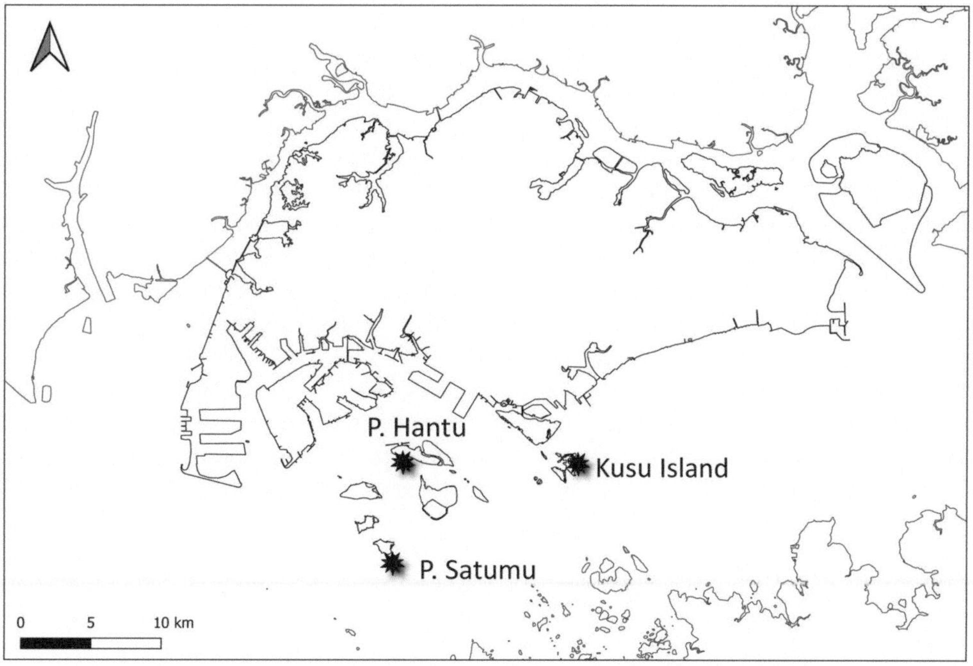

Fig. 2.1 Map of Singapore showing the three fringing reef sites at Pulau Hantu, Kusu Island and Pulau Satumu. Base map from the Tropical Marine Science Institute, National University of Singapore

interest. The locally refined and aligned Singapore Regional Model (SRMRA), developed using Delft3D-Flow, provided good representation of tidal and seasonal forcing for Singapore and the surrounding domain. A local model, referred to as Singapore Local Model (SLM), was nested within the previously calibrated SRMRA. The SRMRA was built using a curvilinear grid system, with varying resolutions from 150 m in the Singapore Straits to two-kilometre resolution at the boundaries of the domain. The SLM was built on a curvilinear grid, with varying resolutions from 25 m in the Johor Straits to approximately 300 m near the boundaries. Nesting allows the smaller model (SLM) to capture interactions that develop between the seasonal and tidal flows that occur outside the local coastal waters of Singapore before they propagate into the domain of the SLM. Both these models are described in detail in Hasan et al. (2011, 2016) and demonstrated very good representation of tidal and seasonal forcing in the Singapore coastal waters (see Hasan et al. 2016). Hydrodynamic model outputs in 10-min time steps from September 2011 to September 2012 were extracted for two reference locations, named West and East, located at 103.668°E, 1.194°N and 103.961°E, 1.242°N, respectively. The hydrodynamic output at the West location was used in conjunction with SSC data for Pulau Hantu and Pulau Satumu, while the hydrodynamic output generated for the East location was used with the SSC data from Kusu Island.

Data on SSC were obtained from September 2011 to September 2012 by in situ data loggers (Infinity CLW data logger, JFE Advantech Co. Ltd.) that were deployed at Pulau Hantu, Kusu Island and Pulau Satumu. The in situ loggers were deployed on a semi-permanent frame on the reef crest at each location, approximately 50–70 cm above the seabed. The reef crests were situated approximately 3–4 meters below mean sea level. Data acquisition was set as a burst sampling of 120 samples over 1 min, at 5-min intervals. The Infinity loggers provided measures of turbidity (in Formazin Turbidity Units, FTU), which was then calibrated with gravimetric analyses of 1 L water samples taken from each reef to derive high-frequency SSC data. Water samples for the calibration were obtained at the location of the frame housing the in situ loggers, on a fortnightly sampling frequency. Derived (Fitted) SSC data were generated from a linear model which included location and month as factors with gravimetric measures of SSC as the dependent variable and turbidity as the predictor variable. Overall, the predicted SSC based on turbidity measures showed a good fit to the measured gravimetric SSC, with an adjusted-R^2 of 0.86 (Fig. 2.2).

2.2.1 Current Velocity and Suspended Sediment Concentrations

The computed flow direction and speed from the hydrodynamic models showed higher current velocities at the East reference location than the West reference location (Fig. 2.3). The computed current velocities were up to 0.79 m s^{-1} at the West reference location and 1.1 m s^{-1} at the East reference location. For both East and West reference locations, the computed current velocities in the eastward direction were much higher than current velocities in the westward direction (Fig. 2.3). Current velocities were generally slower between March and April and again between September and October. These periods coincided with the inter-monsoon periods. Relatively higher current velocities were noted from May to August and between November and February, which coincided with the Southwest monsoon and the Northeast monsoon, respectively (Fig. 2.3). The intra-annual patterns in current velocities were consistent for both reference locations (Fig. 2.3).

To elucidate relationships between suspended sediment concentration and hydrodynamic forcing, median concentration of suspended sediment was plotted against current velocity for individual reef locations to identify threshold velocities. Threshold velocities here are defined as the magnitude of current speed beyond which the suspended sediment concentration–velocity relationship changes. In general, the relationship between flow velocity and SSC was highly site-dependent and temporally variable (Fig. 2.4). Among sites, the correlation between flow velocity and sediment concentration was most apparent and most consistent at Pulau Hantu. The data demonstrated a weak relationship between velocity and suspended sediment concentration for velocities lower than 0.4 m s^{-1}. At velocities greater than 0.4 m s^{-1}, the suspended sediment concentration–current velocity relationship was stronger. The threshold velocities in the sediment concentration–current velocity relationships were more apparent during eastward flows than westerly flows at Pulau Hantu and Pulau Satumu (Fig. 2.4). Threshold velocity of approximately 0.59 m s^{-1} was identified at Pulau Satumu for flow in the eastward direction. A sharp increase in sediment concentration was associated with current velocity that was greater than 0.59 m s^{-1}. Sediment concentration observed at Pulau Satumu appeared invariant with increasing current velocity in the westward direction. Threshold velocity of approximately 0.68 m s^{-1} was identified at Kusu Island for flow in the westward direction, which was when increased SSC was observed (Fig. 2.4).

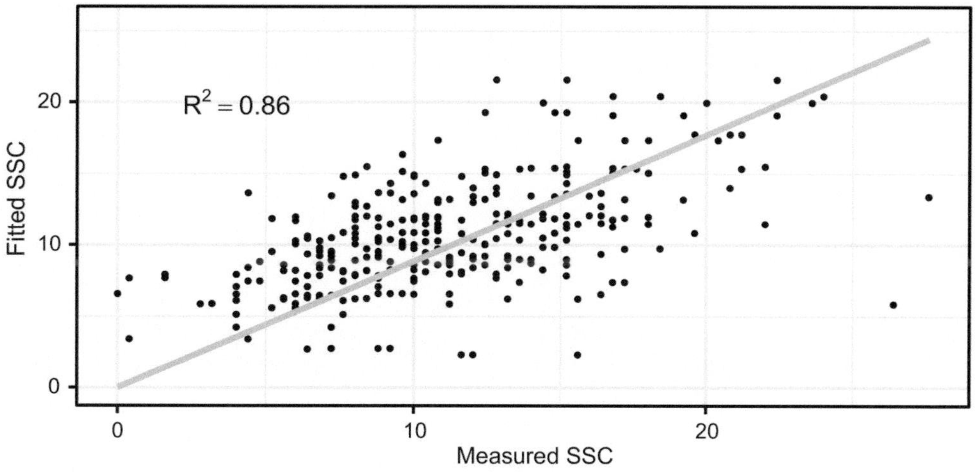

Fig. 2.2 Comparison of measured suspended sediment concentration (measured SSC) with suspended sediment concentration estimated from calibrated turbidity (fitted SSC)

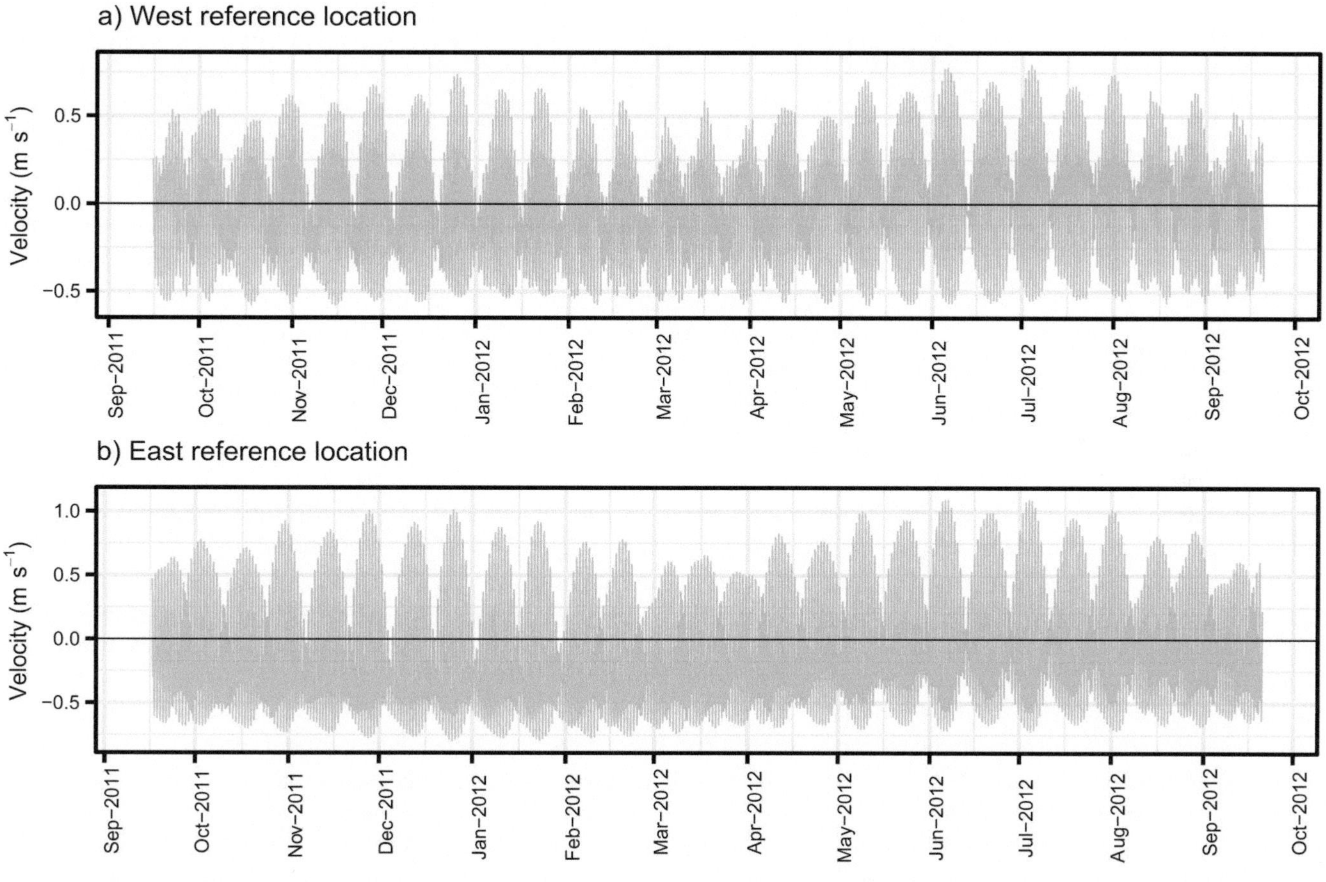

Fig. 2.3 Computed current velocities (m s^{-1}) in 10-min time steps generated from the Singapore Local Model (SLM) nested within the Singapore Regional Model (SRMRA) for a) West and b) East reference locations. Negative current velocity values are indicative of flows in the westward direction and positive current velocity values are indicative of flows in the eastward direction

2.2.2 Frequencies of Variation in Suspended Sediment Concentration

Natural frequencies of variation in sediment concentrations were identified using spectral analysis computed through the 'spec.pgram' routine in the package 'stats' (R Core Team 2024). The spec.pgram calculates the periodogram utilising a fast Fourier transformation, which then smooths the periodogram through a series of Daniell smoothers. Daniell smoothers are smoothers of moving averages that give a half weight to the end values of a given span. A span of three was used in the Daniell smoothers that was applied to this data

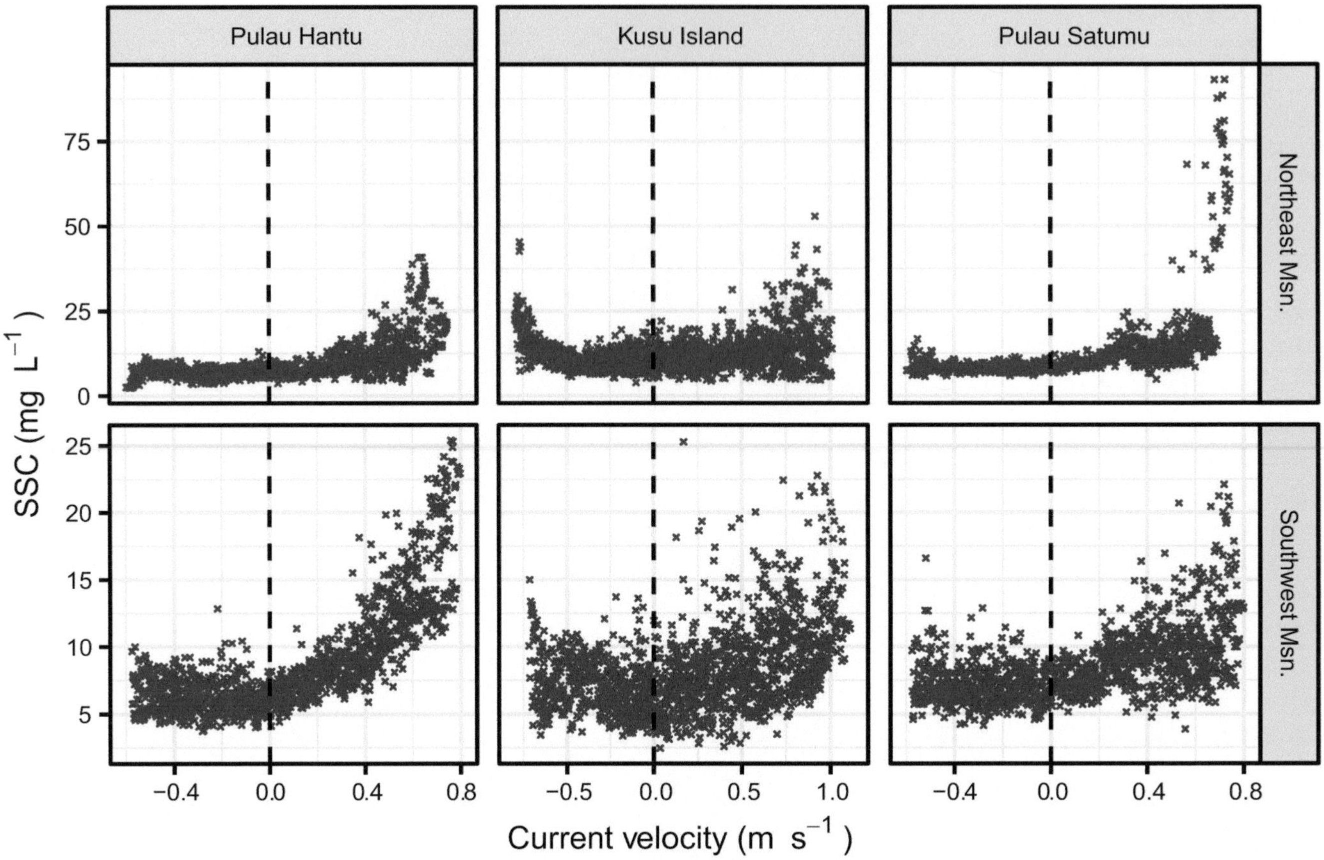

Fig. 2.4 Scatterplots of median suspended sediment concentration (SSC, mg L^{-1}) against flow velocity (m s^{-1}) at Pulau Hantu, Kusu Island and Pulau Satumu for the northeast and southwest monsoons. Negative current velocity values are indicative of flows in the westward direction and positive current velocity values are indicative of flows in the eastward direction

set. The spectral analyses of SSC derived from the in situ turbidity loggers identified high-intensity peaks of SSC in the periodogram that corresponded with the spring-neap tidal cycle (at ~13.5 days) and possibly to the monsoon seasons (>150 days, Fig. 2.5a–c). The intra-tidal spectral peaks of SSC were much weaker and occurred at periods that did not indicate vertical mixing for tidal signals (Fig. 2.5d–f). If tidally induced vertical mixing was an important mechanism in the semi-diurnal tidal environment characteristic for Singapore, suspended sediment peaks would be expected at 6 h. However, the periodograms did not indicate such and pointed towards advection as a more important mechanism in controlling suspended sediment concentrations (Fig. 2.5d–f). Based on these results, the variation in SSC observed at the three reefs appeared to be less influenced by mechanisms of the daily ebb and flood of water currents, but more of the spring-neap oscillations in tidal intensity and potentially monsoonal forcings in the delivery of sources of sediment. This suggests that advection potentially plays a more important role than resuspension on the sediment dynamics at the reef scape.

The influence of spring-neap tidal variations in current velocities on SSC was examined over a representative fortnight cycle from 21 December 2011 to 3 January 2012 (Fig. 2.6). Spring tides are characterised by increased tidal range in water level, occurring every 14.7 days, while neap tides are typically of smaller tidal ranges occurring between the spring cycles (Pugh 1987; Kvale 2006). Computed current velocities during the spring tide were higher than during the neap tide, and at the same time, higher SSC were generally recorded during the spring tide cycle and lowest during the neap tides (Fig. 2.6). At both Pulau Hantu and Pulau Satumu, higher SSC were recorded during maximum flow velocities in the eastward direction. Variations in SSC over the representative spring-neap cycle observed at Kusu Island were greater and showed a less consistent trend than those observed at Pulau Hantu or Pulau Satumu. Given its location, the high variability in SSC at Kusu Island may reflect the influence of the seasonal monsoonal wind-driven impact on the SSC variation during this period. The reefs at Pulau Hantu and Pulau Satumu are relatively sheltered from predominant Northeast monsoon winds, as these reefs are located on the south/south-western part of the islands, which

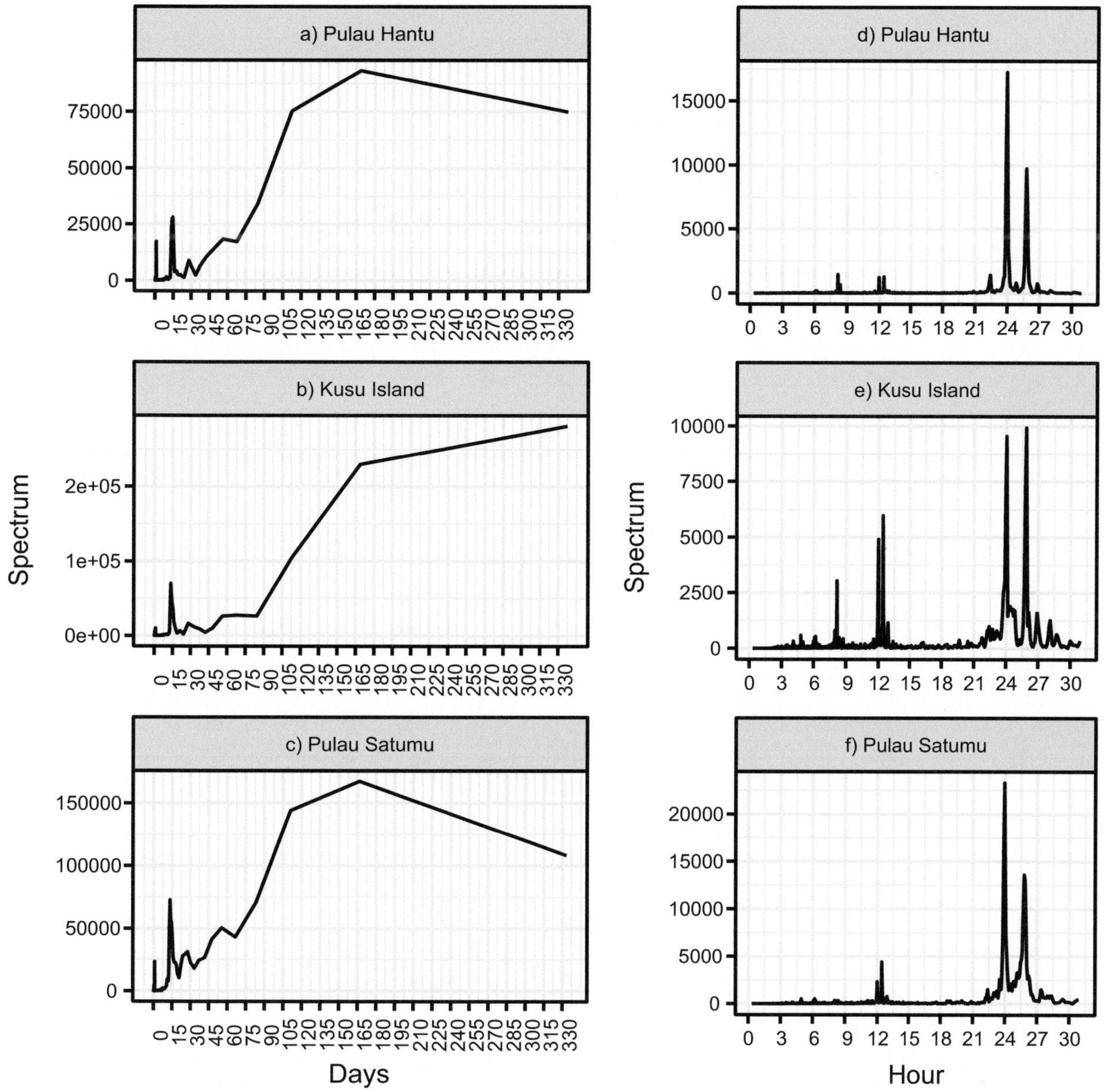

Fig. 2.5 Spectral density of suspended sediment concentrations (mg L^{-1}) against frequencies (in days) over long (**a**, **b**, **c**) and short (**d**, **e**, **f**) temporal scales at Pulau Hantu (**a** and **d**), Kusu Island (**b** and **e**) and Pulau Satumu (**c** and **f**)

are leeward to the predominant wind direction. The reef studied at Kusu Island is windward and faces the direction of prevailing Northeast monsoon winds, which may have increased the influence of wind-waves along the reef, resulting in higher mixing in the vicinity of Kusu Island.

While different patterns in sediment concentration and current velocities at the three reefs were observed, some generalities in the SSC patterns with current velocity were present. These were (1) higher SSC during eastward flows and (2) extremely poor relationships between SSC and current velocity at low tidal amplitudes. At all reef sites, the lower SSC during the westward flow may be related to the slightly lower peak current velocities during the westward flows. Although relatively similar in overall magnitude, peak current velocities are marginally greater during eastward than during the westward flow (Fig. 2.6). Lower current velocities during westward flows are likely to result in lower resuspension rates, especially when current velocities do not exceed critical velocities that would result in sediment resuspension for bed sediments.

Time-series plots are presented for SSC and computed current velocities from September 2011 to September 2012 (Figs. 2.7 and 2.8) to elucidate intra-annual variations in current velocity on SSC. To extract seasonal trends, hydrodynamic and SSC were detided by applying a cascading moving average (Godin 1972). The Godin-filtered data

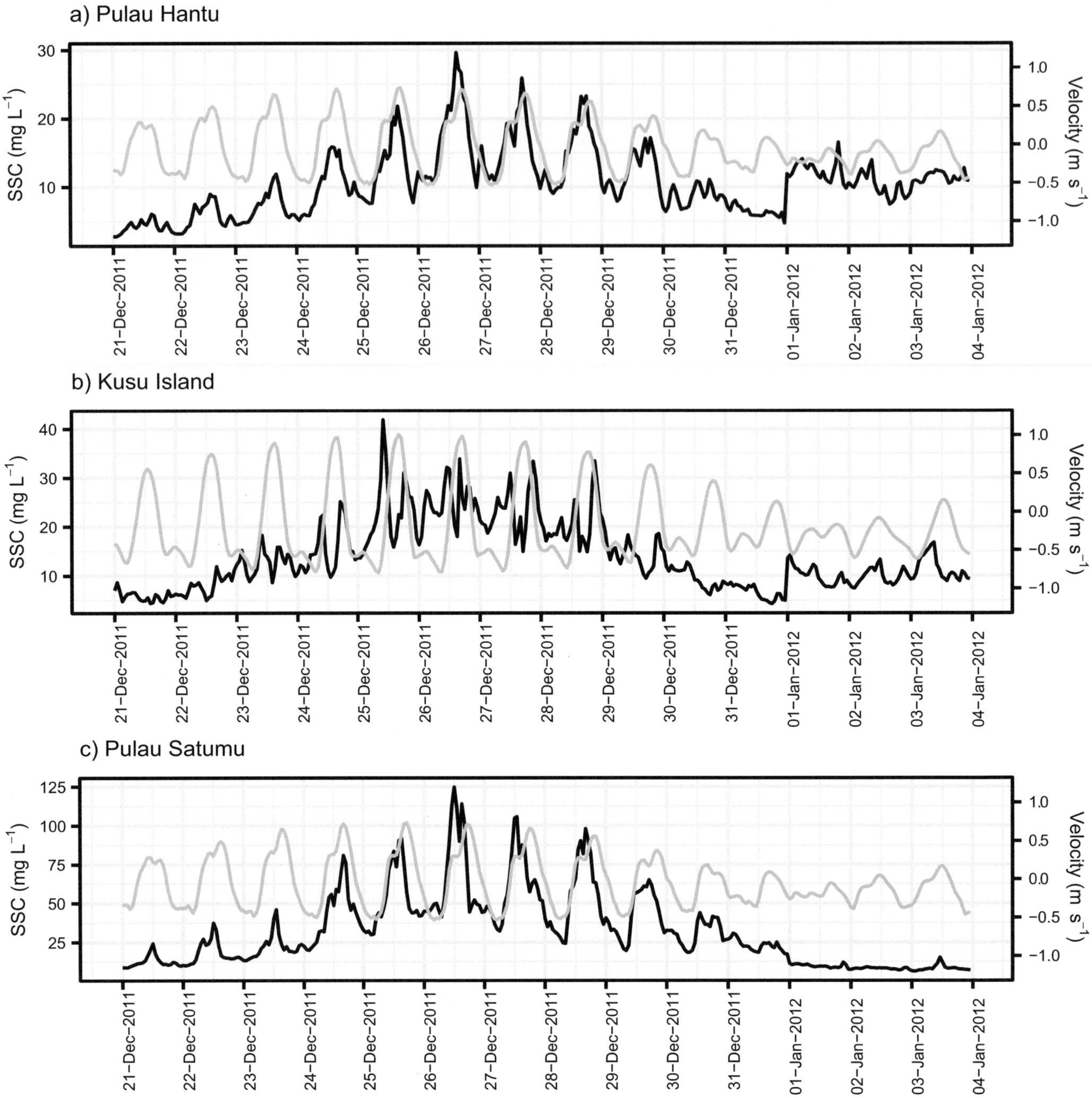

Fig. 2.6 Variations in suspended sediment concentration (SSC, black line) and east-west current velocity (grey line) over a representative 14-day cycle observed at (**a**) Pulau Hantu, (**b**) Kusu Island and c) Pulau Satumu. Negative current velocity values are indicative of flows in the westward direction and positive current velocity values are indicative of flows in the eastward direction. Note unequal axis scales for SSC

were then superimposed onto the plots of the raw data for interpretation. The detided suspended sediment data closely corresponded to the intra-annual patterns in SSC at all locations (Figs. 2.7 and 2.8). The detided data revealed two periods of elevated SSC. Markedly elevated SSC occurred from late December to February at Pulau Hantu (Fig. 2.7) and from late December to March at Kusu Island (Fig. 2.8). An elevation of smaller magnitude occurred later in the year but was asynchronous among sites (Figs. 2.7 and 2.8). The second SSC peak at Pulau Satumu was relatively short-lived, occurring over only part of July (Fig. 2.7). At Pulau Hantu and Kusu Island, increases in SSC commenced in July, but extended through to September (Figs. 2.7 and 2.8). Elevated SSC co-occurred with greater current velocities for the year, often during periods associated with the Northeast and Southwest monsoons. Seasonal influences in increases in

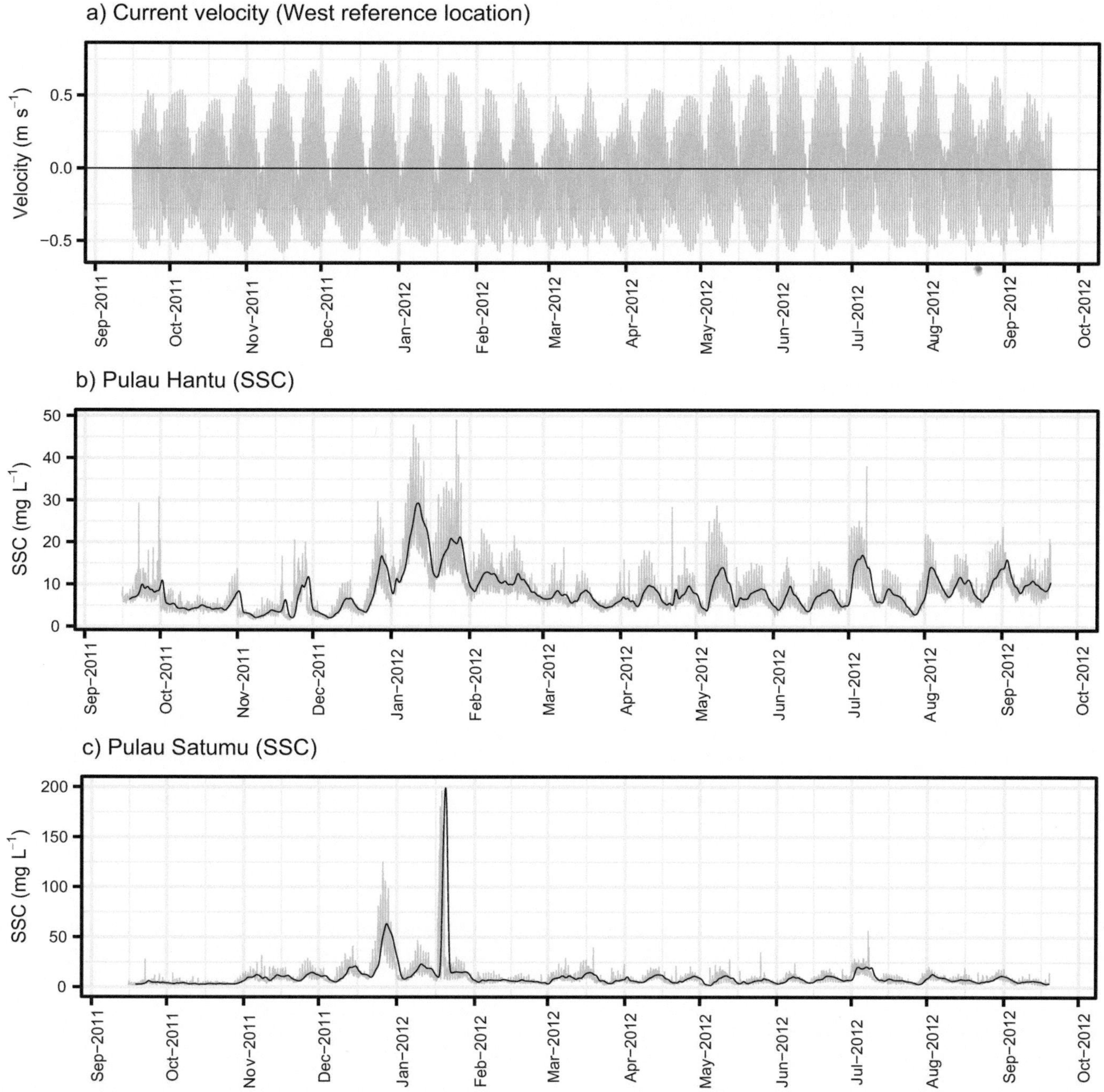

Fig. 2.7 Time-series plots of computed current velocity (in m s^{-1}) modelled at (**a**) West reference location and suspended sediment concentration (SSC) measured at (**b**) Pulau Hantu and (**c**) Pulau Satumu. SSC data are presented as raw values (grey line) and Godin-filtered (black line). Negative current velocity values are indicative of flows in the westward direction and positive current velocity values are indicative of flows in the eastward direction. Raw sediment concentration data for Pulau Satumu are truncated. The peak sediment concentrations occurred on 19th January 2012, at 420 mg L^{-1}. Note different y-axis ranges for Pulau Hantu and Pulau Satumu

SSC measured between September 2011 and September 2012 were reflected in peak spectral densities in SSC, which occurred at 167–169 days (Fig. 2.5). Throughout the year, high-sediment concentration was strongly associated with main flow direction. At Pulau Satumu and Pulau Hantu, elevated SSC consistently occurred during flow from the west to the east (Fig. 2.7). This pattern was less obvious at Kusu Island, where flows were predominantly westwards. However, maxima in SSC at Kusu Island occurred when westward flow was weakest (Fig. 2.8). This observation may be in part due to the location of the reef along the eastern coast of Kusu Island, which places it in the lee of eastward currents. The results suggest that the SSC variability may be

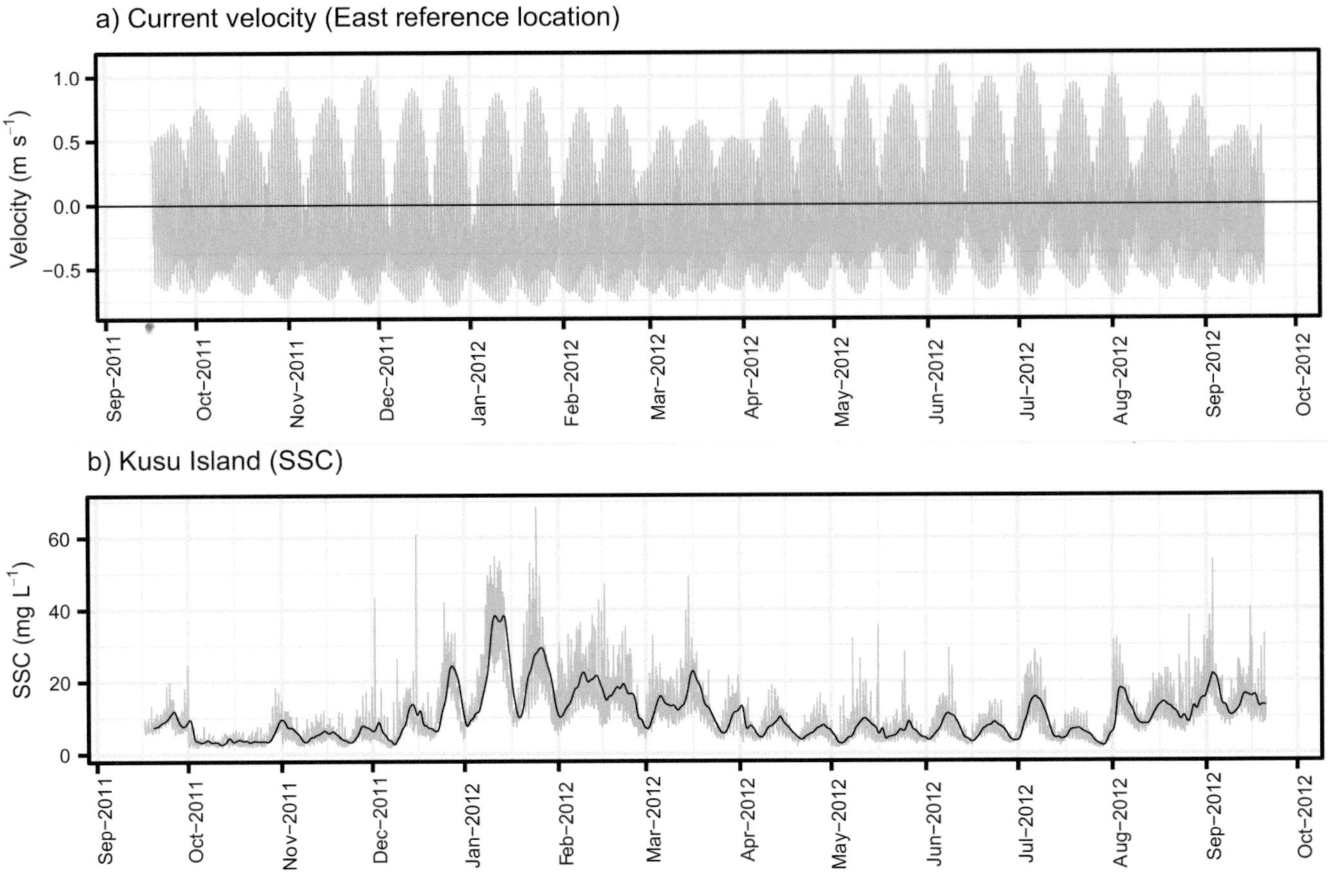

Fig. 2.8 Time-series plots of computed (**a**) current velocity (in m s^{-1}) modelled at East reference location and suspended sediment concentration (mg L^{-1}) measured at (**b**) Kusu Island. Suspended sediment concentration (SSC) data are presented as raw values (grey line) and Godin-filtered (black line). Negative current velocity values are indicative of flows in the westward direction and positive current velocity values are indicative of flows in the eastward direction

driven by localised effects, where increased velocity potentially has increased resuspension of seabed sediments.

2.3 Sediment Flux on the Reef

Sediment flux can play a significant role in controlling the distribution of coral reefs. Net sediment flux is the interplay between sediment resuspension and sediment accumulation, the two main processes which determine how elevated sediment load affects coral reefs. Accumulated sediments can smother live corals, causing mortality and blanket the substrata, making it unsuitable for coral spat colonisation, while suspended sediments can increase turbidity and reduce light penetration that is required for photosynthesis. (Rogers 1990). It is therefore important to quantify sediment flux to better understand how organisms respond to the different sediment regimes. There have been various studies documenting sediment flux over coral reefs (e.g. Larcombe and Woolfe 1999; Storlazzi et al. 2004; Browne et al. 2012), and there have also been efforts to model flow over and around coral reefs (e.g. Gourlay and Colleter 2005; Monismith 2007). However, information regarding sediment regimes around reefs in Singapore is limited (but see recent study by Morgan et al. (2020) that examined SSC and sediment accumulation at two reef sites).

This section explores the influence of finer-scale (or localized) hydrodynamics on suspended sediments through a dedicated measurement campaign to record reef bathymetry, current velocity profiles, and suspended sediment concentrations. As net sediment accumulation was not evaluated here, only sediment resuspension/advection is described and discussed. Two separate surveys were conducted at each of the three reef sites, Pulau Hantu, Kusu Island and Pulau Satumu. All surveys were conducted during the spring tide cycle to capture maximum flow velocities. The bathymetric survey of three islands and their surrounding coral reef areas was conducted using a multi-beam sonar system, GeoSwath Plus, integrated with additional peripheral sensors to obtain accurate seabed morphology. GeoSwath Plus was used to map the areas of interest as it can produce accurate digital terrain maps (DTM) in water depths up to 100 m. The GeoSwath Plus operates on 250 kHz frequency, relying on two trans-receivers, each consisting of a transmitter and four receivers. The trans-receivers were each mounted on the starboard and port sides of the research vessel. Acoustic beams of 0.75° azimuth, with up to 50 pings per second, were transmitted simultaneously on both the starboard and port sides of the vessel, with swath width set at 40–60 m. The trans-receivers were deployed at depths between 12 and 50 m, towed at vessel speeds of 2–3 knots.

Current velocity profiles were obtained along a series of transects using an Acoustic Doppler Current Profiler (ADCP) or a Pulse Coherent Acoustic Doppler Profiler (PC-ADP), based on the maximum water depth at each of the reef sites. The PC-ADP was deployed at Pulau Hantu in water depths which were less than 20 m. The ADCP was utilised at Kusu Island and Pulau Satumu, where water depths were greater than 20 m. All transects were conducted, as close as possible, parallel to the reef and the shoreline around the reefs of interest (Fig. 2.9). Both the ADCP and the PC-ADP, with the sensor located 0.8 m from the water line, were deployed downward-looking, which emits acoustic pings for averaging intervals of 30 s. The 1.5 MHz PC-ADP transmitted pings along three beams at an angle of 15° to the vertical axis, equally spaced at 120° relative azimuth angles. The 500 kHz ADCP transmitted pings along three beams oriented at 25° off the vertical axis, equally spaced at 120° relative azimuth angles. Current velocity information was obtained at vertical intervals of 0.25 m and 1 m, for the PC-ADP and ADCP, respectively. The ADCP/PC-ADCP data were corrected for motion of the vessel (pitch, yaw and roll) and the post-processing of the acoustic signal produced velocity information in the East-West direction, North-South direction, as well as in a vertical orientation (up-down in the water column). A resultant vector was calculated from these three measures to yield a resultant velocity for each depth layer.

The in situ SSC along the water column was measured with a Laser In-Situ Scattering and Transmissometry equipment (LISST, Sequioa Scientific, Inc.) at various locations in each study area. The calibrated LISST was deployed simultaneously from a non-moving vessel with the ADCP or PC-ADP at these individual locations along the transect tracks. Concurrent measures of SSC from the LISST and acoustic backscattering signal from the ADCP/PC-ADP were then obtained for the water column. The acoustic backscattering signal was then correlated with the LISST-measured SSC to derive SSC estimates for the water column. Acoustic signal and LISST-measured SSC were strongly correlated ($R > 0.9$; Fig. 2.10), with different equations describing the linear relationship between the acoustic signal and LISST-measured SSC for different survey days and sites. Therefore, different conversions were used for each of the reef sites, and for each specific survey day to derive SSC for the water column.

The data presented in the graphs for this section were taken from two or three representative transects for each of

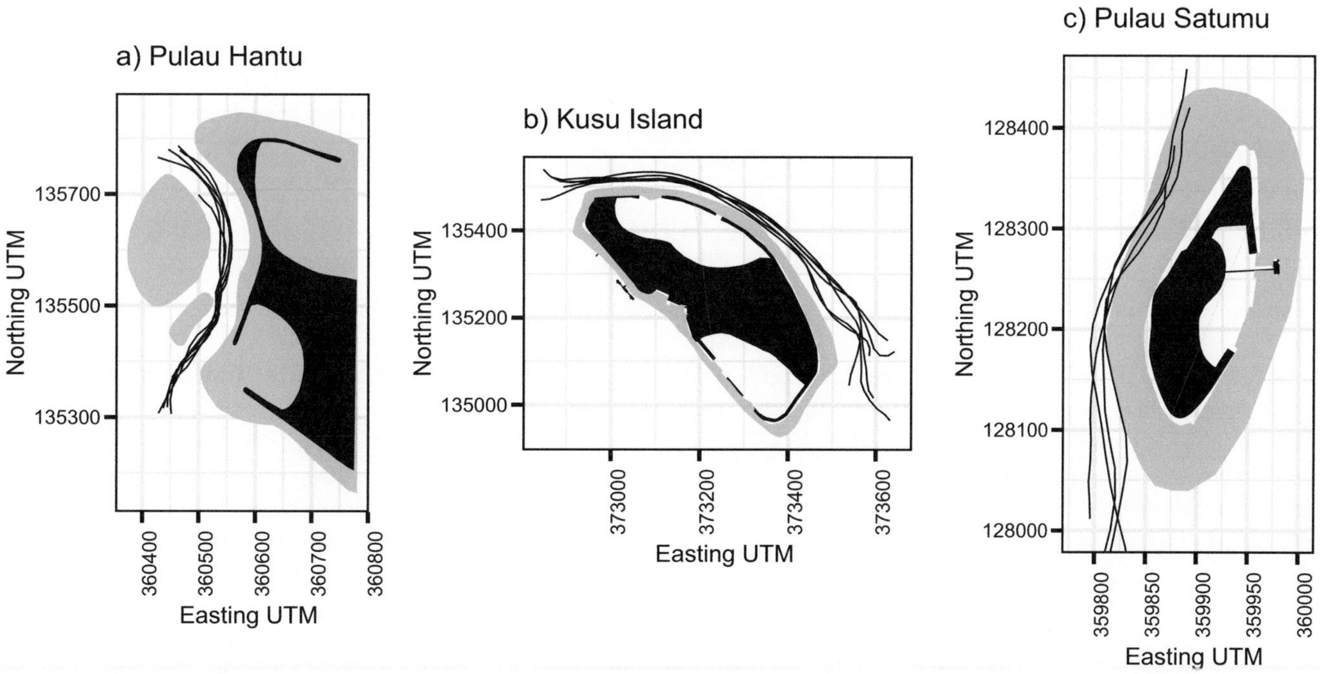

Fig. 2.9 Representative transects for the bathymetry and current velocity profile surveys conducted at (**a**) Pulau Hantu, (**b**) Kusu Island and (**c**) Pulau Satumu. Reef areas are denoted in light grey and land areas are denoted in black

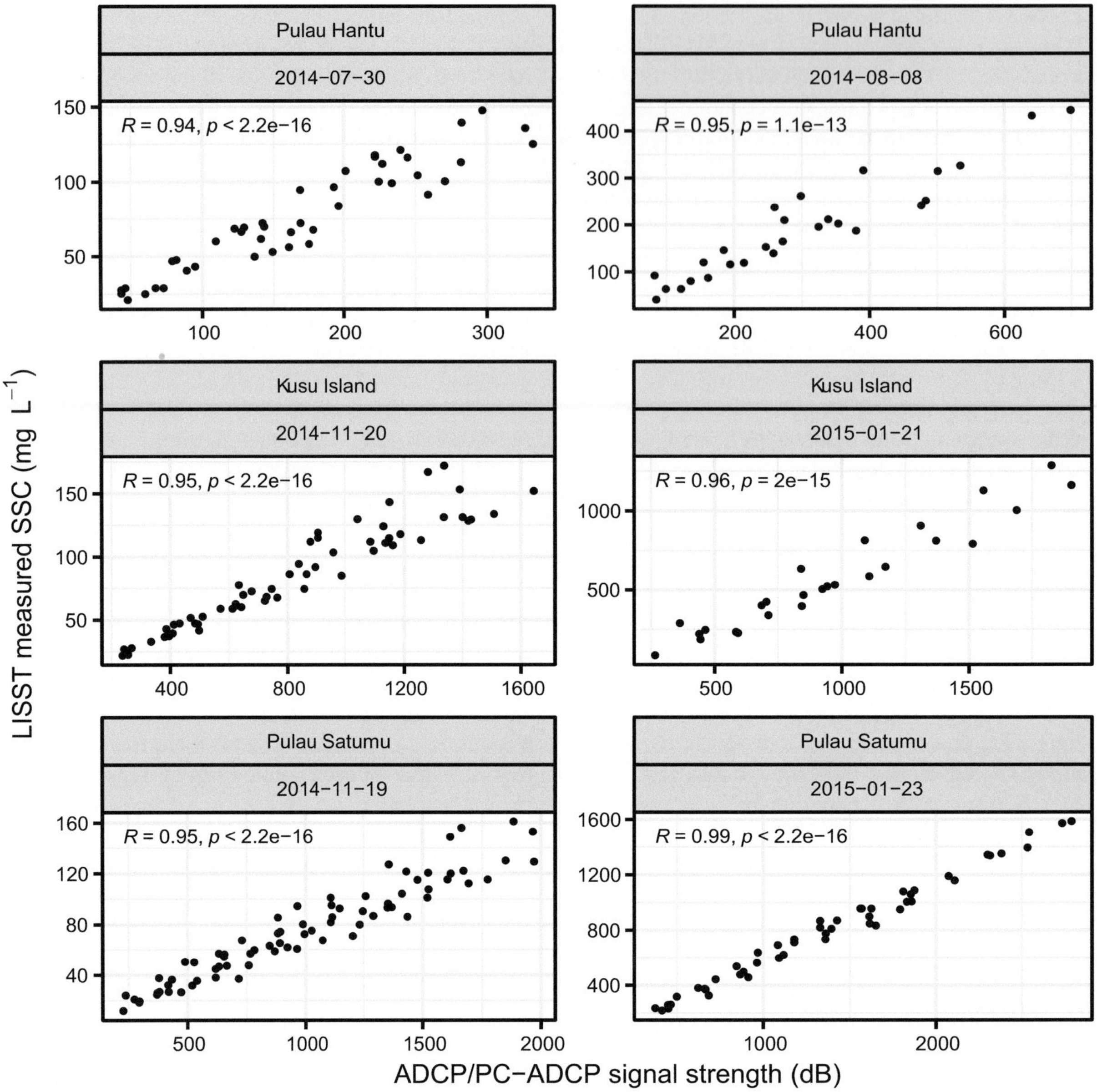

Fig. 2.10 Relationship between acoustic signal strength (dB) obtained from the Acoustic Doppler Current Profiler (ADCP) and the Pulse Coherent Acoustic Doppler Profiler (PC-ADCP) and suspended sediment concentration (SSC, mg L^{-1}) measured from the laser in situ scattering and transmissiometry equipment (LISST)

the reef sites and do not represent the full dataset. General summaries are discussed for the entire dataset, but visual representations are only provided for the representative transects. Surveys at Pulau Hantu were conducted in July and August 2014, while surveys for Kusu Island and Pulau Satumu were conducted in November 2014 and January 2015. There is an expectation that the different survey times will have an impact on the current velocity profiles. Both surveys at Pulau Hantu are conducted during the Southwest monsoon, while the surveys at Kusu Island and Pulau Satumu were conducted during the Northeast monsoon.

2.3.1 Depth-Averaged Water Flow Patterns

The depth-averaged flow patterns observed at Pulau Hantu were different between the two surveys (Fig. 2.11). During the July survey, currents along the fringing reef flowed in a north-south direction, while currents nearer the patch reef to the west flowed in a westerly direction. Current flow patterns during the August survey, on the other hand, generally moved in a westward direction, away from the fringing reef. These differences may be due to the tide stages when the transects were conducted; the July survey was conducted during the

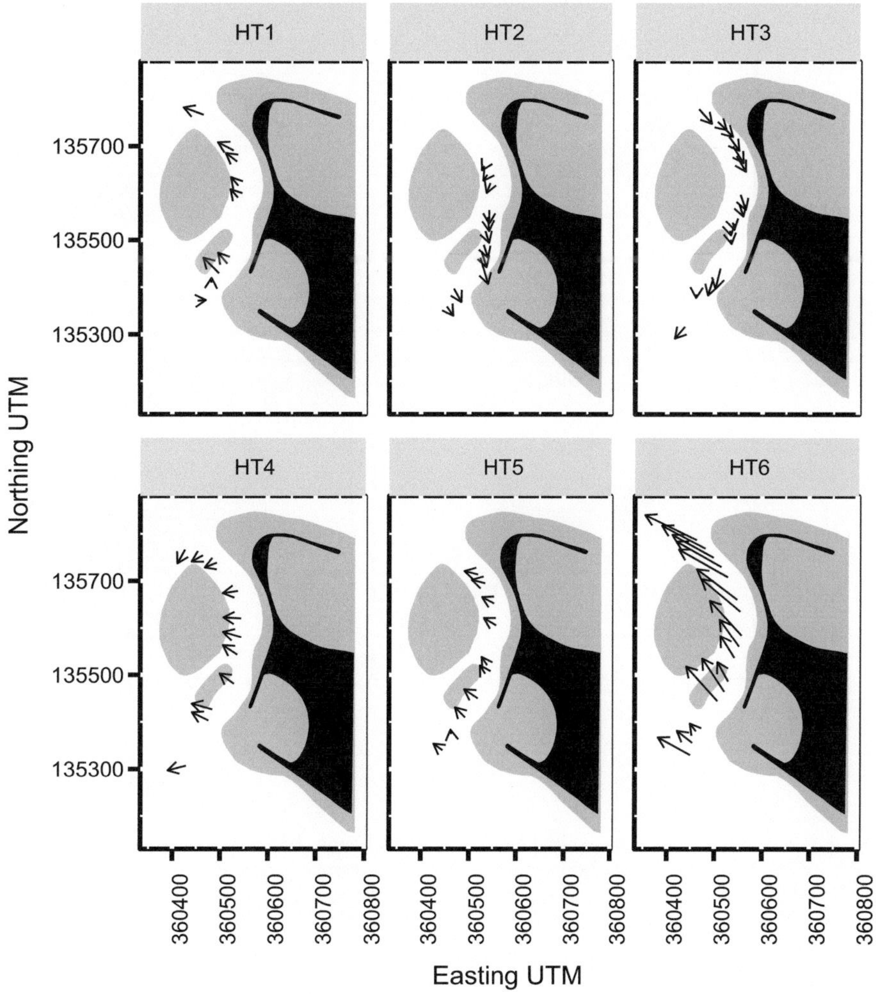

Fig. 2.11 Depth averaged water flow patterns at Pulau Hantu. Bathymetry and current velocity profile surveys at HT1-3 were conducted in July 2014, and HT4-6 were conducted in August 2014. Reef areas are denoted in grey and land areas are denoted in black

rising tide and the August survey occurred on the ebbing tide.

The depth-averaged flow fields at Kusu Island were more consistent between the two surveys (Fig. 2.12), as both surveys were conducted during the ebbing tide phase, where generally, water will be moving out of the Singapore Straits. The flow fields at Kusu Island indicated water flowing in a south-westerly direction, washing onto the fringing reef on the north-east of the island. Water flow patterns at Pulau Satumu did not seem to be affected by tide stage, but potentially more by the bathymetry (Fig. 2.13). All transects at Pulau Satumu were conducted during the rising tide, except for RL2. Generally, water at the edge of the reef (at the reef slope) moved away from the reef, while water directly over the reef flowed towards the island in an eastward flow.

2.3.2 Vertical Profiles in Current Velocity and SSC

Not surprisingly, the current velocities observed at the three reefs were markedly different. Current velocity range observed at Pulau Hantu was largest, between 0.001 m s^{-1} and 4.63 m s^{-1}, between 0.002 m s^{-1} and 0.85 m s^{-1} at Pulau Kusu, and between 0.002 m s^{-1} and 1.84 m s^{1} at Pulau Satumu. The vertical profiles in current velocity and SSC were also markedly different among the three reef sites. At Pulau Hantu, the vertical profile in current velocity was least variable among the three reef sites. Current velocity at Pulau Hantu was observed to increase with increasing water depth (Fig. 2.14). Near-bed flows at Pulau Hantu were as high as 4.7 m s^{-1} (data not shown) despite near-surface flows close to 0 m s^{-1}. This phenomenon is contrary to theory for shallow,

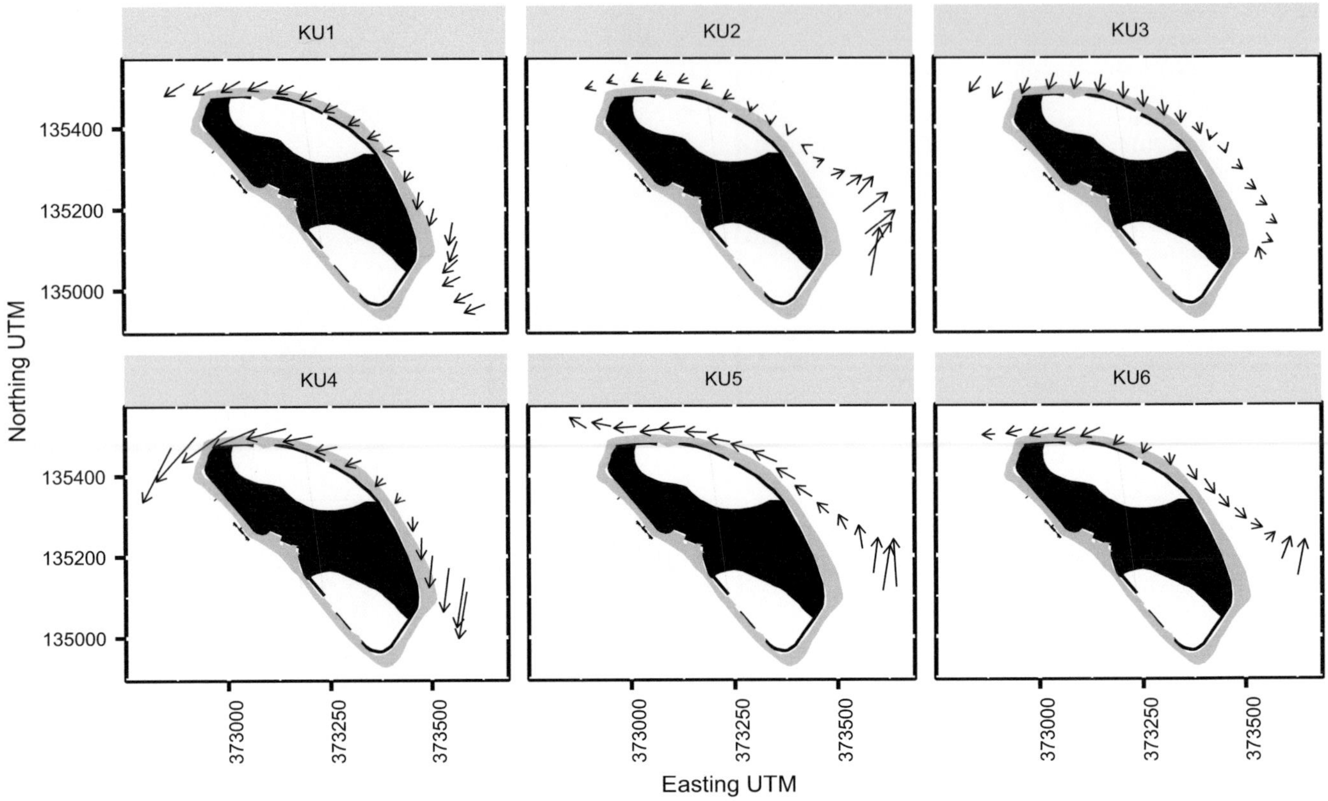

Fig. 2.12 Depth averaged water flow patterns at Kusu Island. Bathymetry and current velocity profile surveys at KU1-3 were conducted in Nov 2014, and KU4-6 were conducted in January 2015. Reef areas are denoted in grey and land areas are denoted in black

coastal waters, where the water speed should decrease with increasing velocity due to bed friction. With the increased bed friction, the flow of water is retarded and thus should be slower at the seabed than at the surface. Similar to the current velocity, SSC was observed to increase with depth (Fig. 2.15) and with increasing velocity (Fig. 2.16). The higher SSC at depth, at increased current velocities, may be due to increased turbulence near the seabed that resulted in events of local resuspension of sediments in the vicinity.

Vertical profiles of current velocity were highly variable at Kusu Island (Fig. 2.17). where it was apparent that the upper layers were dominated by highly mixed and turbulent flow and may reflect the more exposed nature of the location of this reef. The survey at Kusu Island was conducted during the Northeast monsoon, and the reef is exposed to the predominant monsoon winds, which may have resulted in increased and turbulent water movement. Measured SSC at Kusu Island showed variable patterns with depth for the two different sampling campaigns. During the November 2014 survey, SSC measured at Kusu Island was invariable through the water column; however, in January 2015, clear depth variations were noted. Higher SSC at Kusu Island was observed nearer to the seabed, with a difference of more than 30 mg L^{-1} between surface SSC and near-bed SSC (Fig. 2.18). The relationship between SSC and current velocity did not show any clear relationships and did not indicate resuspension events (Fig. 2.19). Measured velocities were commonly <0.5 m s^{-1}. A threshold velocity of 0.68 m s^{-1} to resuspend sediments at Kusu Island was earlier identified. It is likely that the flow conditions during the survey campaign did not capture the current velocity range that would result in resuspension of sediment. The results nonetheless demonstrate that at low current velocities, local advection from the external sources or increased turbulence from wind waves or ship wakes may be an important mechanism in the SSC dynamics at Kusu Island.

Vertical profiles of current velocity at Pulau Satumu were also highly variable (Fig. 2.20). Like Kusu Island, measured SSC at Pulau Satumu also showed variable patterns with depth for the two separate surveys (Fig. 2.21). Some indications of depth differences were observed during the November survey, where concentrations increased with depths up to 15 m but rapidly decreased after. No clear pattern in SSC with depth was observed for the January survey. There was also no clear relationship between SSC and current velocity, with no clear indication of resuspension occurring (Fig. 2.22); however, the velocities measured during the sampling campaign were considerably lower than the thresh-

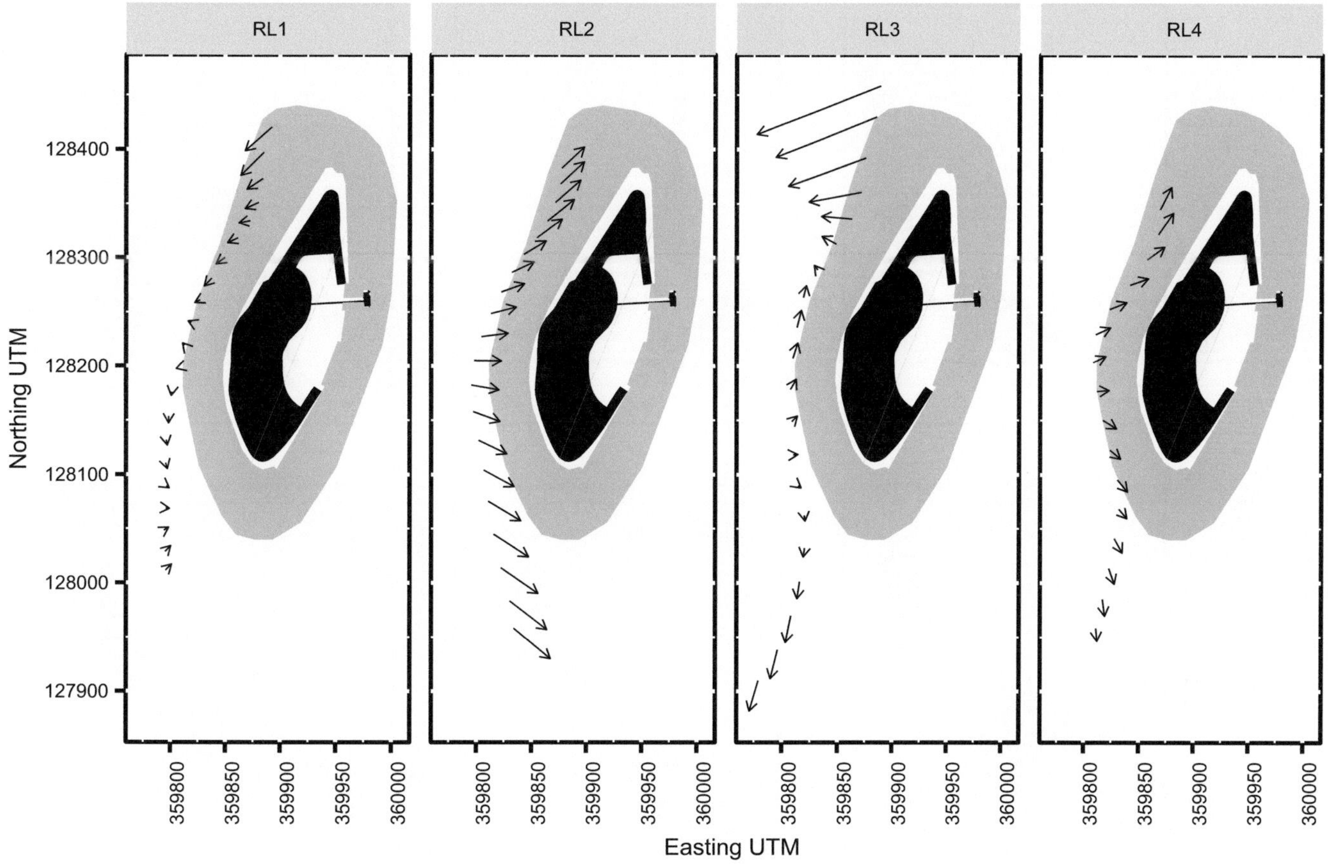

Fig. 2.13 Depth averaged water flow patterns at Pulau Satumu. Bathymetry and current velocity profile surveys at RL1-2 were conducted in Nov 2014, and RL3-4 were conducted in January 2015. Fringing reef is denoted in grey and land areas are denoted in black

old velocity previously identified, of 0.59 m s^{-1}. It is likely that advection is a more likely mechanism in the determination of SSC observed at Pulau Satumu.

These spatial differences in SSC and current velocity may arise from both orientation of the reef to flow and complexity of surrounding environments and/or the bathymetry. The reefs fringing both Pulau Satumu and Pulau Hantu are located on the westward sides of the island, i.e. in the lee of flows in the westward direction, which may be a reason why threshold velocity (based on computed velocity) and SSC relationships at these two reefs (Fig. 2.4) were more apparent with the eastward direction. Similarly, Kusu Island is sheltered from flows from the westward direction by the Pulau Seringat Complex, which is an amalgamation of Pulau Seringat, Lazarus Island and a newly reclaimed patch reef. In addition, Pulau Satumu and Kusu Island are proximal to the main shipping channels and deeper waters and may face increased turbulence in the water column due to ship wakes and wind waves.

2.4 Long-Term Trends in Suspended Sediments

Aside from the influence of tidally driven flow on suspended sediment dynamics, sources of input of sediments will also affect the concentration of suspended sediments in an environment. In Singapore, sediment has been introduced into the shallow-water coastal areas from extensive land reclamation that has taken place in the past and which is still a continuing coastal activity. Studies documenting decadal changes in water clarity in Singapore are few. There have been several studies in the past that have examined the status of water quality (including suspended sediments) in the reef environments in Singapore. Many of these studies only cover limited time spans. Data coverage in these studies was as short as once-off surveys, to spans of 3–12 months (Basheer et al. 2002, 2004; Gin et al. 2003; Dikou and van Woesik 2006; Browne et al. 2015; Morgan et al. 2020). Few studies exceeded a year's worth of data acquisition. To fill this data gap, a dedicated long-term sampling programme that moni-

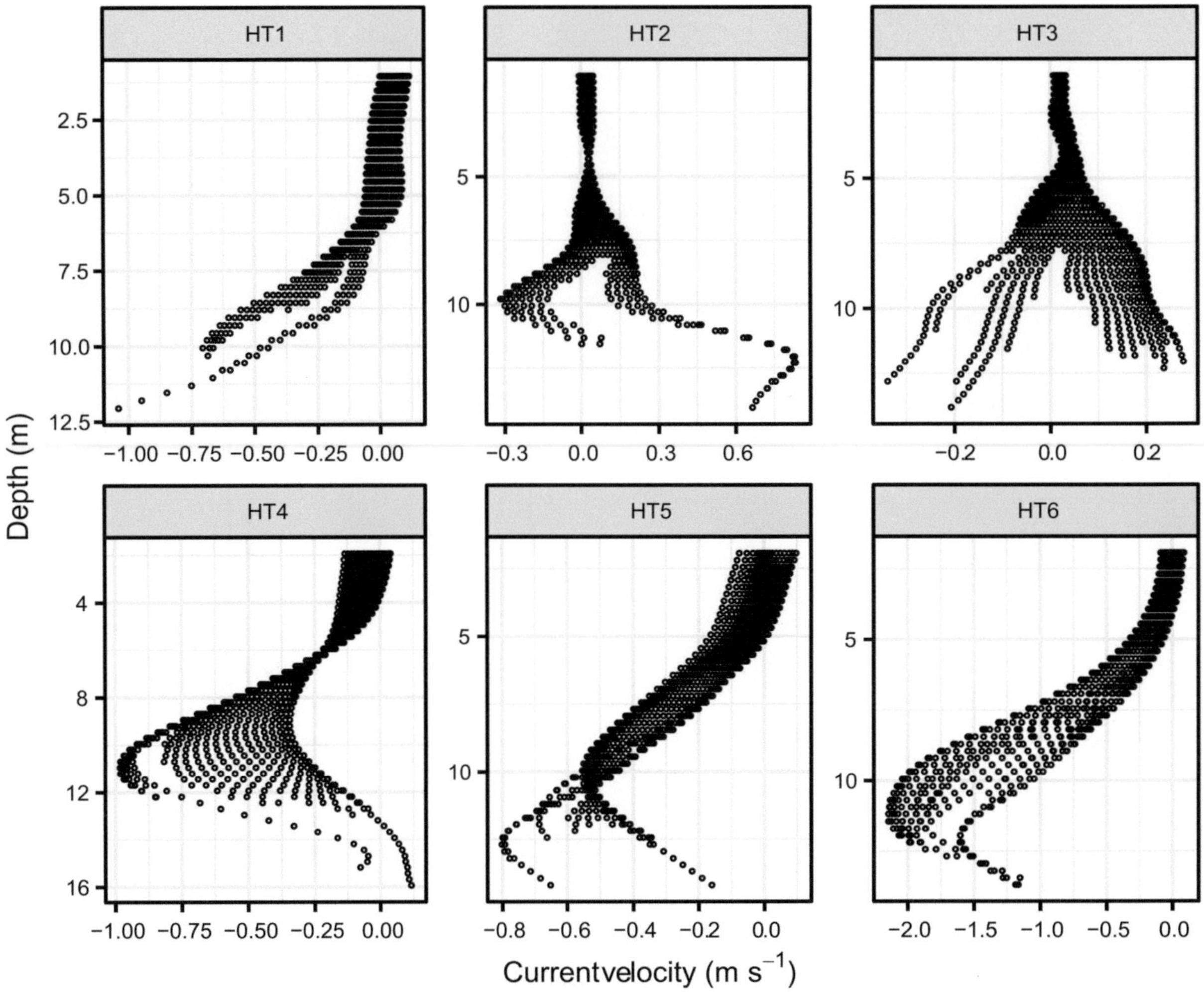

Fig. 2.14 Current velocity profile with depth for representative transects at Pulau Hantu. Bathymetry and current velocity profile surveys at HT1-3 were conducted in July 2014, and HT4-6 were conducted in August 2014

tored SSC on reefs was conducted. Concentrations of suspended sediments on three reefs (Pulau Hantu, Kusu Island and Pulau Satumu) were measured monthly from 2012 to present. Triplicate samples were obtained at the reef crest, at approximately 3 m below mean sea level, using 1 L water samplers for SSC. Samples were vacuum filtered in the laboratory on pre-weighed glass fibre filter paper (Millepore AP400, pore size 0.7 μm), dried at 105 °C for at least 24 h and weighed. The drying process was continued until a stable weight was obtained.

Long-term trends in SSC at the reefs appeared consistent for the three reef sites sampled, with no clear spatial differences among reef sites (Fig. 2.20). Measured SSC values were between 1 mg L^{-1} and 195 mg L^{-1}, which are similar to previous studies (e.g. Low and Chou 1994; Dikou and van Woesik 2006; Browne et al. 2015; Morgan et al. 2020). The greatest variations in measured SSC were observed between distinct time periods in the long-term trends (Fig. 2.23). The differences in SSC between time periods may be associated with a specific coastal activity at that point in time. Between 2012 and present, the most significant coastal development projects included the building of the Pasir Panjang Container Terminal (Phases 3 and 4) and the initiation of the Tuas Mega Port. Pasir Panjang Container Terminal phases 3 and 4 were completed in 2015, and the Tuas Mega Port Phase 1 was between 2015 and 2021 and Phase 2, which commenced in 2018. The elevated SSC observed between 2013 and 2015 and between mid-2018 and 2021 may be linked to the reclamation that occurred during that time.

Aside from local sources, such as those resulting from land reclamation activities, sediments could enter the Singapore domain from a variety of external sources. Sediment input into the Singapore Strait may potentially be influenced by the predominant flow of currents into the region, driven by monsoonal current flows. Monsoon-driven flow can potentially deliver sediment sources from the South China Sea during the Northeast monsoon and from the Java Sea and Malacca Straits during the Southwest monsoon.

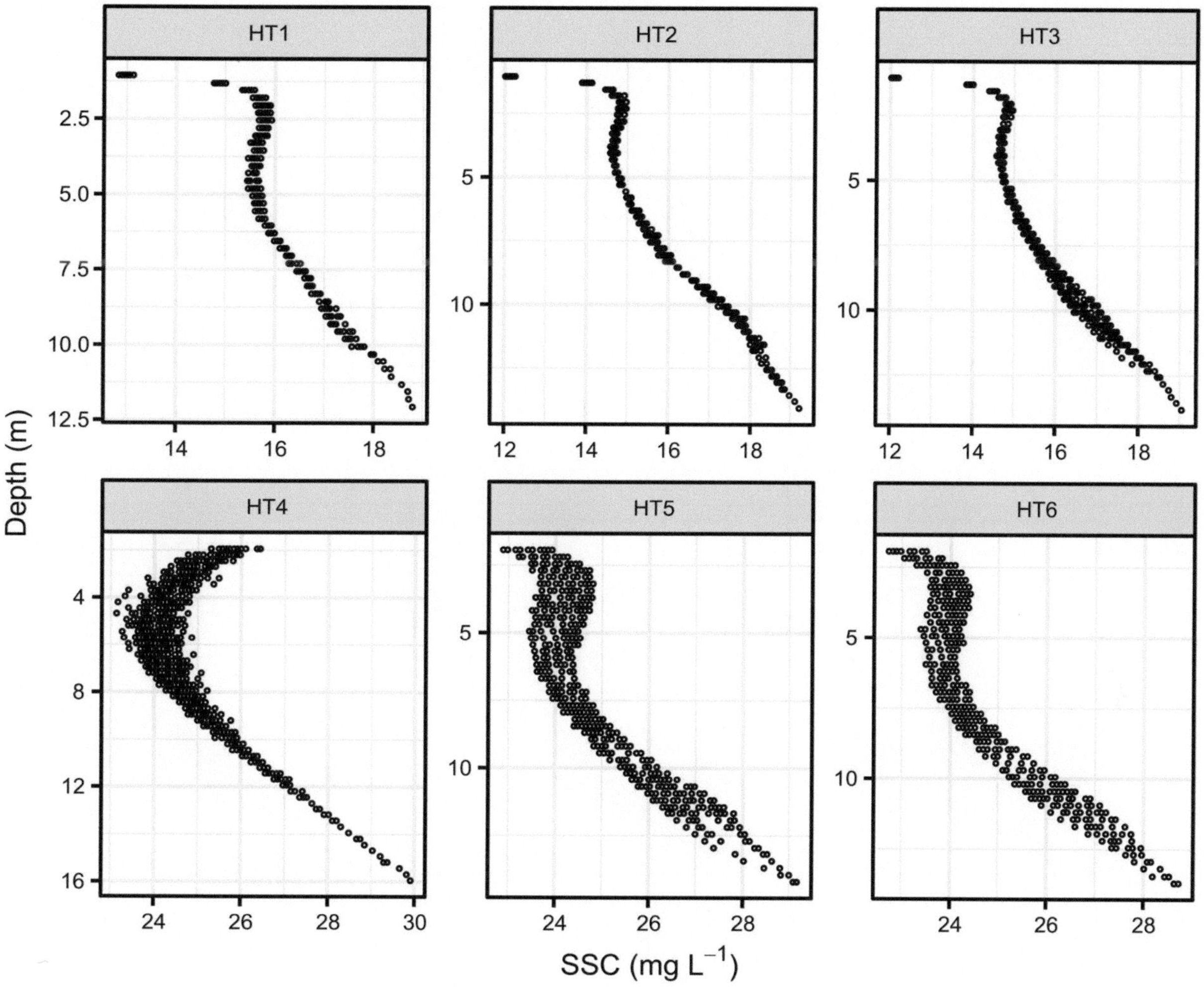

Fig. 2.15 Variation of suspended sediment concentration (SSC, mg L^{-1}) with depth for representative transects at Pulau Hantu. Bathymetry and current velocity profile surveys at HT1-3 were conducted in July 2014, and HT4-6 were conducted in August 2014

However, no distinct seasonal variation in SSC was observed (Fig. 2.24), suggesting that sediment inputs into the Singapore domain were not driven by monsoonal flows. In contrast, previous studies have reported elevated suspended sediment concentrations during the Southwest monsoon (Browne et al. 2015) or during the inter-monsoon periods (Dikou and van Woesik 2006). Discrepancy in the findings between previous studies and the long-term trends presented here could be a result of the data collection period. The long-term data have likely captured the larger variation in SSC that occurred among different time periods and masked the smaller signal that was driven by monsoon periods. It is therefore important to note that making generalisations on seasonal influences based on 12 months of data can lead to misleading conclusions, as there is no replication of the seasonal influences. Minimally 24 months of data would be required for making inferences on seasonal influences. However, these narrow datasets do provide some insights into the intra-annual variability in water quality conditions and should not be discounted entirely.

2.5 The Murky Future

The results from the various surveys and studies presented here demonstrate the importance of current velocity in determining sediment flux, which will also correspond to the frequency of the spring-neap cycles of the semi-diurnal tide. The intra-tidal (flood-ebb) effects on sediment flux were spatially variable. Literature has suggested that reefs in Singapore have experienced elevated sedimentation and correspondingly, an increased accumulation of fine sediments on the seabed. Fine sediments are acknowledged as the dominant abiotic component on coral reefs in Singapore (e.g. Chou and Tun 2012). A demonstration of high mobility in sediments is provided in this study, with some indication for local resuspension of bed sediments. Moreover, the impor-

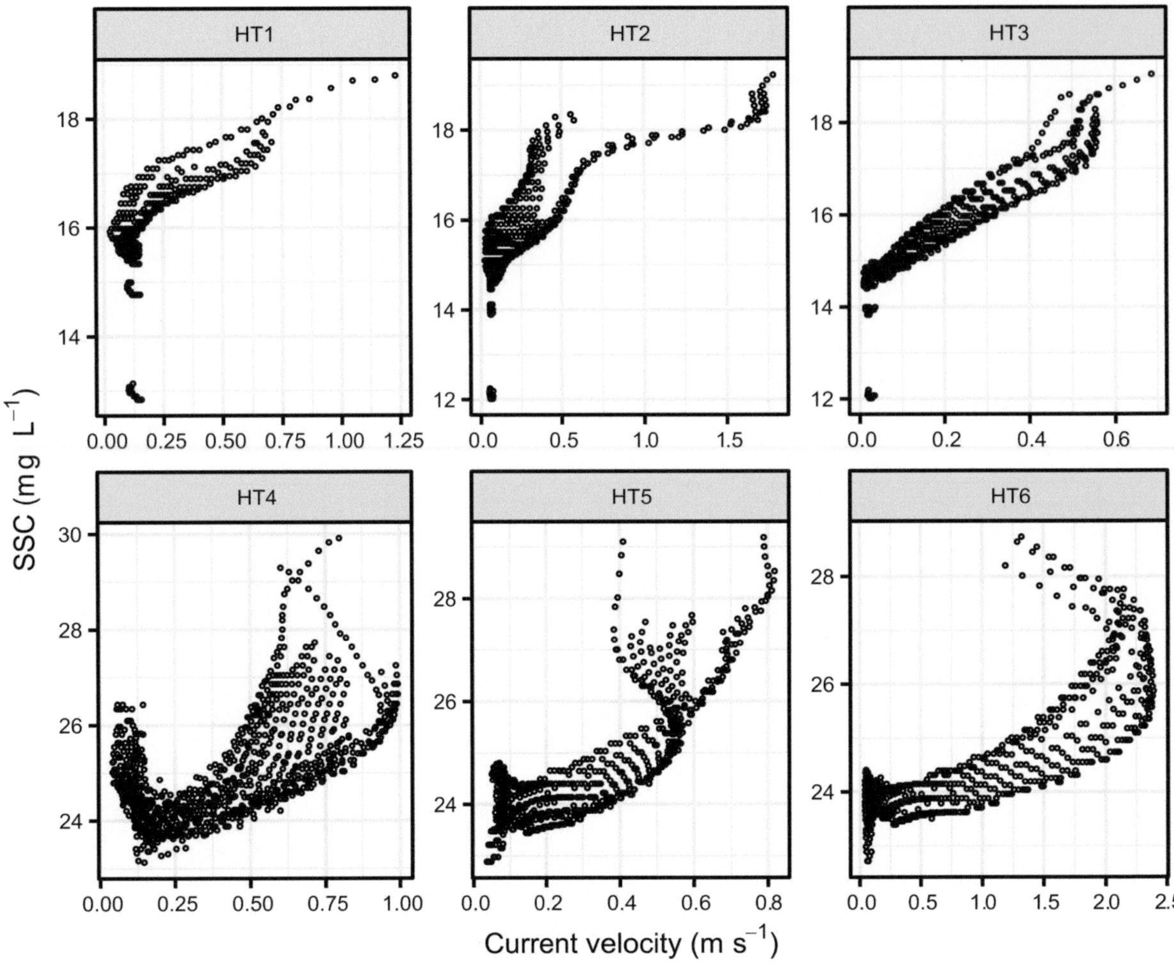

Fig. 2.16 Relationship between suspended sediment concentration (SSC, mg L^{-1}) and current velocity (m s^{-1}) for representative transects at Pulau Hantu. Bathymetry and current velocity profile surveys at HT1–3 were conducted in July 2014, and HT4–6 were conducted in August 2014

tance of orientation of the reef to main flow direction on suspended sediment is demonstrated, with evidence for resuspension due to island wakes. Reef environments in Singapore may appear to be both sources and sinks of suspended sediments, depending on the different hydrodynamic conditions.

The strong association of sediment flux with tidal currents is unusual for shallow reef environments; the prevailing theory is that wind and swell waves are dominant causes of high suspended sediment concentrations (Larcombe et al. 1995, 2001). However, there has been some evidence for flow processes influencing suspended sediment dynamics of shallow water islands by resuspension and/or upwelling processes (Wolanski et al. 1984, 1996; Ingram and Chu 1987), which may be relevant in Singapore. Complex eddies are formed in the wake of shallow-water islands (Hammer and Hauri 1977; Wolanski et al. 1984), creating bottom friction, which can generate strong upwelling (Wolanski et al. 1984). Island-wake effects were demonstrated at Pulau Satumu, which were clearly indicated by the flow patterns captured during the flooding tide. Some indications of island wake are also observed at Kusu Island, where higher current speeds were recorded when flows 'wrapped' around the island (A. C. Lee, pers. obs). However, as measurements were not obtained for the area in the lee of the flow direction, further conclusions cannot be made.

Wave-related resuspension is an important source of sediments over many coral reefs (Larcombe et al. 2001; Douillet et al. 2001; Ouillon et al. 2020). However, wave effects on sediment resuspension are frequently ignored in Singapore as the wave climate has been described as a low energy regime (Chia et al. 1988) and therefore not thought to contribute significantly to sediment dynamics (Zoelaeha 2010). An influencing factor to sediment dynamics to consider in Singapore is the presence of ship wakes. Singapore, as a large shipping port, is subject to high shipping traffic volume ranging from large shipping vessels, such as crude carriers, to high-speed ferries. The wash wakes from large vessels and

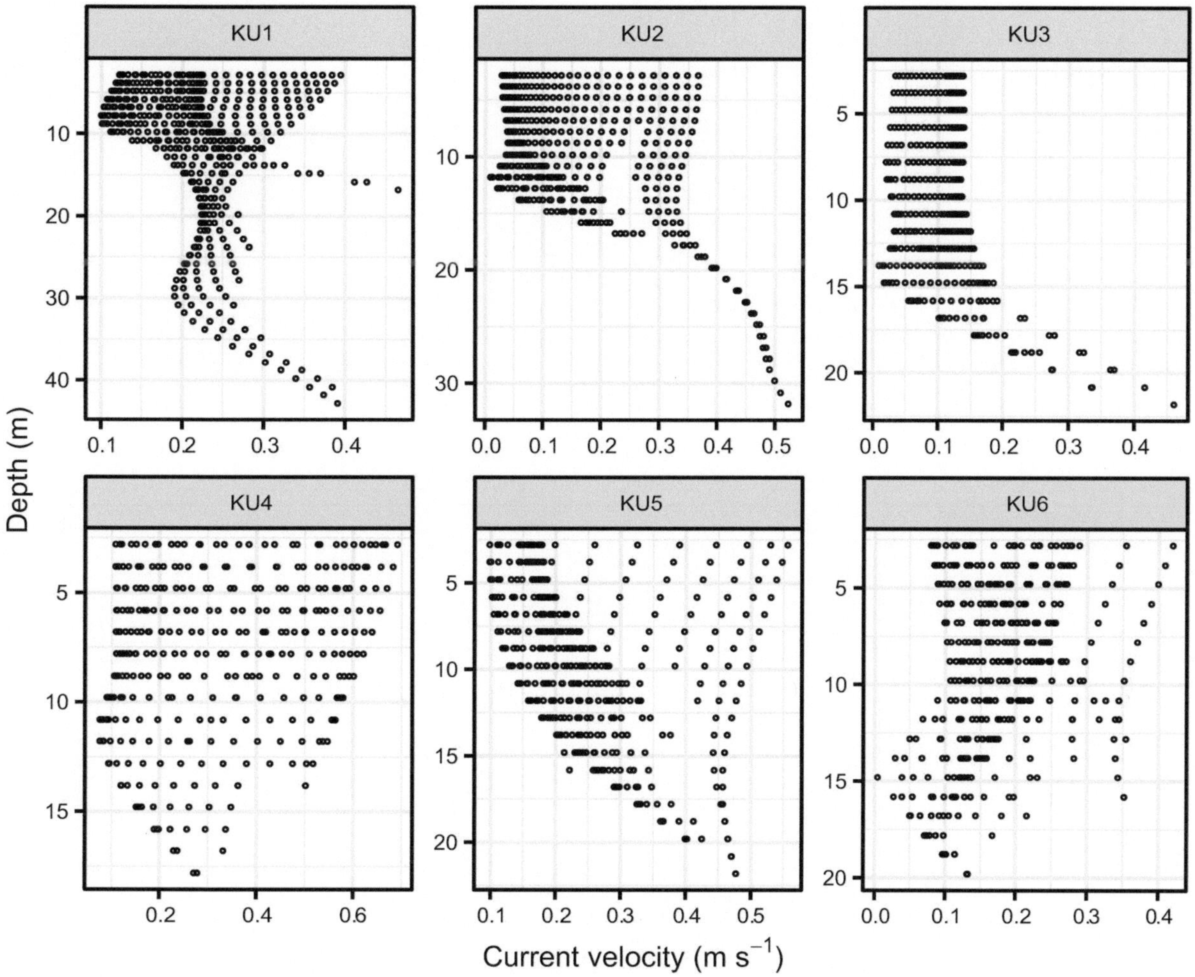

Fig. 2.17 Current velocity profile with depth for representative transects at Kusu Island. Bathymetry and current velocity profile surveys at KU1-3 were conducted in Nov 2014, and KU4-6 were conducted in January 2015

high-speed ferries have been shown to significantly impact resuspension of bottom sediments and the transport of the resuspended sediments (Erm et al. 2011; Houser 2011). In a numerical study that evaluated the impact of ship wakes from fast ferries on two near-shore areas in Singapore, Wong (2018) demonstrated that ship wakes can generate wave heights as high as 0.5 m and are likely to affect Singapore's coastal hydrodynamics. While there are no specific studies on the effect of ship-wakes on near-shore environments in Singapore, Browne et al. (2017) have made some anecdotal observations on the frequency (8–15 ship wakes during daylight hours), size (>0.4 m) and the duration of increased sediment resuspension (10–30 min). Similarly, over the long-term monitoring programme, frequent occurrences of pulse increase of sediments in the water column due to the passing of high-speed vessels have been observed at local reefs (A. C. Lee, pers. obs.). During the ship-wake events, not only does visibility deteriorate noticeably, often reducing horizontal underwater visibility to less than 10 cm, but the underwater conditions become highly chaotic. Sometimes these conditions will dissipate within 10–20 min, while at other times, conditions do not improve noticeably over time (A. C. Lee, pers. obs.). The bathymetry and configuration of Singapore reefs may further influence the behaviour of ship wakes and their breaking characteristics and their effects on water oscillations, which in turn affect the sediment dynamics over the reef. The reef areas in this study consisted of steep slopes with typical gradients of 1:3 or steeper for most parts of the reef (A. C. Lee, pers. obs.). Steep slopes encourage the formation of plunging waves, characterised by the release of wave energy in a violent way (USACE 1984). The steep slopes of local reefs may encourage breaking of ship wakes, while the reflectance of the wake energy off the armoured backreef may also change sediment resuspension characteristics. However, it is unknown how the interaction between ship wakes and the characteristics of the reef and armoured backreef can influence sediment resuspension. This will require further study that was not done here.

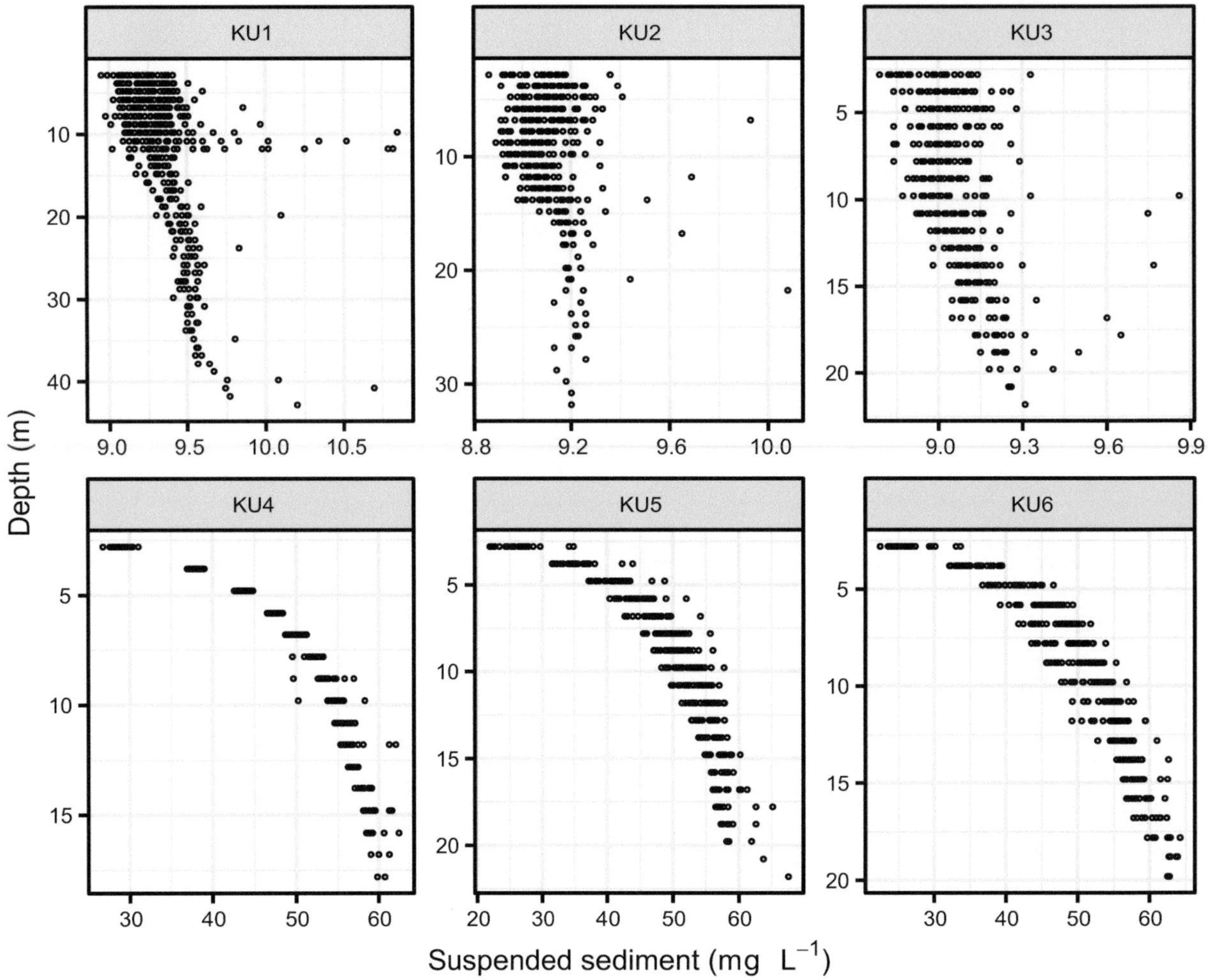

Fig. 2.18 Variation of suspended sediment concentration (SSC, mg L^{-1}) with depth for representative transects at Kusu Island. Bathymetry and current velocity profile surveys at KU1-3 were conducted in November 2014, and KU4-6 were conducted in January 2014

Entrainment of suspended sediments over coral reefs may also be promoted by coastal modifications. In Singapore, a large amount of the coastline has been structurally armoured, with breakwaters constructed shoreward of most fringing reefs. A typical backreef (on the leeward) consists of lagoons or reef flats, which promote energy dissipation of oncoming waves and are natural sinks for sediments (Lugo-Fernandez et al. 1998a, b). The armoured seawall at the back of reefs creates energy reflectance. Increased energy reflectance has been demonstrated for steep-faced seawalls (USACE 1984). When reef-impounded water can only be discharged through the seaward side of the reef, wave-setup would be at its highest (Gourlay 1996). This will result in higher mean water level in the leeward side of the reef. However, in Singapore, where armouring of backreefs is evident, the increased wave set-up caused by the higher mean water level could potentially affect sediment dynamics.

It is undeniable that the condition of coral reefs in Singapore has deteriorated over the last few decades (Chou et al. 2004; Hoeksema and Koh 2009). Enhanced coastal development (locally and distantly), land reclamation, dredging for shipping channel maintenance and a busy port, with its traffic of container vessels, tankers, bulk carriers and fast ferries, are all factors that influence the sediment dynamics in the reef environment. Globally, increased sediment stress has been identified as a major cause of decline in shallow-water, light-sensitive ecosystems including coral reefs and seagrass beds (Fabricius 2005; Erftemeijer and Lewis 2006). Increased sediment stress currently ranks below only direct habitat destruction as the most important driver of coral reef

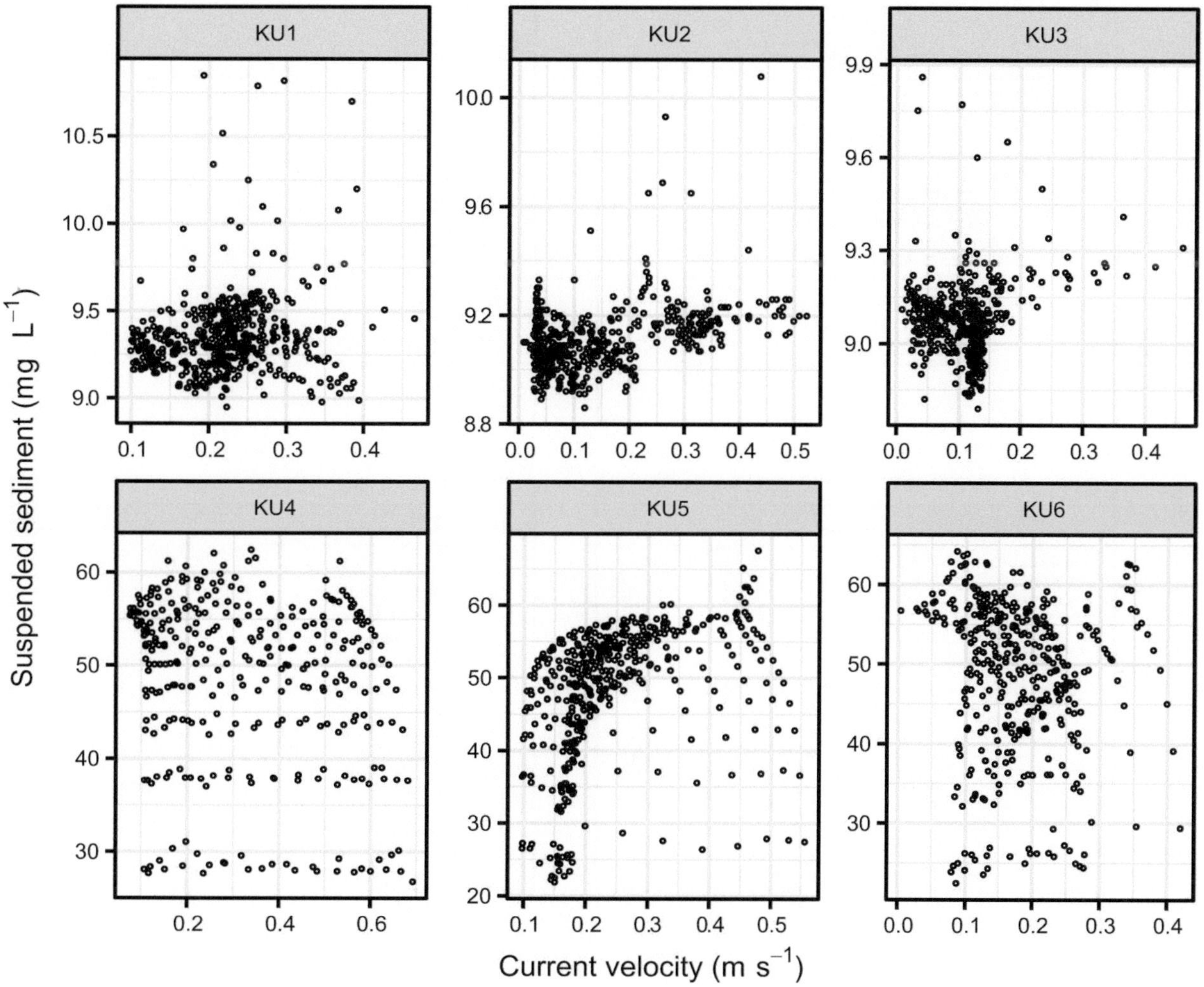

Fig. 2.19 Relationship between suspended sediment concentration (SSC, mg L^{-1}) and current velocity (m s^{-1}) for representative transects at Kusu Island. Bathymetry and current velocity profile surveys at KU1-3 were conducted in November 2014, and KU4-6 were conducted in January 2015

condition (Fabricius 2005; Burke et al. 2002, 2011, 2012). Singapore reefs are not exempt from this. Understanding how sediments behave in a topographically complex seabed is imperative in the management of the health of our reefs.

To fully comprehend the dynamics of suspended sediment behaviour on coral reefs in Singapore waters, there is a need to be able to better quantify both local and external sources of sediment delivered into the system. Furthermore, it is important to not only focus on sediment dynamics directly on the reefs but also at the adjacent off-reef areas too. As a transitional reef (sensu Zweifler et al. 2021), it is possible that the corals in Singapore have developed adaptive strategies to deal with sediment stress over the last six decades, particularly from large land reclamation projects (Chou 2023). Sediment stress in Singapore is unlikely to reduce in the future, however, given that there have been no alarming declines in hard coral cover on most of the reefs in Singapore in the last 15 years (Chan et al. 2024), there is a high potential that coral reefs in Singapore have become better adapted to high sediment load conditions and may show even greater resilience than before.

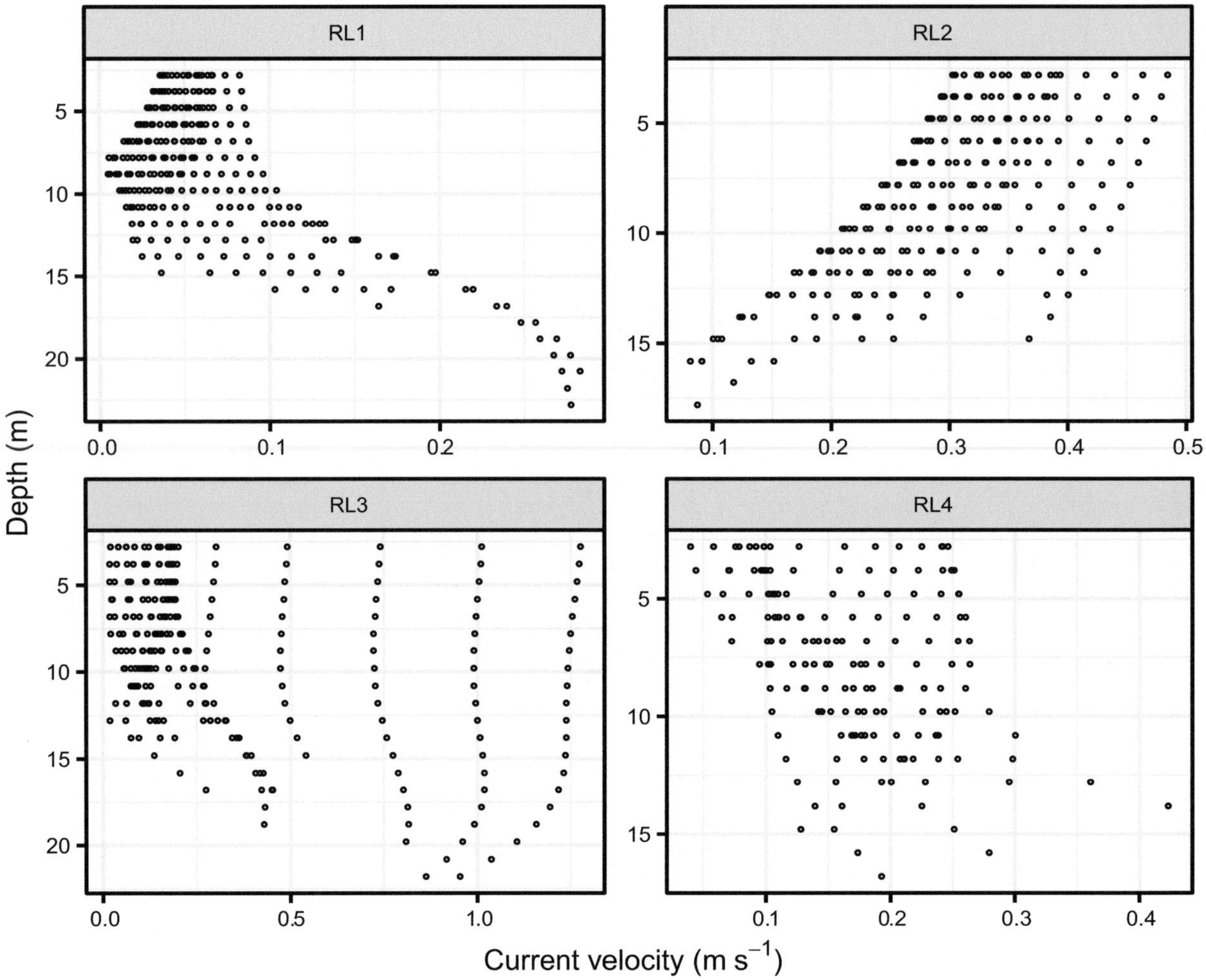

Fig. 2.20 Current velocity profile with depth for representative transects at Pulau Satumu. Bathymetry and current velocity profile surveys at RL1-2 were conducted in Nov 2014, and RL3-4 were conducted in January 2015

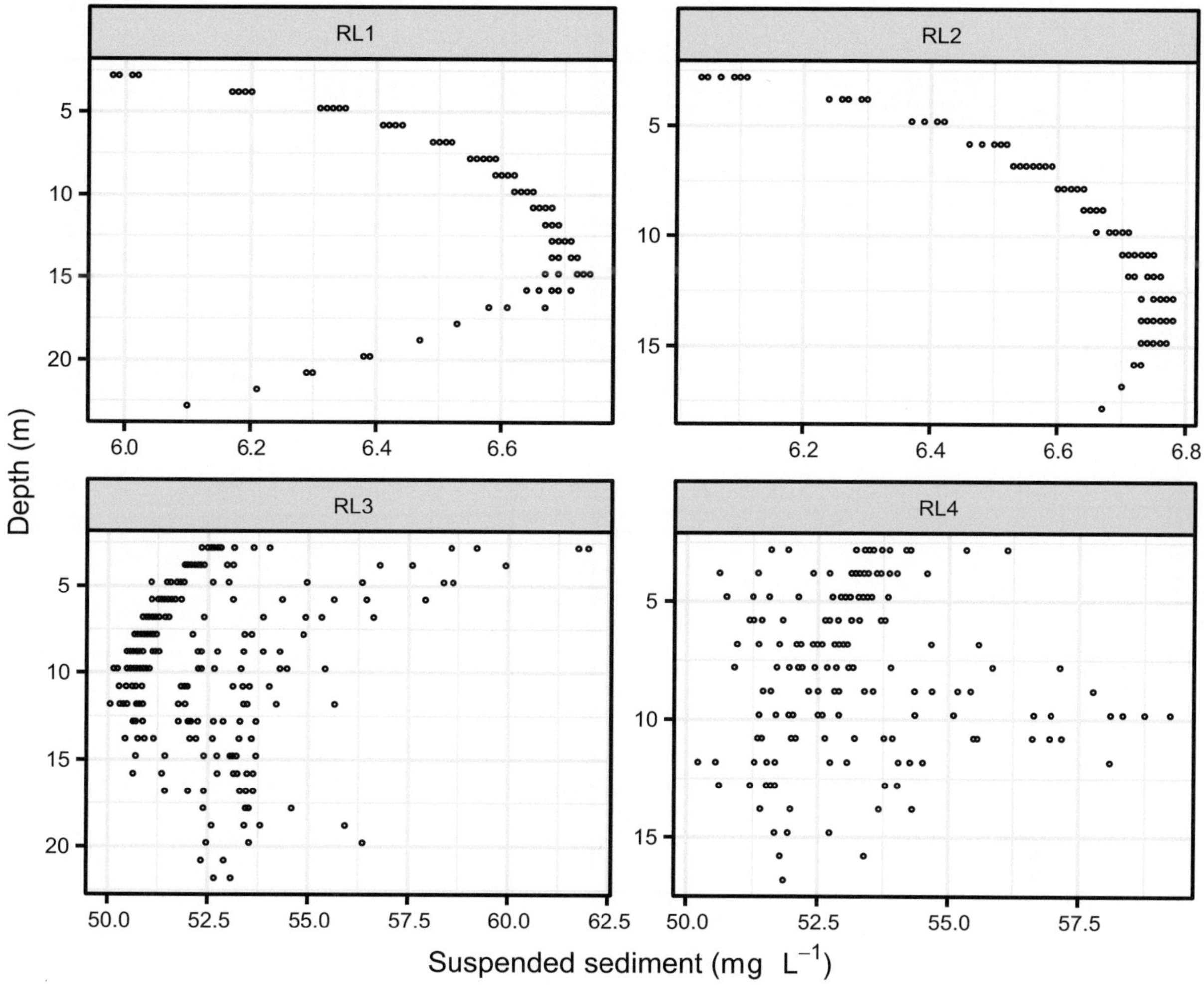

Fig. 2.21 Variation of suspended sediment concentration (SSC) (mg L^{-1}) with depth for representative transects at Pulau Satumu. Bathymetry and current velocity profile surveys at RL1-2 were conducted in November 2014, and RL3-4 were conducted in January 2014

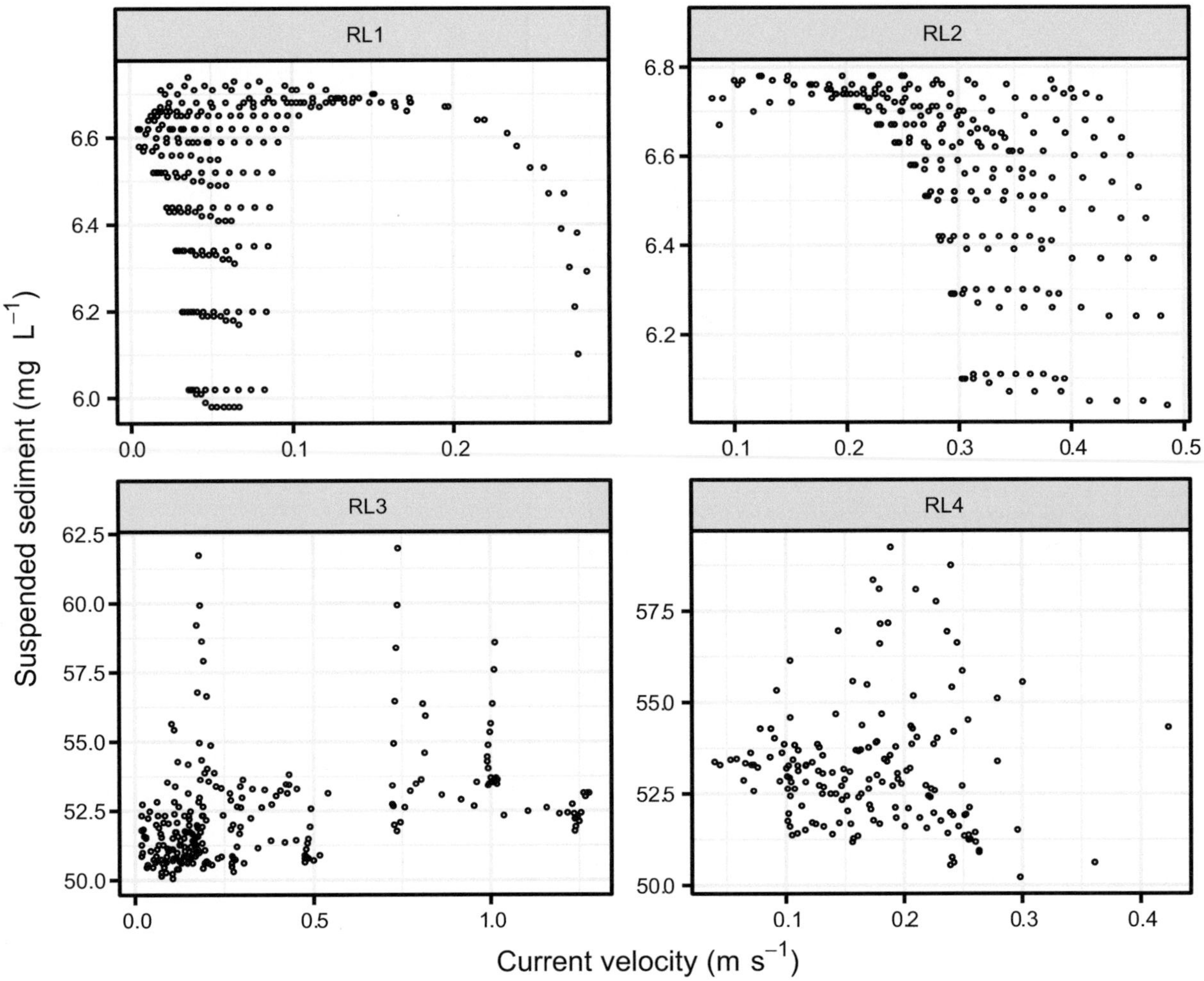

Fig. 2.22 Relationship between suspended sediment concentration (SSC) (mg L^{-1}) and current velocity (m s^{-1}) for representative transects at Pulau Satumu. Bathymetry and current velocity profile surveys at RL1-2 were conducted in November 2014, and RL3-4 were conducted in January 2015

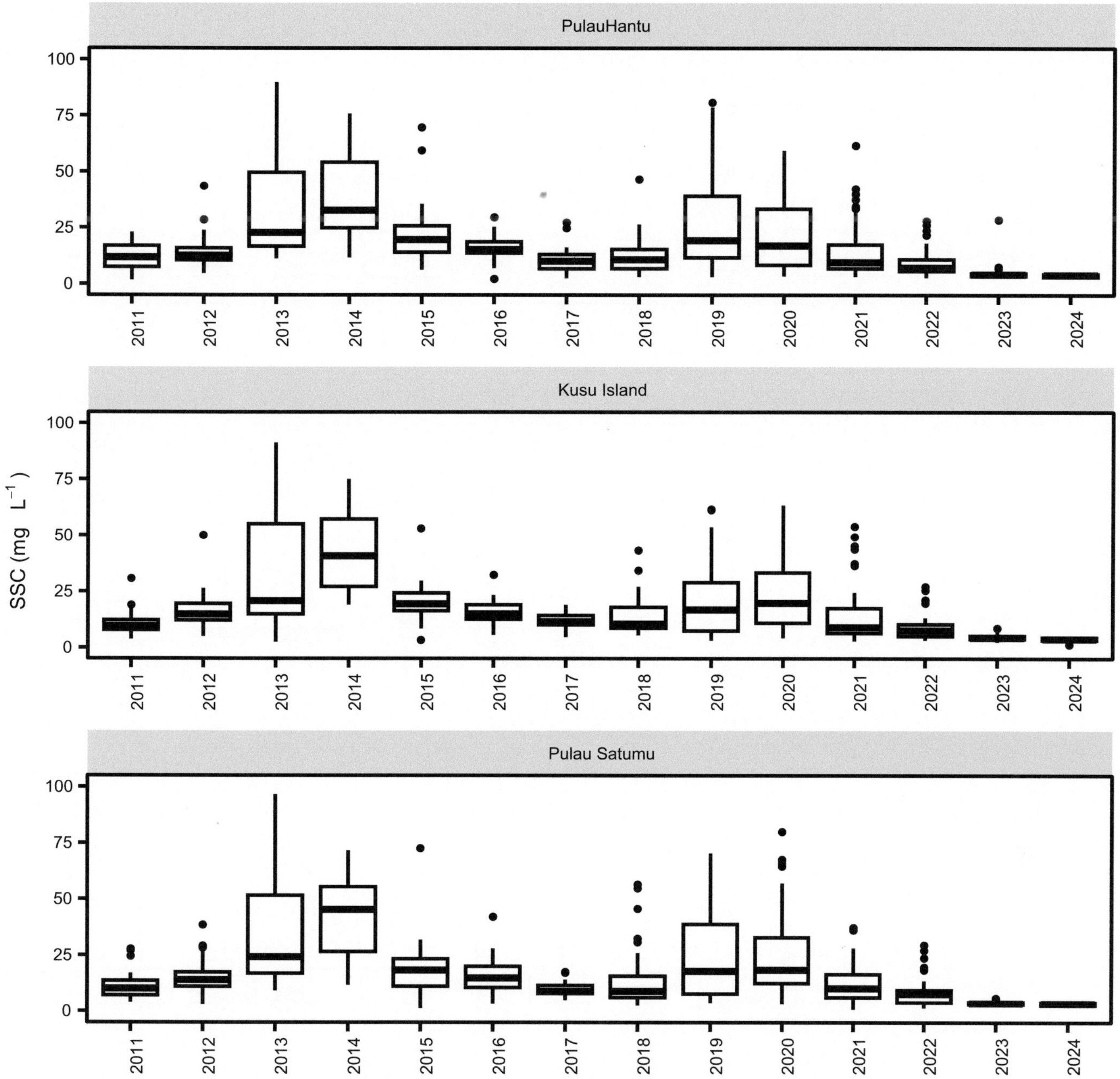

Fig. 2.23 Long-term trends in suspended solid concentration (SSC, mg L^{-1}) at Pulau Hantu, Kusu Island and Pulau Satumu

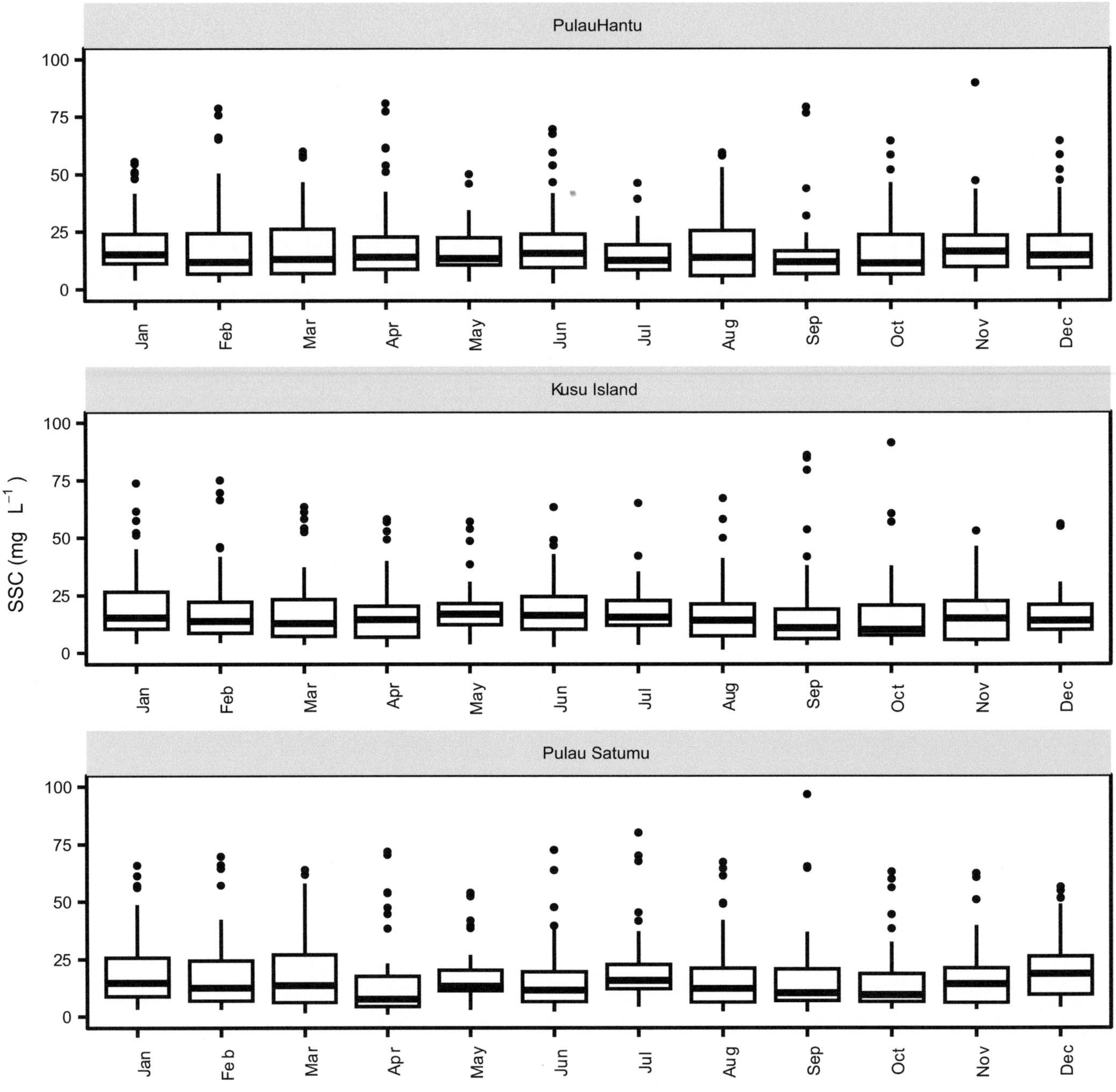

Fig. 2.24 Monthly variation in suspended solid concentration (SSC, mg L^{-1}) at Pulau Hantu, Kusu Island and Pulau Satumu

References

Basheer C, Tan KS, Lee HK (2002) Organotin and Irgarol-1051 contamination in Singapore coastal waters. Mar Pollut Bull 44(7):697–703

Basheer C, Lee HK, Tan KS (2004) Endocrine disrupting alkylphenols and bisphenol-a in coastal waters and supermarket seafood from Singapore. Mar Pollut Bull 48:1161–1167

Browne NK, Smithers SG, Perry CT, Ridd PV (2012) A field-based technique for measuring sediment flux on coral reefs: application to turbid reefs on the Great Barrier Reef. J Coast Res 284:1247–1262

Browne NK, Yaakub SM, Tay, Tay JK, Todd PA (2017) Recreating the shading effects of ship wake induced turbidity to test acclimation responses in the seagrass Thalassia hemprichii Estuarine Coastal and Shelf Science 19987–95. https://doi.org/10.1016/j.ecss.2017.09.034

Browne NK, Tay JKL, Low J, Larson O, Todd PA (2015) Fluctuations in coral health of four common inshore reef corals in response to seasonal and anthropogenic changes in water quality. Mar Environ Res 105:39–52

Burke L, Selig E, Spalding M (2002) Reefs at risk in Southeast Asia. World Resources Institute, 74 pp

Burke L, Reytar K, Spalding M, Perry A (2011) Reefs at risk: revisited. World Resources Institute, Washington, DC, 130 pp

Burke L, Reytar K, Spalding M, Perry A (2012) Reefs at risk: revisited in the coral. World Resources Institute, Triangle, 86 pp

Chan YKS, Ng CSL, Tun KPP, Chou LM, Huang D (2024) Decadal decline of functional diversity despite increasing taxonomic and phylogenetic diversity of coral reefs under chronic urbanisation stress. Ecol Indic 164:112143

Chia LS, Habibullah K, Chou LM (1988) The coastal environmental profile of Singapore. International Center for Living Aquatic Resources Management, Manila, 92 pp

Chou LM (2006) Marine habitats in one of the world's busiest Harbours. In: Wolanski E (ed) The environment in Asia Pacific Harbours. Springer, pp 377–391

Chou LM, Chia LS (1991) The marine environment. In: Chia LS, Rahman A, Tay D (eds) The biophysical environment of Singapore. Singapore University Press, Singapore

Chou LM, Tun K (2012) Status and challenges of coral reef monitoring in East Asia: Singapore. In: 8th ICRI East Asia Regional Workshop, 3–5th September 2012, Jeju, Republic of Korea

Chou ML (2023) Progress in Sustainable Development Sustaining marine biodiversity in Singapore's heavily urbanized coast Elsevier 265–282

Chou LM, Yu JY, Loh T (2004) Impacts of sedimentation on soft-bottom benthic communities in the southern islands of Singapore. Hydrobiologia 515:91–106

Chuang SH (1977) Ecology of Singapore and Malayan coral reefs – preliminary classification. In: Proceedings of the third international coral reef symposium, vol 1, pp 55–61

Cooper TF, Uthicke S, Humphrey C, Fabricius KE (2007) Gradients in water column nutrients, sediment parameters, irradiance and coral reef development in the Whitsunday Region, central Great Barrier Reef Est Coast Shelf. Science 74:458–470

Dikou A, van Woesik R (2006) Survival under chronic stress from sediment load: spatial patterns of hard coral communities in the southern islands of Singapore. Mar Pollut Bull 52(11):1340–1354

Douillet P, Ouillon S, Cordier E (2001) A numerical model for fine suspended sediment transport in the southwest lagoon of New Caledonia. Coral Reefs 20(4):361–372

Dronkers J (1986) Tide-induced residual transport of fine sediment. In: van de Kreeke J (ed) Physics of Shalow estuaries and bays. Springer-Verlag, Berlin, pp 228–244

Dyer KR (1986) Coastal and estuarine sediment dynamics. Wiley, Great Britain, 342 pp

Erftemeijer PL, Lewis RR (2006) Environmental impacts of dredging on seagrasses: a review. Mar Pollut Bull 52(12):1553–1572

Erftemeijer PLA, Riegl B, Hoeksema BW, Todd PA (2012) Environmental impacts of dredging and other sediment disturbances on corals: a review. Mar Pollut Bull 64(9):1737–1765

Erm A, Alari V, Lips I, Kask J (2011) Resuspension of sediment in a semi-sheltered bay due to wind waves and fast ferry wakes. Boreal Environ Res 16:149–163

Fabricius KE (2005) Effects of terrestrial runoff on the ecology of corals and coral reefs: review and synthesis. Mar Pollut Bull 50(2):125–146

Fabricius KE, Wolanski E (2000) Rapid smothering of coral reef organisms by muddy marine snow. Estuar Coast Shelf Sci 50:115–120

Fabricius KE, Wild C, Wolanski E, Abele D (2003) Effects of transparent exopolymer particles and muddy terrigenous sediments on the survival of hard coral recruits. Estuar Coast Shelf Sci 57:613–621

Gin KYH, Koh ST, Lin II (2003) Spectral irradiance profiles of suspended marine clay for the estimation of suspended sediment concentration in tropical waters. Int J Remote Sens 24(16):3235–3245

Godin G (1972) The analysis of tides. University of Toronto Press, 264 pp

Gourlay MR (1996) Wave set-up on coral reefs. 1. Set-up and wave-generated flow on an idealised two dimensional horizontal reef. Coast Eng 27(3–4):161–193

Gourlay MR, Colleter G (2005) Wave-generated flow on corals reefs – an analysis for two-dimensional horizontal reef-tops with steep faces. Coast Eng 52(4):353–387

Hammer WM, Hauri IR (1977) Fine-scale surface currents in the Whitsunday Islands, Queensland, Australia: effect of tide and topography. Mar Freshw Res 28(3):333–359

Hasan GMJ, van Maren DS, Cheong HF (2011) Improving hydrodynamic modeling of an estuary in a mixed tidal regime by grid refining and aligning. Ocean Dyn 62(3):395–409

Hasan GMJ, van Maren DS, Ooi SK (2016) Hydrodynamic modeling of Singapore's coastal waters: nesting and model accuracy. Ocean Model 97:141–151

Hoeksema BW, Koh EG (2009) Depauperation of the mushroom coral fauna (Fungiidae) of Singapore (1860s-2006) in changing reef conditions. Raffles Bull Zool Suppl 22:91–101

Hoitink AJF, Hoekstra P (2003) Hydrodynamic control of the supply of reworked terrigenous sediment to coral reefs in the Bay of Banten (NW Java, Indonesia). Estuar Coast Shelf Sci 58:743–755

Houser C (2011) Sediment resuspension by vessel-generated waves along the Savannah River, Georgia. J Waterw Port Coast Ocean Eng 137(5):246–257

Ingram RG, Chu VH (1987) Flow around islands in Rupert Bay: an investigation of the bottom friction effect. J Geophys Res 92(C13):14521

Kirk JTO (1985) Effects of suspensoids (turbidity) on penetration of solar radiation in aquatic ecosystems. Hydrobiologia 125:195–208

Koh T, Lin J (2006) The land reclamation case: thoughts and reflections. Singapore Year Book Int Law 10:1

Kurniawan A, Ooi SK, Hummel S, Gerritsen H (2011) Sensitivity analysis of the tidal representation in Singapore regional waters in a data assimilation environment. Ocean Dyn 61(8):1121–1136

Kvale EP (2006) The origin of neap-spring tidal cycles. Mar Geol 235:5–18

Lai S, Loke LHL, Hilton MJ, Bouma TJ, Todd PA (2015) The effects of urbanisation on coastal habitats and the potential for ecological engineering: a Singapore case study. Ocean Coastal Manag 103:78–85

Larcombe P, Ridd PV, Prytz A, Wilson B (1995) Factors controlling suspended sediment on inner-shelf coral reefs, Townsville, Australia. Coral Reefs 14:163–171

Larcombe P, Costen A, Woolfe KJ (2001) The hydrodynamic and sedimentary setting of nearshore coral reefs, central great barrier reef shelf, Australia: Paluma shoals, a case study. Sedimentology 48(4):811–835

Lee AC, Tan KS, Sin TM (2009) Intertidal assemblages on coastal defence structures in Singapore I: faunal study. Raffles Bull Zool Suppl 22:237–254

Lim TS (2017) Land from Sand: Singapore's Reclamation Story. https://biblioasia.nlb.gov.sg/vol-13/issue-1/apr-jun-2017/land-from-sand/#fn:3. Accessed 1 Oct 2024

Low JKY, Chou LM (1994) Sedimentation rates in Singapore waters. In: Sudara S, Wilkinson CR, Chou LM (eds) Third ASEAN-Australia symposium on living coastal resources. 2. Chulalongkorn University, Bangkok, pp 697–701

Lugo-Fernandez A, Roberts HH, Suhayda JN (1998a) Wave transformations across a Caribbean fringing-barrier Coral Reef. Cont Shelf Res 18:1099–1124

Lugo-Fernandez A, Roberts HH, Wiseman WJJ (1998b) Tide effects on wave attenuation and wave set-up on a Caribbean coral reef. Estuar Coast Shelf Sci 47:385–393

Monismith S (2007) Hydrodynamic of coral reefs. Ann Rev Fluid Mech 39(1):37–55

Morgan KM, Moynihan MA, Sanwlani N, Switzer AD (2020) Light limitation and depth-variable sedimentation drives vertical reef compression on turbid coral reefs. Front Mar Sci 7:571256

MPA (2023) Maintenance dredging works for existing Dock 1 & Dock 2 Gate Entrance and Q2 at Benoi Yard – Port Marine Notice Number 125 of 2023. https://www.mpa.gov.sg/media-centre/details/port-marine-notice-no.-125-of-2023-maintenance-dredging-works-for-existing-dock-1-dock-2-gate-entrance-and-q2-at-benoi-yard

MPA (2024) Dredging works in Temasek fairway – Port Marine Notice Number 37 of 2024. https://www.mpa.gov.sg/media-centre/details/port-marine-noticeno-37-of-2024-dredging-works-in-temasek-fairway

Ouillon S, Douillet P, Lefebvre JP, Le Gendre R, Jouon A, Bonneton P, Fernandez JM, Chevillon C, Magand O, Lefèvre J, Lehir P, Laganier R, Dumas F, Marchesiello P, Bel Madani A, Andréfouët S, Panché JY, Fichez R (2020) Circulation and suspended sediment transport in a coral reef lagoon: the south-west lagoon of New Caledonia. Mar Pollut Bull 61:269–296

Presto MK, Ogston AS, Storlazzi CD, Field ME (2006) Temporal and spatial variability in the flow and dispersal of suspended sediment on a fringing reef flat, Molokai, Hawaii. Estuar Coast Shelf Sci 67:67–81

Pugh DT (1987) Tides, surges and mean sea level. Wiley, New York, 472 p

R Core Team (2024) R: a language and environment for statistical computing. R Foundation for Statistical Computing, Vienna. https://www.R-project.org

Rogers CS (1990) Responses of coral reefs and reef organisms to sedimentation. Mar Ecol Prog Ser 62:185–202

Sin TM, Ang HP, Buurman J, Lee AC, Leong YL, Ooi SK, Steinberg P, Teo SLM (2016) The urban marine environment of Singapore. Reg Stud Mar Sci 8:331–339

Singapore Department of Statistics (2024) https://www.singstat.gov.sg/find-data/search-by-theme/society/environment/latest-data. Accessed 1 Jan 2025

Storlazzi CD, Ogston AS, Bothner MH, Field ME, Presto MK (2004) Wave- and tidally-driven flow and sediment flux across a fringing coral reef: Southern Molokai, Hawaii. Cont Shelf Res 24(12):1397–1419

Tay JYL, Wong SKM, Chou LM, Todd PA (2018) Land reclamation and the consequent loss of marine habitats around the Ayer Islands, Singapore. Nat Singapore 11:1–5

Te FT (1997) Turbidity and its effect on corals: a model using the extinction coefficient (K) of photosynthetic active radiance (PAR). In: 8th international coral reef symposium. Smithsonian Tropical Research Institute, Panama

USACE (1984) Shore protection manual. U.S. Army Coastal Engineering Research Center, U.S. Army Engineer Waterways Experiment Station, Vicksburg

van Maren DS, Gerritsen H (2012) Residual flow and tidal asymmetry in the Singapore Strait, with implications for resuspension and residual transport of sediment. J Geophys Res 117(C4):C04021

Wolanski E, Imberger J, Heron ML (1984) Island wakes in shallow coastal waters. J Geophy Res 89(C6):10553–10569

Wolanski E, Asaeda T, Tanaka A, Deleersnijder E (1996) Three-Dimensional Island wakes in the field, laboratory experiments and numerical models. Cont Shelf Res 16(11):1437–1452

Wolanski E, Fabricius K, Spagnol S, Brinkman R (2005) Fine sediment budget on an inner-shelf coral-fringed Island, great barrier reef of Australia. Estuar Coast Shelf Sci 65(1–2):153–158

Wong PY (2018) Evaluating impact of passing ship-vessel on coastal hydrodynamics using XBeach. Master's dissertation. National University of Singapore

Woolfe KJ, Larcombe P (1999) Terrigenous sedimentation and coral reef growth: a conceptual framework Marine Geology 155(3-4):331-345. https://doi.org/10.1016/S0025-3227(98)00131-5

Yong KY, Lee SL, Karunaratne GP (1991) Coastal reclamation in Singapore: a review. In: Chia LS, Chou LM (eds) Urban coastal area management: the experience of Singapore. ICLARM conference proceedings 25. International Center for Living Aquatic Resources Management, Philippines, pp 59–67

Zoelaeha BD (2010) Fine sediment transport near coral reefs islands in the Singapore Strait. Master's dissertation. Delft University of Technology.

Zweifler A, O'Leary M, Morgan K, Browne NK (2021) Turbid coral reefs: past, present and future—a review. Diversity 13(6):251.https://doi.org/10.3390/d13060251

Impacts of Sedimentation on Coral Communities

3

Stacia Loong, Min Choo, Yong Kit Samuel Chan, and Danwei Huang

Abstract

Urban coral reefs experience chronic disturbances from anthropogenic activities that result in environmental effects such as sedimentation, which can often be traced to coastal development, land reclamation, and runoff. Prolonged sedimentation can reduce light availability for autotrophic corals and physically smother coral species and trigger stress responses that impact reef health. These effects have led to notable changes and declines of coral biodiversity in Singapore, where high sedimentation rates have shifted coral composition and reduced species diversity. Here, we review how different coral species, growth forms and reef communities respond to sedimentation. We also examine coral–sediment interactions in Singapore by surveying ten contemporary reef sites to quantify patterns of sediment accumulation on scleractinian corals. Results reveal that certain coral species and growth forms—particularly the laminar, massive and encrusting forms—accumulate higher sediment loads. Understanding sedimentation patterns across Singapore's reefs can enhance our responses to ongoing reef degradation and inform future conservation efforts. The complex interactions between physical sedimentation, turbidity and light attenuation and their combined impacts on coral health remain unclear. Future research should establish sedimentation and turbidity thresholds that cause significant stress to Singapore's coral communities, providing essential data for informed reef management practices.

S. Loong (✉)
Department of Biological Sciences, National University of Singapore, Singapore, Singapore

Tropical Marine Science Institute, National University of Singapore, Singapore, Singapore
e-mail: stacialwy@gmail.com

M. Choo
Department of Biological Sciences, National University of Singapore, Singapore, Singapore

Y. K. S. Chan
Department of Biological Sciences, National University of Singapore, Singapore, Singapore

AIMS@JCU, Division of Research and Innovation, James Cook University, Townsville, Australia

Australian Institute of Marine Science, Townsville, Australia

D. Huang
Department of Biological Sciences, National University of Singapore, Singapore, Singapore

Tropical Marine Science Institute, National University of Singapore, Singapore, Singapore

Lee Kong Chian Natural History Museum, National University of Singapore, Singapore, Singapore

Centre for Nature-based Climate Solutions, National University of Singapore, Singapore, Singapore

Keywords

Biodiversity · Coral morphology · Reef survey · Scleractinia · Southeast Asia · Urbanisation

3.1 Introduction

Stony corals (Cnidaria: Anthozoa: Scleractinia) are ecosystem engineers that contribute immensely to diverse ecosystem and structural functions on coral reefs that, in turn, provide an extensive range of ecosystem services such as coastal protection and seafood supply (Buss and Jackson 1979; Jackson 1991; Jones et al. 1994; Moberg and Folke 1999). However, increasing human stressors on coral reefs such as pollution, overfishing and sedimentation together with climate change have been impairing coral survivability and reef functions worldwide (Bellwood et al. 2004; Hughes et al. 2010, 2017). This global decline has led to unrelenting losses in the diversity and abundance of corals (Pandolfi

L. M. Chou, D. Huang (eds.), *Coral Reefs of Singapore's Urbanised Sea*, Coral Reefs of the World 23,
https://doi.org/10.1007/978-981-95-0903-4_3

et al. 2003; Jackson et al. 2001), trends which have grave knock-on effects on reef-associated fish and invertebrate assemblages, further impacting ecosystem resilience and functional redundancy (Pratchett et al. 2018; Hoey and Bellwood 2009). Although some reef systems may show recovery and resilience to the myriad of disturbances, the varying levels of global and localised perturbations complicate findings on whether recovery of reef assemblages is heading towards a pre-disturbance state or phase-shifting to a new, less diverse steady state (Aronson et al. 2004; Connell 1997; Halford et al. 2004).

The complexity in unravelling these shifts is often emphasised as being a result of the 'shifting baseline syndrome', where the lack of long-term baselines established before anthropogenic impact produces biases in our current knowledge of ecosystem condition (Jackson et al. 2001; Hughes et al. 2010; Pandolfi 2002). Long-term ecological studies have provided detailed monitoring records of coral community structure. For example, multi-decadal scientific records of coral reefs in Jakarta Bay, Indonesia, have shown how prolonged anthropogenic impact has altered coral community structure, with sharp decreases in species from Acroporidae and Milleporidae (van der Meij and Suharsono 2010). Similarly, a 91-year study of coral reefs at Low Island, Great Barrier Reef has shown constant degradation due to multiple natural and anthropogenic stressors, with coral species richness decreasing to about one-third of initial records in 1928 and community structure shifting from branching to domed and encrusting forms (Fine et al. 2019).

3.2 Coral Reefs and Sedimentation

Sedimentation can originate from both natural and anthropogenic sources. Naturally occurring sediments play a crucial role in coral reefs and are produced even in undisturbed watersheds with corals historically facing varying levels of sedimentation (Hubbard 1997; Potts and Jacobs 2000; Rogers and Ramos-Scharrón 2022). For instance, over 90% of the sediment on natural coral reefs are carbonates derived from the breakdown of calcifying reef organisms (Erftemeijer et al. 2012; Rogers and Ramos-Scharrón 2022). Additionally, terrigenous sediment from terrestrial runoff can also contribute to the sediment load in reefs (Rogers and Ramos-Scharrón 2022). The amount of sediment from terrestrial sources fluctuates over time, as it is influenced by seasonal changes and weather conditions (Otaño-Cruz et al. 2017). Natural processes such as wave action and tidal currents also determine the rate, mode, and direction of sedimentation (Tebbett et al. 2023).

Conversely, anthropogenic activities—including deforestation, dredging, coastal development, pollution and ship wakes—often intensify natural sedimentation by increasing sediment input and/or resuspension through various mechanisms (Erftemeijer et al. 2012; Otaño-Cruz et al. 2017; Rogers and Ramos-Scharrón 2022). For instance, deforestation can increase terrigenous sediment input into coastal areas due to enhanced erosion (Otaño-Cruz et al. 2017) while dredging operations can increase the breakdown of carbonates and promote the resuspension of particulates from the seafloor. Dredging has frequently been documented as a contributor of coral reef loss, either directly (e.g. through removal or burial) or indirectly (e.g. by increasing sedimentation and turbidity in the vicinity) (Erftemeijer et al. 2012). The interaction between natural and anthropogenic sources is particularly evident on wave-protected nearshore fringing reefs, where sedimentation rates are typically higher, especially in locations closest to urban areas (Wolanski et al. 2005). Moreover, the temporality and duration of sedimentation are likely to be further influenced by anthropogenic climate change such as altered rainfall patterns (Nearing et al. 2004).

Sedimentation is known to affect coral health through a range of effects. Physical contact of coral tissue with sediment can stress corals, resulting in coral bleaching or even tissue death (Burke et al. 2011; Howarth et al. 2011; Weber et al. 2006). Turbid reef conditions due to increased suspended sediment loads can reduce light intensity, leading to a decrease in photosynthetic activity of the coral endosymbiont Symbiodiniaceae that can result in starvation and subsequent coral bleaching (Brown 1997; Glynn 1996). Reduced light availability can also lead to reef compression, which is the narrowing of the depth range where photosynthetic corals can thrive and reef construction can take place (Heery et al. 2018; Morgan et al. 2020). While Indo-Pacific coral reefs typically extend below 30 m in depth, reefs within 20 km of major urban centers often have a maximum depth of less than 10 m (Baum et al. 2015; Heery et al. 2018). Minor increases in suspended sediment, which may have minimal direct physiological effects on corals, can also significantly affect coral distribution by further limiting light penetration and shallowing the photic zone on already light-limited reefs (Morgan et al. 2020).

High sediment levels can result in the smothering of coral colonies as energy normally meant for metabolic processes such as growth or tissue regeneration is redirected to the clearing of sediment through mechanisms such as mucus production and ciliary action (Fabricius 2005; Riegl and Branch 1995; Riegl et al. 1996; Meesters et al. 1992). While corals may consume sediment particles to compensate for the reduction in autotrophic food production, most would stop heterotrophic feeding during high sediment load scenarios, with excessive sediment smothering hindering tentacular action for feeding (Erftemeijer et al. 2012; Rosenfeld et al. 1999).

Sedimentation can affect reproductive success and larval recruitment, primarily by impacting larval survivability in open water and depressing larval settlement rates from the reduction of available substrates (Babcock and Smith 2000; Birrell et al. 2005; Gilmour 1999; Goh and Lee 2008). Highly turbid conditions could also affect fertilisation success (Te 1992; Erftemeijer et al. 2012).

Furthermore, nutrient-rich sediment can result in increased microbial activity, leading to possible coral disease and necrosis of coral tissue smothered by such sediment (Pollock et al. 2014; Weber et al. 2012). Sediment exposed to industrial runoff can carry considerable amounts of heavy metals such as lead and copper (Goh and Chou 1997), the latter of which has been shown to inhibit coral development and reproductive processes (Amin et al. 2009; Reichelt-Brushett and Harrison 2005; Todd et al. 2010).

More broadly, sedimentation impacts a variety of organisms in coral reef ecosystems, influencing their health and the overall integrity of the reefs. Crustose coralline algae (CCA), which are crucial for reef calcification and promoting coral larvae settlement, are particularly sensitive to increased sedimentation (Kendrick 1991). Sediment also interacts with turf algae, which are highly effective at trapping sediment (Purcell 2000). The accumulation of sediment-laden turf algae can disrupt ecosystem functioning and resilience by hindering essential processes like the removal of algae by herbivorous fish, reducing nutrient availability for detritivores and competing for space with hard corals and macroalgae (Ricardo et al. 2017; Tebbett et al. 2017, 2022; Speare et al. 2019).

Similarly, macroalgal species that thrive on the nutrients in sediment can form dense mats, which overgrow and damage large coral areas (Fabricius 2005). These mats trap additional sediment, restrict gas exchange and create anoxic conditions as they age and collapse. An increase in sedimentation above natural ranges also puts sedentary or slow-moving organisms at risk of smothering if they are unable to survive or escape from the accumulating sediments (Newell et al. 2004; Hendrick et al. 2016). The smothering of habitats and shifts in species assemblages due to sedimentation often lead to changes in community structure, composition and faunal abundance (Miller et al. 2002; Pineda et al. 2017). These changes can extend far beyond the immediate site of sediment deposition, sometimes affecting areas kilometers away (Stronkhorst et al. 2003). In Singapore, these effects were evident on both nearshore and offshore reef sites. An algal-sediment-dominated state was identified, characterised by high cover of sediment, macroalgae, turf algae and low abundance of hard coral, secondary consumers and sea urchins (Loong et al. 2025). Sedimentation also led to a decline in echinoderm species as well as a shift towards more resilient echinoderm species (Lim and Chou 1988).

Sediment accumulation in coral reefs, defined as the downward movement of particles suspended in the water column, has been measured to quantify sedimentation through various techniques (Thomas and Ridd 2005; Morgan et al. 2020). These techniques include discontinuous methods, such as rulers, sediment tube traps and SedPods, as well as continuous methods like optical backscatter point sensors and video cameras (Thomas and Ridd 2005). Among these, sediment tube traps and SedPods are most frequently used to measure sediment accumulation rates (Field et al. 2013; Morgan et al. 2020; Ng et al. 2022).

Sediment tube traps are containers placed in the water to capture all particles, including those that merely pass over the reef or are briefly deposited. This method can sometimes lead to overestimation because corals can mechanically remove sediment, and not all sediments may be permanently deposited (Field et al. 2013). Additionally, there is no universal standard for the height of trap openings, but it is suggested to use two traps with a height difference of 20–30 cm (Ng et al. 2022). SedPods are PVC pipes containing concrete. They collect sediment from multiple deposition and erosion events, and are useful proxies for sediment accumulation on encrusting and massive corals (Field et al. 2013).

Depending on the research objectives, the choice of method is also dependent on the site, as wave actions, currents and human activities can influence sedimentation rates and the effectiveness of different measurement methods. The use of a combination of methods can provide a more accurate and nuanced understanding of sediment dynamics in reef ecosystems. When SedPods are used alongside traditional tube traps, they can offer a more comprehensive understanding of sedimentation processes on coral reefs (Field et al. 2013; Morgan et al. 2020).

3.3 Responses of Corals to Sedimentation

Corals exhibit various active and passive responses to sedimentation that determine their ability to tolerate chronically high sediment levels (Erftemeijer et al. 2012). Active sediment removal mechanisms include polyp expansion, movement and tentacular action (Bongaerts et al. 2012; Riegl and Branch 1995; Stafford-Smith and Ormond 1992). Other mechanisms for sediment rejection include mucus sheet production (Coffroth 1990; Brown and Bythell 2005; Stafford-Smith 1993). The sediment rejection efficacy of coral species is also determined by sediment load as it is energetically costly (Lee et al. 2021). Studies have shown that hemispherical and columnar coral growth forms demonstrate effective passive sediment shedding due to the combination of gravitational force and wave energy (Bak and Elgershuizen 1976; Dodge and Vaisnys 1977; Stafford-Smith 1993; Riegl and Branch 1995). With these variable responses of different corals to sedimentation conditions, the coral community as a whole is shaped by these processes in the long term. Consequently, coral morphology has been used to infer sedimentation conditions in paleoecological studies (Plusquellec et al. 1999; Sanders and Baron-Szabo 2005).

Sedimentation can affect the quality of light reaching coral reefs, leading corals to adapt through photoadaptation (Rogers 1990; Browne et al. 2019). This process involves corals shifting their energy allocation by reducing growth rates to increase productivity (Anthony and Fabricius 2000; Tuttle et al. 2020; Tuttle and Donahue 2022). The magnitude of the decrease in growth rate is often dependent on both the environment and life-history traits of the species such as sediment exposure and coral morphology (Humanes et al. 2017a, b; Tuttle et al. 2020; Tuttle and Donahue 2022). Conversely, some coral species have demonstrated notable increases in respiration and decreases in net photosynthesis as a result of increased metabolic activities when exposed to different types of sediment (Rogers 1990; Riegl and Branch 1995; Philipp and Fabricius 2003). Through reduced autotrophy brought about by low-light conditions, corals can also undergo hypoxic events (Jones et al. 2016).

Hypoxia can be caused by both physical (i.e. direct smothering of corals that prevents gaseous exchange) and biological (i.e. increased oxygen demand, inducing bacterial action in the sediment and death of underlying tissue) processes (Silveira et al. 2019; Rushmore et al. 2021; Pezner et al. 2023). Additionally, organic matter from sediment can drive microbial processes that create anoxic environments with low pH, subsequently causing tissue degradation (Weber et al. 2012). Nutrient loading from sedimentation can exacerbate this by stimulating microbial respiration and algal bloom. The subsequent decay of organic matter further depletes oxygen levels and may produce toxic hydrogen sulfide, further adding stress to corals (Weber et al. 2012; Johnson et al. 2021).

In response to hypoxia, hard corals can activate specific molecular pathways. For instance, corals have demonstrated the ability to upregulate the genes involved in anaerobic metabolism, such as tauropine dehydrogenase, which supports energy production when oxygen is limited (Bollati et al. 2022). There is also evidence of genes for hydrogen sulfide metabolism, such as persulfide dioxygenase (ETHE1), which when activated would allow for corals to metabolise toxic sulfide compounds that build up during hypoxia (Bollati et al. 2022).

Hypoxic events can also stimulate the coral antioxidant defence system, which, if overwhelmed, leads to cellular and tissue damage (Hughes et al. 2020). DNA damage has been documented in *Acropora* spp. exposed to nocturnal hypoxia, further compromising coral health (Deleja et al. 2022). Episodic anthropogenic hypoxia, as seen in the Caribbean, has led to bleaching and coral mortality, highlighting the destructive potential of sudden low-oxygen events (Johnson et al. 2018). Regions such as the northern Gulf of Mexico and East China Sea show a concerning trend of expanding hypoxic zones due to nutrient loading, exacerbating the threat to coral reefs (Hughes et al. 2020).

Tropical coral taxa are particularly vulnerable to hypoxia compared to their temperate counterparts, with hypoxia often leading to reduced species diversity and abundance (Hughes et al. 2020; Johnson et al. 2021). Under chronic hypoxic conditions, coral communities may undergo structural shifts, with branching corals being replaced by more spherical or encrusting forms that are better adapted to low-oxygen environments (Hughes et al. 2020). These shifts can have lasting consequences, as post-hypoxia benthic composition and coral cover may not return to their original state (Johnson et al. 2021). However, there is evidence suggesting that corals from chronically turbid reefs may exhibit a degree of adaptation to hypoxic events, potentially making them more resilient in such conditions (Guest et al. 2016b; Morgan et al. 2017; Zweifler et al. 2021; Zweifler et al. 2024).

3.4 Sedimentation on Coral Reefs in Singapore

In Singapore, rapid urbanisation through extensive coastal development, land reclamation projects and shipping activity has led to the loss of about 65% of reef area (Chou et al. 2019; Lai et al. 2015). Consequentially, corals in Singapore's reefs experience chronic anthropogenic disturbances, especially relating to the high-sedimentation rates, with recorded levels of suspended sediment solids exceeding optimal threshold values for typical tropical reefs (Dikou and van Woesik 2006; Erftemeijer et al. 2012; Rogers 1990). Compounding these chronic stressors on Singapore's reefs are thermal anomalies leading to mass coral bleaching (loss of algal endosymbiont Symbiodiniaceae from coral cells), such as the 2010 global event that resulted in the bleaching of two-thirds of corals on Singapore's most diverse reefs at Raffles Lighthouse (Pulau Satumu), as well as bleaching-associated mortality observed across survey sites (Guest et al. 2012, 2016b; Tun et al. 2011; see also Ng et al. 2020).

Singapore's urban reefs are characterised by elevated sedimentation rates of typically 5–35 mg/cm^2/day, a consequence of extensive coastal development and land reclamation (Chou 2006; Dikou and van Woesik 2006). These processes have led to reduced light penetration in reef habitats, impacting coral growth and reef health. Sedimentation rates across Singapore's reefs are strongly influenced by current dynamics, such as the dominant east-west currents, and the history of urbanisation, including land reclamation (Low and Chou 1994). For example, reefs located farther from the mainland, such as those near Raffles

Lighthouse, typically experience lower sedimentation rates due to their reduced exposure to reclamation activities and greater exposure to open ocean conditions, which promote higher wave energy and stronger currents that mitigate sediment accumulation (Chou et al. 2017). Similarly, Kusu Island, despite its proximity to the mainland, benefits from open ocean exposure, resulting in relatively low turbidity levels (Morgan et al. 2020).

Reefs closer to the mainland, such as those at Pulau Semakau and Pulau Hantu, typically face higher sedimentation rates. Pulau Semakau's elevated rates are primarily due to land-use changes following the development of the Semakau Landfill in 1999 (Low and Chou 1994; Chow et al. 2019). Pulau Hantu also experiences high sedimentation, likely because its sheltered location reduces current strength, making it less effective at preventing sediment buildup (Low and Chou 1994; Morgan et al. 2020). Despite efforts to mitigate the impacts of reclamation on Singapore's reefs, sedimentation rates have continued to rise since the 1980s (Low and Chou 1994; Chow et al. 2019; Todd et al. 2013; Morgan et al. 2020).

As a result, sediment-tolerant coral taxa are dominant on Singapore's coral reefs, which are usually limited to growth only down to ~6 m depth due to reduced light penetration (Chow et al. 2019; Guest et al. 2016a; Heery et al. 2018; Tun et al. 1994; Wong et al. 2018). While shallow coral communities continue to persist and there are no clear data to show a significant decline in species diversity, such anthropogenic stressors have influenced shifts in coral species composition and structure to assemblages which are more tolerant to sedimentation and bleaching (Chan et al. 2024; Chow et al. 2019; Ng et al. 2020; Wong et al. 2018). Sediment-tolerant taxa often dominate urban reefs due to their higher tolerance to sediment influx and other environmental stressors (Yeemin et al. 2013; Guest et al. 2016a; Browne et al. 2019; Chan et al. 2024). These taxa typically possess life-history traits such as slow growth, longer generation times, and large corallites, which optimise energy resources amidst disturbances (Darling et al. 2012; van Woesik et al. 2022; Browne et al. 2019).

Few studies have quantified local extinction of scleractinian corals, in part due to the challenges in evaluating species observations and ranges (Carlton et al. 1999; Dulvy et al. 2003), with only two studies to date by Hoeksema and Koh (2009) and Poquita-Du et al. (2019b) documenting declines in species richness among Fungiidae and Pocilloporidae corals, respectively. Other studies show the acclimatisation of corals following acute heat stress events, with implications for coral resilience in Singapore (Guest et al. 2016b; Poquita-Du et al. 2019a). Despite the chronic sedimentation faced by coral reefs in Singapore, scleractinian species richness here remains comparable to locations further from anthropogenic activity in the Southeast Asian region (Affendi et al. 2005; Heery et al. 2018; Huang et al. 2009, 2015).

Although overall species richness may remain stable, chronic sedimentation imposes selective pressure that inhibits the recruitment of sediment-sensitive coral taxa, such as Acroporidae (Bauman et al. 2015). Sedimentation particularly impacts the early life stages of corals, affecting the settlement, survival and growth of competitive coral larvae, which are vital for the recruitment of new individuals to the reef (Humanes et al. 2017a, b; Babcock and Smith 2000). In contrast, stress-tolerant corals typically have higher fecundity than competitive species, and these long-lived, stress-tolerant corals can persist for decades without replenishment (Darling et al. 2012; Chan et al. 2024).

3.5 Ecology of Coral–Sediment Interactions in Singapore

3.5.1 Assemblage Responses of Corals to Sedimentation

Clearly, prolonged sedimentation exposure and accumulation can lead to declines in coral cover and significant shifts in coral taxonomic community structure, favouring sediment-tolerant species over time. These shifts typically involve a transition from fast-growing, competitive species to slow-growing, stress-tolerant species (Browne et al. 2019; Chan et al. 2023; Chan et al. 2024). However, there has been limited evidence of such shifts in Singapore since the 1980s, partly due to a lack of data prior to that period and the lower taxonomic resolution in the early stages of ecological monitoring (Guest et al. 2016a; Chan et al. 2023). Since monitoring began, sediment-tolerant taxa such as *Montipora*, *Porites* and *Galaxea*, along with generalist coral taxa like *Pachyseris* and *Merulina*, have consistently dominated Singapore's shallow reefs, with recorded shifts occurring between these taxa (e.g. from *Montipora* to *Goniastrea* and *Pachyseris*) (Guest et al. 2016a). In contrast, competitive coral taxa such as *Acropora* are more prevalent on reefs farther from the mainland, where sedimentation is less intense (Darling et al. 2012; Guest et al. 2016a; Zweifler et al. 2021; Chan et al. 2024).

In addition to changes in species composition, sedimentation drives shifts in coral growth forms. Sediment-laden environments favour more massive, structurally simple, and volumetrically compact morphologies over branching forms. Long-term analyses on data extending back to the 2000s by Guest et al. (2016a), and Chan et al. (2024) have highlighted a marked decline in branching corals in favour of massive forms. While decreases in laminar taxa have been observed, laminar corals still dominate Singapore's reefs alongside massive corals, due to their ability to optimise light capture in highly sedimented waters, and slow growth rates (Guest et al. 2016a; Morgan et al. 2020; Chan et al. 2024).

Changes in coral assemblages due to sedimentation also extend to their symbiotic relationships. Despite the high diversity of coral species in Singapore, this diversity is not reflected in their symbiont communities (Tanzil et al. 2016; Smith et al. 2020). Sediment stress has been associated with a shift in Symbiodiniaceae diversity, favouring more stress-tolerant clades (Poquita-Du et al. 2020; Smith et al. 2020). In particular, two genera associated with stressed reefs—*Cladocopium* and *Durusdinium*—accounted for 91% of the symbionts found in five coral species: *Pectinia alcicornis*, *Merulina ampliata*, *Platygyra daedalea*, *Podabacia crustacea* and *Pachyseris speciosa* (Smith et al. 2020). Although *Cladocopium* was the dominant symbiont in these five species, it exhibited low community diversity compared to other studies (Hume et al. 2020; Smith et al. 2020). *Durusdinium* was not dominant but featured more diverse assemblages, likely encompassing more stress-tolerant species (Smith et al. 2020). In other coral species such as *Diploastrea heliopora* and *Pocillopora acuta*, however, *Durusdinium* was the dominant symbiont community (Tanzil et al. 2016; Smith et al. 2020; Ong et al. 2022), reflecting the stressors (i.e. high sedimentation and turbidity) characteristic of Singapore's urban reefs.

The patterns described above strongly suggest that sediment regimes influence variation in coral community assemblages among sites, driven by differential sediment tolerance among species. Thus, this study aims to quantify differences in sediment accumulation on corals surveyed in situ, where it is hypothesised that variation in sediment cover on coral colonies would be associated with coral colony characteristics and species identity. Here, we establish relationships between sediment cover on coral species and colony characteristics such as size and growth form. Findings on sedimentation patterns among Singapore's reefs can further our understanding of current reef degradation and guide future conservation efforts.

3.5.2 Characterisation of Coral–Sediment Interactions in Singapore

To characterise sediment accumulation on living coral communities, subtidal surveys using the line intercept transect (LIT) method (English et al. 1994) were carried out from September 2019 to February 2020 along the reef crest of ten sites (nine fringing reefs and one patch reef). These represent two main reef types in Singapore—fringing reefs which are found near the coastlines of the southern islands of Singapore, and patch reefs such as Terumbu Pempang Tengah (Fig. 3.1). These sites experience a range of anthropogenic and environmental conditions due to variations in anthropogenic influence, sediment regime and wave exposure (Browne 2012; Dikou and van Woesik 2006).

Using SCUBA, five 20-m transect tapes were placed per site along the reef crest at a depth of 3 m below chart datum (Table 3.1). Planar, scaled photographs of every live scleractinian coral colony on the transect were taken using an Olympus TG4 digital camera; the camera was positioned parallel to the colony to reduce parallax error. For larger sized coral colonies, multiple images were taken of the colony from the same distance and stitched together in Adobe Photoshop CS6.

Sediment accumulation on coral colonies was quantified by the surface area and perimeter of sediment coverage on each coral colony. Sediment coverage was defined as sediment found on the surface of a colony that was surrounded by live coral and was quantified using ImageJ (Abràmoff et al. 2004; Schneider et al. 2012). Sediment with macroalgal growth was also included in the surface area calculation of sediment coverage. Dead corals covered with sediment that were not surrounded by live coral cover (but could have been part of the live coral colony) were not included in the surface area calculation, as it cannot be determined if the death of the coral was due to sediment smothering, or if the sediment covered the coral only after it died. Sediment coverage was estimated visually in some larger colonies (≥50 cm in diameter) as it cannot be calculated in ImageJ due to site visibility or limitations relating to photo-stitching. Sediment coverage area was normalised by the surface area of each colony, estimated with the geometric models for various coral growth forms. We used the halved geometric mean diameter (GMD) as a proxy to determine coral colony radius following McWilliam et al. (2018) (Table 3.2).

All statistical analyses were performed in R version 3.6.3 and RStudio version 1.2.5033 (RStudio Team 2019). To analyse sediment covering corals, species were classified according to their life-history strategies based on Darling et al. (2012), with observed percentage cover of sediment $\left(100 \times \frac{\text{Sediment Area}}{\text{Corrected Colony Area}}\right)$ on all coral colonies and sediment and coral cover data in the ten sites plotted using the 'ggplot2' R package. The fixed effects of coral colony characteristics (coral species, growth form and colony size) on sediment cover on corals were analysed using linear mixed effects models implemented in the R package 'nlme'. Quantitative variables such as coral colony size and the area of sediment cover were log-transformed for normality. Only species with sediment data for at least ten colonies per species were used in the model to adequately represent species effects. To account for site- and transect-level effects that were not measured, random effects of Site, Transect and Transect nested within Site were included to account for transect and site-level variation that was not measured.

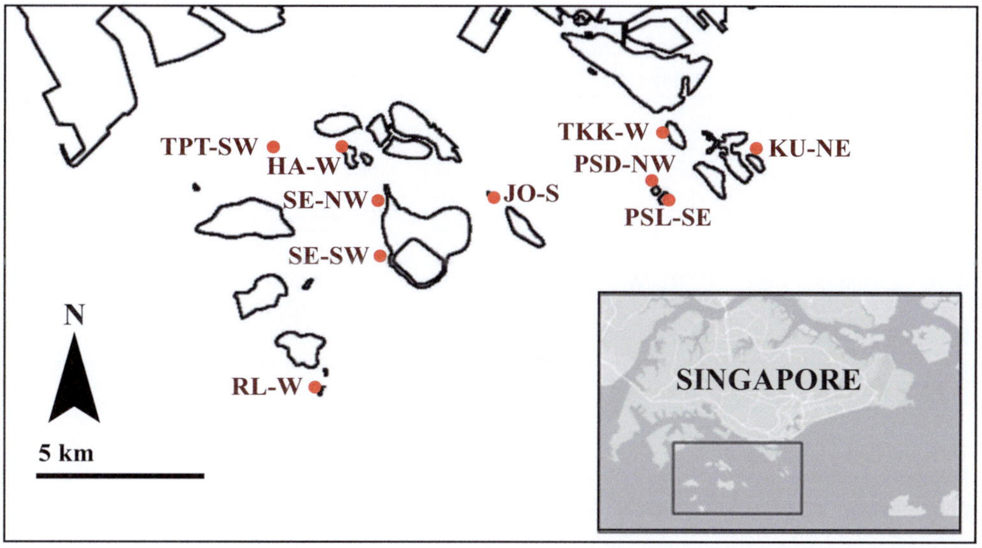

Fig. 3.1 Map showing survey sites (red circles) and site codes from Table 3.1 at the southern islands of Singapore

Model simplification was carried out through the stepwise removal of variables with the best-fit model chosen based on the Akaike Information Criterion (AIC).

Generalised linear mixed models (GLMMs), which can be used with dependent proportion data such as the percentage of sediment cover, were used to analyse similar fixed effects as above on the percentage sediment cover on coral colonies, using the R package 'glmmADMB' with the Gaussian family to account for proportional data in percentage cover. The same random effects were used in the GLMM, with model simplification carried out as above.

Checking of final models for assumptions of data normality and homogeneity of variances was carried out based on graphical validation with diagnostic plots, where fixed and random effects residuals were fitted against predicted values.

3.5.3 Effects of Species, Growth Form and Colony Size

A total of 1252 scleractinian coral colonies were recorded at 10 reef sites, encompassing 47 genera and 115 species, making up about 70% of coral species recorded in present-day Singapore (Wong et al. 2018; Chan et al. 2024). Sediment cover across sites showed some variation, with average sediment cover of 14.22% across all sites and a range of 8.8–20.44% (Fig. 3.2). Jong (South) and Semakau (NW) showed a greater spread of sediment cover. Sediment cover did not vary considerably with distance from the nearest shore (mainland Singapore), although sites closer to the mainland shores typically experience greater anthropogenic impact. The distance-from-shore trend was similar for coral cover, but with greater variation in the range of coral cover (Fig. 3.3).

When sediment cover on coral colonies were grouped according to species life-history strategies based on compiled species traits as proposed in Darling et al. (2012), the most common life-history strategy for all 115 surveyed species were stress-tolerant (27.8%), followed by generalists (6.96%), competitive (3.48%) and weedy (1.74%) (Fig. 3.4). A sizeable number of species (60%) did not have any life-history strategy associated with their species coral trait. Competitive species such as *Acropora valida* and *Turbinaria mesenterina* had low sediment cover, with median sediment cover below 5%. Stress-tolerant and generalist species showed no clear distinction in observed sediment cover. Species were also classified based on their water clarity preference derived from textual descriptions of preferred habitats where coral species can be grouped into 'turbid', 'clear' or 'both' (Diaz and Madin 2011; Veron 2000). Similar to the grouping of observed sediment cover in coral colonies based on life-history strategy, there was no clear distinction in the amount of sediment cover according to water clarity preference.

To examine the influence of coral colony characteristics and coral species on the area of sediment cover on corals, sediment cover area was modelled as a function of coral colony growth form, coral colony size and coral species and their interactions (Table 3.3). Models with corrected colony surface area did not have the lowest AIC value and were not used for analysis of the area of sediment cover on corals. The corrected colony surface area was only used in the GLMM analysis.

In the best-fit model, the effect of colony size on sediment cover was not significant, and only significant when accounting for the size–growth-form interaction, showing that size was not a significant predictor of sediment cover area (Table 3.3 and Fig. 3.5). The branching growth form followed by columnar and foliose growth forms showed the

Table 3.1 Survey sites for coral–sediment interactions

Site	Site code	Date	GPS coordinates	Distance from mainland (km)	Average coral cover (%)	Average sediment cover (%)
Kusu Island Northeast	KU-NE	16 Sep 2019	1.225682°, 103.859902°	4.01	35.74	17.66
Pulau Semakau Northwest	SE-NW	17 Sep 2019	1.211533°, 103.755917°	5.85	6.98	21.4
Pulau Hantu West	HA-W	18 Sep 2019	1.227255°, 103.746892°	4.99	51.06	12.29
Raffles Lighthouse West	RL-W	19 Sep 2019	1.160556°, 103.740556°	11.63	63.39	10.04
Pulau Jong South	JO-S	12 Oct 2019	1.214029°, 103.787490°	4.57	31.69	20.44
Pulau Subar Laut Southeast	PSL-SE	8 Nov 2019	1.212479°, 103.836210°	5.57	40.67	15.67
Terumbu Pempang Tengah Southwest	TPT-SW	22 Nov 2019	1.227453°, 103.727479°	6.39	53.25	13.84
Pulau Tekukor West	TKK-W	29 Jan 2020	1.231172°, 103.836508°	3.80	16.92	12.44
Pulau Subar Darat Northwest	PSD-NW	30 Jan 2020	1.216500°, 103.831500°	5.24	45.46	8.89
Pulau Semakau Southwest	SE-SW	13 Feb 2020	1.202560°, 103.754875°	6.76	31.08	9.55

Table 3.2 Geometric models used to determine surface area of coral colonies

Coral morphology	Surface area formula
Massive (CM)	$2\pi r^2$
Foliose (CF)	$4\pi^2 r^2$
Laminar (CL)	$2\pi r\sqrt{(r+20)}$
Branching (ACE/ACB/ACT/ACD/ACS/CB)	$0.31\pi^2 r^2$
Encrusting (CE)	πr^2
Columnar (CS)	$0.31\pi^2 r^2$

Adapted from McWilliam et al. (2018)

lowest effect of sediment cover per unit increase in colony area.

Of the 31 species included in the model, only 15 showed significant species-level effect on the area of sediment cover. Species which had higher sediment cover per unit area were mainly massive and encrusting forms, with *Cyphastrea chalcidicum* showing the greatest effect at 1.657 ± S.E. 0.4002. This was followed by a mix of species comprising predominantly foliose and laminar forms such as *Pectinia crassa* (1.181 ± S.E. 0.3756) and *Echinopora gemmacea* (0.8914 ± S.E. 0.3688). *Porites lutea* had relatively lower sediment cover per unit area (0.8092 ± S.E. 0.4015) compared to other massive species.

3.5.4 Discussion

Corals often experience sediment smothering associated with anthropogenic activities such as dredging and natural resuspension events, although the former is more prevalent for coral assemblages in urbanised reefs (Chou 2006; van Maren et al. 2014; Larcombe et al. 1995). Regardless of the sediment source, corals may exhibit 'passive' and/or 'active' processes for sediment clearance. 'Passive' processes are typically associated with coral morphology, while 'active' processes involve strategies including polyp expansion, mucus production and ciliary action and can show variation across species (Duckworth et al. 2017; Stafford-Smith and Ormond 1992), potentially leading to the patterns demonstrated in this study.

Analyses here show that laminar, massive and encrusting forms exhibit the greatest amount of sediment cover per unit increase in surface area. Laminar forms appear well adapted to sedimentation, where funnel shapes in laminar species such as *Turbinaria* amass sediment at the base of the funnel as a sacrificial sediment trap to keep the rest of colony surface area clear of sediment (Riegl et al. 1996). In high wave energy conditions, this design also helps in self-cleaning through vortices which replace water in the funnel, consequently removing accumulated sediment (Riegl et al. 1996; Duckworth et al. 2017). Thus, the increased dominance of laminar growth forms over the last few decades may coincide with the increase in anthropogenic activity, thereby conferring a competitive advantage in response to sediment smothering. Interestingly, such laminar forms which tend to be found at deeper parts of reefs due to their flat, spreading forms that maximise light capture have been found in shallower depths in Singapore, likely a result of high turbidity but it may also support the case for their selective advantage in present reefs (Chow et al. 2019; Dikou and van Woesik 2006; Hoogenboom et al. 2008). Compared to branching, foliose and columnar species, massive and encrusting spe-

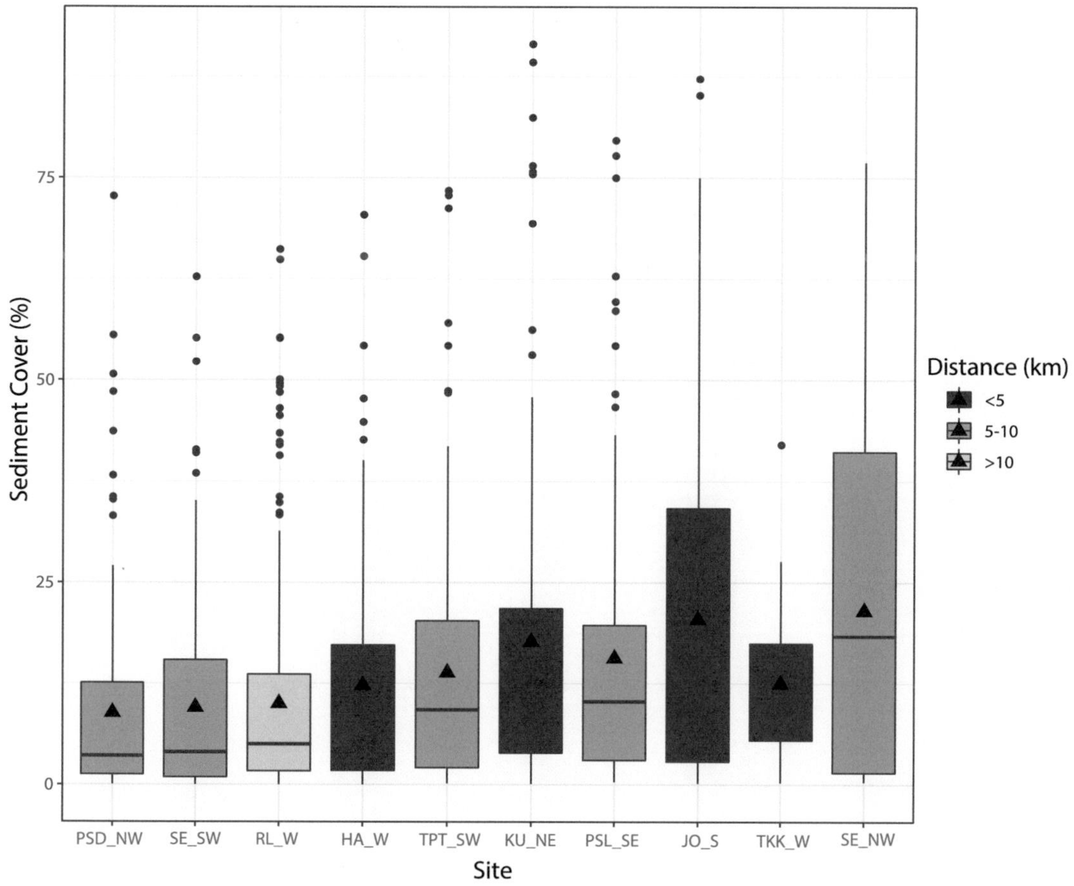

Fig. 3.2 Box plots of percentage sediment cover on corals at all 10 survey sites, in order of increasing median sediment cover. Average sediment cover represented by triangles, and ranges of distance from mainland indicated by box shading

cies benefit less from passive clearance of sediment due to gravitational effects combined with wave energy, since there are less vertical surfaces for sediment to slide off and be passively cleared (Duckworth et al. 2017). Thus, if relying on passive sediment clearing mechanisms alone, massive and encrusting morphologies would be more prone to higher sediment cover on colonies (Brown and Howard 1985). However, corals with such morphologies appear to be more tolerant of turbid environments, suggesting species-level controls of feeding behaviour and/or sediment rejection (Darling et al. 2012; Nugues and Roberts, 2003). Thus, despite showing trends of higher sediment cover, these growth forms continue to persist in Singapore's reefs.

The results show no clear distinctions in observed sediment cover across all surveyed species based on their life-history histories or water clarity preference. However, the models highlight significant variation in sediment cover among the 15 species analysed. *Cyphastrea chalcidicum*, which is typically of an encrusting or massive growth form, exhibits the greatest sediment cover per unit area, with species exhibiting similar encrusting or massive morphologies such as *Favites abdita*, *Favites pentagona* and *Dipsastraea speciosa* also showing the highest species effect on observed sediment cover.

As active sediment clearance mechanisms typically include metabolically taxing responses, variation in species-level sediment cover might be explained by their life-history strategies (Telesnicki and Goldberg 1995). With most of these species adopting stress-tolerant life-history strategies, lower metabolic energy spent on sediment clearance might be an adaptive strategy, leading to higher sediment cover observed (Darling et al. 2012). Furthermore, the smaller corallite sizes of *Cyphastrea chalcidicum* might reduce the effectiveness of sediment clearance through ciliary action, yet the persistence of *Cyphastrea* on Singapore's reefs suggests that active sediment rejection is not the only requirement for survival in turbid conditions (Stafford-Smith 1993; Guest et al. 2016a). Other traits which determine life-history strategies, such as growth and reproductive traits and physiological tolerances, may reduce the need for high sediment rejection to ensure long-term survival (Stafford-Smith and Ormond 1992). Additionally, stress from sedimentation alone might not be sufficient to invoke significant responses from coral, but a combination of stressors such as heat

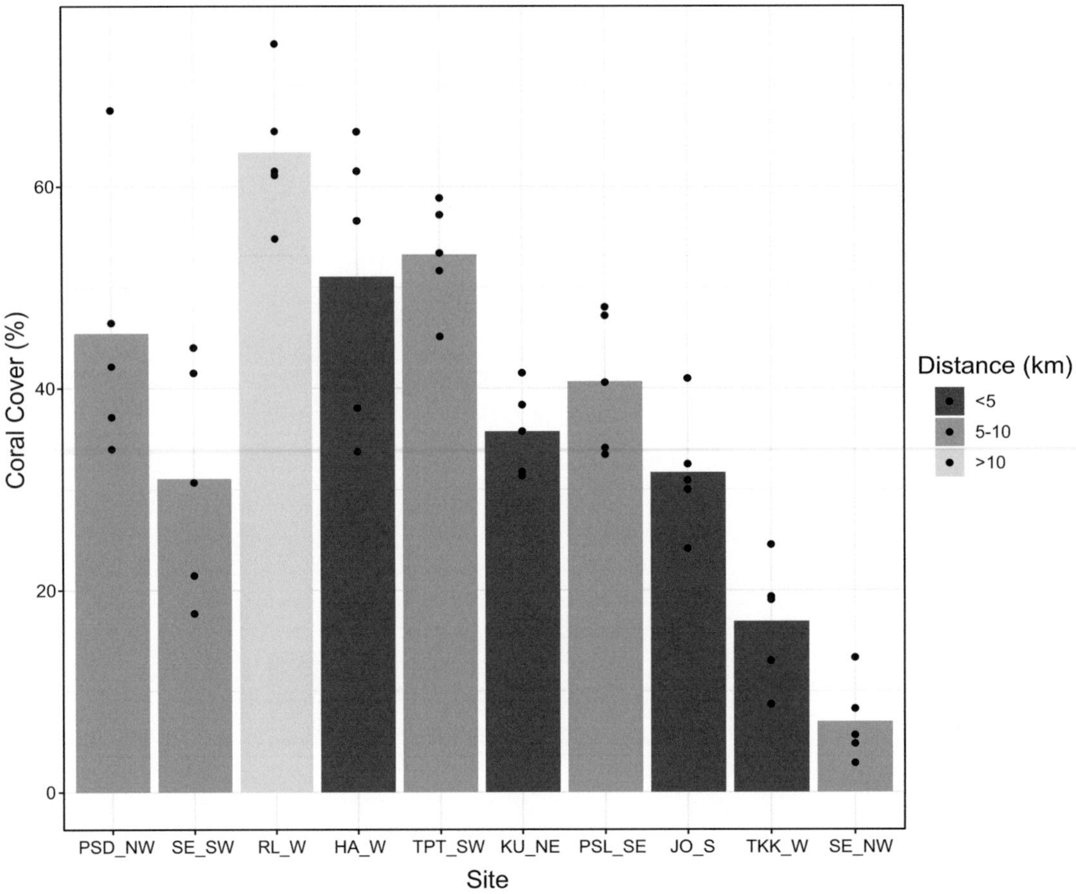

Fig. 3.3 Average coral cover at all 10 survey sites. Transect coral cover (n = 5) at each site represented by dots, and ranges of distance from mainland indicated by box shading

and sedimentation could synergistically reduce photophysiological performance and interfere with gene regulatory processes involving assembly and disassembly of cilia (Poquita-Du et al. 2019a).

Interestingly, laminar species such as *Echinopora gemmacea*, *Merulina ampliata* and *Mycedium elephantotus* that show mid-range sediment cover effect are considered generalist species (Darling et al. 2012). This is indicative of the effectiveness of laminar growth forms that balance between sediment clearance and tolerating being covered by sediment (Riegl et al. 1996). *Mycedium elephantotus*, which has been recorded to form mucus sheets along with strong ciliary mechanisms to remove sediment, could potentially explain why it exhibits the lowest sediment cover effect compared to the other plating species (Stafford-Smith and Ormond 1992). While stress-tolerant *Porites lutea* shows lower sediment cover despite having a massive morphology, this could be attributed to the species-specific production of mucus sheets during instances of high sediment, thereby resulting in higher sediment clearance rates and lower observed sediment cover (Jones et al. 2016).

3.6 Knowledge Gaps and Challenges

While this study offers valuable insights, several knowledge gaps and challenges persist. The specific ways in which physical sedimentation and turbidity interact to impact coral health and community structure remain poorly understood (Tuttle and Donahue 2022). While suspended solids alone may not harm corals, their presence in turbid waters—which are primarily caused by suspended solids in Singapore's reefs (Gin et al. 2000; Gin et al. 2002; van Maren et al. 2014)—can exacerbate the negative impact of turbidity (Bessell-Browne et al. 2017). This close link between suspended particles and light attenuation complicates efforts to determine whether coral responses are due to one or both of these stressors (Bessell-Browne et al. 2017), with most studies still attempting to disentangle the effects of deposited sediment and suspended sedimentation/turbidity (Tuttle and Donahue 2022; Browne et al. 2012; Morgan et al. 2020). Due to the interconnectedness of the three stressors, the realistic thresholds at which coral communities in Singapore begin to experience significant stress from deposited sediment, sus-

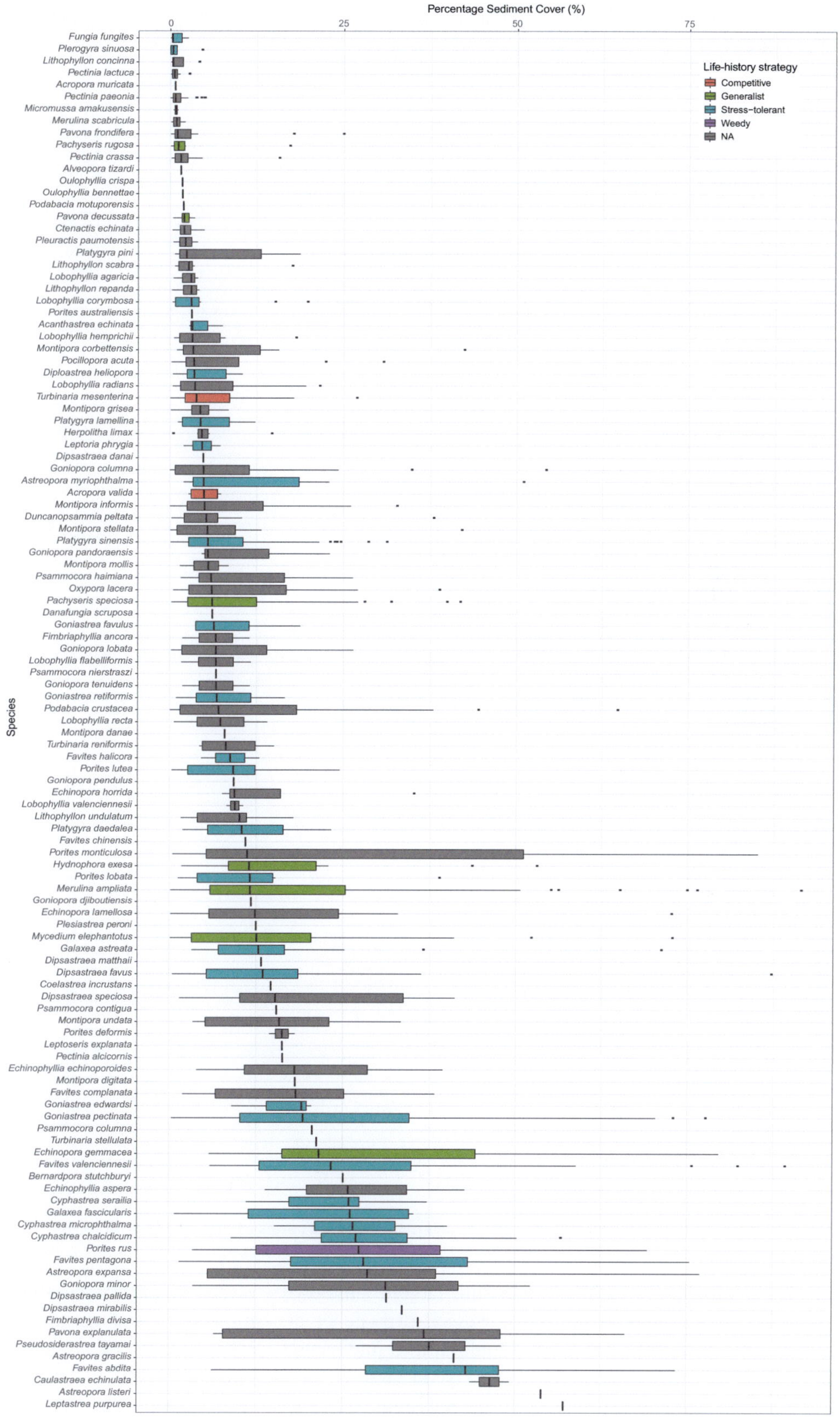

Fig. 3.4 Box plots of percentage sediment cover of all 115 coral species recorded, in order of increasing median sediment cover

Table 3.3 Best linear model with significant fixed effects of colony characteristics (site as random effect) on sediment cover (log-transformed)

Fixed effects		Estimate	S.E.
Species	*Cyphastrea chalcidicum****	1.657	0.400
	*Favites abdita****	1.646	0.414
	*Favites valenciennesi****	1.448	0.372
	*Favites pentagona****	1.437	0.361
	*Goniastrea pectinata****	1.361	0.316
	*Dipsastraea speciosa***	1.236	0.378
	*Pectinia crassa***	1.181	0.376
	*Dipsastraea favus***	1.150	0.400
	*Pavona frondifera***	1.010	0.373
	*Echinopora gemmacea**	0.891	0.369
	*Merulina ampliata****	0.844	0.247
	*Merulina scabricula**	0.828	0.395
	*Porites lutea**	0.809	0.402
	*Lithophyllon undulatum**	0.723	0.364
	*Mycedium elephantotus**	0.630	0.290
Growth form	CE*	−3.075	1.544
Growth form: colony size	CE: Colony radius**	1.498	0.537
	CM: Colony radius**	1.225	0.476
	CL: Colony radius*	1.190	0.469
	CF: Colony radius*	0.994	0.476

*p-value <0.05; ** p-value <0.01; *** p-value <0.001

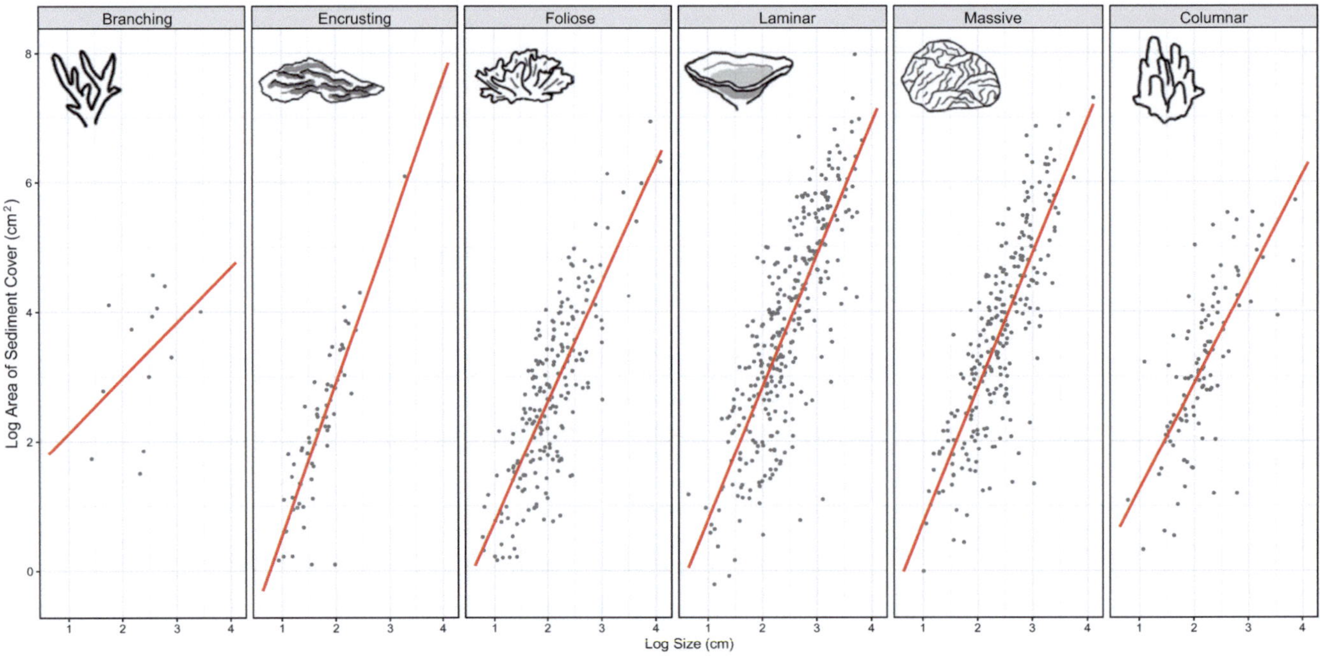

Fig. 3.5 Relationships between colony size and sediment cover on corals of different growth forms

pended sediment or turbidity are still largely unknown (Browne 2012; Jones et al. 2016; Bessell-Browne et al. 2017). Therefore, future research should aim to understand the effects of sedimentation and turbidity on various coral communities and identify which of these pressures most significantly affects different coral taxa in Singapore. These thresholds are crucial for guiding coastal management practices and urban development near coral reefs (Browne 2012).

As Singapore continues to develop its coastlines, the impacts of sedimentation on reefs are likely to intensify, compounded by future climate changes. Ongoing and planned coastal projects are projected to escalate sediment loads, particularly from terrestrial sources (Zweifler et al. 2021). This increased sedimentation could trigger nutrient loading and eutrophication, further stressing coral communities. Climate-related stressors, such as rising sea temperatures and ocean acidification, are expected to compound these challenges, making corals more vulnerable to sedimentation's detrimental effects (Brunner et al. 2021; Zweifler et al. 2021). The synergistic interaction of these stressors—exacerbated by sea-level rise and moderated tidal currents in urbanised settings—could lead to the transformation of coral habitats and further declines in reef biodiversity (Morgan et al. 2020; Zweifler et al. 2021). Additionally, the predicted increase in the frequency and intensity of extreme weather events due to climate change could trigger massive pulses of sedimentation, posing acute threats to coral reefs (Perry et al. 2018). Moreover, increased wave energy and sediment resuspension due to climate change or coastal development can create chronic turbidity, reducing light availability crucial for coral survival (Storlazzi et al. 2015). It is essential to recognise that these stressors will likely interact in complex ways, creating synergistic effects that further challenge the resilience of Singapore's coral reefs.

Long-term spatiotemporal monitoring using innovative techniques, such as environmental DNA (eDNA) analysis, is becoming more essential for identifying areas most vulnerable to sedimentation and tracking changes in reef health over time (Muenzel et al. 2024). This approach, particularly valuable in turbid waters like those surrounding Singapore, can complement traditional visual surveys and enable the detection of shifts in coral community composition and other reef fauna (Ip et al. 2021a, b; Yip et al. 2025). Given the anticipated increase in turbidity due to ongoing coastal development and climate change, understanding the impacts of sedimentation on both corals and reef-associated organisms is vital for effective conservation.

3.7 Conclusion

This study reviews the impacts of anthropogenic and natural sedimentation on coral reefs, particularly on coral–sediment interactions within Singapore's urban reefs and demonstrates that sediment cover on corals in Singapore is influenced by a complex interplay of factors, including species-specific traits, growth forms and colony size. Life-history strategies may partially explain the observed variation in sediment tolerance among corals. However, further research is needed to explore the influence of sediment characteristics, such as grain size and nutrient content, on coral responses. This will deepen our understanding of sediment regimes and their implications for coral communities over the longer term (Weber et al. 2006).

By identifying coral species and growth forms that are particularly vulnerable to sedimentation, conservation strategies can be tailored to prioritise the protection of these sensitive taxa (Davison et al. 2024). Additionally, more detailed spatial mapping of sediment stress can inform the selection of suitable sites for coral restoration, focusing efforts on areas with lower sedimentation levels. This knowledge also facilitates the development of more effective restoration techniques, such as selecting resilient coral species or implementing sediment control measures to enhance the success of transplantation efforts (Boström-Einarsson et al. 2020). By integrating these findings into management practices, Singapore can take proactive steps to safeguard its reefs in the face of anthropogenic and climate-related pressures.

Acknowledgements We thank all past and present members of the Reef Ecology Study Team for their tireless efforts in studying Singapore's coral reef biodiversity and health. This research is supported by the National Research Foundation and National Parks Board, Singapore, under the Marine Climate Change Science programme (Award No.: NRF-MCCS21-1-5-0001).

References

Abràmoff MD, Magalhães PJ, Ram SJ (2004) Image processing with ImageJ. Biophoton Int 11(7):36–42

Affendi YA, Tajuddin BH, Lee YL, Alfian AKA, Yusuf Y (2005) Scleractinian coral diversity of Kg. Tekek, Pulau Tioman Marine Park. In: Second regional symposium on environment and natural resources, Kuala Lumpur, pp 20–31

Amin B, Ismail A, Arshad A, Yap CK, Kamarudin MS (2009) Anthropogenic impacts on heavy metal concentrations in the coastal sediments of Dumai, Indonesia. Environ Monit Assess 148:291–305. https://doi.org/10.1007/s10661-008-0159-z

Anthony KR, Fabricius KE (2000) Shifting roles of heterotrophy and autotrophy in coral energetics under varying turbidity. J Exp Mar Biol Ecol 252(2):221–253. https://doi.org/10.1016/s0022-0981(00)00237-9

Aronson RB, Macintyre IG, Wapnick CM, O'Neill MW (2004) Phase-shifts, alternative states, and the unprecedented convergence of two reef systems. Ecology 85:1876–1891. https://doi.org/10.1890/03-0108

Babcock R, Smith L (2000) Effects of sedimentation on coral settlement and survivorship. Proceedings of the Ninth International Coral Reef Symposium, Bali, Indonesia 1:245–248

Bak RPM, Elgershuizen JHBW (1976) Patterns of oil sediment rejection in corals. Mar Biol 37:715–730. https://doi.org/10.1007/BF00389121

Baum G, Januar HI, Ferse SCA, Kunzmann A (2015) Local and regional impacts of pollution on coral reefs along the Thousand Islands north of the megacity Jakarta, Indonesia. PLoS One 10(9). https://doi.org/10.1371/journal.pone.0138271

Bauman AG, Guest JR, Dunshea G, Low J, Todd PA, Steinberg PD (2015) Coral settlement on a highly disturbed equatorial reef system. PLoS One 10(5):e0127874. https://doi.org/10.1371/journal.pone.0127874

Bellwood DR, Hughes TP, Folke C, Nyström M (2004) Confronting the coral reef crisis. Nature 429:827–833. https://doi.org/10.1038/nature02691

Bessell-Browne P, Negri AP, Fisher R, Clode PL, Jones R (2017) Impacts of light limitation on corals and crustose coralline algae. Sci Rep 7:11783. https://doi.org/10.1038/s41598-017-11783-z

Birrell CL, McCook JL, Willis BL (2005) Effects of algal turfs and sediment on coral settlement. Mar Pollut Bull 51:408–414. https://doi.org/10.1016/j.marpolbul.2004.10.022

Bollati E, Rosenberg Y, Simon-Blecher N, Tamir R, Levy O, Huang D (2022) Untangling the molecular basis of coral response to sedimentation. Mol Ecol 31(3):884–901. https://doi.org/10.1111/mec.16263

Bongaerts P, Hoeksema BW, Hay KB, Hoegh-Guldberg O (2012) Mushroom corals overcome live burial through pulsed inflation. Coral Reefs 31:399

Boström-Einarsson L, Babcock RC, Bayraktarov E, Ceccarelli D, Cook N, Ferse SCA, Hancock B, Harrison P, Hein M, Shaver E, Smith A, Suggett D, Stewart-Sinclair PJ, Vardi T, McLeod IM (2020) Coral restoration – a systematic review of current methods, successes, failures and future directions. PLoS One 15:e0226631. https://doi.org/10.1371/journal.pone.0226631

Brown BE, Howard LS (1985) Assessing the effects of "stress" on reef corals. In: Blaxter JHS, Russell F, Yonge M(eds) Advances in Marine Biology. Academic Press, pp 1–63. https://doi.org/10.1016/S0065-2881(08)60049-8

Brown B (1997) Coral bleaching: causes and consequences. Coral Reefs 16:S129–S138

Brown BE, Bythell JC (2005) Perspectives on mucus secretion in reef corals. Mar Ecol Prog Ser 296:291–309

Browne NK (2012) Spatial and temporal variations in coral growth on an inshore turbid reef subjected to multiple disturbances. Mar Environ Res 77:71–83

Browne N, Smithers S, Perry C, Ridd P (2012) A field-based technique for measuring sediment flux on coral reefs: application to turbid reefs on the Great Barrier Reef. J Coast Res 28(5):1247–1262

Browne N, Braoun C, Mcilwain J, Nagarajan R, Zinke J (2019) Borneo coral reefs subject to high sediment loads show evidence of resilience to various environmental stressors. PeerJ 7:e12167

Brunner CA, Uthicke S, Ricardo GF, Hoogenboom MO, Negri AP (2021) Climate change doubles sedimentation-induced coral recruit mortality. Sci Total Environ 768:143897. https://doi.org/10.1016/j.scitotenv.2020.143897

Burke L, Reytar K, Spalding M, Perry A (2011) Reefs at risk revisited. World Resources Institute, Washington, DC. https://www.wri.org/research/reefs-risk-revisited

Buss L, Jackson J (1979) Competitive networks: nontransitive competitive relationships in cryptic coral reef environments. Am Nat 113(2):223–234

Carlton JT, Geller JB, Reaka-Kudla ML, Norse EA (1999) Historical extinctions in the sea. Annu Rev Ecol Syst 30:515–538

Chan YKS, Affendi YA, Ang PO, Baria-Rodriguez MV, Chen CA, Chui APY, Giyanto GM, Huang H, Kuo C-Y, Kim SW, Lam VYY, Lane DJW, Lian JS, Lin SMNN, Lunn Z, Nañola CL, Nguyen VL, Park HS, Suharsono SM, Vo ST, Vibol O, Waheed Z, Yamano H, Yeemin T, Yong E, Kimura T, Tun K, Chou LM, Huang D (2023) Decadal stability in coral cover could mask hidden changes on reefs in the East Asian Seas. Commun Biol 6:630. https://doi.org/10.1038/s42003-023-05000-z

Chan YKS, Ng CSL, Tun KPP, Chou LM, Huang D (2024) Decadal decline of functional diversity despite increasing taxonomic and phylogenetic diversity of coral reefs under chronic urbanisation stress. Ecol Indic 164:112143. https://doi.org/10.1016/j.ecolind.2024.112143

Chou LM (2006) Marine habitats in one of the world's busiest harbours. In: Wolanski E (ed) The environment in Asia Pacific harbours. Springer, pp 377–391

Chou LM, Toh TC, Ng CSL (2017) Effectiveness of reef restoration in Singapore's rapidly urbanizing coastal environment. Int J Environ Sci Dev 8:576–580. https://doi.org/10.18178/ijesd.2017.8.8.1018

Chou LM, Huang D, Tan KS, Toh TC, Goh BPL, Tun K (2019) Singapore. In: Sheppard CRC (ed) World seas: an environmental evaluation. Volume II: the Indian Ocean to the Pacific. Academic, London, pp 539–558

Chow GSE, Chan YKS, Jain SS, Huang D (2019) Light limitation selects for depth generalists in urbanised reef coral communities. Mar Environ Res 147:101–112. https://doi.org/10.1016/j.marenvres.2019.04.010

Coffroth MA (1990) Mucous sheet formation on poritid corals: effects of altered salinity and sedimentation. Proceedings of the Fifth International Coral Reef Congress, Tahiti 4:165–171

Connell J (1997) Disturbance and recovery of coral assemblages. Coral Reefs 16:S101–S113

Darling E, Alvarez-Filip L, Oliver T, Mcclanahan T, Côté I, Bellwood D (2012) Evaluating life-history strategies of reef corals from species traits. Ecol Lett 15:1378–1386

Davison GWH, Gan JWM, Huang D, Hwang WS, Lum SKY, Yeo DCJ (2024) The Singapore red data book: red lists of Singapore biodiversity, 3rd edn. National Parks Board, Singapore, 688 pp

Deleja M, Paula JR, Repolho T, Franzitta M, Baptista M, Lopes V, Simão S, Fonseca VF, Duarte B, Rosa R (2022) Effects of hypoxia on coral photobiology and oxidative stress. Biology 11:1068. https://doi.org/10.3390/biology11071068

Díaz M, Madin J (2011) Macroecological relationships between coral species' traits and disease potential. Coral Reefs 30:73–84.https://doi.org/10.1007/s00338-010-0668-4

Dikou A, van Woesik R (2006) Survival under chronic stress from sediment load: spatial patterns of hard coral communities in the southern islands of Singapore. Mar Pollut Bull 52:1340–1354

Dodge RE, Vaisnys JR (1977) Coral populations and growth patterns: responses to sedimentation and turbidity associated with dredging. J Mar Res 35:715–730

Duckworth A, Giofre N, Jones R (2017) Coral morphology and sedimentation. Mar Pollut Bull 125:289–300

Dulvy NK, Sadovy Y, Reynolds JD (2003) Extinction vulnerability in marine populations. Fish Fish 4:25–64

English S, Wilkinson C, Baker V (1994) Line intercept transect. In: English S, Wilkinson C, Baker V (eds) Survey manual for tropical marine resources. Australian Institute of Marine Science, Townsville, pp 34–51

Erftemeijer P, Riegl B, Hoeksema B, Todd P (2012) Environmental impacts of dredging and other sediment disturbances on corals: a review. Mar Pollut Bull 64:1737–1765

Fabricius KE (2005) Effects of terrestrial runoff on the ecology of corals and coral reefs: review and synthesis. Mar Pollut Bull 50:125–146

Field ME, Chezar H, Storlazzi CD (2013) SedPods: a low-cost coral proxy for measuring net sedimentation. Coral Reefs 32:155–159. https://doi.org/10.1007/s00338-012-0953-5

Fine M, Hoegh-Guldberg O, Meroz-Fine E, Dove S (2019) Ecological changes over 90 years at Low Isles on the Great Barrier Reef. Nat Commun 10:4409

Gilmour J (1999) Experimental investigation into the effects of suspended sediment on fertilisation, larval survival, and settlement in a scleractinian coral. Mar Biol 135:451–456

Gin KYH, Lin X, Zhang S (2000) Dynamics and size structure of phytoplankton in the coastal waters of Singapore. J Plankton Res 22(8):1465–1484

Gin KYH, Koh ST, Lin II, Chan ES (2002) Application of spectral signatures and colour ratios to estimate chlorophyll in Singapore's coastal waters. Estuar Coast Shelf Sci 55:719–728

Glynn PW (1996) Coral reef bleaching: facts, hypotheses, and implications. Glob Chang Biol 2:495–509

Goh BPL, Lee CS (2008) A study of the effect of sediment accumulation on the settlement of coral larvae using conditioned tiles. In: Proceedings of the Eleventh International Coral Reef Symposium, Ft. Lauderdale, Florida, pp 1235–1239

Goh B, Chou L (1997) Heavy metal levels in marine sediments of Singapore. Environ Monit Assess 44:67–80.https://doi.org/10.1023/A:1005763918958

Guest J, Baird A, Maynard J, Muttaqin E, Edwards A, Campbell S, Yewdall K, Yang A, Affendi Y, Chou L (2012) Contrasting patterns of coral bleaching susceptibility in 2010 suggest an adaptive response to thermal stress. PLoS One 7:e33353

Guest J, Tun K, Low J, Vergés A, Marzinelli E, Campbell A, Bauman AG, Feary DA, Chou LM, Steinberg P (2016a) 27 years of benthic and coral community dynamics on turbid, highly urbanised reefs off Singapore. Sci Rep 6:36260

Guest JR, Low J, Tun K, Wilson B, Ng C, Raingeard D, Ulstrup KE, Tanzil JTI, Todd PA, Toh TC, McDougald D, Chou LM, Steinberg PD (2016b) Coral community response to bleaching on a highly disturbed reef. Sci Rep 6:20717

Halford A, Cheal AJ, Ryan D, Williams D (2004) Resilience to large-scale disturbance in coral and fish assemblages on the Great Barrier Reef. Ecology 85:1892–1905. https://doi.org/10.1890/03-4017

Heery E, Hoeksema B, Browne N, Reimer J, Ang P, Huang D, Chou L, Loke L, Saksena-Taylor P, Alsagoff N, Yeemin T, Sutthacheep M, Vo ST, Bos A, Gumanao G, Syed Hussein MA, Waheed Z, Lane D, Johan O, Friess D (2018) Urban coral reefs: degradation and resilience of hard coral assemblages in coastal cities of East and Southeast Asia. Mar Pollut Bull 135:654–681. https://doi.org/10.1016/j.marpolbul.2018.07.041

Hendrick VJ, Hutchison ZL, Last KS (2016) Sediment burial intolerance of marine macroinvertebrates. PLoS One 11(2). https://doi.org/10.1371/journal.pone.0149114

Hoeksema B, Koh E (2009) Depauperation of the mushroom coral fauna (Fungiidae) of Singapore (1860s–2006) in changing reef conditions. Raffles Bull Zool Suppl 22:91–101

Hoey AS, Bellwood DR (2009) Limited functional redundancy in a high diversity system: single species dominates key ecological process on coral reefs. Ecosystems 12:1316–1328

Hoogenboom MO, Connolly SR, Anthony K (2008) Interactions between morphological and physiological plasticity optimize energy acquisition in corals. Ecology 89(4):1144–1154

Howarth R, Chan F, Conley DJ, Garnier J, Doney SC, Marino R, Billen G (2011) Coupled biogeochemical cycles: eutrophication and hypoxia in temperate estuaries and coastal marine ecosystems. Front Ecol Environ 9(1):18–26. https://doi.org/10.1890/100008

Huang D, Tun KPP, Chou LM, Todd PA (2009) An inventory of zooxanthellate scleractinian corals in Singapore, including 33 new records. Raffles Bull Zool Suppl 22:69–80

Huang D, Licuanan WY, Hoeksema BW, Chen CA, Ang PO, Huang H, Lane DJW, Vo ST, Waheed Z, Affendi YA, Yeeemin T, Chou LM (2015) Extraordinary diversity of reef corals in the South China Sea. Mar Biodivers 45:157–168. https://doi.org/10.1007/s12526-014-0236-1

Hubbard DK (1997) Reefs as dynamic systems. In: Birkeland C (ed) Life and death of coral reefs. Chapman & Hall, New York, pp 43–67. https://doi.org/10.1007/978-1-4615-5995-5_3

Hughes TP, Graham NA, Jackson JB, Mumby PJ, Steneck RS (2010) Rising to the challenge of sustaining coral reef resilience. Trends Ecol Evol 25:633–642

Hughes TP, Kerry JT, Álvarez-Noriega M, Álvarez-Romero JG, Anderson KD, Baird AH, Babcock RC, Beger M, Bellwood DR, Berkelmans R, Bridge TC, Butler IR, Byrne M, Cantin NE, Comeau S, Connolly SR, Cumming GS, Dalton SJ, Diaz-Pulido G, Eakin CM, Figueira WF, Gilmour JP, Harrison HB, Heron SF, Hoey AS, Hobbs J-PA, Hoogenboom MO, Kennedy EV, Kuo C-Y, Lough JM, Lowe RJ, Liu G, McCulloch MT, Malcolm HA, McWilliam MJ, Pandolfi JM, Pears RJ, Pratchett MS, Schoepf V, Simpson T, Skirving WJ, Sommer B, Torda G, Wachenfeld DR, Willis BL, Wilson SK (2017) Global warming and recurrent mass bleaching of corals. Nature 543:373–377

Hughes DJ, Alderdice R, Cooney C, Kühl M, Pernice M, Voolstra CR, Suggett DJ (2020) Coral reef survival under accelerating ocean deoxygenation. Nat Clim Chang 10:296–307. https://doi.org/10.1038/s41558-020-0737-9

Humanes A, Fink A, Willis BL, Fabricius KE, de Beer D, Negri AP (2017a) Effects of suspended sediments and nutrient enrichment on juvenile corals. Mar Pollut Bull 125(1–2):166–175. https://doi.org/10.1016/j.marpolbul.2017.08.003

Humanes A, Ricardo G, Willis BL, Fabricius KE, Negri AP (2017b) Cumulative effects of suspended sediments, organic nutrients and temperature stress on early life history stages of the coral *Acropora tenuis*. Sci Rep 7:44101. https://doi.org/10.1038/srep44101

Hume BCC, Mejia-Restrepo A, Voolstra CR, et al (2020) Fine-scale delineation of Symbiodiniaceae genotypes on a previously bleached central Red Sea reef system demonstrates a prevalence of coral host-specific associations. Coral Reefs 39:583–601. https://doi.org/10.1007/s00338-020-01917-7

Ip YCA, Chang JJM, Lim KKP, Jaafar Z, Wainwright BJ, Huang D (2021a) Seeing through sedimented waters: environmental DNA reduces the phantom diversity of sharks and rays in turbid marine habitats. BMC Ecol Evol 21:166. https://doi.org/10.1186/s12862-021-01895-6

Ip YCA, Tay YC, Chang JJM, Ang HP, Tun KPP, Chou LM, Huang D, Meier R (2021b) Seeking life in sedimented waters: environmental DNA from diverse habitat types reveals ecologically significant species in a tropical marine environment. Environ DNA 3:654–668. https://doi.org/10.1002/edn3.162

Jackson JBC (1991) Adaptation and diversity of reef corals. Bioscience 41:475–482

Jackson JBC, Kirby MX, Berger WH, Bjorndal KA, Botsford LW, Bourque BJ, Bradbury RH, Cooke R, Erlandson J, Estes JA, Hughes TP, Kidwell S, Lange CB, Lenihan HS, Pandolfi JM, Peterson CH, Steneck RS, Tegner MJ, Warner RR (2001) Historical overfishing and the recent collapse of coastal ecosystems. Science 293:629–638

Johnson MD, Rodriguez LM, Altieri AH (2018) Shallow-water hypoxia and mass mortality on a Caribbean coral reef. Bull Mar Sci 94:143–144

Johnson MD, Scott JJ, Leray M, Lucey N, Bravo LMR, Wied WL, Altieri AH (2021) Rapid ecosystem-scale consequences of acute deoxygenation on a Caribbean coral reef. Nat Commun 12:4522. https://doi.org/10.1038/s41467-021-24777-3

Jones C, Lawton J, Shachak M (1994) Organisms as ecosystem engineers. Oikos 69(3):373–386. https://doi.org/10.2307/3545850

Jones R, Bessell-Browne P, Fisher R, Klonowski W, Slivkoff M (2016) Assessing the impacts of sediments from dredging on corals. Mar Pollut Bull 102(1):9–29. https://doi.org/10.1016/j.marpolbul.2015.10.049

Kendrick GA (1991) Recruitment of coralline crusts and filamentous turf algae in the Galapagos Archipelago: effect of simulated scour, erosion and accretion. J Exp Mar Biol Ecol 147:47–63. https://doi.org/10.1016/0022-0981(91)90179-D

Lai S, Loke LHL, Hilton MJ, Bouma TJ, Todd PA (2015) The effects of urbanization on coastal habitats and the potential for ecological engineering: a Singapore case study. Ocean Coast Manag 103:78–85

Larcombe P, Ridd P, Prytz A, Wilson B (1995) Factors controlling suspended sediment on inner-shelf coral reefs, Townsville, Australia. Coral Reefs 14:163–171

Lee DL, Poquita-Du RC, Todd PA (2021) Active sediment rejection efficiency of five scleractinian coral species in Singapore. Nature in Singapore 14:e2021051. https://doi.org/10.26107/NIS-2021-0051

Lim GSY, Chou LM (1988) The echinoderm fauna of sediment stressed reefs in Singapore. Proc Sixth Int Coral Reef Symposium 2:245–250

Loong S, Taira D, Chan YKS, et al (2025) Covariations of coral, macroalgal, macroinvertebrate and fish communities in an urbanised tropical reef environment. Coral Reefs https://doi.org/libproxy1.nus.edu.sg/10.1007/s00338-025-02723-9

Low JKY, Chou LM (1994) Coral reef fish communities in a sediment stressed environment. In: Fourth LIPI-JSPS Joint Seminar on Marine Science, Jakarta, pp 91–99

McWilliam M, Hoogenboom MO, Baird AH, Kuo C, Madin JS, Hughes TP (2018) Biogeographical disparity in the functional diversity and redundancy of corals. Proc Natl Acad Sci USA 115(12):3084–3089. https://doi.org/10.1073/pnas.1716643115

Meesters EH, Bos A, Gast GJ (1992) Effects of sedimentation and lesion position on coral tissue regeneration. Proceedings of the Seventh International Coral Reef Symposium, Guam, 2:671–678

Miller DC, Muir CL, Hauser OA (2002) Detrimental effects of sedimentation on marine benthos: what can be learned from natural processes and rates? Ecol Eng 19(3):211–232. https://doi.org/10.1016/S0925-8574(02)00081-2

Moberg F, Folke C (1999) Ecological goods and services of coral reef ecosystems. Ecol Econ 29(2):215–233

Morgan KM, Perry CT, Johnson JA, Smithers SG (2017) Nearshore turbid-zone corals exhibit high bleaching tolerance on the Great Barrier Reef following the 2016 ocean warming event. Front Mar Sci 4:224. https://doi.org/10.3389/fmars.2017.00224

Morgan KM, Moynihan MA, Sanwlani N, Switzer AD (2020) Light limitation and depth-variable sedimentation drives vertical reef compression on turbid coral reefs. Front Mar Sci 7:571256. https://doi.org/10.3389/fmars.2020.571256

Muenzel D, Bani A, Brauwer MD, Stewart E, Djakiman C, Halwi, Purnama R, Yusuf S, Santoso P, Hukom FD, Struebig M, Jompa J, Limmon G, Dumbrell A, Beger M (2024) Combining environmental DNA and visual surveys can inform conservation planning for coral reefs. Proc Natl Acad Sci USA 121:e2307214121. https://doi.org/10.1073/pnas.2307214121

Nearing MA, Pruski FF, O'Neal MR (2004) Expected climate change impacts on soil erosion rates: a review. J Soil Water Conserv 59(1):43–50

Newell RC, Seiderer LJ, Simpson NM, Robinson JE (2004) Impacts of marine aggregate dredging on benthic macrofauna off the south coast of the United Kingdom. J Coast Res 20(1):115–125. https://doi.org/10.2112/1551-5036(2004)20[115:IOMADO]2.0.CO;2

Ng CSL, Huang D, Toh KB, Sam SQ, Kikuzawa YP, Toh TC, Taira D, Chan YKS, Hung LZT, Sim WT, Rashid AR, Afiq-Rosli L, Ng NK, Chou LM (2020) Responses of urban reef corals during the 2016 mass bleaching event. Mar Pollut Bull 154:111111. https://doi.org/10.1016/j.marpolbul.2020.111111

Ng MS, Teo A, Todd PA (2022) Sediment trap height affects mass, particle size, and biogeochemical composition of material collected in an equatorial coral reef. Mar Pollut Bull 183:114086. https://doi.org/10.1016/j.marpolbul.2022.114086

Nugues MM, Roberts CM (2003) Partial mortality in massive reef corals as an indicator of sediment stress on coral reefs. Mar Pollut Bull 46:314–323

Ong JH, Wainwright BJ, Jain SS, Afiq-Rosli L, Lee JN, Huang D (2022) Species and spatio-environmental effects on coral endosymbiont communities in Southeast Asia. Coral Reefs 41:1131–1145. https://doi.org/10.1007/s00338-022-02254-7

Otaño-Cruz A, Montañez-Acuña AA, Torres-López V, Hernández-Figueroa EM, Hernández-Delgado EA (2017) Effects of changing weather, oceanographic conditions, and land uses on spatio-temporal variation of sedimentation dynamics along near-shore coral reefs. Front Mar Sci 4:249. https://doi.org/10.3389/fmars.2017.00249

Pandolfi J (2002) Coral community dynamics at multiple scales. Coral Reefs 21:13–23

Pandolfi JM, Bradbury RH, Sala E, Hughes TP, Bjorndal KA, Cooke RG, McArdle DA, McClenachan LE, Newman MJH, Paredes G, Warner RR, Jackson JBC (2003) Global trajectories of the long-term decline of coral reef ecosystems. Science 301:955–958

Perry CT, Alvarez-Filip L, Graham NAJ, Mumby PJ, Wilson SK, Kench PS, Manzello DP, Morgan KM, Slangen ABA, Thomson DP, Januchowski-Hartley F, Smithers SG, Steneck RS, Carlton R, Edinger EN, Enochs IC, Estrada-Saldívar N, Haywood MDE, Kolodziej G, Murphy GN, Pérez-Cervantes E, Suchley A, Valentino L, Boenish R, Wilson M, Macdonald C (2018) Loss of coral reef growth capacity to track future increases in sea level. Nature 558:396–400. https://doi.org/10.1038/s41586-018-0194-z

Pezner AK, Courtney TA, Barkley HC, Chou W-C, Chu H-C, Clements SM, Cyronak T, DeGrandpre MD, Kekuewa SAH, Kline DI, Liang Y-B, Martz TR, Mitarai S, Page HN, Rintoul MS, Smith JE, Soong K, Takeshita Y, Tresguerres M, Wei Y, Yates KK, Andersson AJ (2023) Increasing hypoxia on global coral reefs under ocean warming. Nat Clim Chang 13:403–409. https://doi.org/10.1038/s41558-023-01619-2

Philipp E, Fabricius K (2003) Photophysiological stress in scleractinian corals in response to short-term sedimentation. J Exp Mar Biol Ecol 287(1):57–78. https://doi.org/10.1016/S0022-0981(02)00495-1

Pineda MC, Strehlow B, Sternel M, Duckworth A, den Haan J, Jones R, Webster NS (2017) Effects of sediment smothering on the sponge holobiont with implications for dredging management. Sci Rep 7:5156. https://doi.org/10.1038/s41598-017-05243-x

Plusquellec Y, Webb G, Hoeksema BW (1999) Automobility in Tabulata, Rugosa, and extant scleractinian analogues: stratigraphic and paleogeographic distribution of Paleozoic mobile corals. J Paleontol 73(6):985–1001

Pollock FJ, Lamb JB, Field SN, Heron SF, Schaffelke B, Shedrawi G, Bourne DG, Willis BL (2014) Sediment and turbidity associated with offshore dredging increase coral disease prevalence on nearby reefs. PLoS One 9:e102498

Poquita-Du RC, Huang D, Chou LM, Mrinalini, Todd PA (2019a) Short-term exposure to heat and sediment triggers changes in coral gene expression and photo-physiological performance. Front Mar Sci 6:121

Poquita-Du RC, Quek Z, Jain S, Schmidt-Roach S, Tun K, Heery E, Chou L, Todd PA, Huang D (2019b) Last species standing: loss of Pocilloporidae corals associated with coastal urbanization in a tropical city-state. Mar Biodivers 49:1727–1741. https://doi.org/10.1007/s12526-019-00939-x

Poquita-Du RC, Huang D, Chou LM, Todd PA (2020) The contribution of stress-tolerant endosymbiotic dinoflagellate *Durusdinium* to *Pocillopora acuta* survival in a highly urbanized reef system. Coral Reefs 39:745–755. https://doi.org/10.1007/s00338-020-01902-0

Potts D, Jacobs J (2000) Evolution of reef-building scleractinian corals in turbid environments: a paleo-ecological hypothesis. Proceedings of the Ninth International Coral Reef Symposium, Bali, Indonesia, 23–27 October 2000, pp 249–254

Pratchett MS, Thompson CA, Hoey AS, Cowman PF, Wilson SK (2018) Effects of coral bleaching and coral loss on the structure and function of reef fish assemblages. In: van Oppen MJH, Lough J (eds) Coral bleaching: patterns, processes, causes and consequences. Springer, Basel, pp 265–293

Purcell SW (2000) Association of epilithic algae with sediment distribution on a windward reef in the northern Great Barrier Reef, Australia. Bull Mar Sci 66:199–214

Reichelt-Brushett AJ, Harrison PL (2005) The effect of selected trace metals on the fertilization success of several scleractinian coral species. Coral Reefs 24:524–534

Ricardo GF, Jones RJ, Nordborg M, Negri AP (2017) Settlement patterns of the coral *Acropora millepora* on sediment-laden surfaces. Sci Total Environ 609:277–288. https://doi.org/10.1016/j.scitotenv.2017.07.153

Riegl B, Branch GM (1995) Effects of sediment on the energy budgets of four scleractinian (Bourne 1900) and five alcyonacean (Lamouroux 1816) corals. J Exp Mar Biol Ecol 186:259–275

Riegl B, Heine C, Branch GM (1996) The function of funnel-shaped coral growth in a high-sedimentation environment. Mar Ecol Prog Ser 145:87–94

Rogers CS (1990) Responses of coral reefs and reef organisms to sedimentation. Mar Ecol Prog Ser 62:185–202

Rogers CS, Ramos-Scharrón CE (2022) Assessing effects of sediment delivery to coral reefs: a Caribbean watershed perspective. Front Mar Sci 8:773968. https://doi.org/10.3389/fmars.2021.773968

Rosenfeld M, Bresler V, Abelson A (1999) Sediment as a possible source of food for corals. Ecol Lett 2:345–348

RStudio Team (2019) RStudio: integrated development for R. RStudio, Inc, Boston. http://www.rstudio.com/

Rushmore ME, Ross C, Fogarty ND (2021) Physiological responses to short-term sediment exposure in adults of the Caribbean coral *Montastraea cavernosa* and adults and recruits of *Porites astreoides*. Coral Reefs 40:1579–1591. https://doi.org/10.1007/s00338-021-02156-0

Sanders D, Baron-Szabo R (2005) Scleractinian assemblages under sediment input: their characteristics and relation to the nutrient input concept. Palaeogeogr Palaeoclimatol Palaeoecol 216:139–181

Schneider CA, Rasband WS, Eliceiri KW (2012) NIH image to ImageJ: 25 years of image analysis. Nat Methods 9(7):671–675

Silveira CB, Luque A, Roach TNF, Villela H, Barno A, Green K, Reyes B, Rubio-Portillo E, Le T, Mead S, Hatay M, Vermeij MJA, Takeshita Y, Haas A, Bailey B, Rohwer F (2019) Biophysical and physiological processes causing oxygen loss from coral reefs. eLife 8:e49114

Smith EG, Gurskaya A, Hume BC, Voolstra CR, Todd PA, Bauman AG, Burt JA (2020) Low Symbiodiniaceae diversity in a turbid marginal reef environment. Coral Reefs 39(3):545–553. https://doi.org/10.1007/s00338-020-01956-0

Speare KE, Duran A, Miller MW, Burkepile DE (2019) Sediment associated with algal turfs inhibits the settlement of two endangered coral species. Mar Pollut Bull 144:189–195. https://doi.org/10.1016/j.marpolbul.2019.04.066

Stafford-Smith MG (1993) Sediment-rejection efficiency of 22 species of Australian scleractinian corals. Mar Biol 115:229–243

Stafford-Smith M, Ormond R (1992) Sediment-rejection mechanisms of 42 species of Australian scleractinian corals. Mar Freshw Res 43:683–705

Storlazzi CD, Norris BK, Rosenberger KJ (2015) The influence of grain size, grain color, and suspended-sediment concentration on light attenuation: why fine-grained terrestrial sediment is bad for coral reef ecosystems. Coral Reefs 34:967–975. https://doi.org/10.1007/s00338-015-1268-0

Stronkhorst J, Ariese F, van Hattum B, Postma JF, de Kluijver M, Den Besten PJ, Bergman MJN, Daan R, Murk AJ, Vethaak AD (2003) Environmental impact and recovery at two dumping sites for dredged material in the North Sea. Environ Pollut 124(1):17–31. https://doi.org/10.1016/S0269-7491(02)00430-X

Tanzil JTI, Ng A, Tey YQ, Tan YB, Yun ET, Huang D (2016) A preliminary characterisation of *Symbiodinium* diversity in some common corals from Singapore. Cosmos 12:1–13. https://doi.org/10.1142/S0219607716500014

Te FT (1992) Response to higher sediment loads by *Pocillopora damicornis* planulae. Coral Reefs 11:131–134

Tebbett SB, Goatley CHR, Bellwood DR (2017) The effects of algal turf sediments and organic loads on feeding by coral reef surgeonfishes. PLoS One 12(1):e0169479. https://doi.org/10.1371/journal.pone.0169479

Tebbett SB, Morais J, Bellwood DR (2022) Spatial patchiness in change, recruitment, and recovery on coral reefs at Lizard Island following consecutive bleaching events. Mar Environ Res 173:105537. https://doi.org/10.1016/j.marenvres.2021.105537

Tebbett SB, Schlaefer JA, Bowden CL, Collins WP, Hemingson CR, Ling SD, Morais J, Morais RA, Siqueira AC, Streit RP, Swan S, Bellwood DR (2023) Bio-physical determinants of sediment accumulation on an offshore coral reef: a snapshot study. Sci Total Environ 895:165188. https://doi.org/10.1016/j.scitotenv.2023.165188

Telesnicki G, Goldberg W (1995) Effects of turbidity on the photosynthesis and respiration of two South Florida reef coral species. Bull Mar Sci 57:527–539

Thomas S, Ridd P (2005) Field assessment of innovative sensor for monitoring of sediment accumulation at inshore coral reefs. Mar Pollut Bull 51(1–4):470–480. https://doi.org/10.1016/j.marpolbul.2004.10.026

Todd PA, Ong X, Chou LM (2010) Impacts of pollution on marine life in Southeast Asia. Biodivers Conserv 19:1063–1082

Todd PA, Zhao W, Chou LM (2013) Environment-induced colour changes in the massive corals *Diploastrea speciosa* and *Diploastrea heliopora*. Nat Singapore 6:173–178

Tun KPP, Cheshire AC, Chou LM (1994) Photosynthetic production of four scleractinian corals from Singapore. In: Sudara S, Wilkinson C, Chou LM (eds) Third ASEAN-Australia Symposium on Living Coastal Resources. Chulalongkorn University, Bangkok, pp 69–77

Tun KPP, Chou LM, Low JKY, Yeemin T, Phongsuwan N, Setiasih N, Wilson J, Affendi YA, Kee Alfian AA, Lane D, van Bochove J-W, Kluskens B, Nguyen VL, Vo ST, Gomez E (2011) A regional review on the 2010 coral bleaching event in Southeast Asia. In: Status of coral reefs in East Asian Seas region: 2010. Tokyo, Ministry of the Environment, pp 9–27

Tuttle LJ, Donahue MJ (2022) Effects of sediment exposure on corals: a systematic review of experimental studies. Environ Evid 11:4. https://doi.org/10.1186/s13750-022-00256-0

Tuttle LJ, Johnson C, Kolinski S, Minton D, Donahue MJ (2020) How does sediment exposure affect corals? A systematic review protocol. Environ Evid 9:17. https://doi.org/10.1186/s13750-020-00200-0

van der Meij SET, Suharsono HBW (2010) Long-term changes in coral assemblages under natural and anthropogenic stress in Jakarta Bay (1920–2005). Mar Pollut Bull 60:1442–1454

van Maren DS, Liew SC, Hasan GMJ (2014) The role of terrestrial sediment on turbidity near Singapore's coral reefs. Cont Shelf Res 76:75–88. https://doi.org/10.1016/j.csr.2013.12.001

van Woesik R, Shlesinger T, Grottoli AG, et al (2022) Coral-bleaching responses to climate change across biological scales. Glob Change Biol 28:4229–4250. https://doi.org/10.1111/gcb.16192

Veron JEN (2000) Corals of the world. Australian Institute of Marine Science, Townsville

Weber M, Lott C, Fabricius KE (2006) Sedimentation stress in a scleractinian coral exposed to terrestrial and marine sediments with contrasting physical, organic, and geochemical properties. J Exp Mar Biol Ecol 336:18–32

Weber M, de Beer D, Lott C, Polerecky L, Kohls K, Abed RMM, Ferdelman TG, Fabricius KE (2012) Mechanisms of damage to corals exposed to sedimentation. Proc Natl Acad Sci USA 109:E1558–E1567

Wolanski E, Fabricius K, Spagnol S, Brinkman R (2005) Fine sediment budget on an inner-shelf coral-fringed island, Great Barrier Reef of Australia. Estuar Coast Shelf Sci 65(1–2):153–158. https://doi.org/10.1016/j.ecss.2005.06.003

Wong JSY, Chan YKS, Ng CSL, Tun KPP, Darling ES, Huang D (2018) Comparing patterns of taxonomic, functional, and phylogenetic diversity in reef coral communities. Coral Reefs 37:737–750. https://doi.org/10.1007/s00338-018-1698-6

Yeemin T, Pengsakun S, Yucharoen M, Klinthong W, Sangmanee K, Sutthacheep M (2013) Long-term changes in coral communities under stress from sediment. Deep Sea Res Part II Top Stud Oceanogr 96:32–40.https://doi.org/10.1016/j.dsr2.2013.04.019

Yip ZT, Quek ZBR, Huang D (2025) Spatiotemporal eDNA monitoring of marine biodiversity in a hyperurbanised coastal environment. Environmental DNA 7:e70174. https://doi.org/10.1002/edn3.70174

Zweifler A, O'Leary M, Morgan K, Browne NK (2021) Turbid coral reefs: past, present and future—a review. Diversity 13(6):251. https://doi.org/10.3390/d13060251

Zweifler A, Dee S, Browne NK (2024) Resilience of turbid coral communities to marine heatwave. Coral Reefs 43:1303–1315. https://doi.org/10.1007/s00338-024-02538-0

4 Energetic Needs of Corals Living in Singapore's Extreme Reef Environment

Jani Thuaibah Isa Tanzil, Wei Long Ow Yong, Yan Xiang Ow, Wai Leong, Chin Soon Lionel Ng, Jonas Chee Haw Khaw, and Karenne Tun

Abstract

Singapore's corals live in an extreme reef environment characterised by high-sedimentation rates, turbidity and nutrient levels, and low light. Proximity to coastal development and other anthropogenic impacts over several decades has resulted in a compressed zone of coral cover up to 6–7 m depth, beyond which live coral cover declines rapidly, with the benthos primarily composed of rubble and silt. Despite the numerous environmental challenges, Singapore's remaining reefs still host high coral species diversity and exhibit relatively rapid growth. Here, we review the physiological performance of corals living in such harsh conditions from an energetics approach to shed light on their persistence and provide perspectives on the challenges that could jeopardise their future existence. Past and recent studies on coral photo-physiology, heterotrophy, respiration, growth and reproduction are discussed and synthesised to obtain the daily energy budgets of some common corals from Singapore's reefs. Overall, corals at shallower reef depths (~2–3 m) showed positive energy balances, while those in deeper waters (~6–7 m) acquired just enough to sustain their daily metabolic needs with little to no scope for growth or reproduction. These findings help explain the current vertical distribution of corals on Singapore's reefs and underscore the increasing role trophic plasticity is expected to play in the resilience of Singapore's corals.

J. T. I. Tanzil (✉) · W. L. O. Yong · Y. X. Ow · W. Leong · C. S. L. Ng · J. C. H. Khaw
Tropical Marine Science Institute, National University of Singapore, Singapore, Singapore
e-mail: jani.tanzil@nus.edu.sg

K. Tun
National Biodiversity Centre, National Parks Board, Singapore, Singapore

Keywords

Photo-physiology · Coral energetics · Energy budget · Heterotrophy · Extreme reef · Trophic plasticity

4.1 Singapore's Extreme Reefs

'Extreme reefs' exist in suboptimal conditions characterised by high-sedimentation rates, turbidity and nutrient levels, as well as low light regimes that are typically associated with dynamic coastal conditions further affected, or not, by other anthropogenic disturbances (Kleypas et al. 1999; Perry and Larcombe 2003; Schoepf et al. 2023). Singapore's coral reefs are an example of such 'extreme' reefs. After losing >60% of her coral reefs from decades of extensive coastal urbanisation, Singapore's remaining 10 km^2 of reefs are mainly found as fringing or patch reefs around her southern offshore islands co-existing alongside industries and a busy shipping port (Burke et al. 2002; Tun 2012; Tan et al. 2023). Singapore's corals survive in a very light-limited environment with high light attenuation (K_d of 0.4–0.6 m^{-1}) after being exposed to five decades of increased sedimentation that has reduced underwater visibility from >10 m in the early 1960s to ~2 m today (Tun et al. 1994; Dikou and van Woesik 2006; Tun 2012; Morgan et al. 2020). Depending on the proximity to mainland and monsoonal-influenced seasonality, Photosynthetically Active Radiation (PAR) levels can fluctuate seasonally from as low as ~50 µmol m^{-2} day to up to >500 µmol m^{-2} day^{-1}, and extreme PAR regimes can last for up to 2–3 months (Browne et al. 2015a; Tanzil et al. 2019). Today, Singapore's reef photic zones are typically limited to ~6–7 m depths at mid-tide with highest percent live coral cover found at ~2–3 m (Chou 2010; Guest et al. 2016; Chow et al. 2019; Chan et al. 2024).

The vertical truncation of Singapore's habitable reef zones that used to extend to >10 m depths is a consequence of the high sedimentation and turbidity generated by decades

L. M. Chou, D. Huang (eds.), *Coral Reefs of Singapore's Urbanised Sea*, Coral Reefs of the World 23,
https://doi.org/10.1007/978-981-95-0903-4_4

of extensive land reclamation, heavy shipping traffic and regular dredging of shipping channels (Browne et al. 2014; Chou et al. 2017). Despite these challenging conditions, Singapore's reefs still support highly diverse coral communities (Huang et al. 2009; Ng et al. 2021), with some sites supporting high live coral cover of between 50% and 70% (Guest et al. 2016; Wong et al. 2018; Chan et al. 2024), and rapid coral growth (Tanzil et al. 2013). While various factors could contribute to the persistence and resilience of Singapore's corals, this chapter aims to provide insights from an energetics perspective. In particular, how corals living on Singapore's sedimented reefs obtain energy (and to what extent), their basal metabolic needs and other energetic investments such as growth and reproduction will be reviewed and discussed. By estimating the daily energy budgets of some common corals, we can shed light on the energy balancing act that corals need to perform to live in such a harsh environment.

4.2 Coral Energy Dynamics

For all organisms, energy flow serves as an important currency for physiological performance, including basic metabolic needs, recovery, growth and reproduction—all of which have implications for survival and fitness (Maltby 1999). The dynamics of energy availability, acquisition and reserves can therefore greatly influence the risk of mortality or chances of survival during period/s of resource scarcity. Whether a given stress response ultimately leads to mortality depends not only on the duration and intensity of the stress event, but also on the organisms' energy status, which is a function of (i) the current and past energy intakes and losses, and (ii) energy allocation for key physiological processes (i.e. maintenance, growth and reproduction) (Koiijman 2000).

Hermatypic reef-building corals are polytrophic: while energy acquisition is largely determined by rate of photosynthetic carbon fixation by their symbiotic algae (zooxanthellae) and the rate of carbon transfer by these photo-endosymbionts (Muscatine et al. 1981; Davies 1984; Grottolli et al. 2006), the host can also acquire energy through heterotrophy (feeding) (Sebens et al. 1996; Anthony and Fabricius 2000; Houlbrèque et al. 2004; Grottolli et al. 2006; Borell et al. 2008). Under optimum conditions for photosynthesis in nutrient poor environments, corals are principally autotrophic as their zooxanthellae can supply in excess of the coral host's energetic requirements (Muscatine et al. 1981; Grottolli et al. 2006). However, in sub-optimal environments, such as low light or fluctuating light regimes characteristic of extreme reefs (Todd 2008; Cacciapaglia and van Woesik 2016; Tanzil et al. 2019), much lower quantities of photosynthates are produced and translocated. To cope, corals can employ various photo-acclimation strategies to maximise autotrophy (Hennige et al. 2008), including regulating their endosymbiont and chlorophyll populations (Muscatine 1967; Drew 1972; Muscatine 1980; Fitt 2000; Warner et al. 2002; Lesser and Farrell 2004) and augmenting their skeletal morphology to maximise light capture (Ow and Todd 2010). However, corals also exhibit nutritional plasticity and can increase heterotrophy to supplement energy inputs (Houlbrèque and Ferrier-Pagès 2009; Santos et al. 2021). For example, some coral species can double their rates of particle feeding under prolonged shading (Anthony and Fabricius 2000; Tremblay et al. 2015), and display markedly different $\partial^{13}C$ and $\partial^{15}N$ in their tissue relative to their symbionts, indicating potential shifts from autotrophy to heterotrophy (Conti-Jerpe et al. 2020; Santos et al. 2021). In terms of main energy expenditures, the coral's key needs include daily respiration, growth, reproduction (gamete production), mucus production and other maintenance/repair processes (Edmunds and Davies 1989; Koiijman 2000). Corals are also capable of storing surplus energetic carbon as lipid reserves, to be catabolised at times of need to offset shorter term energy deficits. Better understanding of the energy budgets (Fig. 4.1) of corals living on extreme reefs, such as those found in Singapore, may shed light on how corals may be able to persist in such harsh environments and help better predict their chances for survival where environmental quality is anticipated to further deteriorate with climate change and increasing coastal urbanisation.

The following sections will review past studies on the main energetic inputs and outputs of corals living in an extreme reef environment. Findings from photo-physiology and feeding experiments conducted between 2018 and 2019 at the St. John's Island National Marine Laboratory (SJINML) are also presented and discussed. These experiments focused on three coral species common to Singapore's turbid reefs—*Merulina ampliata, Pachyseris speciosa* and *Porites lutea*. Based on these past and present works, the daily energy budgets of the three coral species under different light scenarios were then estimated. *Merulina ampliata, Pachyseris speciosa* and *Porites lutea* rank in the top ten for percent cover not only on reefs in Singapore but also around Southeast Asia (Heery et al. 2018). As such, greater understanding on how these reef-building corals are able to persist on Singapore's reefs may help predict responses of other regional reefs that will be or are already facing similar extreme environmental conditions.

4.3 Photosynthesis and Respiration

Studies that have aimed to quantify autotrophic inputs and the photic needs of Singapore's corals are limited (Tun et al. 1994; Dikou and van Woesik 2006; Browne et al. 2014, 2015a; b; Chow et al. 2019; Morgan et al. 2020; Ow Yong et al. (*in review*)). Of these, only Tun et al. (1994) had mea-

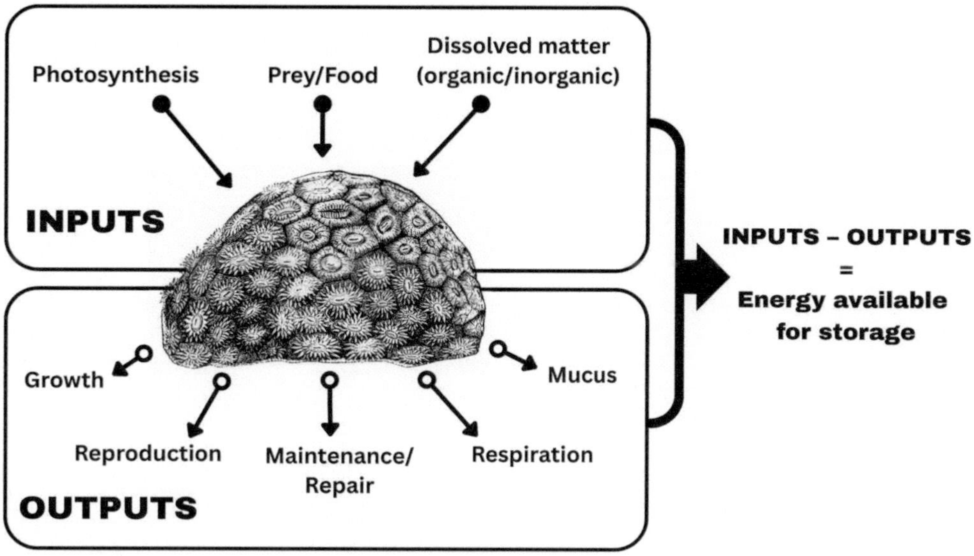

Fig. 4.1 Simple coral energy budget schematic, involving inputs such as photosynthesis by coral endosymbionts and heterotrophic feeding, and outputs such as growth, reproduction, repair, respiration and mucus production. Excess energy is stored away, e.g. as lipids in the corals—important surplus for periodic rather than continuous energy demands

sured the relationship between photosynthesis and respiration vs. light to obtain photokinetic parameters—maximum net photosynthesis rate ($P_{net\text{-}max}$), maximum gross photosynthesis rates ($P_{max\text{-}gross}$), irradiance at maximum photosynthesis (I_{max}), saturating light irradiance (I_k), light compensation point (I_c), maximum light conversion efficiency (α) and average dark respiration rate (R_d)—of four genera of corals (*Heliofungia*, *Fungia*, *Platygyra* and *Goniopora*). Acquired through photosynthesis-irradiance (P-E) curves, photokinetic parameters provide information about the acclimation of the coral photo-endosymbionts' photosynthetic apparatus to different light levels, including how efficiently they can absorb photons. P-E curves are typically determined through oxygen respirometry done under a range of irradiance (Rex et al. 1995; Sakshaug et al. 1997). More recently, optical techniques based on active chlorophyll fluorescence have also been used to generate P-E curves, e.g. rapid light curves (RLC) using pulse-amplitude modulated (PAM) fluorometry (Ralph and Gademann 2005). While these two methods are able to provide photosynthetic parameters, results may not be completely interchangeable and discrepancies can exist (Herlory et al. 2007; Wientjes et al. 2017).

To quantify photo-responses of common corals living in Singapore's turbid reefs, photosynthesis-respiration trials for *Merulina ampliata, Pachyseris speciosa* and *Porites lutea* were conducted in October 2018 at the SJINML. Oxygen respirometry methods were used to generate P-E curves to allow for more comparable results with previous studies in Singapore and elsewhere. Briefly, for each species, replicate coral fragments obtained from three colonies (i.e. total 9 fragments per species) sampled from Kusu Island were placed in transparent watertight acrylic chambers (600 ml) filled with 50-μm filtered seawater. Magnetic stirrers ensured water mixing in each chamber, and a water bath maintained temperatures at 29.5 ± 0.5 °C. Light shading caused by the acrylic chambers were compensated by increasing source light levels, and the corals were exposed to five PAR levels—0 μmol m^{-2} s^{-1} (dark), 60 ± 20 μmol m^{-2} s^{-1}, 120 ± 20 μmol m^{-2} s^{-1}, 250 ± 20 μmol m^{-2} s^{-1} and 340 ± 20 μmol m^{-2} s^{-1}. All corals were kept in the dark prior to respiration trials. Exposure time for each irradiance treatment was set at 20 min, with a 20-min dark period in between to relax the corals' photoreceptors and reduce risk of light over-saturation (Schrameyer et al. 2014).

The P-E curves for the three study species are shown in Fig. 4.2, and a summary of coral photo-physiology data from Singapore is presented in Table 4.1. Results from the experiment show that saturating light irradiance (E_k) occurred at ~120 μmol m^{-2} s^{-1} for *M. ampliata* and *Pac. speciosa* and ~ 160 μmol m^{-2} s^{-1} for *Por. lutea* (Fig. 4.2), and $P_{max\text{-}gross}$ occurred at ~260 μmol m^{-2} s^{-1} for *M. ampliata* and *Pac. speciosa* and ~300 μmol m^{-2} s^{-1} for *Por. lutea*. These saturating light irradiance (E_k) values are substantially lower than previously found elsewhere—E_k of ~400 μmol m^{-2} s^{-1} for *Porites lutea* from Wakatobi (Hennige et al. 2008), E_k of ~200 μmol m^{-2} s^{-1} for *Pachyseris speciosa* from Great Barrier Reef (DiPerna et al. 2018)—suggesting that corals living in Singapore's turbid and dynamic reefs have relatively high light conversion rates and require less light to achieve maximum photosynthesis rates, and may be already acclimated to lower light conditions. The E_k values for *Por. lutea*, M. *ampliata* and *Pac. speciosa* here are also lower than those for *Fungia* (E_k = 337 μmol m^{-2} s^{-1}), *Goniopora* (E_k = 467 μmol m^{-2} s^{-1}), *Heliofungia* (E_k = 267 μmol m^{-2} s^{-1}) and *Platygyra* (E_k = 326 μmol m^{-2} s^{-1}) from Pulau Hantu in the 1990s, which were measured as whole colony in situ with ambient circulated seawater exchanged every 15 min (Tun et al. 1994). However, whether these differences indicate species differences in photo-physiology, methods of measurement (in situ vs. ex situ respiration trials), and/or reflect

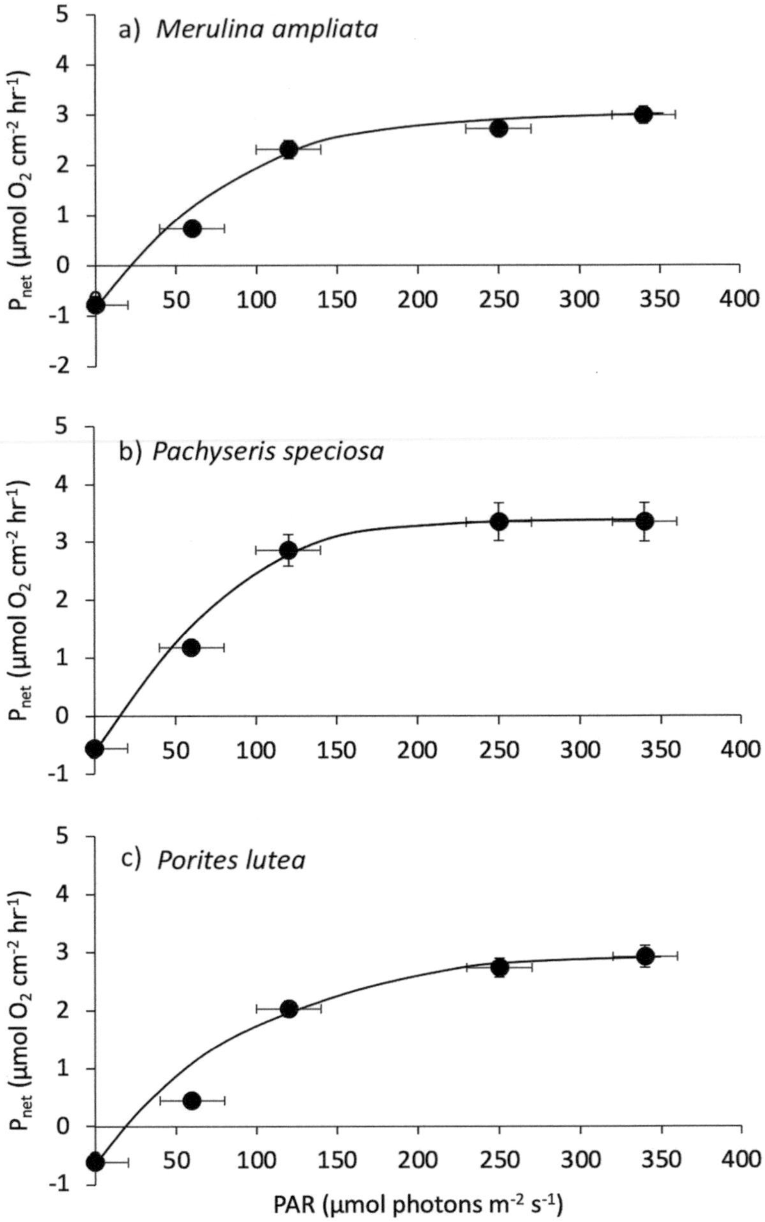

Fig. 4.2 Photosynthesis (net photosynthesis) (μmol O_2 cm^{-2} h^{-1}) vs. photosynthetically active radiation (PAR) (μmol photons m^{-2} s^{-1}) for (**a**) *Merulina ampliata*, (**b**) *Pachyseris speciosa* and (**c**) *Porites lutea*. Corals were exposed to respiration trials under five photosynthetically active radiation (PAR) levels, i.e. 0 μmol m^{-2} s^{-1} (Dark), 60 ± 20 μmol m^{-2} s^{-1}, 120 ± 20 μmol m^{-2} s^{-1}, 250 ± 20 μmol m^{-2} s^{-1} and 340 ± 20 μmol m^{-2} s^{-1}

any changes in baseline environmental conditions in the >25 years between the two studies cannot be ascertained without further investigation. Nevertheless, the maximum gross photosynthesis rates obtained in this study (3.6–4 μmol O_2 cm^{-2} h^{-1}) are within range of what were previously reported (3.3–13.8 μmol O_2 cm^{-2} h^{-1}) by Tun et al. (1994), suggesting somewhat comparable estimations of autotrophic energy inputs for Singapore's corals.

Given 'normal' environmental and biological states (e.g. temperature, zooxanthellae and chlorophyll abundances), it is possible to then estimate the energetic inputs from photosynthesis under different light conditions from the P-E relationships found. Here, photosynthesis rate is measured as the rate of production of O_2 assuming a mole-to-mole relationship of CO_2 and O_2. Conversion to energy in joules (J) assumes 6 mol of O_2 results in formation of 1 mol glucose = 2817 kJ (1 calorie = 4.185 J). Photosynthetically active radiation (PAR) at ~2–3 m depth in Singapore average ~ 200 μmol m^{-2} s^{-1} when measured from 7 am to 7 pm, and ~ 300 μmol m^{-2} s^{-1} when measured at mid-day from 11 am to 2 pm (Tanzil et al. 2019; Fig. 4.3). At this light level, assuming an energy transfer quotient of 0.78 from zooxanthellae to coral animal (Edmunds and Davies 1989), the values for energy per day produced through photosynthesis for *Pac. speciosa*, *M. ampliata* and *Por. lutea* are 18.44 J cm^{-2} day^{-1}, 16.61 J cm^{-2} day^{-1} and 14.44 J cm^{-2} day^{-1}, respectively. Similar estimations can be generated for exposure under other light conditions, e.g. at ~6–7 m where aver-

Table 4.1 Summary of photokinetic parameters (averages) from past and present photophysiological studies on corals living in Singapore's turbid reefs

Taxon	F_v/F_m	$P_{max\text{-}gross}$	R_d	α	E_k	E_c	E_{95}	Location	Source
Pocillopora acuta	0.65–0.7							Pulau Hantu	Poquita-Du et al. (2020)
Fungia		5.2	1.0	0.015	337	67.9	617	Pulau Hantu	Tun et al. (1994)
Goniopora		6.6	1.3	0.014	467	96.8	856	Pulau Hantu	Tun et al. (1994)
Heliofungia		13.8	3.0	0.052	267	73.5	489	Pulau Hantu	Tun et al. (1994)
Platygyra		3.3	0.8	0.01	326	80.8	597	Pulau Hantu	Tun et al. (1994)
Pachyseris speciosa		4.0	0.55	0.038	120	17	260	Kusu Island	This study
Merulina ampliata		3.6	0.78	0.037	120	26	260	Kusu Island	This study
Porites lutea		3.6	0.62	0.026	160	20	300	Kusu Island	This study
Merulina ampliata	0.57		1.15					Labrador Park, Kusu Island, Pulau Hantu	Browne et al. (2014)
Pachyseris speciosa	0.53		0.72					Labrador Park, Kusu Island, Pulau Hantu	Browne et al. (2014)
Platygyra sinensis	0.59		1.44					Labrador Park, Kusu Island, Pulau Hantu	Browne et al. (2014)
Merulina ampliata	0.58							Labrador Park, Kusu Island, Pulau Hantu	Browne et al. (2015a, 2015b)
Pachyseris speciosa	0.55							Labrador Park, Kusu Island, Pulau Hantu	Browne et al. (2015a, 2015b)
Platygyra sinensis	0.63							Labrador Park, Kusu Island, Pulau Hantu	Browne et al. (2015a, 2015b)
Goniastrea pectinata	~0.66–0.74							St. John's island	Bollati et al. (2022)
Mycedium elephantotus	~0.63–0.67							St. John's island	Bollati et al. (2022)

F_v/F_m maximum photosynthetic yield, $P_{max\text{-}gross}$ maximum gross photosynthesis (μmol O_2 cm^{-2} h^{-1}), R_d dark respiration rate (μmol O_2 cm^{-2} h^{-1}), α maximum light conversion efficiency, E_k saturating light irradiance

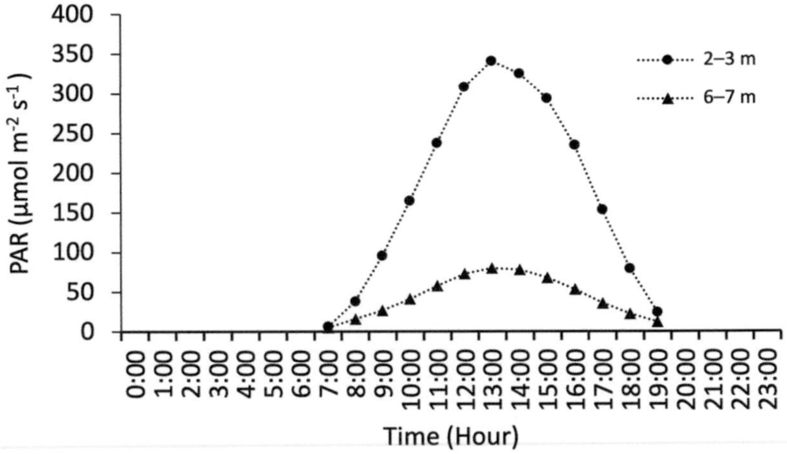

Fig. 4.3 Average hourly photosynthetically active radiation (PAR) measured at 2–3 m and 6–7 m depths at Kusu Island, Singapore, over the period December 2018–July 2020

age 12-h PAR = ~50 µmol m^{-2} s^{-1} and average mid-day PAR (11 am–2 pm) = ~80 µmol m^{-2} s^{-1}. At this depth and light condition, autotrophic energy inputs decrease to 8.49 J cm^{-2} day^{-1}, 7.57 J cm^{-2} day^{-1} and 5.82 J cm^{-2} day^{-1} for *Pac. speciosa*, *M. ampliata* and *Por. lutea*, respectively.

Similarly, energy consumed for daily coral respiration can be calculated. While it is possible to use only dark respiration rates to estimate daily respiratory demands, it has been shown that respiration rates after or during exposure to light can be significantly elevated (Edmunds and Davies 1989; Schrameyer et al. 2014). Light respiration can increase 11–25 times higher than steady-state dark respiration rates (Schrameyer et al. 2014) as the activity of light-driven metabolic pathways increase with increasing irradiance, such as due to demands on host-supplied carbon with irradiance, or extra demand for carbon in light-enhanced calcification.

Quantifying light respiration rates is challenging—the exchange of respiratory gases (O_2 and CO_2) in photosynthetic symbioses is difficult to study in the light because respiratory O_2 uptake is masked by the O_2 production from photosynthesis. However, it is still possible to obtain estimates by measuring enhanced post-illumination dark respiration (EPIR), which is the respiratory activity measured just after transition from light to darkness. It should be noted that EPIR likely still underestimates light respiration as EPIR is measured in the dark and therefore would not account for production of any reducing agents during photosynthetic light reactions. Nevertheless, EPIR values can be used to quantify the relationship between respiration and irradiance (R-E) (Fig. 4.4), which can be used to provide potentially better estimates for daily respiratory needs. Therefore, given the light regimes for ~2–3 m depths on Singapore's reefs (Fig. 4.3), the 24-h energy consumption based on dark and light respiration for *M. ampliata*, *Pac. speciosa* and *Por. lutea* can be estimated to be 9.05 J cm^{-2} day^{-1}, 7.42 J cm^{-2} day^{-1} and 10.94 J cm^{-2} day^{-1}, respectively. This estimation also assumes that respiratory substrate used is lipid (Patton et al. 1977), with a lipid oxyjoule coefficient of 19.63 J ml^{-1} O_2. At ~6–7 m depths, where light levels are lower, daily estimated respiratory needs were estimated to be 6.81 J cm^{-2} day^{-1} and 8.68 J cm^{-2} day^{-1} for *Pac. speciosa* and *Por. lutea,* respectively. In addition to lower respiration rates based on EPIR values, corals also tend to conserve energy under low light by reducing the excretion of organic matter and rates of coral and zooxanthellae respiration (Bessell-Browne et al. 2017; Brunner et al. 2022).

The average ratio of energy obtained through daily photosynthesis and consumed through daily respiration (P:R) suggests that while some corals living on Singapore's extreme reefs can still maintain positive production at light conditions experienced at 6–7 m, e.g. *Pac. speciosa* (P:R = 1.25), others such as *M. ampliata* (P:R = 0.84) and *Por. lutea* (P:R = 0.67) may already be at their energetic limits. Similar conclusions were drawn by Tun et al. (1994) who predicted that corals on Pulau Hantu would be net consumers below ~4 m depth, and obtain <50% of their daily carbon requirements from photosynthesis at 6 m. These estimates, however, have yet to account for heterotrophic inputs (feeding) as well as other energy outputs (e.g. growth, mucus production).

4.4 Feeding and Heterotrophy

The coastal environment of Singapore is rich in suspended particulate organic material (Tham et al. 1970; Schmoker et al. 2014; Tanzil et al. 2019). Organic content of monthly total suspended solids measured over the period 2010–2015 averaged 22–23 ± 10% (in total suspended solids ~22 mg l^{-1}) (Tanzil et al. 2019). Seawater chlorophyll-a concentration, an indicator of the amount of phytoplankton over this same period averages ~ 1.7 µg l^{-1}, an order of magnitude higher than concentration found in the coastal waters of the Red Sea (0.13 µg l^{-1}) (Fahmy 2003). High plankton abundance as well as high densities of suspended particles that contain a wide array of organic matter, including microbes and coral

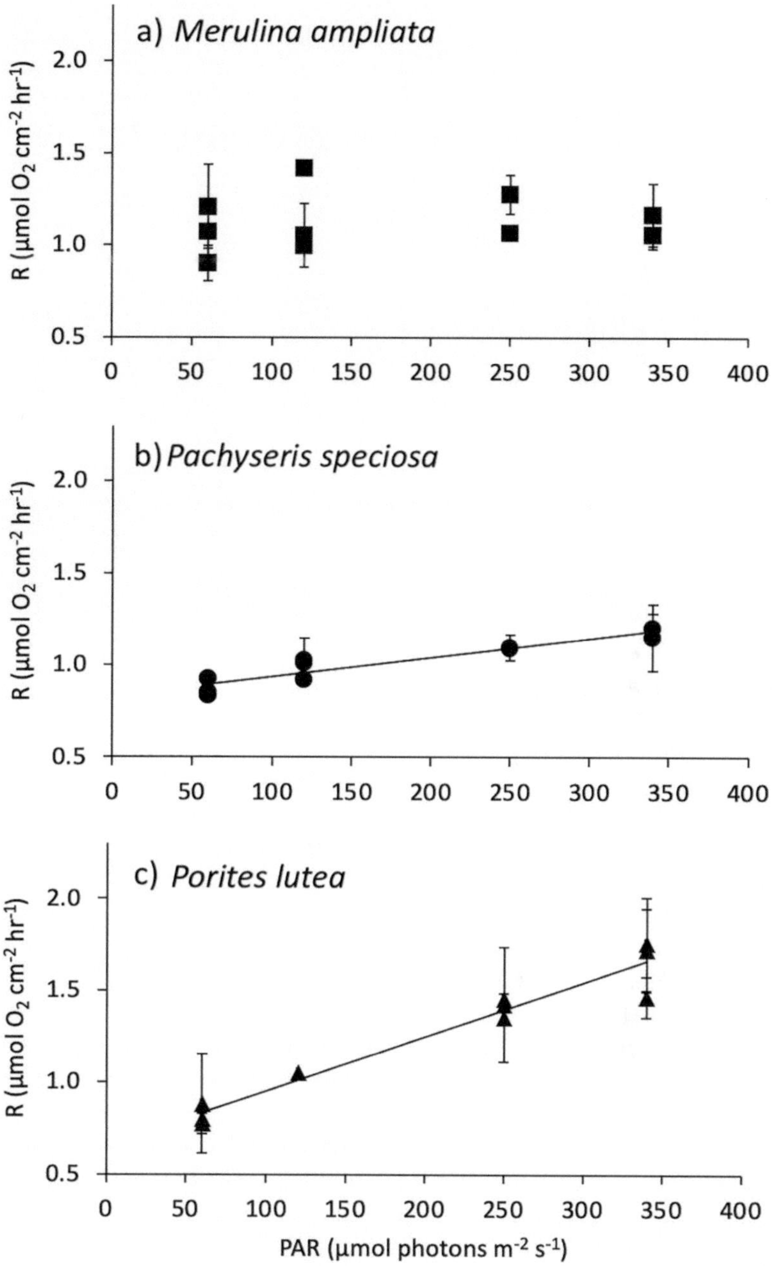

Fig. 4.4 Relationship between respiration (R) (as post-illumination dark respiration) and photosynthetically active radiation (PAR) (μmol photons m^{-2} s^{-1}) for (**a**) *Merulina ampliata*, (**b**) *Pachyseris speciosa* and (**c**) *Porites lutea*. Fitted lines shown only for significant ($p < 0.05$) linear relationships

mucus (Lopez and Levinton 1987), constitute rich potential food sources to corals in nutrient-rich waters. In addition to being voracious predators of zooplankton—corals can ingest 0.5 to >2 prey items per polyp per hour of ingestion depending on prey size and density (Sebens et al. 1996; Toh et al. 2013; Tanzil and Ow, *pers. obs.*)—many corals also feed on fine particles caught in mucus films or strands, which are drawn by cilia into the polyps' mouth (Anthony and Larcombe 2000; Fig. 4.5). Heterotrophic consumption of zooplankton can account for up to 60% of coral's daily metabolic requirements (Falkowski et al. 1984; Grottolli et al. 2006; Houlbrèque and Ferrier-Pagès 2009), rising to 100% in the event of bleaching (Grottolli et al. 2006; Seemann et al. 2012; Bessell-Browne et al. 2014). Energy from supplemental feeding can be extremely beneficial to corals (Toh et al. 2014; Conlan et al. 2017). Toh et al. (2014) found that *Pocillopora acuta* juveniles (parent colonies from Kusu Island, Singapore) fed with day-old *Artemia* nauplii increased their ecological volume by >74 times their initial size.

Although it has been assumed that corals feed primarily at night, with tentacle expansion of several coral species observed to occur only at night when zooplankton densities tend to be highest, day-time polyp extension has been observed for a number of coral species on Singapore's reefs (Tanzil, *pers. obs.*). Therefore, in the highly turbid, low-light, nutrient-rich environment of Singapore's reefs, where dis-

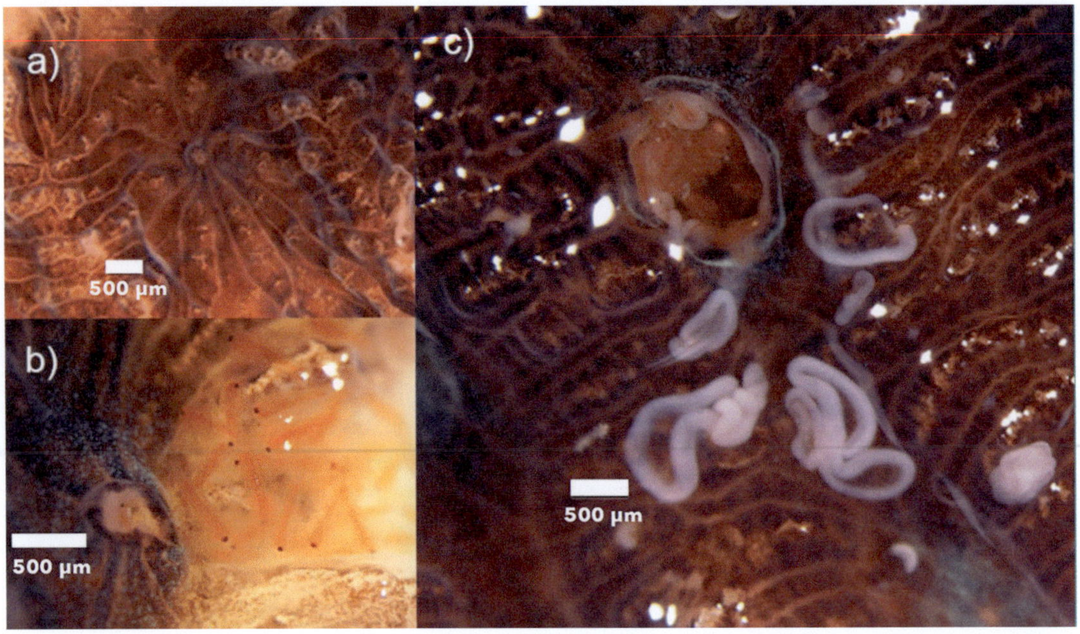

Fig. 4.5 Images taken under microscope following feeding with *Artemia* in aquarium conditions. (**a**) *Merulina ampliata* polyp, (**b**) *Merulina ampliata* polyp ~20 min after being fed with *Artemia* (polyp with *Artemia* prey captured on mucus string near mouth), (**c**) *Merulina ampliata* corallites after ~1 h of feeding (~15–20 *Artemia* individuals in the polyp with mouth expanded and protrusions of digestive filaments)

solved and particulate nutrients are abundant, heterotrophy could account for a larger than expected energy source. No previous study has quantified the feeding rates of corals from Singapore's reefs. Here, we present results from a feeding trial conducted at SJINML in August 2019 on *M. ampliata*, *Pac. speciosa* and *Por. lutea*. To estimate total daily feeding rates, day (~1 pm) as well as night (~8 pm) feeding trials were conducted for the three species. Briefly, pre-counted (concentration ranged from 1500 to 2000 individuals per litre) freshly hatched *Artemia salina* nauplii were fed to coral fragments ($n = 4$ for a total of 12 fragments from 3 colonies per species) each housed in chambers filled with 500 ml 0.5 µm-filtered seawater. Corals were allowed to feed for a measured time between 60 and 90 min, gently removed and rinsed, and the concentration of *Artemia* remaining in the chamber fluid was quantified post-feeding. While results are available for day vs. night feeding (Hickey 2020), for the purposes of this work, we present only average daily feeding rates that account for diurnal feeding potential. Average daily *Artemia* nauplii clearance rates for *M. ampliata*, *Pac. speciosa* and *Por. lutea* were 22.38 ± SE 3.47 ind cm^{-2} h^{-1}, 22.10 ± SE 2.85 ind cm^{-2} h^{-1} and 18.76 ± SE 4.86 ind cm^{-2} h^{-1}, respectively. These values are comparable to those found in other studies for other small polyp coral species such as *Pocillopora damicornis, Acropora* spp. and *Hydnophora rigida* (Kuanui et al. 2016; Tagliafico et al. 2018).

Using these coral feeding rates and assuming (a) feeding occurs throughout day and night, (b) digestion time of captured prey of ~2 h (Kuanui et al. 2016) (i.e. 12 units of feed capture per day) and (c) energy estimated from consumption of plankton or other organic substance is similar to the energy content of freshly hatched *Artemia salina* nauplii (0.00924 J per individual) (Paffenhöfer 1967), estimates of energy from heterotrophy can be calculated. Heterotrophic inputs for *M. ampliata*, *Pac. speciosa* and *Por. lutea* from Singapore's reefs are estimated to be 2.48 J cm^{-2} day^{-1}, 2.48 J cm^{-2} day^{-1} and 2.08 J cm^{-2} day^{-1}, respectively. While these estimates do not account for nutrition, variability in prey types and digestibility of food sources, they are likely conservative values given the organic-rich suspended particulates in Singapore's waters (Tanzil et al. 2019).

4.5 Growth

Despite living in challenging environmental conditions, corals in Singapore still have high growth rates comparable to corals living in clearer waters elsewhere. For example, *Porites lutea* from several of Singapore's reefs at ~2–3 m at mid-tide demonstrate skeletal calcification rates of up to ~1.6 g cm^{-2} $year^{-1}$ (Tanzil et al. 2013, 2016), comparable to those found in the Great Barrier Reef (~1.7 g cm^{-2} $year^{-1}$) (Lough and Barnes 1997; De'ath 2009) and elsewhere around Southeast Asia (Tanzil et al. 2013, 2016). However, skeletal growth requires significant energy investment—laying down 1 mg dry weight of $CaCO_3$ is estimated to require 0.152 J (Zubay 1993; McConnaughey and Whelan 1997;

Allemand et al. 1998; Anthony and Fabricius 2000; Anthony and Larcombe 2000). This energy cost of calcification is estimated based on the predicted expenditure of 1 mole of ATP (30,500 J) for every 2 moles of Ca^{2+} (amounts to 2×10^5 mg dry weight of $CaCO_3$) transported across the membrane of calicoblastic cells to the site of calcification, and assuming the cost of producing organic skeletal matrix is negligible. For slow-growing massive *Porites* corals, calcification at a rate of 1.6 g cm^{-2} $year^{-1}$ would therefore require ~0.67 J cm^{-2} day^{-1} of energy, while faster growing species such as *Merulina ampliata* (average ~ 3.86 g cm^{-2} $year^{-1}$ from Lane 1991; Dikou 2009; Madin et al. 2016; Toh et al. 2017; Ng et al. 2021) and *Pachyseris speciosa* (~3.79 g cm^{-2} $year^{-1}$ from Lane 1991; Afiq-Rosli et al. 2017; Ng et al. 2021; Khaw unpublished data) require higher energy costs of ~1.61 and ~ 1.58 J cm^{-2} day^{-1}. It is no surprise then that during periods of extreme stress, such as during severe coral bleaching caused by anomalous sea warming events, when corals lose their zooxanthellae—and therefore their main source of energy—coral calcification is significantly reduced or effectively ceased (Goreau and Macfarlane 1990; Leder et al. 1991; Tudhope et al. 1992; Mendes and Woodley 2002; Tanzil 2012).

In addition to skeletogenesis, however, there are also energy costs associated with growth of the coral-animal tissue and growth of their associated zooxanthellae. Edmunds and Davies (1986) estimated that with a constant zooxanthellae to coral-tissue biomass ratio, 4.35×10^5 zooxanthellae are added daily to a coral with 1 g of dry tissue for the coral *Porites*. Currently, there have been no studies to investigate such relationships for corals in Singapore. However, based on limited studies elsewhere (Edmunds and Davies 1989; Brandtneris et al. 2016), the average energy invested in zooxanthellae and tissue growth can be estimated as ~1% and 5%, respectively, of energy obtained through photosynthesis.

4.6 Reproduction

Expenditure on reproduction is a significant energy investment for corals. Lipids constitute ~60–70% of the dry weight of coral eggs (Arai et al. 1993), of which 70–80% are wax esters, 1–10% TGA and ~ 10% phospholipids. Phospholipids mainly control the structural function and cell membrane formation (Lee et al. 2006; Harii et al. 2007) while wax esters are primarily responsible for the buoyancy of eggs (Arai et al. 1993). Wax esters and TGA are the main energy source for eggs during the planktonic stage, with TGA used as shorter term energy source that can be rapidly hydrolysed (Villinski et al. 2002; Lee et al. 2006; Harii et al. 2007). Based on total energy equivalents of per cm^2 coral tissue before and after a spawning event at the Great Barrier Reef, energy associated with reproductive outputs can range between 132 and 384 J cm^{-2}, with smaller polyp corals such as *Montipora digitata* and *Acropora tenuis* having lower reproductive energy outputs compared to species with larger polyps (e.g. *Lobophyllia hemprichii* and *Galaxea fascicularis*) (Leuzinger et al. 2003). While there have been studies on coral spawning and reproduction potential (fecundity) for Singapore's corals (Guest et al. 2002; Guest 2005), as well as monitoring efforts led by the National Parks Board Singapore together with partners, there have been no studies quantifying the energetic costs associated with gamete production. As Singapore's mass coral spawning period is well-documented and can be accurately predicted (i.e. onset of spawning a few days after the full moon around March/April) (Guest 2005), it would be possible to sample corals just before and after spawning to obtain more accurate estimates of energy outputs associated with reproduction.

4.7 Other Energetic Losses

While growth and reproduction constitute significant energy investments for corals, other sizeable energetic losses can also be associated with maintenance or repair. Corals produce mucus as a defense against desiccation and pathogens (Shnit-Orland and Kushmaro 2009) and as primary mechanism for sediment rejection/clearance (Stafford-Smith and Ormond 1992). Mucus production is thought to account for up to 40–60% of unaccounted carbon released in coral energy budgets (Crossland 1987; Edmunds and Davies 1989; Coffroth 1990; Wild et al. 2005; Nakajima et al. 2009) and can cost corals up to three times their daily energy budgets (Riegl and Branch 1995). However, mucus comprises a variety of types—fluid mucus is significantly more expensive energetically compared to mucus sheets or tunics—such as those seen around *Porites* corals (Coffroth 1990). Time of day, stress intensity and duration and exposure to air can also affect mucus production rates and therefore energy outputs. For instance, mucus production at night is 30% that of daytime rates (Crossland 1987). Given the high sedimentation rates (~8–15 mg cm^{-2} day^{-1}) and total suspended solids (~22 mg l^{-1}) on Singapore's reefs (Tanzil et al. 2019), we can expect higher energetic expenditure associated with mucus production and other sediment removal mechanisms (e.g. ciliary movement). A study by Browne et al. (2014) for *Merulina ampliata* and *Pachyseris speciosa* from Singapore's reefs found elevated respiration rates by ~30% and ~ 10%, respectively when subjected to sedimentation rates above 9.9 mg l^{-1}.

Wild et al. (2005) estimated that mucus release by *Acropora aspera* and *A. millepora* under submersed conditions was equivalent to ~7–10 mg C m^{-2} coral surface area h^{-1}, while Coffroth et al. (1990) quantified the energy content of mucus released by *Porites asteroides* to be 20.9 J mg^{-1} mucus ash-free dry weight (AFDW). Based on these numbers and excluding other losses associated with increased sedimentation and/or other stresses, it can be estimated that, at minimum, mucus production per cm^{-2} of coral expends ~2.77 J of energy.

4.8 Daily Energy Budgets of Singapore's Corals

Above, we have discussed the photo-physiology and other biological processes that constitute key input and output parameters in a coral energy budget model (Fig. 4.6). Using the state variables and rates obtained through aquaria-based experiments presented here for *Merulina ampliata, Pachyseris speciosa* and *Porites lutea* together, with consolidated data from other studies to estimate other energy inputs, outputs and transfer quotients, the energy budget models for these corals under different light regime scenarios can be developed. The estimated energy in J cm^{-2} day^{-1} for the three coral species living at either the reef crest (~2–3 m depth) or reef slope (~6–7 m) is presented in Table 4.2. Polytrophic energy inputs are balanced against main energy outputs estimated for daily respiration, growth of tissue and zooxanthellae, and mucus production to determine the remaining energy available as scope for growth and reproduction.

From the energy budgets of corals living at the shallower reef crest, results show positive scope for growth and reproduction with relatively high energy balances and P:R ratios for all three species (Table 4.2). Energy available after accounting for the key physiological outputs would be more than sufficient to meet the energy investments needed for average calcification rates (0.67–2.2 J cm^{-2} day^{-1}). Taking aside energy required for skeletal growth, remaining energy estimated would still be sufficient to scope for investing in reproduction. For example, approximating energetic requirement of ~140 J cm^{-2} for gamete production for *M. ampliata* based on other corals with similar-sized polyps (Leuzinger et al. 2003), *M. ampliata* living at this depth in Singapore's turbid reef environment would need a ~ 30-day equivalent of ~4.5 J cm^{-2} day^{-1} of surplus energy. However, the same cannot be said for corals living at the reef slope (6–7 m)—negative energy balances and P:R ratios <1 were found for *M. ampliata* and *Por. lutea.* Although P:R ratio and energy balances were positive for *Pac. speciosa* in this lower light regime scenario, these values would not be sufficient to allow for average skeletal growth rates, much less reproduction.

The lack of scope for growth and reproduction for corals living in deeper waters may explain why there have been no observations or records of corals spawning beyond the reef crest and flat (Tun, *pers. obs.*). Energy budgets may also help explain the vertical distribution of corals living in Singapore's reef environment (Chow et al. 2019). Based on the energy budget models presented here, with the assumption that there are no other stressors that incur further energetic expenses or increase/decrease feeding rates, *Por. lutea* is predicted to be a net consumer at depths below ~4 m, *M. ampliata* at depths below ~5 m and *Pac. speciosa* at depths below 7 m. At Pulau Hantu, a nearer shore reef of similar distance from mainland Singapore as Kusu Island, the lower depth range of *Pac. speciosa* extended to >6 m while that of *M. ampliata* and *Por. lutea* was ~4.5 m (Chow et al. 2019). It should be noted that the median depth range for all three species was found to be ≤4 m (Chow et al. 2019). Visual observations at Kusu Island and Sisters' Island by the authors note similar distributions where larger *Por. lutea* colonies were less abundant at deeper

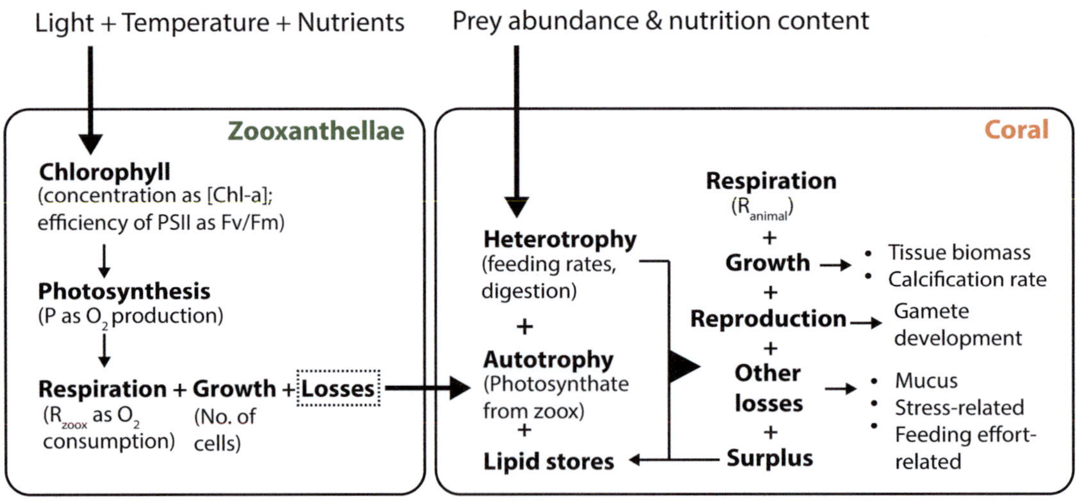

Fig. 4.6 Schematic diagram showing the partitioning of energy budget into zooxanthellae and coral components. State variables and rates were obtained either from aquarium experiments conducted or estimated based on other published studies from Singapore or elsewhere (see Table 4.1)

Table 4.2 Predicted daily (24 h) energy budgets (J cm^{-2} day^{-1}) for *Merulina ampliata*, *Pachyseris speciosa* and *Porites lutea* on the reef crest (~2–3 m depth) and slope (~6–7 m depth)

Reef zone (depth)	Species	Inputs (J cm^{-2} day^{-1})		Outputs (J cm^{-2} day^{-1})			Inputs-outputs (J cm^{-2} day^{-1})	
		P_{gross}	Feeding	R	Growth (tissue and zoox.)	Mucus production	Scope for growth and reproduction	P:R ratio
Reef crest (2–3 m)	*Merulina ampliata*	16.61	2.48	9.05	0.86	2.77	6.41	1.84
	Pachyseris speciosa	18.44	2.45	7.42	0.98	2.77	9.72	2.49
	Porites lutea	14.44	2.08	10.94	0.74	2.77	2.07	1.32
Reef slope (6–7 m)	*Merulina ampliata*	7.57	2.48	9.05	0.33	2.77	−2.1	0.84
	Pachyseris speciosa	8.49	2.45	6.81	0.39	2.77	0.97	1.25
	Porites lutea	5.82	2.08	8.68	0.23	2.77	−3.78	0.67

P_{gross} Gross photosynthesis, *R* respiration, *P:R* photosynthesis to respiration ratio. For light regimes at these two depths, see Fig. 4.3

depths of >4 m, while colonies of *Pac. speciosa* were still found beyond 6 m depths. Foliose coral species, such as *Pachyseris*, are known to be depth generalists—able to maximise light capture and survive in shallow, as well as the deep mesophotic light environment (Sinniger et al. 2022; Hoogenboom et al. 2008; Ohara et al. 2013). While efficiency in light capture and photosynthesis would increase energy inputs, reduced energy outputs would also allow for larger energy balances. Respiration trials presented here found *Pac. speciosa* to have the lowest respiration rates (Fig. 4.4) compared to *M. ampliata* and *Por. lutea*. High photo-efficiency combined with low energetic demands could further explain *Pac. speciosa* abundance and distribution on Singapore's turbid reefs.

4.9 Conclusions

While the coral energy budget models presented are far from comprehensive, they have provided insights into how some corals are able to persist and grow in Singapore's turbid, light-limited reef environment. Energetic perspectives may also shed light on other environmental limits for coral growth—important for developing site- and species-specific environmental thresholds for coral health that can be applied to assess environmental risks (Lesser 2013). Such approaches may be useful as part of environmental impact assessments (EIAs) or environmental management and mitigation plans (EMMPs) for coral reefs. For example, specific environmental quality objectives (EQOs) with differing limits of turbidity and sedimentation can be imposed on different stages of projects to ensure that the energy budgets of corals are not compromised, particularly during ecologically sensitive periods such as marine heat waves or ecologically significant events such as coral spawning. Additionally, there is potential to apply energetic approaches for evaluating suitability of coral candidates for reef restoration purposes. Energy budget models allow for scenario testing at the species and individual levels to determine likelihoods of transplant survival at potential recipient sites. Models may also be useful for identifying the most appropriate period/s for transplantation efforts given the coral's energetic state and environmental conditions. For example, quantitative energy budgets could help identify coral species and genotypes suitable for restoring Singapore's deeper (>6 m) degraded reef slopes where live coral cover has decreased from up to 45% in the 1980s to <5–16% today, with no signs of recovery (Guest et al. 2016; Chan et al. 2024).

Based on Singapore's future plans for large-scale coastal development as part of coastal protection measures against sea-level rise (URA 2024), it is expected that Singapore's reefs will continue to experience turbid and low-light conditions. Low-light penetration may also further exacerbate impacts of rising sea levels on coral reefs by further limiting the amount of light reaching corals (Law and Huang 2023). Given anticipated environmental challenges with climate change and increasing coastal urbanisation, as well as disturbances from other human activities (e.g. sea-based aquaculture, maritime activities), improving the use of models to predict outcomes related to effects of environmental changes may help secure coral reefs' survival in a sea of change.

Acknowledgements We would like to acknowledge the St. John's Island National Marine Laboratory (SJINML) for providing the necessary infrastructure for conducting our research. SJINML is a National Research Infrastructure under the National Research Foundation, Singapore. This work would also not have been possible if not for the assistance of various student interns and colleagues, in particular Clare Ho, Denise Yu, Wilson Yeo and Wong Tung Yee. We also thank the SJINML lab and aquarium managers for their assistance. Part of the research presented here was funded by the National Parks Board Singapore under the Technical Committee for the Coastal and Marine Environment (TCCME) as part of the project 'Development of an adaptive framework for assessing risk to environmental stress: an energetics approach to safeguarding Singapore's corals', and under the WWF-Singapore project 'A pilot of site-specific restoration of degraded reef areas in Singapore: light limited reef zones'. We also thank the National Parks Board Singapore for allowing this research (permit no. RP19-018).

References

Afiq-Rosli L, Taira D, Loke HX, Toh TC, Tok KB, Ng CSL, Cabaitan PC, Chou LM, Song T (2017) In situ nurseries enhance coral transplant growth in sedimented waters. Mar Biol Res 13(8):878–887

Allemand D, Furla P, Bénazet-Tambutté S (1998) Mechanisms of carbon acquisition for endosymbiont photosynthesis in Anthozoa. Can J Bot 76(6):925–941

Anthony KRN, Fabricius K (2000) Shifting roles of heterotrophy and autotrophy in coral energetics under varying turbidity. J Exp Mar Biol Ecol 252:221–253

Anthony KRN, Larcombe P (2000) Coral reefs in turbid waters: sediment-induced stresses in corals and likely mechanisms of adaptation. In: Proceedings of the Ninth International Coral Reef Symposium, Bali, Indonesia, 23–27 Oct 2000

Arai I, Kato M, Heyward A, Ikeda Y, Iizuka T, Maruyama T (1993) Lipid composition of positively buoyant eggs of reef building corals. Coral Reefs 12(2):71–75

Bessell-Browne P, Stat M, Thomson D, Clode PL (2014) *Coscinaraea marshae* corals that have survived prolonged bleaching exhibit signs of increased heterotrophic feeding. Coral Reefs 33:795–804

Bessell-Browne P, Fisher R, Duckworth A, Jones R (2017) Mucous sheet production in *Porites*: an effective bioindicator of sediment related pressures. Ecol Indic 77:276–285

Bollati E, Rosenberg Y, Simon-Blecher N, Tamir R, Levy O, Huang D (2022) Untangling the molecular basis of coral response to sedimentation. Mol Ecol 31:884–901

Borell EM, Yuliantri AR, Bischof K, Richter C (2008) The effects of heterotrophy on photosynthesis and tissue composition of two scleractinian corals under elevated temperature. J Exp Mar Biol Ecol 364:116–123

Brandtneris VW, Brandt ME, Glynn PW, Gyory J, Smith T (2016) Seasonal variability in calorimetric energy content of two Caribbean mesophotic corals. PLoS One 11(4):e0151953

Browne NK, Precht E, Last KS, Todd PA (2014) Photo-physiological costs associated with acute sediment stress events in three near-shore turbid water corals. Mar Ecol Prog Ser 502:129143

Browne NK, Tay JKL, Low J, Larson O, Todd PA (2015a) Fluctuations in coral health of four common inshore reef corals in response to seasonal and anthropogenic changes in water quality. Mar Environ Res 105:3952

Browne NK, Tay JKL, Todd PA (2015b) Recreating pulsed turbidity events to determine coral-sediment thresholds for active management. J Exp Mar Biol Ecol 466:98–109

Brunner CA, Ricardo GF, Uthicke S, Negri AP, Hoogenboom MO (2022) Effects of climate change and light limitation on coral recruits. Mar Ecol Prog Ser 690:65–82

Burke L, Selig E, Spalding M, McManus J (2002) Reefs at risk in South East Asia. World Resources Institute

Cacciapaglia C, van Woesik R (2016) Climate-change refugia: shading reef corals by turbidity. Glob Chang Biol 22:1145–1154

Chan YKS, Ng CSL, Tun KPP, Chou LM, Huang D (2024) Decadal decline of functional diversity despite increasing taxonomic and phylogenetic diversity of coral reefs under chronic urbanisation stress. Ecol Indic 164:112143.

Chou LM (2010) Regional coral reef monitoring and reporting through an informal mechanism – the experience of Southeast Asia. In: Status of coral reefs in East Asian Seas region: 2010. Ministry of the Environment, Tokyo, pp 117–118

Chou LM, Toh TC, Ng CSL (2017) Effectiveness of reef restoration in Singapore's rapidly urbanizing coastal environment. Int J Environ Sci Dev 8(8):576–580

Chow GSE, Chan YKS, Jain SS, Huang D (2019) Light limitation selects for depth generalists in urbanised reef coral communities. Mar Environ Res 147:101–112

Coffroth MA (1990) Mucous sheet formation on poritid corals: an evaluation of coral mucus as a nutrient source on reefs. Mar Biol 105(1):39–49

Conlan JA, Humphrey CA, Severati A, Francis DS (2017) Influence of different feeding regimes on the survival, growth and biochemical composition of *Acropora* coral recruits. PLoS One 12(11):e0188568

Conti-Jerpe IE, Thompson PD, Wong CWM, Oliveira NL, Duprey NN, Moynihan MA, Baker DM (2020) Trophic strategy and bleaching resistance in reef-building corals. Sci Adv 6(15):eaaz5443

Crossland CJ (1987) In situ release of mucus and DOC-lipid from the corals *Acropora variabilis* and *Stylophora pistillata* in different light regimes. Coral Reefs 6(1):35–42

Davies PS (1984) The role of zooxanthellae in the nutritional energy requirements of *Pocillopora eydouxi*. Coral Reefs 2:181–186

De'ath G, Lough JM, Fabricius KE (2009) Declining coral calcification on the Great Barrier Reef. Science 323(5910):116–119

Dikou A (2009) Skeletal linear extension rates of the foliose scleractinian coral *Merulina ampliata* (Ellis & Solander, 1786) in a turbid environment. Mar Ecol 30(4):405–415

Dikou A, van Woesik R (2006) Survival under chronic stress from sediment load: spatial patterns of hard coral communities in the southern islands of Singapore. Mar Pollut Bull 52:7–21

Diperna S, Hoogenboom M, Noonan S, Fabricius K (2018) Effects of variability in daily light integrals on the photophysiology of the corals *Pachyseris speciosa* and *Acropora millepora*. PLoS One 13(9):e0203882

Drew EA (1972) The biology and physiology of alga-invertebrates symbioses. II. The density of symbiotic algal cells in a number of hermatypic hard corals and alcyonarians from various depths. J Exp Mar Biol Ecol 9(1):71–75

Edmunds PJ, Davies PS (1986) An energy budget for *Porites porites* (Scleractinia). Mar Biol 92:339–347.

Edmunds PJ, Davies PS (1989) An energy budget for *Porites porites* (Scleractinia), growing in a stressed environment. Coral Reefs 8:37–43

Fahmy M (2003) Water quality in the Red Sea coastal waters (Egypt): analysis of spatial and temporal variability. Chem Ecol 19(1):67–77

Falkowski PG, Dubinsky Z, Muscatine L, Porter JW (1984) Light and the bioenergetics of a symbiotic coral. Bioscience 34(11):705–770

Fitt WK (2000) Seasonal patterns of tissue biomass and densities of symbiotic dinoflagellates in reef corals and relation to coral bleaching. Limnol Oceanogr 45(3):677–685

Goreau TJ, Macfarlane AH (1990) Reduced growth rate of *Montastrea annularis* following the 1987–1988 coral-bleaching event. Coral Reefs 8:211–215

Grottolli AG, Rodrigues L, Palardy J (2006) Heterotrophic plasticity and resilience in bleached corals. Underw Nat 440:1186–1189

Guest JR (2005) Reproductive patterns of scleractinian corals on Singapore's reefs. Honours Thesis, Newcastle University

Guest JR, Baird AH, Goh BPL, Chou LM (2002) Multispecific, synchronous coral spawning in Singapore. Coral Reefs 21(4):422–423

Guest JR, Tun K, Low J, Verges A, Marzinelli EM, Campbell AH, Baumann AG, Feary DA, Chou LM, Steinberg PD (2016) 27 years of benthic and coral community dynamics on turbid, highly urbanised reefs off Singapore. Sci Rep 6:36260

Harii S, Nadaoka K, Yamamoto M, Iwao K (2007) Temporal changes in settlement, lipid content and lipid composition of larvae of the

spawning hermatypic coral *Acropora tenuis*. Mar Ecol Prog Ser 346:89–86
Heery EC, Hoeksema BW, Browne NK, Reimer JD, Ang PO, Huang D, Friess DA, Chou LM, Loke LHL, Saksena-Taylor P, Alsagoff N, Yeemin T, Sutthacheep M, Vo ST, Bos AR, Gumanao GS, Hussein MAS, Waheed Z, Lane DJW, Johan O, Todd PA (2018) Urban coral reefs: degradation and resilience of hard coral assemblages in coastal cities of East and Southeast Asia. Mar Pollut Bull 135:654–681
Hennige SJ, Smith DJ, Perkins R, Consalvey M, Paterson DM, Suggett DJ (2008) Photoacclimation, growth and distribution of massive coral species in clear and turbid waters. Mar Ecol Prog Ser 369:77–88
Herlory O, Richard P, Blanchard GF (2007) Methodology of light response curves: application of chlorophyll fluorescence to microphytobenthic biofilms. Mar Biol 153:91–101
Hickey V (2020) Overeating in a 'buffet of turbidity': a preliminary insight into heterotrophy among four common coral species from Singapore's reefs. MSc thesis, University College Cork, Ireland
Hoogenboom MO, Connolly SR, Anthony KRN (2008) Interactions between morphological and physiological plasticity optimize energy acquisition in corals. Ecology 89(4):1144–1154
Houlbrèque F, Ferrier-Pagès C (2009) Heterotrophy in tropical scleractinian corals. Biol Rev 85:1–17
Houlbrèque F, Tambutte E, Allemand D, Ferrier-Pagès C (2004) Interactions between zooplankton feeding, photosynthesis and skeletal growth of the scleractinian coral *Stylophora pistillata*. J Exp Biol 207:1461–1469
Huang D, Tun K, Chou LM, Todd PA (2009) An inventory of zooxanthellate scleractinian corals in Singapore including 33 new records. Raffles Bull Zool 22:69–80
Kleypas JA, McManus JW, Menez LAB (1999) Environmental limits to coral reef development: where do we draw the line? Am Zool 39:146–159
Koiijman SALM (2000) Dynamic energy and mass budgets in biological systems. Cambridge University Press, Cambridge
Kuanui P, Chavanich S, Viyakarn V, Park HS, Omori M (2016) Feeding behaviours of three tropical scleractinian corals in captivity. Trop Zool 29(1):1–9
Lane DJW (1991) Growth of scleractinian corals on sediment-stressed reefs at Singapore. In: Proceedings of the regional symposium on living resources in coastal areas. University of The Philippines, Quezon City, 30 Jan–1 Feb 1989
Leder JJ, Szmant AM, Swart PK (1991) The effect of prolonged "bleaching" on skeletal banding and stable isotopic composition in *Montastrea annularis*. Coral Reefs 10:19–27
Lee RF, Hagen W, Kattner G (2006) Lipid storage in marine zooplankton. Mar Ecol Prog Ser 307:273–306
Lesser MP (2013) Using energetic budgets to assess the effects of environmental stress on corals: are we measuring the right things? Coral Reefs 32(1):25–33
Lesser MP, Farrell JH (2004) Exposure to solar radiation increases damage to both host tissues and algal symbionts of corals during thermal stress. Coral Reefs 23(3):367–377
Leuzinger S, Anthony KR, Willis BL (2003) Reproductive energy investment in corals: scaling with module size. Oecologia 136(4):524–531
Lopez GR, Levinton JS (1987) Ecology of deposit-feeding animals in marine sediments. Q Rev Biol 62:235–260
Lough JM, Barnes DJ (1997) Several centuries of variation in skeletal extension, density and calcification in massive *Porites* colonies from the Great Barrier eef: a proxy for seawater temperature and a background of variability against which to identify unnatural change. J Exp Mar Biol Ecol 221(1):29–67
Madin JS, Hoogenboom MO, Connolly SR, Darling ES, Falster DS, Huang D, Keith SA, Mizerek T, Pandolfi JM, Putnam HM, Baird AH (2016) A trait-based approach to advance coral reef science. Trends Ecol Evol 31(6):419–428
Maltby L (1999) Studying stress: the importance of organism-level responses. Ecol Appl 9(2):431–440
McConnaughey TA, Whelan JF (1997) Calcification generates protons for nutrient and bicarbonate uptake. Earth Sci Rev 42:95–117
Mendes JM, Woodley JD (2002) Effect of the 1995–1996 bleaching event on the polyp tissue depth, growth, reproduction and skeletal band formation in *Montastrea annularis*. Mar Ecol 235:93–102
Morgan KM, Moynihan MA, Sanwlani N, Switzer AD (2020) Light limitation and depth-variable sedimentation drives vertical reef compression on turbid coral reefs. Front Mar Sci 7:571256
Muscatine L (1967) Glycerol excretion by symbiotic algae from corals and tridacna and its control by host. Science 156(3774):516–519
Muscatine L (1980) Productivity of zooxanthellae. In: Falkowski PG (ed) Primary productivity in the sea. Springer Science & Business Media, pp 381–402
Muscatine L, McCloskey LR, Marian RE (1981) Estimating the daily contribution of carbon from zooxanthellae to coral animal respiration. Limnol Oceanogr 26(4):601–611
Nakajima R, Yoshida T, Azman BA, Zaleha K, Othman BH, Toda T (2009) In situ release of coral mucus by *Acropora* and its influence on the heterotrophic bacteria. Aquat Ecol 43(4):815–823
Ng CSL, Chan YKS, Nyuyen NTH, Kikuzawa YP, Sam SQ, Toh TC, Mock AYJ, Chou LM, Huang D (2021) Coral community composition and carbonate production in an urbanized seascape. Mar Environ Res 168:105322
Ohara T, Fuiji T, Kawamura I, Miyuzama M, Montenegro J, Shikiba H, White KN, Reimer JD (2013) First record of a mesophotic *Pachyseris foliosa* reef from Japan. Oceanarium 43:71–72
Ow YX, Todd PA (2010) Light-induced morphological plasticity in the scleractinian coral Goniastrea pectinata and its functional significance. Coral Reefs 29(3):797–808
Paffenhöfer GA (1967) Caloric content of larvae of the brine shrimp *Artemia salina*. Helgoländer Meeresun 16(1–2):130–135
Patton JS, Abraham S, Benson AA (1977) Lipogenesis in the intact coral *Pocillopora capitata* and its isolated zooxanthellae: evidence for a light-driven carbon cycle between symbiont and host. Mar Biol 44(3):235–247
Perry CT, Larcombe P (2003) Marginal and non-reef-building coral environments. Coral Reefs 22(4):427–432
Poquita-Du RC, Goh YL, Huang D, Chou LM, Todd PA (2020) Gene expression and photophysiological changes in *Pocillopora acuta* coral holobiont following heat stress and recovery. Microorganisms 8(8):1227
Ralph PJ, Gademann R (2005) Rapid light curves: a powerful tool to assess photosynthetic activity. Aquat Bot 82(3):222–237
Rex A, Montebon F, Yap HT (1995) Metabolic responses of the scleractinian coral *Porites cylindrica* Dana to water motion. I. Oxygen flux studies. J Exp Mar Biol Ecol 186(1):33–52
Riegl B, Branch GM (1995) Effects of sediment on the energy budgets of four scleractinian (Bourne 1900) and five alcyonacean (Lamouroux 1816) corals. J Exp Mar Biol Ecol 186(2):259–275
Sakshaug E, Bricaud A, Dandonneau Y, Falkowski PG, Kiefer DA, Lengendre L, Morel A, Parslow J, Takahashi M (1997) Parameters of photosynthesis: definitions, theory and interpretation of results. J Plankton Res 19(11):1637–1670
Santos MEA, Baker DM, Conti-Jerpe IE, Reimer JD (2021) Populations of a widespread hexacoral have trophic plasticity and flexible syntrophic interactions across the Indo-Pacific Ocean. Coral Reefs 40:543–558
Schmoker C, Mahjoub MS, Calbet A, Hsiao SH, Russo F, Larsen O, Trottet A, Drillet G (2014) A review of the zooplankton in Singapore waters. Raffles Bull Zool 62:726–749
Schoepf V, Baumann JH, Barshis DJ, Browne NK, Camp EF, Comeau S, Cornwall C, Guzmán HM, Riegl B, Rodolfo-Metalpa R, Sommer

B (2023) Corals at the edge of environmental limits: a new conceptual framework to re-define marginal and extreme coral communities. Sci Total Environ 884:163688

Schrameyer V, Wangpraseurt D, Hill R, Kühl M, Larkum AW, Ralph PJ (2014) Light respiratory processes and gross photosynthesis in two scleractinian corals. PLoS One 9(10):e110814

Sebens KP, Vandersall KS, Savina LA, Graham KR (1996) Zooplankton capture by two scleractinian corals, *Madracis mirabilis* and *Montastrea cavernosa*, in a field enclosure. Mar Biol 127:303–317

Seemann J, Carballo-Balanos R, Berry KL, Gonzalez CT, Richter C, Leinfelder RR (2012) Importance of heterotrophic adaptation of corals to maintain energy reserves. In: Proceedings of the 12th International Coral Reef Symposium, Cairns, Australia, 9–13 Jul 2012

Shnit-Orland M, Kushmaro A (2009) Coral mucus-associated bacteria: a possible first line of defense. FEMS Microbiol Ecol 67(3):371–380

Sinniger F, Albelda RL, Prasetia R, Rouzé H, Sitorus ED, Harii S (2022) Overview of the mesophotic coral ecosystems around Sesoko Island, Okinawa, Japan. Galaxea J Coral Reef Stud 24(1):69–76

Stafford-Smith MG, Ormond RFG (1992) Sediment-rejection mechanisms of 42 species of Australian scleractinian corals. Mar Freshw Res 43(4):683

Tagliafico A, Rangel S, Kelaher B, Christidis L (2018) Optimizing heterotrophic feeding rates of three commercially important scleractinian corals. Aquaculture 483:96–101

Tan YHJ, Tham JKQ, Paul A, Rana U, Ang HP, Nguyen NTH, Yee ATK, Leong BPI, Drummond S, Tun KPP (2023) Remote sensing mapping of the regeneration of coastal natural habitats in Singapore: implications for marine conservation in tropical cities. Singap J Trop Geogr 44:130–148

Tanzil JTI (2012) Bleaching susceptibility and growth characteristics of *Porites lutea* from the Andaman Sea, South Thailand. Res Bull Phuket Mar Biol Cent 71:49–56

Tanzil JTI, Brown BE, Dunne RP, Lee JN, Kaandorp JA, Todd PA (2013) Regional decline in growth rates of massive *Porites* corals in Southeast Asia. Glob Chang Biol 19:3011–3023

Tanzil JTI, Lee JN, Brown BW, Quax R, Kaandorp JA, Lough JM, Todd PA (2016) Luminescence and density banding patterns in massive *Porites* corals around the Thai-Malay Peninsula, Southeast Asia. Limnol Oceanogr 61:2003–2226

Tanzil JTI, Goodkin NF, Sin TM, Chen ML, Fabbro GN, Boyle EA, Lee AC, Toh KB (2019) Multi- colony coral skeletal Ba/ca from Singapore's turbid urban reefs: relationship with contemporaneous in-situ seawater parameters. Geochim Cosmochim Acta 250:191–208

Tham AK, Tan CM, Gan MY, Tan CS, Foo HT, Gan JL (1970) The distribution of plankton in Singapore Straits for the period from April 1968 to March 1969. In: Proceedings of the 2nd CSK symposium, Tokyo, Japan, 28 September–1 October 1970

Todd PA (2008) Morphological plasticity in scleractinian corals. Biol Rev 83(3):315–337

Toh TC, Peh JWK, Chou LM (2013) Heterotrophy in recruits of the scleractinian coral *Pocillopora damicornis*. Mar Freshw Behav Physiol 46:313–320

Toh TC, Ng CSL, Peh JW, Toh KB, Chou LM (2014) Augmenting the post-transplantation growth and survivorship of juvenile scleractinian corals via nutritional enhancement. PLoS One 9(6):e98529

Toh TC, Ng CSL, Loke HX, Taira D, Toh KB, Afiq-Rosli L, Poquita-Du RC, Cabaitan P, Sam SQ, Kikuzawa YP, Chou LM, Song T (2017) A cost-effective approach to enhance scleractinian diversity on artificial shorelines. Ecol Eng 99:349–357

Tremblay P, Maguer JF, Grover R, Ferrier-Pages C (2015) Trophic dynamics of scleractinian corals: stable isotope evidence. J Exp Biol 218:1223–1234

Tudhope AW, Allison N, Le Tissier MDA, Scoffin TP (1992) Growth characteristics and susceptibility to bleaching in massive *Porites* corals, South Thailand. In: Proceedings of the 7th International Coral Reef Symposium, 22–233762

Tun KPP (2012) Optimisation of reef survey methods and application of reef metrics and biocriteria for the monitoring of sediment-impacted reefs. PhD dissertation, National University of Singapore

Tun KPP, Cheshire AC, Chou LM (1994) Photosynthetic production of four scleractinian corals from Singapore. In: Proceedings 3rd ASEAN-Australia symposium on living coastal resources, 16–20 May 1994

URA Singapore (2024) Master plan. https://www.ura.gov.sg/Corporate/Planning/Master-Plan. Accessed 1 December 2024

Villinski JT, Villinski JC, Byrne M, Raff RA (2002) Convergent maternal provisioning and life-history evolution in echinoderms. Evolution 56(9):176–1765

Warner M, Chilcoat G, McFarland W, Fitt W (2002) Seasonal fluctuations in the photosynthetic capacity of photosystem II in symbiotic dinoflagellates in the Caribbean reef-building coral *Montastraea*. Mar Biol 141(1):31–38

Wientjes E, Philippi J, Borst JW, van Amerongen H (2017) Imaging the Photosystem I/Photosystem II chlorophyll ratio inside the leaf. Biochim Biophys Acta Bioenerg 1858(3):259–265

Wild C, Woyt H, Huettel M (2005) Influence of coral mucus on nutrient fluxes in carbonate sands. Mar Ecol Prog Ser 287:87–98

Wong JSY, Chan YKS, Ng CSL, Tun KPP, Darling ES, Huang D (2018) Comparing patterns of taxonomic, functional and phylogenetic diversity in reef coral communities. Coral Reefs 37(3):737–750

Zubay GL (1993) Biochemistry, 3rd edn. Dubuque, Iowa

Coral Recruitment Potential on Singapore's Urban Reefs

5

Kareen Vicentuan, Jani Thuaibah Isa Tanzil, Iris Baula, Pei Ying Liew, and Ai Chin Lee

Abstract

The successful recruitment of coral juveniles to reefs damaged or degraded by disturbance events is a key mechanism of reef recovery and resilience. Coral larvae and newly settled corals are particularly sensitive to external influences, and environmental conditions during critical post-settlement periods play a crucial role in driving recruitment rates. Coral reef communities in Singapore persist in a highly sedimented, urban environment subjected to frequent and multiple anthropogenic and climate change-related disturbances. Here we examined the spatio-temporal patterns of coral recruitment on Singapore's reefs. Recruitment of coral spat observed in this chapter was influenced by density and diversity of the adult coral population in the vicinity, competitive interactions with other benthic species, climate and water conditions. Overall mean recruitment rate of 167.4 ± 8.6 spat m^{-2} y^{-1} was generally higher than previously reported for Singapore's reefs. Highest numbers of recruits were observed at Pulau Satumu, which also had the highest coral cover and diversity of the three reefs examined in this study, and lowest number of recruits was observed at Pulau Hantu. Only small numbers of juvenile colonies have been observed over the years, possibly indicating high post-recruitment mortality caused by suboptimal environmental conditions. From the 15-year ecological monitoring dataset, it was apparent that climate change-induced warming of seawater played a significant role in depressing coral recruitment in Singapore. However, there is a possibility of reef replenishment for as long as the corals continue to produce viable larvae.

K. Vicentuan (✉) · J. T. I. Tanzil · I. Baula · P. Y. Liew · A. C. Lee
Tropical Marine Science Institute, National University of Singapore, Singapore, Singapore
e-mail: tmskvc@nus.edu.sg

Keywords

Coral spat · Ecological monitoring · Reefs · Settlement · Recruitment · Sediment

5.1 Introduction

Coral reefs worldwide have faced significant declines in the past decades because of various local and global-scale threats (Bellwood et al. 2004; Fabricius 2005; Hoegh-Guldberg et al. 2007; De'ath et al. 2012; D'Angelo and Wiedenmann 2014). Pressures from coastal development and sedimentation (Rogers 1990; Fabricius 2005; Risk 2014; Tuttle and Donahue 2022), eutrophication and pollution (Dubinsky and Stambler 1996; Haapkylä et al. 2007; van Dam et al. 2011), boat damage, and irresponsible tourism and fishing practices (Saila et al. 1993; Edinger et al. 1998; Barker and Roberts 2004) are being compounded by global-scale climate change-associated disturbances, such as ocean warming and acidification (Hoegh-Guldberg et al. 2007; Kleypas and Yates 2009; De'ath et al. 2012). These threaten to upset the balance between damage to coral reefs and their recovery potential.

A key recovery mechanism for reefs impacted by disturbance is coral recruitment (Tamelander 2002; Coles and Brown 2007; Sawall et al. 2010). This refers to the successful settlement of sexually produced coral larvae and the formation of new recruits to recolonise damaged reef areas (Caley et al. 1996; Gilmour et al. 2013; Gouezo et al. 2019; Edmunds 2023). Although corals can reproduce asexually (Highsmith 1982; Sammarco 1982; Stoddart 1983; Lirman 2000), coral recruits resulting from sexual reproduction add genetic variation to coral populations, which may increase chances of survival in a rapidly changing environment. Coral recruitment is a process by which coral larvae settle, metamorphose and grow on a reef and eventually contribute to the growth and recovery of coral reefs, thereby influencing the distribution and dynamics of marine communities (Richmond

L. M. Chou, D. Huang (eds.), *Coral Reefs of Singapore's Urbanised Sea*, Coral Reefs of the World 23,
https://doi.org/10.1007/978-981-95-0903-4_5

et al. 2018). It is a complex process that requires the survival of coral offspring throughout sequential life-history stages, from larval availability, successful settlement to post-settlement survivorship and growth (Connell et al. 1997; Ritson-Williams et al. 2009; Gleason and Hofmann 2011; Doropoulos et al. 2017).

Although often used interchangeably, the terms settlement and recruitment are two separate processes (Hodgson 1990). Coral settlement can be defined summarily as the firm attachment of coral larvae and formation of an exoskeleton on the substratum by the larvae (Keough and Downes 1982; Te 1992). Settlement can therefore be said to comprise two further processes: larval metamorphosis and attachment to the substratum. The definition of coral recruitment, however, is still highly debatable. Recruitment interpretations have ranged from corals that survive to reproduce (Cameron and Endean 1985), to corals that are large enough to be detected in the field (Fitzhardinge 1988; Ritson-Williams et al. 2009), to recently settled juveniles that have survived for a significant period of time after settlement, of at least 6 months, which roughly corresponds to growth of the settled juveniles to a minimum of 0.5 cm diameter (Connell 1985). In the study of population dynamics, recruitment is generally termed as the addition of individuals to local populations (Rockwood 2006). In this case, population is defined as 'a group of interbreeding organisms found in the same space at the same time' (Rockwood 2006). Taking this into the context of the coral reef ecosystem, it would imply that the definition of coral recruitment should echo that of Cameron and Endean (1985). However, since the age and size of corals at first reproduction are not known for most corals and can be highly variable between coral species, recruitment taken in this sense would be difficult, if not almost impossible to quantify. Therefore, to facilitate measurement and quantification, most coral recruitment studies have, for now, realistically defined coral recruitment to follow a variation of Connell (1985)—the survival of newly settled corals over a longer period, past a critical post-settlement phase. For the purpose of this work, the term 'recruitment' is defined as the point when the coral larvae permanently attach on the settlement tile, metamorphose and survive for up to 6 months (the period of time the tiles were left on the reef) until the development of secondary corallites is observed.

Coral recruitment has been quantified in different areas across the world (Edmunds et al. 2015; Price et al. 2019) and has been studied for decades (see review Edmunds 2023). From these studies, some showed marked spatial variability in coral recruitment at various scales, from local to regional scales (Adjeroud et al. 2007; Green and Edmunds 2011; Salinas-de-León et al. 2013; Jouval et al. 2019; Edmunds 2023). The first study on recruitment and mortality of juvenile corals on natural reef substrate was carried out by Connell (1973), where permanent quadrats were established in 1963 for long-term monitoring of Heron Island at the Great Barrier Reef. Subsequent studies determined various factors that influence coral recruitment with the main objective to enhance it. Some factors include biotic cues, interactions and substrate characteristics. In particular, Bak and Engel (1979) found crustose (or calcified) coralline algae (CCA) to be in spatial competition with coral juveniles. Morse et al. (1988) subsequently determined that CCA induce coral larvae settlement and metamorphosis. Other studies that followed explored species-specific preferences among coral/CCA pairings, and more recent studies identified *Titanoderma* spp. to be the preferred CCA by some Caribbean and Pacific corals (Price 2010; Ritson-Williams et al. 2016; Abdul Wahab et al. 2023). Another factor that influences coral recruitment is light intensity that attenuates with depth and varies with turbidity and sedimentation. Early studies in Guam showed that coral recruitment was high in crevices and/or vertical surfaces at shallow depths; at greater depths, recruitment was higher on more exposed surfaces as light becomes limited (Birkeland 1977; Birkeland et al. 1981). Similar results were observed in the Caribbean (Edmunds et al. 2004) and Australia (Turner et al. 2018). Hydrodynamic conditions can also affect coral recruitment. Coral larvae have significantly lower probability for successful settlement when exposed to environments with high wave energy (Eagle et al. 2012; Reidenbach et al. 2021).

Recruitment patterns may also vary among coral taxa, in relation to the density of local reef populations (Hughes et al. 2000; Kayal et al. 2015; Bramanti and Edmunds 2016; Edmunds 2017; Gouezo et al. 2020). Several studies also demonstrated that the density of recruits varies among seasons and years (Adjeroud et al. 2007; Bauman et al. 2015; Davidson et al. 2019). Interannual variability is associated with variations in the type and intensity of spatial scale of disturbance such as storms that can alter the substratum, and bleaching events (Connell et al. 1997; Adjeroud et al. 2002; Mallela and Crabbe 2009). Assuming there is a ready supply of viable larvae, continuous effort in understanding the processes and mechanisms that influence the settlement and recruitment of corals at the receiving or 'sink' reefs is important. It enhances the prediction of reef connectivity and recovery, which is critical for sound reef management as well as restoration and rehabilitation strategies (Bode et al. 2006; Tay et al. 2012; Ani et al. 2024; Carlson et al. 2024). Reefs that function as sinks can be more resilient to disturbances and can recover rapidly through influx of new larvae, thereby contributing to the maintenance of coral cover and ecosystem functions (Ani et al. 2024).

Quantification of coral recruitment has traditionally been done with specific substrate material that is placed on the reef, providing a bare surface that will potentially capture coral spat in pre-settlement stages. Settlement tiles have been a widely used method to quantify taxonomic compositions, patterns and abundance of the recruits although there are few studies where recruits have been counted directly on natural substrata (Roth and Knowlton 2009; Martinez and Abelson 2013; Nativ et al. 2021). Various types of settlement tiles are used for coral recruitment studies. Some studies used fabricated artificial substrata that are naturally not present in the reef environment such as terracotta tiles (Mundy 2000; Humanes and Bastidas 2015; Hamizan et al. 2021), ceramic tiles (Fisk and Harriott 1990; Maida et al. 1994; Petersen et al. 2005; Dikou and van Woesik 2006), limestone (Lohr et al. 2017), fibre-cement (Edmunds et al. 2014; dela Cruz and Harrison 2024), polyvinyl chloride plastic (Price 2010; Fabricius et al. 2017) and concrete blocks (Reyes and Yap 2001; Bachtiar 2002). Earlier studies experimented on slices of corals as substrate to enhance recruitment (Harriott and Fisk 1987), but more recent studies moved away from such destructive practices. dela Cruz and Harrison (2024) used dead coral cuttings while more innovative approaches include the use of mineral-infused lime mortar (Levenstein et al. 2022; Yus et al. 2024) and tile surface modifications with different established macroalgal communities (Fong et al. 2024).

5.2 Coral Recruitment on Singapore Reefs

Coral reefs in Singapore are under considerable stress, where well over 60% of the total reef cover has been lost over the years (Hilton and Manning 1995; Chou 2006; Tan et al. 2023). Coral reef loss and degradation are associated with anthropogenic impact, including high sedimentation from land reclamation, seabed dredging and oil and marine-related industries, aquaculture and recreation (Chou et al. 2019a). Since the 1960s, extensive port development and land reclamation have diminished the coral cover and the recurring dredging activities to maintain shipping lanes have resulted in high sediment loads in Singapore waters. Reefs in Singapore are subject to high sediment regimes, with sedimentation rates varying between ~5–45 mg cm^{-2} d^{-1} (Low and Chou 1994) and a turbidity range of 0.06–230 mg l^{-1} (Morgan et al. 2020). Most of the substrata available for settlement on Singapore reefs constitute loose rubble, usually observed to be covered by sediment. Corals on Singapore reefs have experienced increased chronic exposure to high sedimentation that had reduced water visibility from the 10 m recorded in the 1960s to 2 m in recent times (Chou 1996).

High sedimentation has shown to adversely affect coral health and increase mortality rates of many coral species (Tomascik and Sander 1987; Weber et al. 2012; Browne et al. 2014, 2015a; b). Relative coral cover, species richness and diversity in Singapore reefs decrease with depth (Chow et al. 2019), and coral communities are restricted to above 8 m depth (Guest et al. 2016), a phenomenon known as reef compression (Heery et al. 2018). There are 255 reef-building scleractinian coral species recorded historically in Singapore reefs (Huang et al. 2009), and coral mass spawning occurs biannually, with the majority of spawning in March and April (Guest et al. 2005), an event which contributes to the seeding of coral reefs (Tay 2012). Nevertheless, recruitment in Singapore is generally low as compared to reefs in other parts of the world, e.g. Australia (Thomson et al. 2021), Philippines (dela Cruz and Harrison 2024) and Indonesia (Sawall et al. 2013). It is likely that lack of suitable stable substratum and elevated sediment load in Singapore have prevented the successful recruitment of coral spat to the reef (Dikou and van Woesik 2006; Tun 2012; Bauman et al. 2015).

In the past two decades, several studies have quantified coral recruitment on Singapore reefs with emphasis on the effects of sedimentation (Dikou and van Woesik 2006), substratum preferences and modifications (Nues 2005), recruitment on reef enhancement units (REUs) (Loh et al. 2006) and spatial and temporal variability (Bauman et al. 2015). Average coral recruits in Singapore have been recorded to range from 0.10 to 54.74 recruits m^2 (Nues 2005; Dikou and van Woesik 2006; Loh et al. 2006; Bauman et al. 2015). In Singapore, temporal variation in coral recruitment has been based on data collected from 15 to 26 months. Understanding how recruitment varies over longer periods of time is important for effective reef management to detect the fluctuations of natural recruitment patterns on the reef.

In this work, the changes in density and taxa composition of coral recruits across three locations between 2009 and 2023, encompassing three mass bleaching events (2010, 2016, 2020), were examined, thus providing insights on the long-term patterns of recruitment. Here, we describe two different aspects of coral recruitment potential in Singapore's sediment-stressed reefs. First, the findings from experiments testing how to best sample the potential recruitment of coral spat are presented. The second study details long-term trends in recruitment of coral spat observed at three representative reef sites situated in the Southern Islands. This is the first report documenting the results of a 15-year coral recruitment monitoring dataset in the coral reefs south of mainland Singapore.

5.3 Spatial Variations in Potential Recruitment of Corals

The deployment orientation and placement of settlement tiles may have an influence on settlement abundance (Carleton and Sammarco 1987). To provide an accurate evaluation of settlement and/or recruitment potential, it was also necessary to determine appropriate areas on the reef to place the settlement tiles. As such, any differences in settlement potential with respect to the angle at which the settlement tiles were placed across reef zones (i.e. depth) were examined.

Data on settlement potential were collected at the reefs of Pulau Hantu and Pulau Satumu from April to September 2005 (Fig. 5.1). A total of 288 20 × 20 cm unglazed terracotta tiles were deployed at these reefs. Terracotta tiles were used as they were readily available and have previously been successfully used as coral settlement substratum (Mundy 2000; Arnold and Steneck 2011; Green and Edmunds 2011; van Woesik et al. 2014; Harper et al. 2021). In order to examine the effect of settlement substratum placement and inclination, tiles were fixed onto racks in replicate pairs at three different reef zones, i.e. reef flat (~1.5–2.5 m depth at mid-tide), crest (>2.5–5 m at mid-tide) and slope (>5–7.5 m), and at four different substratum inclinations, i.e. 0° (horizontal), 30°, 60° and 90° (vertical). Six replicate racks for each substratum inclination at each of the three reef zones were deployed along a 100 m stretch of reef. These racks were fabricated by first hammering two lengths of iron bars into dead reef substrata followed by bolting a third to the two upright bars at the desired inclination (Fig. 5.2). As the surfaces of the terracotta tiles used here consisted of a smooth surface and a corrugated surface, the smooth surface was consistently used as the front surface, while the other was used as the back surface (attached to the rack). All tiles were conditioned for 2 weeks pre-deployment and positioned on the racks to face seaward. The tiles were left on the reef for 6 months, after which they were removed and processed for examination under a dissecting microscope. Recruitment on all surfaces of the tile (i.e. front, side and under surfaces) was examined and identified to family level according to skeletal characteristics presented in Babcock et al. (2003).

To investigate the effect of settlement substratum placement at various reef depths, recruit count from the whole tile (i.e. front, side and back surfaces) was analysed. A two-way analysis of variance (ANOVA) was used to analyse the data, with 'Zone' (reef zonation by depth) as fixed and 'Reef' as random factors. Spat count from only the front surfaces of the tiles was used to ascertain the effect of substrate inclination ('Angle') on settlement rate. In this case, a three-way ('Angle' × 'Zone' × 'Reef') ANOVA was used, with 'Angle' and 'Zone' considered fixed factors and 'Reef' as random factor. As recruit count on tiles deployed on the same racks may not be independent, analyses of all settlement data were based on averaged counts of the two tiles deployed at each rack (i.e. total $n = 144$). Assumptions of normality and homoscedasticity of data were examined prior to ANOVAs, and, where applicable, a square-root transformation was applied to count data. Any post-hoc differences were tested using the Ryan F-test (Welsch 1977). In case of non-normality post-transformation, a non-parametric Kruskal-Wallis rank test was used to examine variations of recruit

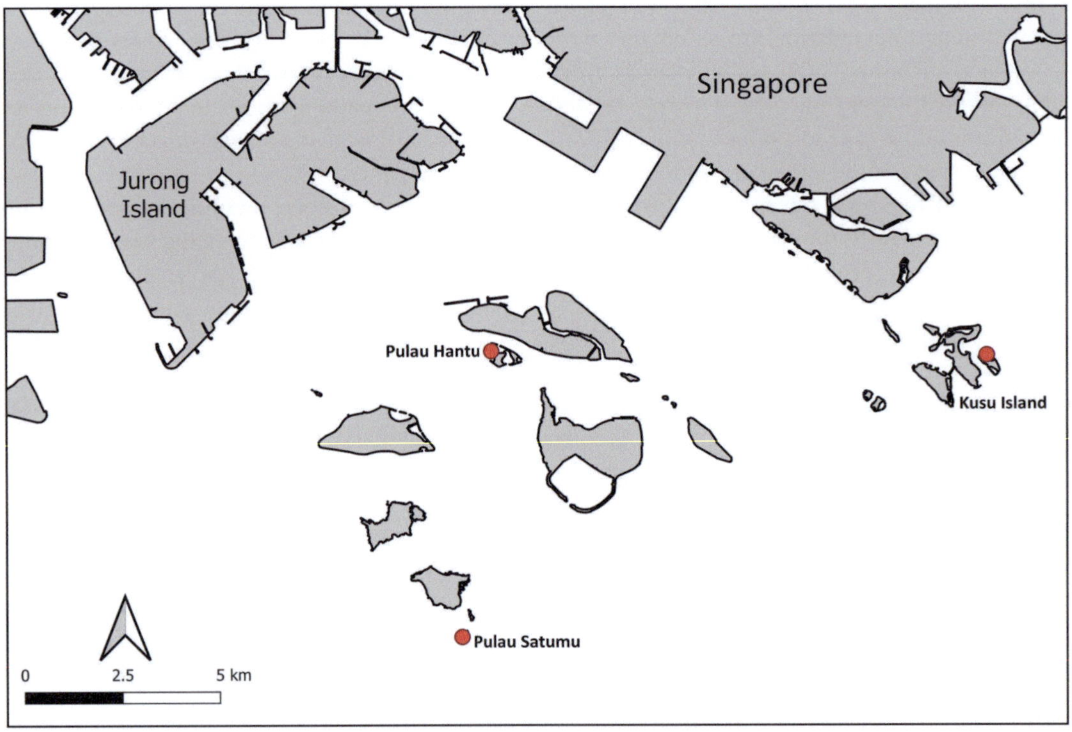

Fig. 5.1 Map of study sites around southern islands of Singapore—Kusu Island (1°13′32.06″ N, 103°51′35.32″ E), Pulau Hantu (1°13′35.38″ N, 103°44′49.23″ E) and Pulau Satumu (1°09′35.43″ N, 103°44′24.48″ E)

Fig. 5.2 Terracotta tiles on angle iron rack inclined at (**a**) 90° (vertical), (**b**) 0° (horizontal) and (**c**) 30° to the horizontal. All scale bars represent 5 cm

counts among sites and years. Dunn's test was used as a post hoc test for multiple comparisons. Analyses were performed using R (R Core Team 2012).

A total of 2350 spat were found from the 288 tiles deployed for 6 months at Pulau Satumu (1895 spat) and Pulau Hantu (455 spat). The majority of spat found (73.1%) belonged to the family Pocilloporidae. Remaining identifiable spat found were from the family Acroporidae (7.8%), Poritidae (3.7%) and Fungiidae (1.2%). Results from the experiment demonstrated differences in recruitment rates at different depths among the reef zones and differences between the two reefs. Two-way ANOVA on square-root transformed spat count revealed a significant interaction between reef sampled and zone ('Reef' × 'Zone'; $F_{[2138]} = 17.60$, $p < 0.001$). Post-hoc analysis showed significantly higher settlement was observed at Pulau Satumu (13.2 ± 1.2 spat per tile; mean ± SE, tile area = 0.04 m^2) compared to Pulau Hantu (3.16 ± 0.6 spat per tile, Ryan F-test, $p < 0.05$) over the study period (Fig. 5.3). At Pulau Satumu, settlement rates were significantly higher on the reef flat and crest, compared to the slope (Ryan F-test, $p < 0.05$). However, settlement rate at Pulau Hantu was lowest at the reef flat compared to the reef crest and slope (Ryan F-test, $p < 0.05$), which did not differ from one another.

Recruitment of coral spat was also different for the different substratum inclinations. Of the 584 spat on upper surfaces of tiles, only 20 were found on those deployed at Pulau Hantu. Due to this disparity in numbers of spat found between reefs, and consequently gross violations in normality and homoscedasticity, statistical analysis to test differences among the different tile inclinations was only performed on spat data from Pulau Satumu. Spat data from Pulau Satumu were found to be normal but heteroscedastic, even after square-root transformation. As such, any differences between the substrate angle and reef zone treatments were compared using the Games-Howell multiple comparison procedure (Games and Howell 1976). Comparing spat densities observed on the top surfaces of the settlement tiles at the different reef zones, settlement rates were significantly higher at the reef flat and crest, compared to the reef slope ($p < 0.05$, Fig. 5.3). At the reef flat and crest, spat densities were significantly higher on tiles deployed at 60° and 90° (vertical), compared to those deployed at 30° and 0° (horizontal) (Games-Howell Q-test, $p < 0.05$, Fig. 5.4). No detectable differences were found for the different substrata angles at the reef slope (Games-Howell Q-test, $p > 0.05$). Notwithstanding post-settlement mortality, our results suggest that recruitment potential on Singapore's reefs would be

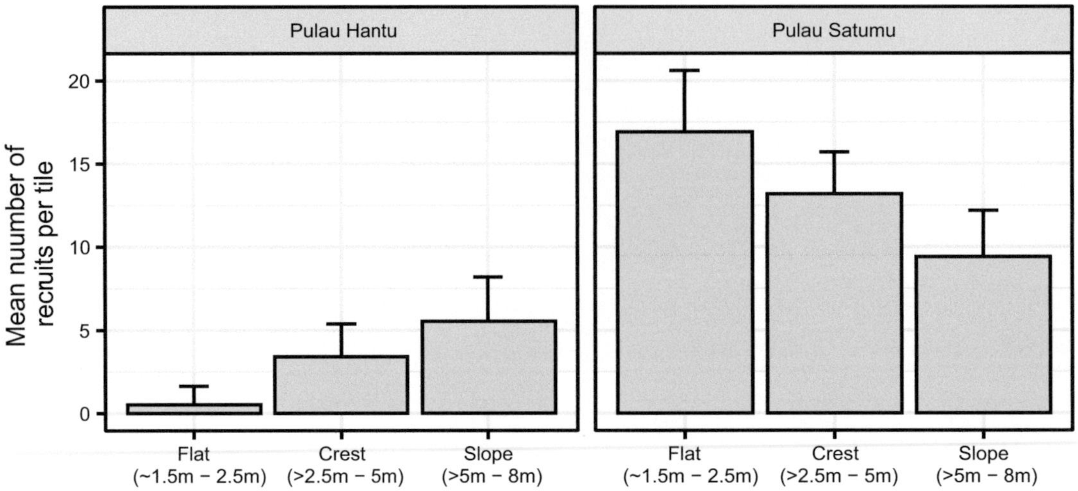

Fig. 5.3 Mean number of recruits (± SE) on all surfaces of tiles deployed at Pulau Hantu and Pulau Satumu at different reef zones (flat—~1.5–2.5 m depth, crest—>2.5–5 m depth and slope—>5–8 m depth)

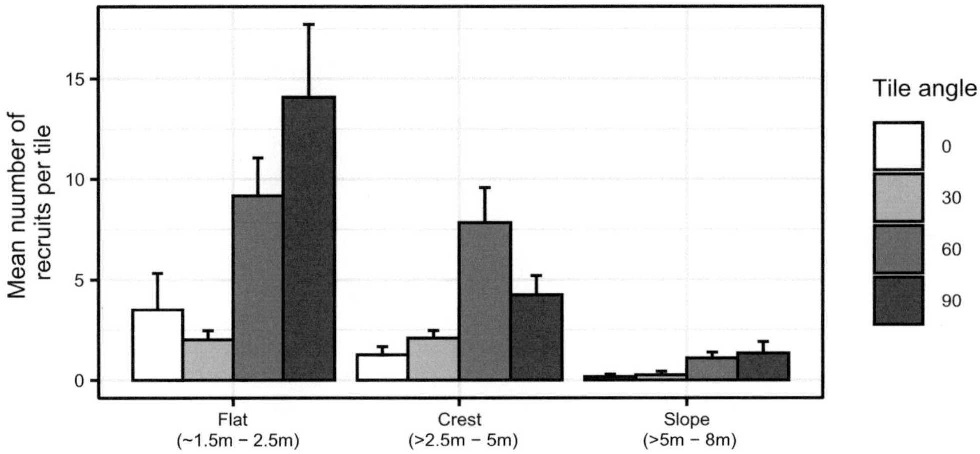

Fig. 5.4 Mean number of recruits (± SE) on upper surfaces of tiles deployed in Pulau Satumu at varying recruitment tile angles (0°—horizontal, 30°, 60° and 90°—vertical) at different reef zones (Flat: ~1.5–2.5 m depth, Crest: >2.5–5 m depth, and Slope: >5–8 m depth)

highest at the reef flat and crest and on substrates orientated at 60° or 90° angles. These findings were used to determine methodologies for the long-term monitoring of recruitment potential on Singapore's reefs.

5.4 Long-Term Variability in Coral Spat Recruitment Rates

The larger-scale spatial and temporal variabilities in recruitment potential of scleractinian corals in Singapore were quantified at Pulau Hantu, Kusu Island and Pulau Satumu from 2009 to 2023. Two sets of tiles were deployed each year, one in March and one in October, to capture the mass spawning events which occur bi-annually in Singapore (Guest et al. 2005). Based on results from the settlement tile placement study above, and taking into consideration logistical and navigational safety constraints, twelve vertical racks were constructed at each site by hammering a stainless-steel angle bar perpendicular to the reef substratum, along the reef crest or upper reef slope at 3–5 m depth (Fig. 5.5a). Unglazed 15 × 15 × 1 cm terracotta tiles were mounted at 50–80 cm above the seabed on each vertical rack. Singapore reefs are exposed to high sediment environment. The seabed is mostly covered with silt, which is very likely to be resuspended with increased turbulent energy from ship wakes especially during low tide when the waves hit the reef (Browne et al. 2015a, 2017). The placement of settlement panels too close to the seafloor or directly on top of the coral reef itself can cause abrasion of the spat, smothering and other physical implications of sediment. As a result, the settlement tiles were secured between 50 cm and 80 cm above the seabed. Each rack bore one hole about 5 mm from the top to allow mounting of terracotta tiles. Tiles were then bolted to each rack via a hole drilled through the center of each tile, with the smooth surface facing towards the reef slope (Fig. 5.5a).

Recruitment tiles were retrieved 2–3 weeks before the predicted spawning event and replaced with new tiles to cap-

Fig. 5.5 (**a**) Recruitment panels and placement of settlement tiles on the reef; (**b**) Damselfish at Pulau Hantu protecting its 'algal farm'; (**c**) Tile covered with bryozoan (Pulau Hantu); (**d**) Sponge encrusting the settlement tile (Pulau Hantu); (**e**) Sea anemone, *Radianthus magnifica*, observed on a tile (Kusu Island). All scale bars represent 5 cm

ture potential recruitment of coral spat for the next spawning season. Upon retrieval, all tiles were then bleached in ~10% sodium hypochlorite solution for 24–48 h, rinsed with flowing tap water and air dried. Recruitment tiles were examined in the laboratory using a Leica S8 APO Stereomicroscope to determine the density of coral settlers on the tile surfaces. Coral spat at any developmental stage were counted and identified to genus level according to skeletal characteristics presented in Babcock et al. (2003) and Mangubhai et al. (2007). Recruits that were not identifiable or damaged were categorised as 'Others'.

Similar to previous coral recruitment studies conducted in Singapore (Nues 2005; Dikou and van Woesik 2006; Loh et al. 2006; Bauman et al. 2015), spatial and temporal variations in recruit density observed at the three studied reefs were evident. Differences in the preference and capabilities of the coral spat, variation in physical conditions (i.e. hydrodynamics, depth gradients, light intensity and its associated sedimentation and composition of available substratum) and biotic interactions (i.e. predation or allelopathy that can alter the species abundance and diversity of scleractinian corals and the co-existing organisms) can greatly influence recruitment of coral spat on the reef. Observations over the 15-year period revealed a considerable degree of natural variation in the density of coral recruits in the Southern Islands of Singapore. A total of 7864 coral recruits settled onto settlement tiles from 2009 to 2023 at the three sites. The density of coral recruits recorded from a single settlement tile ranged from 0 to 3126 recruits m^{-2}, with an overall mean recruitment of 167.4 ± 8.6 recruits m^{-2} y^{-1} (mean ± SE, n = 1080 tiles). Recruit density varied among sites (Fig. 5.6); highest density of recruits was observed for Pulau Satumu (398.3 ± 20.2 recruits m^{-2} y^{-1}), followed by Kusu Island and Pulau Hantu (77.8 ± 5.0 recruits m^{-2} y^{-1} and 26.1 ± 2.2 recruits m^{-2} y^{-1}, respectively). The recruitment rates observed for Pulau Satumu were similar to the results of Tanzil et al. (2008) where ~330 recruits m^{-2} were observed. However, the recruitment at Pulau Hantu from the 2005 study was considerably higher (approximately 79 recruits m^{-2} y^{-1}) than that observed during the long-term monitoring programme.

Recruitment was highly variable over time; however, a general increase in recruit density was observed over the years from 2009 to 2019/2020, which was consistent across all three reef sites (Fig. 5.6). Highest recruit density was recorded in 2020 for Kusu Island and Pulau Satumu (154.2 ± 28.3 recruits m^{-2} y^{-1} and 785.4 ± 90.2 recruits m^{-2} y^{-1}, respectively) and in 2014 for Pulau Hantu at 63.2 ± 15.9 recruits m^{-2} y^{-1}. In spite of that, since 2020, recruit densities

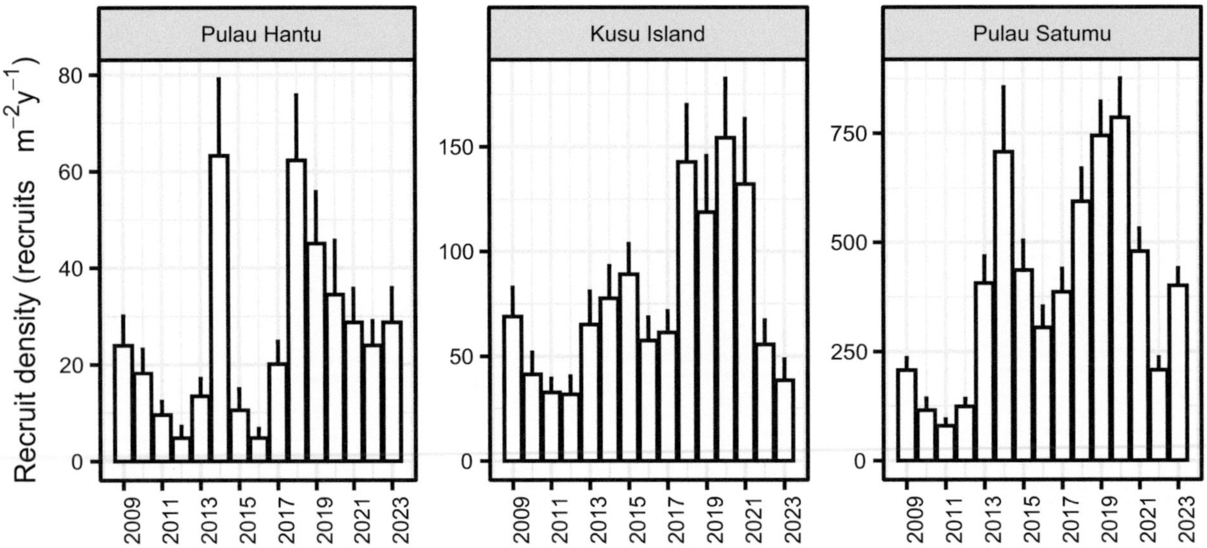

Fig. 5.6 Distribution of total recruit density from 2009 to 2023 across three sites in Singapore. Note the unequal y-axis

at the three reef sites have decreased. Temporal variability in coral recruitment has been widely documented in many reef systems globally (Hughes et al. 2002; Edmunds et al. 2015; Adjeroud et al. 2016; Price et al. 2019; Edmunds 2023). Temporal variability is mainly driven by the change in climatic and oceanographic conditions (Adjeroud et al. 2016), which can be due to weather seasons or global climate change (Price et al. 2019). Subtle changes in the above-mentioned conditions can influence the fecundity, pre-settlement mortality, dispersal and early post-settlement events (Mydlarz et al. 2010; Adjeroud et al. 2016). Furthermore, climate change-induced warming of seawater threatens coral reefs. The effect of thermal bleaching not only exposes corals under compromised health making them more prone to diseases (Bruno et al. 2007; Mydlarz et al. 2010), but it also affects gametogenesis and reduces the number of gametes produced and fertilisation rate (Ritson-Williams et al. 2009), consequently affecting recruitment rates. If the marine heatwaves are protracted, widespread coral mortality is more likely to occur (Browne et al. 2019; Price et al. 2019; Edmunds 2023). The depressed recruitment rates post 2020 may have been a lingering effect from the anomalous sea surface temperature conditions observed that year (ASMC 2021a, b) when coral bleaching was observed (K. Vicentuan, pers. obs.). Between 2020 and 2023, mass coral bleaching has also been reported globally, in the South China Sea (Lyu et al. 2022), Great Barrier Reef (McGowan and Theobald 2023; Henley et al. 2024) and Okinawa (Afzal et al. 2024).

Mean coral recruitment in Singapore observed in the 15-year study was higher than what was previously reported from Singapore reefs (Nues 2005; Dikou and van Woesik 2006; Loh et al. 2006; Bauman et al. 2015). Even then, recruit densities observed from the long-term data were found to be lower than in Australia (600 m^{-2} y^{-1}; Thomson et al. 2021), New Caledonia (1220 m^{-2} y^{-1}; Adjeroud et al. 2022), Mozambique (1130 m^{-2} y^{-1}; Sola et al. 2015), Philippines (1508.4 m^{-2} y^{-1}; dela Cruz and Harrison 2024), Seychelles (595 spat m^{-2} y^{-1}; Chong-Seng et al. 2014), Indonesia (286–705 m^{-2} y^{-1}; Sawall et al. 2013) and Kenya (101–908 m^{-2} y^{-1}; Mangubhai et al. 2007), but was comparable to the reefs at Eilat (190 m^{-2} y^{-1}; Glassom et al. 2006), Solitary Islands (132 spat m^{-2} y^{-1}; Harriott and Banks 1995), Taiwan (111 spat m^{-2} y^{-1}; Soong et al. 2003) and Dubai (121 spat m^{-2} y^{-1}; Bauman et al. 2015). However, it is important to note that these various studies employed different deployment methods and sampling frequencies to what was conducted in this study.

Coral larval availability is naturally related to the density of the adult coral population. Self-recruitment has been identified as an important demographic process (Afiq-Rosli et al. 2021) for the replenishment of various coral taxa, even for highly dispersive corals (Lequeux et al. 2018; Afiq-Rosli et al. 2021; Burt et al. 2024). Hard coral cover at Pulau Satumu was consistently highest among the three reef sites (Fig. 5.7). Having a substantial adult population in the vicinity is likely to be a driver for a higher number of recruits. Connectivity among populations will also play an important role in reef seeding. In a hydrodynamic study of coral larvae sink spots by Tay et al. (2012), Pulau Hantu was predicted to have the least larvae and Pulau Satumu was projected to have the highest number of larvae post-spawning. The spatial variability of our coral recruitment study corresponds with the results of Bauman et al. (2015) that coral recruits were more abundant in Kusu Island and Pulau Satumu compared to all other sites studied.

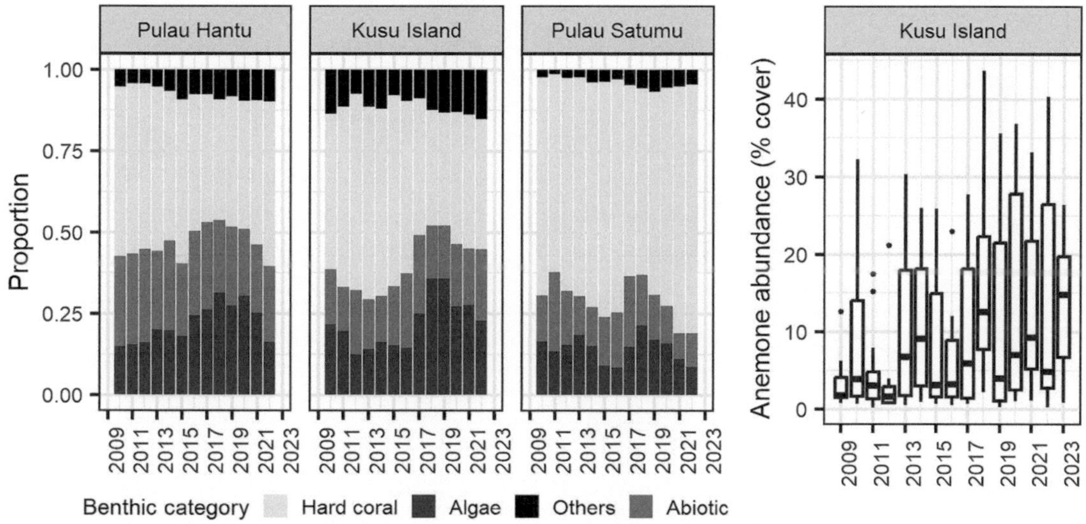

Fig. 5.7 Proportion of cover by the four main benthic categories observed at Pulau Hantu, Kusu Island and Pulau Satumu; and sea anemone abundance observed at Kusu Island, from 2009 to 2023

In addition to larval availability from local adult populations, competitive interactions with other sessile and epibiotic organisms will affect recruitment success of coral larvae. High abundance of epibionts found on the settlement tiles can result in greater competitive interactions between coral spat and the other sessile epibionts, affecting the ability of the coral to recruit to the benthic community (Ritson-Williams et al. 2009; Bento et al. 2017). Although it was not quantified in this study, macroalgae, worms, bryozoans, sponges, tunicates, oysters and barnacles were seen to colonise the tiles, which led to competition for available space for coral spat. High abundance of these sessile epibionts can overgrow the marginal polyps of the coral recruit, reducing coral growth rates (Tanner 1995). Examples of other organisms that were observed from the settlement tiles are provided in Fig. 5.5b–e. Generally, abundances of these other benthic organisms were higher at Kusu Island, followed by Pulau Hantu (Fig. 5.7). Kusu Island is unique, as it hosts a distinctly large population of the magnificent sea anemone, *Radianthus magnifica* (Quoy and Gaimard 1833). Over the years, mean cover by *R. magnifica* at Kusu Island was between 1% and 17%, and as high as 40% of the benthic cover at Kusu Island can consist *of R. magnifica* individuals (Fig. 5.7). It was common for the settlement tiles to be completely covered by the basal disc of numerous *R. magnifica* individuals (Fig. 5.5e), preventing the settlement of coral spat and potentially smothering newly settled larvae, eventually causing recruitment failure.

Very low coral recruitment was consistently observed at Pulau Hantu, and this may be attributed to the presence of territorial turf-farming damselfishes (Family Pomacentridae). These herbivores cultivate dense, relatively fast-growing turf algae patches, which serve as a food source (Ceccarelli et al. 2001). While not quantified, there appeared to be a greater presence of these damselfish at Pulau Hantu than either Pulau Satumu or Kusu Island (A.C. Lee, pers. obs.). Most, if not all, of the settlement tiles that were deployed eventually became 'farming plots' for the damselfishes (Fig. 5.5b). The dense patches of farmed algae would have prevented the settling of coral larvae on the settlement tiles. The farmed algae patches also promote the trapping of sediments that may inhibit settlement of coral larvae and even smothering of settled coral spat (Birrell et al. 2005; Speare et al. 2019). The presence of these herbivores also increases the susceptibility of coral recruits to accidental grazing from the substratum, limiting their survival (Sammarco and Carleton 1981; Christiansen et al. 2009; Baria et al. 2013; Casey et al. 2014). It is very likely that the damselfish present at Pulau Hantu had prevented the settlement of coral larvae, resulting in low recruitment rates. The actual recruitment potential of corals at Pulau Hantu may be higher than what has been recorded from the settlement tiles, as these settlement tiles have been utilised as farming plots by damselfish and may not represent accurate recruitment rates of coral spat.

Observed and identifiable coral recruits from the settlement tiles deployed between 2009 and 2023 consisted of individuals from at least 10 families (Pocilloporidae, Acroporidae, Poritidae, Fungiidae, Merulinidae, Agariciidae, Dendrophylliidae, Euphylliidae, Helioporidae and Lobophylliidae). The dominant taxa composition was mainly from the family Pocilloporidae (84%) (Figs. 5.8 and 5.9). The remaining taxa included were Acroporidae (9%), Poritidae (3%), Fungiidae (1%) and Merulinidae (1%). Representatives of all other families were low in number and only occurred on rare occasions and were grouped together in the 'Others' category, which represented 2% of the total

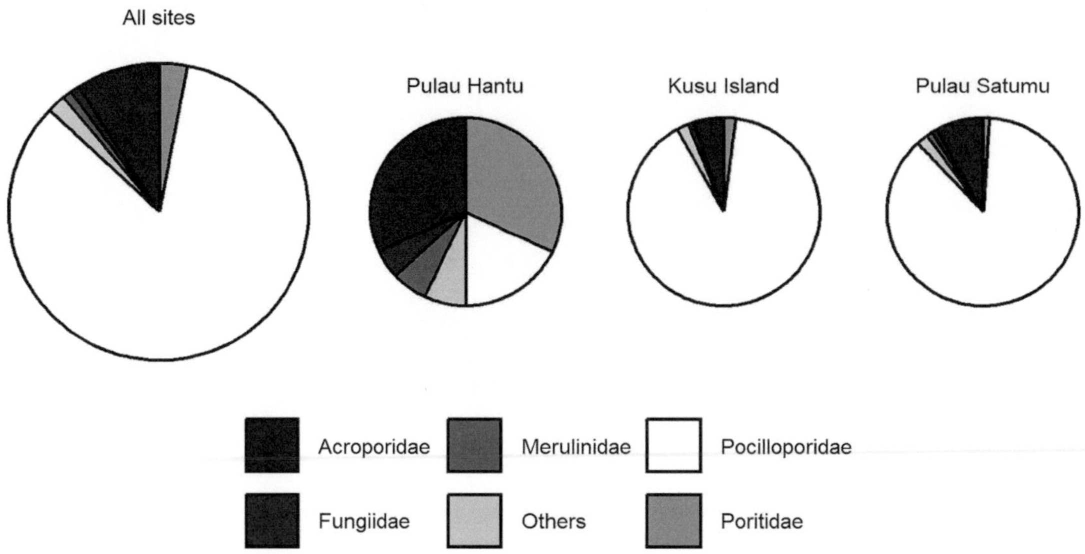

Fig. 5.8 Taxonomic composition of coral recruits by major family groups. Data presented are based on the total recruits observed between 2009 and 2023 for all sites combined, and also for each site

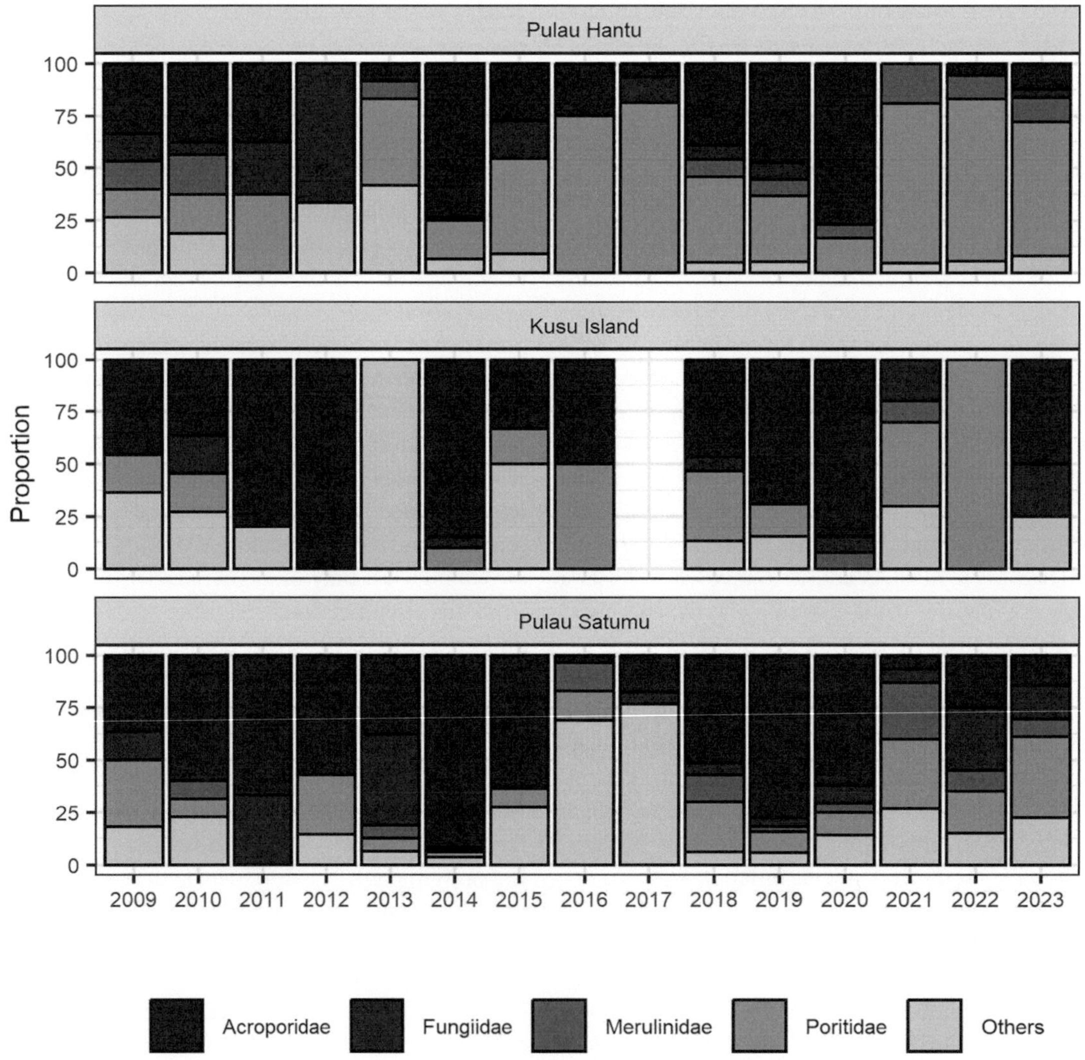

Fig. 5.9 Changes in taxa composition of coral recruits (excluding Pocilloporidae) at Pulau Hantu, Kusu Island and Pulau Satumu from 2009 to 2023

Fig. 5.10 Dominant coral recruits observed under stereo dissecting microscope. (**a**) Pocilloporidae; (**b**, **c**) Acroporidae; (**d**) Poritidae; (**e**, **f**) Merulinidae; (**g**–**i**) others. All scale bars represent 500 μm

recruit communities observed over time. Examples of coral recruits observed on the settlement tiles are presented in Fig. 5.10.

Although all the families identified from the settlement tiles were present at all sites, the taxonomic composition of recruits within each site was highly variable. Pulau Satumu and Kusu Island were consistently dominated by pocilloporids, which were similar to the patterns observed on reefs in nearby regions such as Thailand (Klinthong et al. 2021), Malaysia (Muhammad Faiz et al. 2017), Indonesia (Sawall et al. 2013), Philippines (dela Cruz and Harrison 2024) and Taiwan (Soong et al. 2003), as well as in the Red Sea (Glassom et al. 2004) and Indian Ocean (Jouval et al. 2019). High density of Pocilloporidae recruits, most likely from a single species *Pocillopora acuta* (Poquita-Du et al. 2017, 2019), is possibly a reflection of its life-history characteristic. Pocilloporids are hermaphroditic corals, documented to brood monthly in Singapore, releasing coral larvae, several days just before or after the new moon (Toh et al. 2013; Chou et al. 2019b). For example, larvae of *Pocillopora damicornis* are capable of settling 12–24 h after release (Harrigan 1972). In this study, it is likely that the Pocilloporidae recruits observed on the settlement tiles originated from a process of self-seeding from nearby adult colonies.

In contrast, at Pulau Hantu, the presence of Pocilloporidae recruits was very low compared to Pulau Satumu and Kusu Island. Pulau Hantu tended to be dominated by individuals from Poritidae and Acroporidae. One remarkable observation in Pulau Hantu is the dominance of *Goniopora* corals (Family Poritidae), where large stands of this coral species commonly occur on the upper reef slope. These large columnar colonies form a wide and patchy aggregation in the western side of Pulau Hantu where the tiles were deployed. One congeneric species of *Goniopora*, *Goniopora tenuidens* is known to possess larvatoxic allelochemicals that limit settlement of potential competitors and inhibit larval metamorphosis and post-metamorphic growth of *Pocillopora damicornis* (Fearon and Cameron 1996). Toxin-mediated interactions may influence the ordering of coral assemblage (Endean and Cameron 1990). If hard corals possess toxins

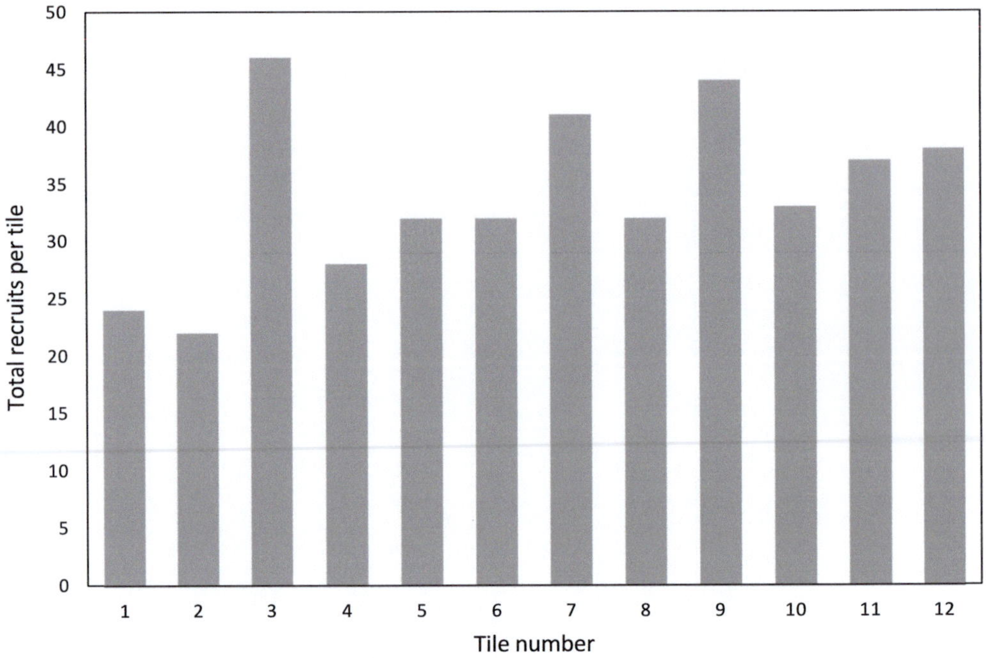

Fig. 5.11 Total recruits observed per tile from 2009 to 2023 at Pulau Hantu

Fig. 5.12 Photo of settlement tile, where, over the 15-year period, *Goniopora* sp. continued to grow around the settlement tile rack. Scale bar represents 5 cm

that restrain settlement of coral species and other organisms, this will affect the determination of community assemblage of the reef. Pulau Hantu had the lowest coral recruit density among sites. It is possible that the presence of *Goniopora* colonies has a negative effect on the recruitment potential in the vicinity of these colonies. In this study, the lowest recruitment density at Pulau Hantu was consistently observed on two specific settlement tiles (Fig. 5.11, tile number 1 and 2) where the adjacent surroundings were overgrown with *Goniopora* colonies (Fig. 5.12), which developed over the 15-year period.

Pulau Satumu is characterised as the site with the highest coral diversity (Huang et al. 2009; Chan et al. 2024) and considered as the most pristine reef in Singapore (Goh and Chou 1993; Loong et al. 2025). Over the years, recruit diversity was generally higher at Pulau Satumu (total of 25 genera identified) than Kusu Island and Pulau Hantu (20 and 15 genera, respectively). The taxonomic composition of recruits, apart from Pocilloporidae, on the settlement tiles from Pulau Hantu, Kusu Island and Pulau Satumu were highly variable over time, with no clear trends in taxa dominance (Fig. 5.9). Excluding the dominant Pocilloporidae recruits, Acroporidae and Poritidae recruits were the more commonly recorded taxa on the settlement tiles at all reef sites examined. Acroporidae are broadcast spawners (Wallace 1985; Baird et al. 2009) and generally exhibit strong recruitment and growth rates (Harriott and Fisk 1988; Morais et al. 2024). Adult colonies form large, branching and/or plating growth forms particularly at shallow depths to capture maximum light that further enhance their fast growth rate. The high surface area-to-volume ratio of Acroporidae makes it susceptible to breakage and dislodgement during disturbances (Woodley et al. 1981; Madin et al. 2014), as well as to sedimentation (Wesseling et al. 1999). Larval settlement success and survivorship of some *Acropora* species are known to be sensitive to increased suspended particles in the water (Moeller et al. 2017; Ricardo et al. 2021). *Acropora* spp. were observed at substantial densities only at Pulau Satumu (Dikou and van Woesik 2006; Huang et al. 2009; Wong et al. 2018; Chan et al. 2024) situated closer to the deeper channels of the Singapore Strait where water currents are less impeded and sedimentation rates are relatively low (Chow et al. 2019; Morgan et al. 2020). The higher water movement may pre-

Fig. 5.13 Coral recruits on the reef. (**a**) *Pectinia* sp.; (**b**) *Pachyseris* sp. (established recruit highlighted in white-bordered insert); (**c**) *Fungia* sp. (left) and *Pectinia* sp. (right); (**d**) Fungiidae (unidentified, possibly *Podabacia* sp.); (**e**) Fungiidae. All scale bars represent 1 cm

vent sediment build-up on the settlement tiles, making it more likely for successful recruitment of coral spat.

5.5 The Future of Coral Recruitment

Over the last 15 years, recruitment potential of corals in Singapore was marginally higher than previously reported recruitment densities for Singapore (Nues 2005; Dikou and van Woesik 2006; Loh et al. 2006; Bauman et al. 2015). The spatial and temporal differences in recruit densities observed here arise from the inherent differences in local adult coral population dynamics, biological interactions with other organisms, physical environment in terms of water flow and anthropogenic inputs. The study design of the long-term monitoring of coral recruitment potential, with how the settlement tiles were placed on the reef, will naturally provide some indication of the influence of local adult populations in the vicinity on the supply and availability of coral larvae. The lower recruit densities that were observed at Pulau Hantu may be a cause of concern for successful reef regeneration and the maintenance of genetic diversity. However, in recent years, juvenile colonies (<5 cm diameter) have been observed on the reefs in Singapore, with highest number of juveniles observed at Pulau Satumu. These consisted of juveniles of plate-forming corals, *Merulina*, *Echinopora*, *Pachyseris* and *Podabacia*, the mushroom coral, *Fungia*, and the brooding species, *Pocillopora acuta* and *Pectinia* (Fig. 5.13). Many of the newly settled juveniles were of the generalist type corals. Generalist corals can be defined as species that possess a mix of weedy (opportunistic), stress-tolerant and competitive abilities (Grime 1977). These corals are species that have high competitive ability and are fast-growing and/or with fast horizontally spreading plating morphology and able to survive in disturbed environments (Darling et al. 2012). This particular life-history characteristic is advantageous in fast-changing and disturbed environments such as Singapore. Albeit in low numbers, the presence of juvenile corals on the reefs in Singapore is a sign of active reef replenishment. This finding is encouraging in light of the various climatic and anthropogenic challenges that the reefs in Singapore must endure.

Acknowledgements The long-term data presented here were derived from ongoing coral reef monitoring programme conducted by Tropical Marine Science Institute, funded by the Public Utilities Board (PUB) of Singapore. The authors would like to thank PUB for the use of their data presented in this work.

References

Abdul Wahab MA, Ferguson S, Snekkevik VK, McCutchan G, Jeong S, Severati A, Randall CJ, Negri AP, Diaz-Pulido G (2023) Hierarchical settlement behaviours of coral larvae to common coralline algae. Sci Rep 13(1):5795. https://doi.org/10.1038/s41598-023-32676-4

Adjeroud M, Augustin D, Galzin R, Salvat B (2002) Natural disturbances and interannual variability of coral reef communities on the outer slope of Tiahura (Moorea, French Polynesia): 1991 to 1997. Mar Ecol Prog Ser 237:121–131. https://doi.org/10.3354/meps237121

Adjeroud M, Penin L, Carroll AG (2007) Spatio-temporal heterogeneity in coral recruitment around Moorea, French Polynesia: implications for population maintenance. J Exp Mar Bio Ecol 341:204–218. https://doi.org/10.1016/j.jembe.2006.10.048

Adjeroud M, Kayal M, Penin L (2016) Importance of recruitment processes in the dynamics and resilience of coral reef assemblages. In: Rossi S, Bramanti L, Gori A, Orejas C (eds) Marine animal forests. Springer, Cham, pp 549–569. https://doi.org/10.1007/978-3-319-17001-5_12-1

Adjeroud M, Peignon C, Gauliard C, Penin L, Kayal M (2022) Extremely high but localized pulses of coral recruitment in the southwestern lagoon of New Caledonia and implications for conservation. Mar Ecol Prog Ser 692:67–79. https://doi.org/10.3354/meps14073

Afiq-Rosli L, Wainwright BJ, Gajanur AR, Lee AC, Ooi SK, Chou LM, Huang D (2021) Barriers and corridors of gene flow in an urbanized tropical reef system. Evol Appl 14(10):2502–2515. https://doi.org/10.1111/eva.13276

Afzal M, Udo T, Ueno M, Nakamura T (2024) Mass coral bleaching and mortality associated with high sea surface temperatures in the summer of 2022 in Sekisei Lagoon, Okinawa, Japan. Galaxea, J Coral Reef Stud 26:20–26. https://doi.org/10.3755/galaxea.G26N-5

Ani CJ, Haller-Bull V, Gilmour JP, Robson BJ (2024) Connectivity modelling identifies sources and sinks of coral recruitment within reef clusters. Sci Rep 14(1):13564. https://doi.org/10.1038/s41598-024-64388-8

Arnold SN, Steneck RS (2011) Settling into an increasingly hostile world: the rapidly closing "recruitment window" for corals. PLoS One 6(12):e28681. https://doi.org/10.1371/journal.pone.0028681

ASMC (2021a) ASEAN Specialised Meteorological Centre (ASMC) Bulletin Issue No. 6. https://www.mss-int.sg/docs/default-source/rcc_bulletin_doc/asmc-bulletin-issue-no%2D%2D6-sep-2020.pdf

ASMC (2021b) ASEAN Specialised Meteorological Centre (ASMC) Bulletin Issue No. 7. https://www.mss-int.sg/docs/default-source/rcc_bulletin_doc/asmc-bulletin-issue-7-mar-2021.pdf

Babcock R, Baird A, Piromvaragorn S, Thomson D, Willis B (2003) Identification of scleractinian coral recruits from Indo-Pacific reefs. Zool Stud 42:211–226

Bachtiar I (2002) Promoting recruitment of scleractinian corals using artificial substrate in the Gili Indah, Lombok Barat, Indonesia. In: Moosa MK, Soemodihardjo S, Soegiarto A et al (eds) Proc 9th Int Coral Reef Symp, Bali, Indonesia, pp 425–430

Baird A, Guest J, Willis B (2009) Systematic and biogeographical patterns in the reproductive biology of scleractinian corals. Annu Rev Ecol Evol Syst 40:551–571. https://doi.org/10.1146/annurev.ecolsys.110308.120220

Bak RPM, Engel MS (1979) Distribution, abundance and survival of juvenile hermatypic corals (Scleractinia) and the importance of life history strategies in the parent coral community. Mar Biol 54(4):341–352. https://doi.org/10.1007/BF00395440

Baria MV, Guest J, Edwards A, Aliño P, Heyward A, Gomez E (2013) Caging enhances post-settlement survival of juveniles of the scleractinian coral *Acropora tenuis*. J Exp Mar Bio Ecol 394:149–153. https://doi.org/10.1016/j.jembe.2010.08.003

Barker NHL, Roberts CM (2004) Scuba diver behaviour and the management of diving impacts on coral reefs. Biol Conserv 120(4):481–489. https://doi.org/10.1016/j.biocon.2004.03.021

Bauman AG, Guest JR, Dunshea G, Low J, Todd PA, Steinberg PD (2015) Coral settlement on a highly disturbed equatorial reef system. PLoS One 10(5):e0127874. https://doi.org/10.1371/journal.pone.0127874

Bellwood DR, Hughes TP, Folke C, Nyström M (2004) Confronting the coral reef crisis. Nature 429(6994):827–833. https://doi.org/10.1038/nature02691

Bento R, Feary DA, Hoey AS, Burt JA (2017) Settlement patterns of corals and other benthos on reefs with divergent environments and disturbances histories around the northeastern Arabian Peninsula. Front Mar Sci 4. https://doi.org/10.3389/fmars.2017.00305

Birkeland C (1977) The importance of rate of biomass accumulation in early successional stages of benthic communities to the survival of coral recruits. In: Taylor DL (ed) Proc 3rd Int Coral Reef Symp, Miami, pp 15–21

Birkeland C, Rowley D, Randall RH (1981) Coral recruitment patterns at Guam. In: Gomez ED, Birkeland CE, Buddemeier RW, Johannes RE, March JA Jr, Tsuda RT (eds) Proc 4th Int Coral Reef Symp, Manila, Philippines, pp 339–344

Birrell CL, McCook LJ, Willis BL (2005) Effects of algal turfs and sediment on coral settlement. Mar Pollut Bull 51(1):408–414. https://doi.org/10.1016/j.marpolbul.2004.10.022

Bode M, Bode L, Armsworth P (2006) Larval dispersal reveals regional sources and sinks in the Great Barrier Reef. Mar Ecol Prog Ser 308. https://doi.org/10.3354/meps308017

Bramanti L, Edmunds PJ (2016) Density-associated recruitment mediates coral population dynamics on a coral reef. Coral Reefs 35:543–553. https://doi.org/10.1007/s00338-016-1413-4

Browne NK, Precht E, Last KS, Todd PA (2014) Photo-physiological costs associated with acute sediment stress events in three near-shore turbid water corals. Mar Ecol Prog Ser 502:129–143. https://doi.org/10.3354/meps10714

Browne NK, Tay J, Todd PA (2015a) Recreating pulsed turbidity events to determine coral–sediment thresholds for active management. J Exp Mar Bio Ecol 466:98–109. https://doi.org/10.1016/j.jembe.2015.02.010

Browne NK, Tay JKL, Low J, Larson O, Todd PA (2015b) Fluctuations in coral health of four common inshore reef corals in response to seasonal and anthropogenic changes in water quality. Mar Environ Res 105:39–52. https://doi.org/10.1016/j.marenvres.2015.02.002

Browne NK, Yaakub SM, Tay JKL, Todd PA (2017) Recreating the shading effects of ship wake induced turbidity to test acclimation responses in the seagrass *Thalassia hemprichii*. Estuar Coast Shelf Sci 199:87–95. https://doi.org/10.1016/j.ecss.2017.09.034

Browne N, Braoun C, McIlwain J, Nagarajan R, Zinke J (2019) Borneo coral reefs subject to high sediment loads show evidence of resilience to various environmental stressors. PeerJ 7:e7382. https://doi.org/10.7717/peerj.7382

Bruno JF, Selig ER, Casey KS, Page CA, Willis BL, Harvell CD, Sweatman H, Melendy AM (2007) Thermal stress and coral cover as drivers of coral disease outbreaks. PLoS Biol 5(6):e124. https://doi.org/10.1371/journal.pbio.0050124

Burt AJ, Vogt-Vincent N, Johnson H, Sendell-Price A, Kelly S, Clegg SM, Head C, Bunbury N, Fleischer-Dogley F, Jeremie M-M, Khan N, Baxter R, Gendron G, Mason-Parker C, Walton R, Turnbull LA

(2024) Integration of population genetics with oceanographic models reveals strong connectivity among coral reefs across Seychelles. Sci Rep 14(1):4936. https://doi.org/10.1038/s41598-024-55459-x

Caley MJ, Carr MH, Hixon MA, Hughes TP, Jones GP, Menge BA (1996) Recruitment and the local dynamics of open marine populations. Annu Rev Ecol Evol Syst 27:477–500. https://doi.org/10.1146/annurev.ecolsys.27.1.477

Cameron AM, Endean R (1985) Do long-lived species structure coral reef ecosystems? In: Proc 5th Int Coral Reef Cong, Tahiti, French Polynesia, pp 211–215

Carleton JH, Sammarco P (1987) Effects of substratum irregularity on success of coral settlement: quantification by comparative geomorphological techniques. Bull Mar Sci 40(1):85–98

Carlson RR, Crowder LB, Martin RE, Asner GP (2024) The effect of reef morphology on coral recruitment at multiple spatial scales. Proc Natl Acad Sci 121(4):e2311661121. https://doi.org/10.1073/pnas.2311661121

Casey J, Choat J, Connolly S (2014) Coupled dynamics of territorial damselfishes and juvenile corals on the reef crest. Coral Reefs 34. https://doi.org/10.1007/s00338-014-1221-7

Ceccarelli DM, Jones GP, McCook LJ (2001) Territorial damselfishes as determinants of the structure of benthic communities on coral reefs. In: Gibson R, Barnes M, Atkinson R (eds) Oceanography and marine biology. An annual review, vol 39. CRC Press, pp 355–389

Chan YKS, Ng CSL, Tun KPP, Chou LM, Huang D (2024) Decadal decline of functional diversity despite increasing taxonomic and phylogenetic diversity of coral reefs under chronic urbanisation stress. Ecol Indic 164:112143. https://doi.org/10.1016/j.ecolind.2024.112143

Chong-Seng K, Graham N, Pratchett M (2014) Bottlenecks to coral recovery in the Seychelles. Coral Reefs 33:449–461. https://doi.org/10.1007/s00338-014-1137-2

Chou L (1996) Response of Singapore reefs to land reclamation. Galaxea 13(1):85–92

Chou L (2006) Marine habitats in one of the world's busiest harbours. In: Wolanski E (ed) The environment in Asia Pacific harbours. Springer, Dordrecht, pp 377–391. https://doi.org/10.1007/1-4020-3655-8_22

Chou LM, Huang D, Tan KS, Toh TC, Goh BPL, Tun K (2019a) Singapore. In: Sheppard C (ed) World seas: an environmental evaluation, vol III: ecological issues and environmental impacts, 2nd edn. Elsevier, pp 539–558. https://doi.org/10.1016/B978-0-08-100853-9.00031-2

Chou LM, Goh BPL, Quek ST (2019b) Planulation in the scleractinian coral *Pocillopora damicornis* in Singapore waters. Int J Environ Sci Dev 10:326–330. https://doi.org/10.18178/ijesd.2019.10.10.1194

Chow GSE, Chan YKS, Jain SS, Huang D (2019) Light limitation selects for depth generalists in urbanised reef coral communities. Mar Environ Res 147:101–112. https://doi.org/10.1016/j.marenvres.2019.04.010

Christiansen N, Ward S, Harii S, Tibbetts I (2009) Grazing by a small fish affects the early stages of a post-settlement stony coral. Coral Reefs 28:47–51. https://doi.org/10.1007/s00338-008-0429-9

Coles SL, Brown EK (2007) Twenty-five years of change in coral coverage on a hurricane impacted reef in Hawai'i: the importance of recruitment. Coral Reefs 26(3):705–717. https://doi.org/10.1007/s00338-007-0257-3

Connell JH (1973) Population ecology of reef-building corals. Biol Geol Coral Reefs 2:205–245. https://doi.org/10.1016/B978-0-12-395526-5.50015-8

Connell JH (1985) The consequences of variation in initial settlement vs. post-settlement mortality in rocky intertidal communities. J Exp Mar Bio Ecol 93(1):11–45. https://doi.org/10.1016/0022-0981(85)90146-7

Connell JH, Hughes TP, Wallace CC (1997) A 30-year study of coral abundance, recruitment, and disturbance at several scales in space and time. Ecol Monogr 67(4):461–488. https://doi.org/10.1890/0012-9615(1997)067[0461:AYSOCA]2.0.CO;2

D'Angelo C, Wiedenmann J (2014) Impacts of nutrient enrichment on coral reefs: new perspectives and implications for coastal management and reef survival. Curr Opin Environ Sust 7:82–93. https://doi.org/10.1016/j.cosust.2013.11.029

Darling ES, Alvarez-Filip L, Oliver TA, McClanahan TR, Côté IM (2012) Evaluating life-history strategies of reef corals from species traits. Ecol Lett 15(12):1378–1386. https://doi.org/10.1111/j.1461-0248.2012.01861.x

Davidson J, Thompson A, Logan M, Schaffelke B (2019) High spatio-temporal variability in Acroporidae settlement to inshore reefs of the Great Barrier Reef. PLoS One 14(1):e0209771. https://doi.org/10.1371/journal.pone.0209771

De'ath G, Fabricius KE, Sweatman H, Puotinen M (2012) The 27–year decline of coral cover on the Great Barrier Reef and its causes. Proc Natl Acad Sci 109(44):17995–17999. https://doi.org/10.1073/pnas.1208909109

dela Cruz DW, Harrison PL (2024) Reef location and season, but not recruitment substrate contour and composition, affect coral recruitment patterns. J Exp Mar Bio Ecol 578:152029. https://doi.org/10.1016/j.jembe.2024.152029

Dikou A, van Woesik R (2006) Survival under chronic stress from sediment load: spatial patterns of hard coral communities in the southern islands of Singapore. Mar Pollut Bull 52(11):1340–1354. https://doi.org/10.1016/j.marpolbul.2006.02.011

Doropoulos C, Roff G, Visser MS, Mumby PJ (2017) Sensitivity of coral recruitment to subtle shifts in early community succession. Ecology 98(2):304–314. https://doi.org/10.1002/ecy.1663

Dubinsky Z, Stambler N (1996) Marine pollution and coral reefs. Glob Chang Biol 2(6):511–526. https://doi.org/10.1111/j.1365-2486.1996.tb00064.x

Eagle JV, Baird AH, Jones GP, Kingsford MJ (2012) Recruitment hotspots: consistent spatial patterns in the relative abundance of coral recruits at One Tree Island, Australia. Galaxea, J Coral Reef Stud 14(1):5–22. https://doi.org/10.3755/galaxea.14.1_5

Edinger EN, Jompa J, Limmon GV, Widjatmoko W, Risk MJ (1998) Reef degradation and coral biodiversity in Indonesia: effects of land-based pollution, destructive fishing practices and changes over time. Mar Pollut Bull 36(8):617–630. https://doi.org/10.1016/S0025-326X(98)00047-2

Edmunds PJ (2017) Unusually high coral recruitment during the 2016 El Niño in Mo'orea, French Polynesia. PLoS One 12(10):e0185167. https://doi.org/10.1371/journal.pone.0185167

Edmunds PJ (2023) Coral recruitment: patterns and processes determining the dynamics of coral populations. Biol Rev 98(6):1862–1886. https://doi.org/10.1111/brv.12987

Edmunds P, Bruno J, Carlon D (2004) Effects of depth and microhabitat on growth and survivorship of juvenile corals in the Florida Keys. Mar Ecol Prog Ser 278:115–124. https://doi.org/10.3354/meps278115

Edmunds PJ, Nozawa Y, Villanueva RD (2014) Refuges modulate coral recruitment in the Caribbean and the Pacific. J Exp Mar Bio Ecol 454:78–84. https://doi.org/10.1016/j.jembe.2014.02.009

Edmunds P, Steneck R, Albright R, Carpenter R, Chui APY, Fan T-Y, Harii S, Kitano H, Kurihara H, Legendre L (2015) Geographic variation in long-term trajectories of change in coral recruitment: a global-to-local perspective. Mar Freshw Res 66(7):609–622. https://doi.org/10.1071/MF14139

Endean R, Cameron A (1990) Trends and new perspectives in coral reef ecology. In: Dubinsky Z (ed) Coral reefs: ecosystems of the world, vol 25. Elsevier, Amsterdam, pp 469–492

Fabricius KE (2005) Effects of terrestrial runoff on the ecology of corals and coral reefs: review and synthesis. Mar Pollut Bull 50(2):125–146. https://doi.org/10.1016/j.marpolbul.2004.11.028

Fabricius KE, Noonan SH, Abrego D, Harrington L, De'Ath G (2017) Low recruitment due to altered settlement substrata as primary constraint for coral communities under ocean acidification. Proc R Soc Lond Ser B Biol Sci 284(1862):20171536. https://doi.org/10.1098/rspb.2017.1536

Fearon RJ, Cameron AM (1996) Larvotoxic extracts of the hard coral *Goniopora tenuidens:* Allelochemicals that limit settlement of potential competitors? Toxicon 34(3):361–367. https://doi.org/10.1016/0041-0101(95)00137-9

Fisk DA, Harriott VJ (1990) Spatial and temporal variation in coral recruitment on the Great Barrier Reef: implications for dispersal hypotheses. Mar Biol 107(3):485–490. https://doi.org/10.1007/BF01313433

Fitzhardinge R (1988) Coral recruitment: the importance of interspecific differences in juvenile growth and mortality. In: Choat JH, Barnes D, Borowitzka MA (eds) Proc 6th Int Coral Reef Symp, Townsville, Australia, pp 673–678

Fong J, Ramsby BD, Flores F, Dada T, Antunes E, Abdul Wahab MA, Severati A, Negri AP, Diaz-Pulido G (2024) Effects of material type and surface roughness of settlement tiles on macroalgal colonisation and early coral recruitment success. Coral Reefs 43(4):1083–1096. https://doi.org/10.1007/s00338-024-02526-4

Games PA, Howell JF (1976) Pairwise multiple comparison procedures with unequal n's and/or variances: a Monte Carlo study. J Educ Stat 1(2):113–125

Gilmour JP, Smith LD, Heyward AJ, Baird AH, Pratchett MS (2013) Recovery of an isolated coral reef system following severe disturbance. Science 340(6128):69–71. https://doi.org/10.1126/science.1232310

Glassom D, Zakai D, Chadwick-Furman N (2004) Coral recruitment: a spatio-temporal analysis along the coastline of Eilat, northern Red Sea. Mar Biol 144:641–651. https://doi.org/10.1007/s00227-003-1243-0

Glassom D, Celliers L, Schleyer M (2006) Coral recruitment patterns at Sodwana Bay, South Africa. Coral Reefs 25(3):485–492. https://doi.org/10.1007/s00338-006-0117-6

Gleason DF, Hofmann DK (2011) Coral larvae: from gametes to recruits. J Exp Mar Bio Ecol 408(1–2):42–57. https://doi.org/10.1016/j.jembe.2011.07.025

Goh NK, Chou L (1993) The coral reef community of Pulau Satumu (Raffles Lighthouse), Singapore, with emphasis on the hard corals. J Singapore Natl Acad Sci 20(21):51–57

Gouezo M, Golbuu Y, Fabricius K, Olsudong D, Mereb G, Nestor V, Wolanski E, Harrison P, Doropoulos C (2019) Drivers of recovery and reassembly of coral reef communities. Proc R Soc Lond Ser B Biol Sci 286:20182908. https://doi.org/10.1098/rspb.2018.2908

Gouezo M, Olsudong D, Fabricius K, Harrison P, Golbuu Y, Doropoulos C (2020) Relative roles of biological and physical processes influencing coral recruitment during the lag phase of reef community recovery. Sci Rep 10:2471. https://doi.org/10.1038/s41598-020-59111-2

Green DH, Edmunds PJ (2011) Spatio-temporal variability of coral recruitment on shallow reefs in St. John, US Virgin Islands. J Exp Mar Bio Ecol 397(2):220–229. https://doi.org/10.1016/j.jembe.2010.12.004

Grime JP (1977) Evidence for the existence of three primary strategies in plants and its relevance to ecological and evolutionary theory. Am Nat 111(982):1169–1194

Guest JR, Baird AH, Goh BPL, Chou LM (2005) Reproductive seasonality in an equatorial assemblage of scleractinian corals. Coral Reefs 24(1):112–116. https://doi.org/10.1007/s00338-004-0433-7

Guest JR, Tun K, Low J, Vergés A, Marzinelli EM, Campbell AH, Bauman AG, Feary DA, Chou LM, Steinberg PD (2016) 27 years of benthic and coral community dynamics on turbid, highly urbanised reefs off Singapore. Sci Rep 6(1):36260. https://doi.org/10.1038/srep36260

Haapkylä J, Ramade F, Salvat B (2007) Oil pollution on coral reefs: a review of the state of knowledge and management needs. Vie Milieu 57(1/2):91–107

Hamizan YM, Saad S, Naim MA, Khodzori MFA, Hanapiah MFM (2021) Comparison of early succession and scleractinian recruitment between artificial live rocks and terracotta tiles. Ocean Sci 56(3):317–325. https://doi.org/10.1007/s12601-021-00028-w

Harper LM, Huebner LK, O'Cain ED, Ruzicka R, Gleason DF, Fogarty ND (2021) Methodological recommendations for assessing scleractinian and octocoral recruitment to settlement tiles. PeerJ 9:e12549. https://doi.org/10.7717/peerj.12549

Harrigan JF (1972) The planula larva of *Pocillopora damicornis*: lunar periodicity of swarming and substratum selection behavior. Ph.D. thesis. University of Hawaii, Hawaii

Harriott VJ, Banks SA (1995) Recruitment of scleractinian corals in the Solitary Islands Marine Reserve, a high latitude coral-dominated community in Eastern Australia. Mar Ecol Prog Ser 123:155–161. https://doi.org/10.3354/meps123155

Harriott VJ, Fisk DA (1987) A comparison of settlement plate types for experiments on the recruitment of scleractinian corals. Mar Ecol Prog Ser 37(2–3):201–208. https://doi.org/10.3354/meps037201

Harriott VJ, Fisk DA (1988) Recruitment patterns of scleractinian corals: a study of three reefs. Mar Freshw Res 39(4):409–416. https://doi.org/10.1071/MF9880409

Heery EC, Hoeksema BW, Browne NK, Reimer JD, Ang PO, Huang D, Friess DA, Chou LM, Loke LHL, Saksena-Taylor P, Alsagoff N, Yeemin T, Sutthacheep M, Vo ST, Bos AR, Gumanao GS, Syed Hussein MA, Waheed Z, Lane DJW, Johan O, Kunzmann A, Jompa J, Suharsono TD, Bauman AG, Todd PA (2018) Urban coral reefs: degradation and resilience of hard coral assemblages in coastal cities of East and Southeast Asia. Mar Pollut Bull 135:654–681. https://doi.org/10.1016/j.marpolbul.2018.07.041

Henley BJ, McGregor HV, King AD, Hoegh-Guldberg O, Arzey AK, Karoly DJ, Lough JM, DeCarlo TM, Linsley BK (2024) Highest ocean heat in four centuries places Great Barrier Reef in danger. Nature 632(8024):320–326. https://doi.org/10.1038/s41586-024-07672-x

Highsmith RC (1982) Reproduction by fragmentation in corals. Mar Ecol Prog Ser 7(2):207–226

Hilton MJ, Manning SS (1995) Conversion of coastal habitats in Singapore: indications of unsustainable development. Environ Conserv 22(4):307–322. https://doi.org/10.1017/S0376892900034883

Hodgson G (1990) Sediment and the settlement of larvae of the reef coral *Pocillopora damicornis*. Coral Reefs 9:41–43. https://doi.org/10.1007/BF00686720

Hoegh-Guldberg O, Mumby PJ, Hooten AJ, Steneck RS, Greenfield P, Gomez E, Harvell CD, Sale PF, Edwards AJ, Caldeira K (2007) Coral reefs under rapid climate change and ocean acidification. Science 318(5857):1737–1742. https://doi.org/10.1126/science.1152509

Huang D, Tun K, Chou L, Todd P (2009) An inventory of zooxanthellate scleractinian corals in Singapore, including 33 new records. Raffles Bull Zool 22:69–80

Hughes TP, Baird AH, Dinsdale EA, Moltschaniwskyj NA, Pratchett MS, Tanner JE, Willis BL (2000) Supply-side ecology works both ways: the link between benthic adults, fecundity, and larval recruits. Ecology 81(8):2241–2249. https://doi.org/10.1890/0012-9658(2000)081[2241:SSEWBW]2.0.CO;2

Hughes TP, Baird AH, Dinsdale EA, Harriott VJ, Moltschaniwskyj NA, Pratchett MS, Tanner JE, Willis BL (2002) Detecting regional variation using meta-analysis and large-scale sampling: latitudinal patterns in recruitment. Ecology 83(2):436–451. https://doi.org/10.1890/0012-9658(2002)083[0436:DRVUMA]2.0.CO;2

Humanes A, Bastidas C (2015) In situ settlement rates and early survivorship of hard corals: a good year for a Caribbean reef. Mar Ecol Prog Ser 539:139–151. https://doi.org/10.3354/meps11501

Jouval F, Latreille AC, Bureau S, Adjeroud M, Penin L (2019) Multiscale variability in coral recruitment in the Mascarene Islands: from centimetric to geographical scale. PLoS One 14(3):e0214163. https://doi.org/10.1371/journal.pone.0214163

Kayal M, Vercelloni J, Wand MP, Adjeroud M (2015) Searching for the best bet in life-strategy: a quantitative approach to individual performance and population dynamics in reef-building corals. Ecol Complex 23:73–84. https://doi.org/10.1016/j.ecocom.2015.07.003

Keough MJ, Downes BJ (1982) Recruitment of marine invertebrates: the role of active larval choices and early mortality. Oecologia 54(3):348–352. https://doi.org/10.1007/BF00380003

Kleypas JA, Yates KK (2009) Coral reefs and ocean acidification. Oceanography 22(4):108–117. https://doi.org/10.5670/oceanog.2009.101

Klinthong W, Yeemin T, Aunkhongthong W, Rongprakhon S (2021) Abundance of coral recruits on settlement plate experiments from Mu Ko Angthong, the Western Gulf of Thailand. Ramkhamhaeng Int J Sci Technol 4(2):19

Lequeux BD, Ahumada-Sempoal M-A, López-Pérez A, Reyes-Hernández C (2018) Coral connectivity between equatorial eastern Pacific marine protected areas: a biophysical modeling approach. PLoS One 13(8):e0202995. https://doi.org/10.1371/journal.pone.0202995

Levenstein MA, Marhaver KL, Quinlan ZA, Tholen HM, Tichy L, Yus J, Lightcap I, Wegley Kelly L, Juarez G, Vermeij MJ (2022) Composite substrates reveal inorganic material cues for coral larval settlement. ACS Sustain Chem Eng 10(12):3960–3971. https://doi.org/10.1021/acssuschemeng.1c08313

Lirman D (2000) Fragmentation in the branching coral *Acropora palmata* (Lamarck): growth, survivorship, and reproduction of colonies and fragments. J Exp Mar Bio Ecol 251(1):41–57. https://doi.org/10.1016/S0022-0981(00)00205-7

Loh T-L, Tanzil JTI, Chou LM (2006) Preliminary study of community development and scleractinian recruitment on fibreglass artificial reef units in the sedimented waters of Singapore. Aquat Conserv 16(1):61–76. https://doi.org/10.1002/aqc.701

Lohr KE, Smith DJ, Suggett DJ, Nitschke MR, Dumbrell AJ, Woodcock S, Camp EF (2017) Coral community structure and recruitment in seagrass meadows. Front Mar Sci 4. https://doi.org/10.3389/fmars.2017.00388

Low J, Chou LM (1994) Sedimentation rates in Singapore waters. Proc 3rd ASEAN-Australian Symp on Living Coral Resources 2:697–701

Lyu Y, Zhou Z, Zhang Y, Chen Z, Deng W, Shi R (2022) The mass coral bleaching event of inshore corals form South China Sea witnessed in 2020: insight into the causes, process and consequence. Coral Reefs 41(5):1351–1364. https://doi.org/10.1007/s00338-022-02284-1

Madin JS, Baird AH, Dornelas M, Connolly SR (2014) Mechanical vulnerability explains size-dependent mortality of reef corals. Ecol Lett 17(8):1008–1015. https://doi.org/10.1111/ele.12306

Maida M, Coll J, Sammarco P (1994) Shedding new light on scleractinian coral recruitment. J Exp Mar Bio Ecol 180(2):189–202. https://doi.org/10.1016/0022-0981(94)90066-3

Mallela J, Crabbe MJC (2009) Hurricanes and coral bleaching linked to changes in coral recruitment in Tobago. Mar Environ Res 68(4):158–162. https://doi.org/10.1016/j.marenvres.2009.06.001

Mangubhai S, Harrison PL, Obura DO (2007) Patterns of coral larval settlement on lagoon reefs in the Mombasa Marine National Park and Reserve, Kenya. Mar Ecol Prog Ser 348:149–159. https://doi.org/10.3354/meps07090

Martinez S, Abelson A (2013) Coral recruitment: the critical role of early post-settlement survival. ICES J Mar Sci 70(7):1294–1298. https://doi.org/10.1093/icesjms/fst035

McGowan H, Theobald A (2023) Atypical weather patterns cause coral bleaching on the Great Barrier Reef, Australia during the 2021–2022 La Niña. Sci Rep 13(1):6397. https://doi.org/10.1038/s41598-023-33613-1

Moeller M, Nietzer S, Schils T, Schupp PJ (2017) Low sediment loads affect survival of coral recruits: the first weeks are crucial. Coral Reefs 36(1):39–49. https://doi.org/10.1007/s00338-016-1513-1

Morais J, Tebbett SB, Morais RA, Bellwood DR (2024) Natural recovery of corals after severe disturbance. Ecol Lett 27(1):e14332. https://doi.org/10.1111/ele.14332

Morgan KM, Moynihan MA, Sanwlani N, Switzer AD (2020) Light limitation and depth-variable sedimentation drives vertical reef compression on turbid coral reefs. Front Mar Sci 7. https://doi.org/10.3389/fmars.2020.571256

Morse DE, Hooker N, Morse ANC, Jensen RA (1988) Control of larval metamorphosis and recruitment in sympatric agariciid corals. J Exp Mar Bio Ecol 116(3):193–217. https://doi.org/10.1016/0022-0981(88)90027-5

Muhammad Faiz MH, Saad S, Mukai Y, Ahmad Z (2017) Relationship between adult abundance and successive coral recruitment density in the Tioman Island Marine Park, Malaysia. Zool Ecol 27(3–4):328–334. https://doi.org/10.1080/21658005.2017.1396793

Mundy C (2000) An appraisal of methods used in coral recruitment studies. Coral Reefs 19:124–131. https://doi.org/10.1007/s003380000081

Mydlarz LD, McGinty ES, Harvell CD (2010) What are the physiological and immunological responses of coral to climate warming and disease? J Exp Biol 213(6):934–945. https://doi.org/10.1242/jeb.037580

Nativ H, Scucchia F, Martinez S, Einbinder S, Chequer A, Goodbody-Gringley G, Mass T (2021) *In situ* estimation of coral recruitment patterns from shallow to mesophotic reefs using an optimized fluorescence imaging system. Front Mar Sci 8:709175. https://doi.org/10.3389/fmars.2021.709175

Nues PCM (2005) Enhancement of coral recruitment. Master's Thesis. National University of Singapore, Singapore

Petersen D, Laterveer M, Schuhmacher H (2005) Innovative substrate tiles to spatially control larval settlement in coral culture. Mar Biol 146(5):937–942. https://doi.org/10.1007/s00227-004-1503-7

Poquita-Du RC, Ng CSL, Loo JB, Afiq-Rosli L, Tay YC, Todd PA, Chou LM, Huang D (2017) New evidence shows that *Pocillopora* '*damicornis*-like' corals in Singapore are actually *Pocillopora acuta* (Scleractinia: Pocilloporidae). Biodivers Data J 5:e11407. https://doi.org/10.3897/BDJ.5.e11407

Poquita-Du RC, Quek ZBR, Jain SS, Schmidt-Roach S, Tun K, Heery EC, Chou LM, Todd PA, Huang D (2019) Last species standing: loss of Pocilloporidae corals associated with coastal urbanization in a tropical city state. Mar Biodivers 49(4):1727–1741. https://doi.org/10.1007/s12526-019-00939-x

Price N (2010) Habitat selection, facilitation, and biotic settlement cues affect distribution and performance of coral recruits in French Polynesia. Oecologia 163(3):747–758. https://doi.org/10.1007/s00442-010-1578-4

Price NN, Muko S, Legendre L, Steneck R, van Oppen MJ, Albright R, Ang P Jr, Carpenter RC, Chui APY, Fan T-Y (2019) Global biogeography of coral recruitment: tropical decline and subtropical increase. Mar Ecol Prog Ser 621:1–17. https://doi.org/10.3354/meps12980

Quoy JRC, Gaimard JP (1833) Voyage de découvertes de l'Astrolabe: exécuté par ordre du roi, pendant les années 1826-1827-1828-1829, sous le commandement de M.J. Dumont d'Urville. Zoologie, Tome quatrième. J. Tastu, Paris, 390 pp. https://www.biodiversitylibrary.org/page/2064331

R Core Team (2012) R: A language and environment for statistical computing. R Foundation for Statistical Computing, Vienna, Austria. https://www.R-project.org/

Reidenbach MA, Stocking JB, Szczyrba L, Wendelken C (2021) Hydrodynamic interactions with coral topography and its impact on larval settlement. Coral Reefs 40:505–519. https://doi.org/10.1007/s00338-021-02069-y

Reyes MZ, Yap HT (2001) Effect of artificial substratum material and resident adults on coral settlement patterns at Danjugan Island, Philippines. Bull Mar Sci 69(2):559–566

Ricardo GF, Harper CE, Negri AP, Luter HM, Abdul Wahab MA, Jones RJ (2021) Impacts of water quality on *Acropora* coral settlement: the relative importance of substrate quality and light. Sci Total Environ 777:146079. https://doi.org/10.1016/j.scitotenv.2021.146079

Richmond R, Tisthammer K, Spies N (2018) The effects of anthropogenic stressors on reproduction and recruitment of corals and reef organisms. Front Mar Sci 5. https://doi.org/10.3389/fmars.2018.00226

Risk MJ (2014) Assessing the effects of sediments and nutrients on coral reefs. Curr Opin Environ Sust 7:108–117. https://doi.org/10.1016/j.cosust.2014.01.003

Ritson-Williams R, Arnold SV, Fogarty N, Steneck R, Vermeij M, Paul V (2009) New perspectives on ecological mechanisms affecting coral recruitment on reefs. Smithson Contrib Mar Sci 38:457. https://doi.org/10.5479/si.01960768.38.437

Ritson-Williams R, Arnold SN, Paul V (2016) Patterns of larval settlement preferences and post-settlement survival for seven Caribbean corals. Mar Ecol Prog Ser 548. https://doi.org/10.3354/meps11688

Rockwood LL (2006) Introduction to population ecology. Blackwell Publishing, UK

Rogers CS (1990) Responses of coral reefs and reef organisms to sedimentation. Mar Ecol Prog Ser 62:185–202. https://doi.org/10.3354/meps062185

Roth M, Knowlton N (2009) Distribution, abundance, and microhabitat characterization of small juvenile corals at Palmyra Atoll. Mar Ecol Prog Ser 376:133–142. https://doi.org/10.3354/meps07787

Saila SB, Kocic VL, McManus JW (1993) Modelling the effects of destructive fishing practices on tropical coral reefs. Mar Ecol Prog Ser 94(1):51–60. https://doi.org/10.3354/meps094051

Salinas-de-León P, Dryden C, Smith D, Bell J (2013) Temporal and spatial variability in coral recruitment on two Indonesian coral reefs: consistently lower recruitment to a degraded reef. Mar Biol 160:97–105. https://doi.org/10.1007/S00227-012-2066-7

Sammarco P (1982) Polyp bailout: an escape response to environmental stress and a new means of reproduction in corals. Mar Ecol Prog Ser 10:57–65. https://doi.org/10.3354/meps010057

Sammarco PW, Carleton JH (1981) Damselfish territoriality and coral community structure: reduced grazing, coral recruitment, and effects on coral spat. In: Gomez ED, Birkeland CE, Buddemeier RW, Johannes RE, March JA Jr, Tsuda RT (eds) Proc 4th Int Coral Reef Symp, Manila, Philippines, pp 525–535

Sawall Y, Phongsuwan N, Richter C (2010) Coral recruitment and recovery after the 2004 Tsunami around the Phi Phi Islands (Krabi Province) and Phuket, Andaman Sea, Thailand. Helgol Mar Res 64(4):357–365. https://doi.org/10.1007/s10152-010-0192-5

Sawall Y, Jompa J, Litaay M, Maddusila A, Richter C (2013) Coral recruitment and potential recovery of eutrophied and blast fishing impacted reefs in Spermonde Archipelago, Indonesia. Mar Pollut Bull 74(1):374–382. https://doi.org/10.1016/j.marpolbul.2013.06.022

Sola E, Marques da Silva I, Glassom D (2015) Spatio-temporal patterns of coral recruitment at Vamizi Island, Quirimbas Archipelago, Mozambique. Afr J Mar Sci 37(4):557–565. https://doi.org/10.2989/1814232X.2015.1113201

Soong K, Chen MH, Chen CL, Dai C-F, Fan T-Y, Li JJ, Fan HM (2003) Spatial and temporal variation of coral recruitment in Taiwan. Coral Reefs 22(3):224–228. https://doi.org/10.1007/s00338-003-0311-8

Speare K, Duran A, Miller M, Burkepile D (2019) Sediment associated with algal turfs inhibits the settlement of two endangered coral species. Mar Pollut Bull 144:189–195. https://doi.org/10.1016/j.marpolbul.2019.04.066

Stoddart JA (1983) Asexual production of planulae in the coral *Pocillopora damicornis*. Mar Biol 76(3):279–284. https://doi.org/10.1007/BF00393029

Tamelander J (2002) Coral recruitment following a mass mortality event. AMBIO J Hum Environ 31(7):551–557. https://doi.org/10.1579/0044-7447-31.7.551

Tan YHJ, Tham JKQ, Paul A, Rana U, Ang HP, Nguyen NTH, Yee ATK, Leong BPI, Drummond S, Tun KPP (2023) Remote sensing mapping of the regeneration of coastal natural habitats in Singapore: implications for marine conservation in tropical cities. Singap J Trop Geogr 44(1):130–148. https://doi.org/10.1111/sjtg.12454

Tanner JE (1995) Competition between scleractinian corals and macroalgae: an experimental investigation of coral growth, survival and reproduction. J Exp Mar Bio Ecol 190(2):151–168. https://doi.org/10.1016/0022-0981(95)00027-O

Tanzil JT, Chou LM, Todd PA (2008) Coral recruitment in sediment-stressed reefs: effect of reef zone and substrate inclination (Poster). In: 11th Int Coral Reef Symp, Fort Lauderdale, Florida

Tay YC (2012) Connectivity of coral populations within the Singapore Strait. PhD thesis. National University of Singapore, Singapore

Tay YC, Todd PA, Rosshaug PS, Chou LM (2012) Simulating the transport of broadcast coral larvae among the Southern Islands of Singapore. Aquat Biol 15:283–297. https://doi.org/10.3354/ab00433

Te FT (1992) Response to higher sediment loads by *Pocillopora damicornis* planulae. Coral Reefs 11(3):131–134. https://doi.org/10.1007/BF00255466

Thomson DP, Babcock RC, Evans RD, Feng M, Moustaka M, Orr M, Slawinski D, Wilson SK, Hoey AS (2021) Coral larval recruitment in North-Western Australia predicted by regional and local conditions. Mar Environ Res 168:105318. https://doi.org/10.1016/j.marenvres.2021.105318

Toh TC, Peh JWK, Chou LM (2013) Heterotrophy in recruits of the scleractinian coral *Pocillopora damicornis*. Mar Freshwat Behav Physiol 46(5):313–320. https://doi.org/10.1080/10236244.2013.832890

Tomascik T, Sander F (1987) Effects of eutrophication on reef-building corals. Mar Biol 94(1):77–94. https://doi.org/10.1007/BF00392901

Tun PPK (2012) Optimisation of reef survey methods and application of reef metrics and biocriteria for the monitoring of sediment-impacted reefs. Ph.D. thesis. National University of Singapore, Singapore

Turner JA, Thomson DP, Cresswell AK, Trapon M, Babcock RC (2018) Depth-related patterns in coral recruitment across a shallow to mesophotic gradient. Coral Reefs 37:711–722. https://doi.org/10.1007/s00338-018-1696-8

Tuttle LJ, Donahue MJ (2022) Effects of sediment exposure on corals: a systematic review of experimental studies. Environ Evid 11(1):4. https://doi.org/10.1186/s13750-022-00256-0

van Dam J, Negri A, Uthicke S, Mueller J (2011) Chapter 9. Chemical pollution on coral reefs: exposure and ecological effects. In: Sanchez-Bayo F, van den Brink P, Mann R (eds) Ecological impacts of toxic chemicals, vol 1. Bentham Science Publisher Ltd., Amsterdam, pp 187–211. https://doi.org/10.2174/978160805121211 01010187

van Woesik R, Scott WJ IV, Aronson RB (2014) Lost opportunities: coral recruitment does not translate to reef recovery in the Florida Keys. Mar Pollut Bull 88(1–2):110–117. https://doi.org/10.1016/j.marpolbul.2014.09.017

Wallace CC (1985) Reproduction, recruitment and fragmentation in nine sympatric species of the coral genus *Acropora*. Mar Biol 88:217–233. https://doi.org/10.1007/BF00392585

Welsch RE (1977) Stepwise multiple comparison procedures. J Am Stat Assoc 72(359):566–575. https://doi.org/10.2307/2286218

Weber M, de Beer D, Lott C, Polerecky L, Kohls K, Abed RMM, Ferdelman TG, Fabricius KE (2012) Mechanisms of damage to corals exposed to sedimentation. Proc Natl Acad Sci 109(24):E1558–E1567. https://doi.org/10.1073/pnas.1100715109

Wesseling I, Uychiaoco A, Aliño PM, Aurin T, Vermaat J (1999) Damage and recovery of four Philippine corals from short-term sediment burial. Mar Ecol Prog Ser 176:11–15. https://doi.org/10.3354/meps176011

Wong JSY, Chan YKS, Ng CSL, Tun KPP, Darling ES, Huang D (2018) Comparing patterns of taxonomic, functional and phylogenetic diversity in reef coral communities. Coral Reefs 37(3):737–750. https://doi.org/10.1007/s00338-018-1698-6

Woodley JD, Chornesky EA, Clifford PA, Jackson JBC, Kaufman LS, Knowlton N, Lang JC, Pearson MP, Porter JW, Rooney MC, Rylaarsdam KW, Tunnicliffe VJ, Wahle CM, Wulff JL, Curtis ASG, Dallmeyer MD, Jupp BP, Koehl MAR, Neigel J, Sides EM (1981) Hurricane Allen's impact on Jamaican coral reefs. Science 214(4522):749–755. https://doi.org/10.1126/science.214.4522.749

Yus J, Nixon EN, Li J, Noriega Gimenez J, Bennett MJ, Flores D, Marhaver KL, Wegley Kelly L, Espinosa-Marzal RM, Wagoner Johnson AJ (2024) Composite substrates for coral larval settlement and reef restoration based on natural hydraulic lime and inorganic strontium and magnesium compounds. Ecol Eng 202:107236. https://doi.org/10.1016/j.ecoleng.2024.107236

Diversity and Assemblages of Symbiodiniaceae Dinoflagellates in Singapore's Coral Reefs: A Case Study on Giant Clams (*Tridacna squamosa* and *T. crocea*)

6

H. R. Samuel Lee, Teresa S. Tay, Yuchen Sun,
Maxine A. D. Mowe, Jing Hui Ong,
Benjamin J. Wainwright, Li Keat Lee, and Mei Lin Neo

Abstract

Symbiotic relationships involving photosynthetic dinoflagellates (Symbiodiniaceae) are integral to the survival and adaptability of reefs globally. Using the ITS2 marker and SymPortal framework, we characterised the endosymbiont communities of two native giant clams, *Tridacna crocea* and *T. squamosa*, across four intertidal sites in Singapore. Environmental parameters (light, temperature, sedimentation and hydrodynamic activity) were also measured to identify associations with endosymbiont assemblages. Additionally, we contextualised these giant clam-based compositions against the endosymbiotic communities observed in other local reef species including hard corals and a nudibranch. Results revealed that giant clams predominantly hosted Symbiodiniaceae of genera *Durusdinium* and *Cladocopium*, with *Symbiodinium* in the background. Between species, *T. crocea* exhibited greater flexibility in its Symbiodiniaceae community. At sites showing marked differences in light intensity and hydrodynamic activity, distinct endosymbiont diversity levels were also observed. Comparisons among reef species similarly suggested a predominance of *Durusdinium* and *Cladocopium*, but the diversity of *in hospite* Symbiodiniaceae communities varied. This study is the first documentation of Symbiodiniaceae diversity in Singapore's giant clams, and includes a reef-wide perspective of the Symbiodiniaceae assemblages in Singapore's reefs. The findings enhance our understanding of the role of Symbiodiniaceae in coral reef resilience.

Keywords

Endosymbiosis · Symbiodiniaceae · Tridacninae · Next-generation sequencing · SymPortal · Zooxanthellae

H. R. S. Lee (✉) · T. S. Tay · Y. Sun · M. L. Neo
Tropical Marine Science Institute, National University of Singapore, Singapore, Singapore
e-mail: slhr@nus.edu.sg; samuel.lhr@u.nus.edu; tmsnml@nus.edu.sg

M. A. D. Mowe
Department of Biological Sciences, National University of Singapore, Singapore, Singapore

J. H. Ong
Tropical Marine Science Institute, National University of Singapore, Singapore, Singapore

Department of Biological Sciences, National University of Singapore, Singapore, Singapore

B. J. Wainwright
Department of Biological Sciences, National University of Singapore, Singapore, Singapore

Yale-NUS College, National University of Singapore, Singapore, Singapore

L. K. Lee
Tropical Marine Science Institute, National University of Singapore, Singapore, Singapore

Institute of Ocean and Earth Sciences, University of Malaya, Kuala Lumpur, Malaysia

6.1 Introduction

6.1.1 Symbiodiniaceae—An Overview

Symbiodiniaceae is a family of dinoflagellates that form important symbiotic relationships in coral reef ecosystems (LaJeunesse et al. 2018; Nitschke et al. 2022). These unicellular photosynthetic microalgae are commonly found in highly evolved mutualistic relationships with a diversity of host species from various taxa such as Cnidaria, Porifera,

L. M. Chou, D. Huang (eds.), *Coral Reefs of Singapore's Urbanised Sea*, Coral Reefs of the World 23,
https://doi.org/10.1007/978-981-95-0903-4_6

Mollusca, Foraminifera and Ciliophora (LaJeunesse et al. 2018). While some host organisms may maintain a single type of Symbiodiniaceae (Baums et al. 2014), other organisms such as corals and giant clams can host multiple species of endosymbionts in a single individual (Saad et al. 2022; Ng et al. 2024). The presence of multiple endosymbiont species could benefit host organisms by enhancing their fitness and survivorship (Baker 2003; Ziegler et al. 2018; Qin et al. 2019), and therefore they exhibit greater tolerance to environmental changes (Finney et al. 2010). For most partnerships, the algal symbionts provide their hosts with up to 100% of their energy requirements in the form of photosynthates such as glycerol, peptides and carbohydrates (Venn et al. 2008; Bleuel et al. 2024). The symbiotic partnership also supports the host's physiological processes such as its respiratory demands, calcification for shelled organisms and UV protection (Chew et al. 2019; Dellaert et al. 2022). In exchange, the microalgae gain inorganic nutrients and respiratory CO_2 from the host taxa and an ideal habitat protected from grazers (Stat et al. 2006; Neo et al. 2015). At reef scales, these synergistic host-Symbiodiniaceae pairings could serve as key drivers of primary productivity in coastal and oceanic waters (Neo et al. 2015), as well as other ecological roles including habitat formation, carbon sequestration and nutrient recycling (Neo et al. 2015; Bleuel et al. 2024).

Identification of Symbiodiniaceae has conventionally relied on molecular techniques as the main method (Sampayo et al. 2009) alongside some phenotypic traits (e.g. cell size and photosynthetic pigment content) to support characterisation (Ziegler et al. 2015; Suggett et al. 2017). The earlier molecular taxonomic research for Symbiodiniaceae had used the highly conserved small ribosomal subunit (SSU) (Rowan and Powers 1991; Carlos et al. 1999) and the variable regions of the large ribosomal subunit (LSU) (Lenaers et al. 1991; Wilcox 1998) for species delineation. These works have collectively highlighted nine major *Symbiodinium* clades: A, B, C (Rowan and Powers 1991), D (Carlos et al. 1999), E (Lajeunesse and Trench 2000; LaJeunesse 2001), F (LaJeunesse 2001), G (Pochon et al. 2001), H (Pochon et al. 2004) and I (Pochon and Gates 2010). Each major clade could further be resolved into diverse genetic subclades using the internal transcribed spacer (ITS) regions as these regions are multicopy and rapidly evolving (LaJeunesse 2001). The latter can help to distinguish Symbiodiniaceae 'types' through examining the fine-scale variations in the intragenomic copies (LaJeunesse 2001). The most recent systematic revision of Symbiodiniaceae proposed that the microalgal taxon be partitioned into multiple genera to better reflect their ecological distinctiveness and long evolutionary histories (LaJeunesse et al. 2018). An extension to the ongoing taxonomic work on Symbiodiniaceae is SymPortal, which is an analytical framework increasingly used to optimise the analysis of Symbiodiniaceae ITS2 data derived through next-generation sequencing (NGS) (Hume et al. 2019). The accuracy of SymPortal database increases with usage by the wider community that provides NGS data of ITS2 rDNA, as the type profiles are continually validated with increased sampling effort (Hume et al. 2019; Soon et al. 2023; Mies et al. 2025), and therefore bypassing the need to use multiple molecular markers to confirm identification. The increased usage of SymPortal also makes future sampling efforts more comparable and maximises efficacy in the classification of global Symbiodiniaceae diversity.

The increased taxonomic resolution of Symbiodiniaceae has advanced knowledge on their diversity, distribution, life-history and ecology, which could reveal their unique physiological attributes and explain their interactions with host species (LaJeunesse et al. 2018; Nitschke et al. 2022). For instance, members of Symbiodiniaceae may exhibit different lifestyles such as being free-living (non-symbiotic), pelagic or benthic in their host taxa, which influence the survival of both host and algal symbionts under fluctuating environmental conditions (Bhattacharya et al. 2024). Given their wide-ranging occurrence across the temperate and tropical regions (LaJeunesse et al. 2018; Nitschke et al. 2022), studies also found that the distinct local environmental conditions could strongly influence the ecological needs of Symbiodiniaceae members (Mies 2019). Broadly, the evolutionary primitive genus *Symbiodinium* (formerly Clade A) is recognised as an opportunist or generalist with a broad global distribution across the Indo-Pacific (LaJeunesse et al. 2018; Mies 2019). They are also mostly adapted to living in shallow-water environments, with high or variable light conditions (Venn et al. 2008). Members of the diverse genus *Cladocopium* (formerly Clade C), on the other hand, are often found in association with a broad range of host taxa such as cnidarians, molluscs and sponges (LaJeunesse et al. 2018). They are physiologically diverse and capable of adapting to a wide range of temperature and irradiance across habitats spanning the intertidal to mesophotic zones (Hume et al. 2015; LaJeunesse et al. 2018). Also known as extremophiles, members of the genus *Durusdinium* (formerly Clade D) are often identified in various host taxa centred in the Indo-West Pacific region (Stat and Gates 2011). They are usually found in symbiosis with hosts living in sub-optimal or stressful conditions including large diurnal or seasonal shifts in temperature and/or broad fluctuations in water turbidity (Tanzil et al. 2016). With the increasing vulnerability of coral reefs to global warming (Thomas et al. 2019; Palacio-Castro et al. 2021), there is a greater imperative to understand Symbiodiniaceae distribution and diversity with respect to their host taxa (LaJeunesse 2002; LaJeunesse et al. 2010).

6.1.2 Symbiodiniaceae Diversity in Giant Clams (Bivalvia: Tridacninae)

Among bivalves, photosymbiotic association was first documented in giant clams (or tridacnines) (Yonge 1936). Many known photosymbiotic bivalves are in the family Cardiidae, where some members of the subfamilies Fraginae (genera *Corculum*, *Fragum* and *Lunulicardia*), Clinocardiinae (genus *Clinocardium*) and all members of subfamily Tridacninae (genera *Hippopus* and *Tridacna*) engage in photosymbiotic associations with Symbiodiniaceae dinoflagellates (Kirkendale and Paulay 2017; Mies 2019). The tridacnines also represent the largest extant marine bivalves distributed across the subtropical and tropical reefs in the Indo-Pacific region (Neo et al. 2017). They serve various ecological roles in the coral reef ecosystem, which include providing suitable habitats and nutrients for algal symbionts (Neo et al. 2015). Giant clams obtain their algal symbionts from the environment independently via horizontal transmission (Mies et al. 2012). They host and store great numbers of endosymbionts in their hypertrophied outer mantle tissues, while the algal symbionts provide the bulk of the host's carbon and nitrogen requirements for growth and metabolism (Ip and Chew 2021). Together with filter-feeding, the combination of phototrophy and heterotrophy is thought to explain why giant clams have rapid growth rates resulting in their large body sizes and longevity in oligotrophic waters (Mies et al. 2017).

Photosymbiosis is considered obligatory for tridacnines' survival; hence, the initiation of symbiosis is considered a pivotal stage in their early life history (Mies et al. 2012). As early as 2 days old, giant clam veligers begin to filter-feed from the water column and acquire algal symbionts, but most of the microalgae absorbed would be digested for nutrition. The assimilation of endosymbionts usually takes place after 6–8 days post-spawning, where a proportion of them translocate to a separate extension, known as the zooxanthellal tubular network, within the host's gut system (Norton et al. 1992). Over time, the pioneer Symbiodiniaceae cells become fully established and functional within this tubular network that extends throughout the mantle tissue of the metamorphosed juveniles. Early studies revealed that the *in hospite* Symbiodiniaceae communities in giant clams are highly diverse (Carlos et al. 2000; Weber 2009). Specifically, tridacnines are mostly associated with Symbiodiniaceae from the genera *Symbiodinium*, *Cladocopium* and *Durusdinium* (DeBoer et al. 2012; Lim et al. 2019; Rossbach et al. 2021). Furthermore, a single giant clam individual can host multiple endosymbiont taxa simultaneously (DeBoer et al. 2012; Mies 2019). Recent studies also found that the same species of giant clams from different geographic locations may host different endosymbiont diversity and communities. For example, the Red Sea populations of *T. maxima* and *T. squamosa* hosted only *Symbiodinium* algal species (Pappas et al. 2017), while other studies observed multiple species of algal symbionts belonging to *Symbiodinium*, *Cladocopium* and *Durusdinium* across various giant clam species in other regions. In particular, at least 11 different Symbiodiniaceae species were observed in the natural populations of *T. crocea*, *T. maxima* and *T. squamosa* from Raja Ampat (DeBoer et al. 2012), at least six species were reported in the cultured *T. crocea* and *T. squamosa* in Ishigaki, Okinawa, Japan (Ikeda et al. 2017), and at least 11 and 13 species were reported in *T. maxima* and *T. noae*, respectively, from Dongsha Atoll (Lim et al. 2019). The compositions of endosymbionts were also highly varied among tridacnine species, suggesting that these observed patterns could be driven by hosts' biology and ecology. For instance, giant clams that co-occur in similar habitats had significantly different endosymbiont community structures and this variation may be explained by their depth occurrences (Ikeda et al. 2017; Lim et al. 2019). Conversely, certain Symbiodiniaceae taxa such as the *Symbiodinium* A1 dominant ITS2 type profiles could influence the phenotypes of host organisms such as their growth rate and tolerance towards climate change-associated stressors (Silverstein et al. 2017; Long et al. 2021; Rossbach et al. 2021). These findings point to the strongly dynamic nature of tridacnine–Symbiodiniaceae relationships across environmental gradients, which may serve as predictors of the resilience of giant clam populations towards unfavourable environmental conditions (Thomas et al. 2019; Palacio-Castro et al. 2021).

6.1.3 Distribution and Diversity of Symbiodiniaceae in Singapore's Coral Reefs

The coral reefs in Singapore can be described as marginal reef systems (Browne and Bauman 2023). These reefs typically thrive under sub-optimal environmental conditions such as low light, chronic sediment inputs and elevated temperatures (Ng et al. 2020; Fong and Todd 2021). Such an environment could provide opportunities to examine hypotheses on the resilience of biodiversity and ecosystem functions of coral reefs in the face of future climate change (Smith et al. 2020; Wall et al. 2020). For example, identifying the presence of stress-tolerant Symbiodiniaceae genera could be associated with the host's enhanced tolerance towards stressors (Mote et al. 2021). In Singapore, following a global bleaching event in 2016, Jain et al. (2020) examined the relative abundances of Symbiodiniaceae in post-bleached *Pachyseris speciosa* colonies. There was an increase in the relative abundance of *Durusdinium* (c.f. *Cladocopium*), where the algal genus is known to occur in hosts living in stressful conditions including thermal stress events (Stat and Gates 2011). Furthermore, another study from Quek et al.

(2023) observed the shuffling of algal symbionts in a *P. speciosa* colony, where the ITS2 type profile transitioned from *Durusdinium* to *Cladocopium* and subsequently back to *Durusdinium* across three time points (i.e. May 2019 to March 2020 and June 2020) coinciding with the development of a minor reef bleaching event. Therefore, the unique reef environment and reef organisms in Singapore could offer novel and useful insights about the capacity of *in hospite* Symbiodiniaceae thriving in such niche environments.

To date, the studies regarding *in hospite* Symbiodiniaceae communities in Singapore have used the ITS2-rDNA sequences to characterise the diversity across various reef taxa. Specifically, these studies have mainly focused on cnidarian species such as zoanthids (Reimer and Todd 2009) and hard corals (Guest et al. 2016; Leveque et al. 2019; Jain et al. 2020, 2021; Smith et al. 2020; Tan et al. 2020; Ong et al. 2022; Quek et al. 2023). In the first local study by Reimer and Todd (2009), they found only members of *Cladocopium* and *Durusdinium* in the zoanthids of genera *Zoanthus* and *Palythoa*. In contrast, the studies on hard corals recovered three Symbiodiniaceae genera, where *Cladocopium* and *Durusdinium* were dominantly present, with much lower abundance of *Symbiodinium*. Only a single study so far has been published for a non-cnidarian taxon, where the common reef nudibranch, *Pteraeolidia semperi*, hosted four genera of Symbiodiniaceae (Soon et al. 2023). Notably, a previously undetected genus *Gerakladium* was identified from a single sample, along with the other common genera: *Cladocopium, Durusdinium* and *Symbiodinium*. Regardless of the host taxa examined, the major Symbiodiniaceae genus observed was either *Cladocopium* or *Durusdinium*, and these genera are typically known for their stress tolerance (Hume et al. 2015; Silverstein et al. 2017; LaJeunesse et al. 2018). Depending on the host taxa, some of these studies identified possible drivers of *in hospite* endosymbiont communities such as fluctuations in seawater temperatures (Jain et al. 2020, 2021; Tan et al. 2020) and high sedimentation loads (Smith et al. 2020; Jain et al. 2021) or host species' ecology (Quek et al. 2023; Soon et al. 2023).

Giant clams are conspicuous photosymbiotic animals on Singapore's coral reefs (Neo and Todd 2013; Neo 2024), but there have been no studies examining their endosymbiont diversity. These symbiotic bivalves are scarce on the reefs because of historical overexploitation (Harrison and Tham 1973; Courtois de Vicose and Chou 1999), followed by habitat loss and degradation due to intensive coastal development in the late 1960s and after (Neo and Todd 2012). The remaining populations of giant clams are at great risk of extirpation because of their extremely low abundance, scattered distribution and poor recruitment (Neo 2024). This current scenario suggests that active intervention is needed to revive and restore giant clam populations in Singapore, as proposed by other studies (Guest et al. 2008; Ow Yong et al. 2022). A knowledge gap that may hamper the management of giant clams in Singapore concerns their resilience while living under sub-optimal conditions. While their current numbers are low, the opportunity to study their resilience is valuable and feasible since the approach of the study is non-lethal. Given the significance of Symbiodiniaceae for these marine bivalves, characterising their endosymbiont communities could offer novel insights into the biological and ecological drivers for symbiosis, as well as a prelude to the resilience of Singapore's giant clam populations for informing future conservation work.

6.1.4 Objectives of the Study

This study aims to examine the patterns of Symbiodiniaceae diversity and communities hosted by the two extant species of giant clams, *T. squamosa* and *T. crocea*, distributed across four intertidal reef sites at the southern islands of Singapore. Here, the distribution of algal symbionts in host taxa across various geographic sites, temperature, light, hydrodynamic activity, sediment load and host species is unknown. Four reef sites were chosen as prior surveys noted the abundant presence of giant clams (i.e. $n \geq 3$ individuals per species) (Neo and Todd 2012). This study will also compare and discuss the *in hospite* Symbiodiniaceae communities between giant clams and the other previously published reef taxa to better understand ITS2 type profile diversity in Singapore's reef environment. Overall, findings from this study will add to our knowledge of Symbiodiniaceae diversity among taxa living in Singapore's coral reefs, as well as the dynamics of the tridacnine–Symbiodiniaceae associations that could drive the persistence of giant clams in marginal reef systems.

6.2 Materials and Methods

6.2.1 Study Sites

Four sites were selected in this study (Fig. 6.1): Pulau Jong (PJ) (1°12′19.7″N, 103°47′12.9″E), Pulau Semakau east (PSE) (1°12′37.6″N, 103°45′54.9″E), Pulau Semakau west (PSW) (1°12′44.8″N, 103°45′27.3″E) and Terumbu Raya (TR) (1°12′52.9″N, 103°45′11.6″E). Broadly, all sites are characterised by large intertidal reef areas that become exposed during low tide periods, and the confirmed presence of wild giant clams (Neo 2024). Furthermore, these sites are readily accessible on foot during low tide windows (≤0.2 chart datum), which made fieldwork possible for non-divers.

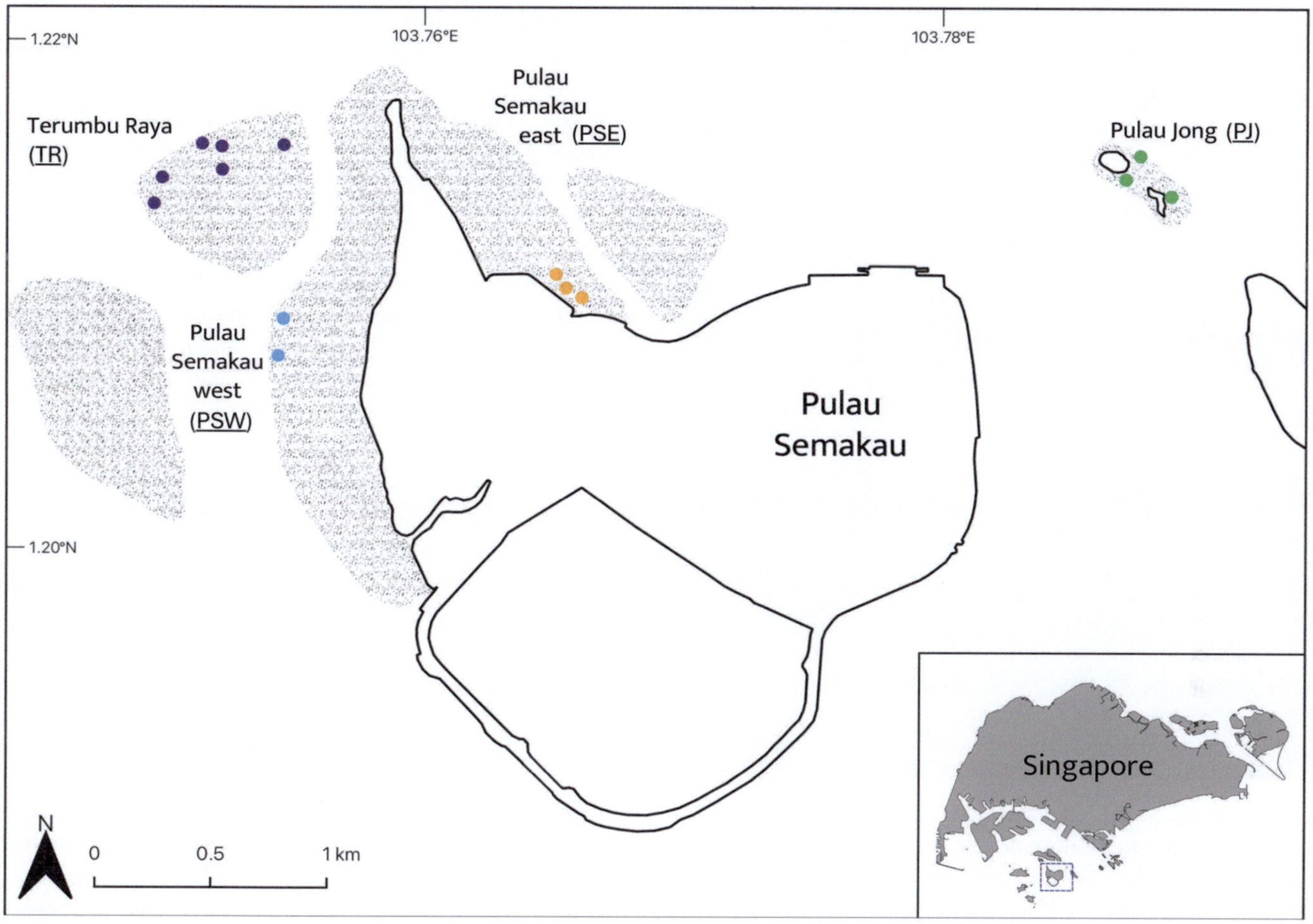

Fig. 6.1 Overview of four study sites (colour-coded) in Singapore (inset): Pulau Jong (PJ), Pulau Semakau east (PSE), Pulau Semakau west (PSW) and Terumbu Raya (TR). Each circle represents the location of a monitoring station, where abiotic parameters (temperature, light, hydrodynamic activity and sedimentation) were measured. The number of monitoring stations per site was proportional to the number of giant clams present at each site. Shaded regions represent the surveyed intertidal reef areas

6.2.2 Measuring Abiotic Parameters

Four abiotic parameters were monitored at the respective study sites: temperature, light (photosynthetically active radiation, PAR), hydrodynamic activity and sediment load. These parameters were chosen as proxies to further assess their influence on the *in hospite* Symbiodiniaceae diversity within native giant clams. Monitoring stations were set up and loggers were deployed over 3 months (July to September 2023), which coincided with the southwest monsoon season of Singapore. The 3 months studied corresponded to the final 3 months of low tides ($\leq$0.2 chart datum) in the annual cycle, which allowed for intertidal fieldwork. The 13 monitoring stations were selected based on their proximity to the native giant clams (i.e. within a radius of $\leq$50 m). The number of monitoring stations is reflected by the number of circles per site in Fig. 6.1. Each monitoring station consisted of one HOBO logger to measure temperature and light intensity, three sediment traps to measure sediment load and two clod cards to measure hydrodynamic activity (Fig. 6.2).

To measure the temperature and light intensity, HOBO loggers were randomly deployed across the monitoring stations at all study sites. Due to insufficient numbers, two models of loggers were deployed: eight units of HOBO UA-002-64 and five units of HOBO MX2202; however, both models offer comparable measurements and wavelength ranges. Before deployment, loggers were normalised against a reference light sensor logger (LI-COR, LI-1500), where a correction equation was obtained and applied to all the recorded photosynthetically active radiation (PAR) measurements. Loggers were set to capture information in intervals of 15 min. Each month, the logger's sensors were manually cleaned to 'restart' measurements, as well as to clean off the fouling algae. For analyses, only data from the first 48 h post-deployment period were used for comparisons, as data after this window were deemed inaccurate due to rapid fouling. The readings of temperature and light encompassed both

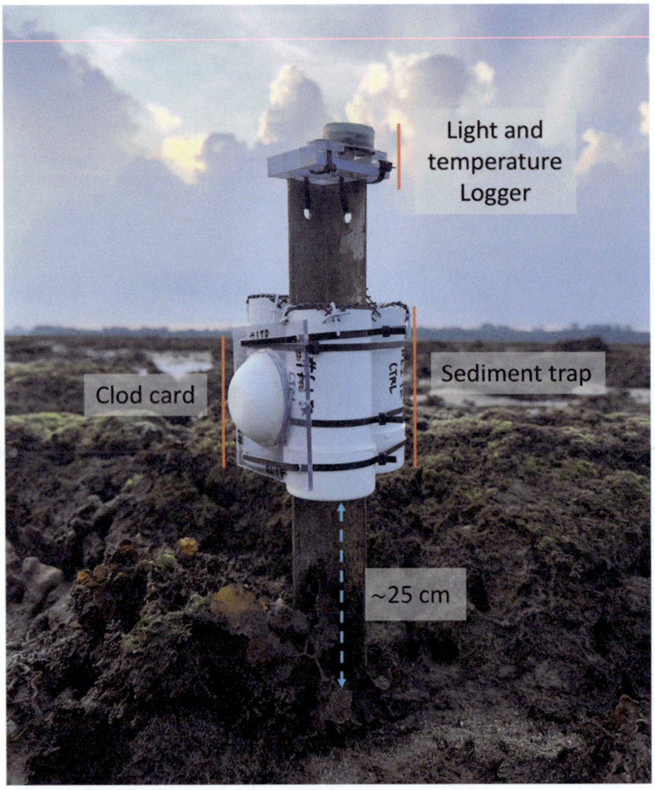

Fig. 6.2 Monitoring stations deployed across all the four study sites. Each station consisted of the following equipment—one HOBO light and temperature logger, two clod cards and three sediment traps. Light and temperature loggers were deployed on-site throughout the entire study period, while clod cards and sediment traps were deployed (ca. 25 cm above the seabed) for 48 h periods twice monthly across all 3 months

the high tide immersion and low tide immersion as the period of recording coincided with low tides. Only the daytime light data between 0700 and 1900 h were accounted for in this analysis, as this period coincides with the maximum time for Symbiodiniaceae photosynthesis. Due to a high logger failure in July, only temperature and light data recorded for August and September were used in subsequent analyses.

To measure the hydrodynamic activity, fresh batches of clod cards (a 1.5:1 ratio of plaster of Paris and water) were produced for each sampling month. The mixture was cast and dried in rubber molds at 60 °C for 48 h, and subsequently filed down to produce hemispheres with a collective mean dry weight of 94.80 g (±2.24 g). Each hemisphere was then adhered to a 15 × 15 cm polyvinylchloride (PVC) plate with silicon and left to cure further for 48 h prior to recording of the dry weight of the entire unit (i.e. a hemisphere attached to a single PVC plate). At each monitoring station, a pair of clod cards was fixed onto iron stakes with their orientation (i.e. seaward or landward) randomised, at ca. 25 cm above seabed level. Clod cards were deployed and retrieved after 48 h, oven dried at 60 °C for 48 h, with the entire unit being subsequently weighed. To validate the consistencies across batches of clod cards, calibration was carried out with three clod cards in still seawater conditions (i.e. 40 L for 48 h), dried and weighed to derive a calibration value specific for the batch (modified from Doty 1971). These calibration values were subsequently used for the computation of the diffusion index (DI) of each deployed clod card, following the equations from Doty (1971).

To measure the sediment load, modified sediment traps were produced by adapting methodologies from English et al. (1997): PVC pipes (diameter × length; 5 × 12 cm) with sealed bottom ends and a top opening lined with plastic mesh (mesh size: 0.5 mm). At each monitoring station, three sediment traps were fixed onto iron stakes at ca. 25 cm above the seabed level. Like the clod card deployment, sediment traps were deployed and retrieved after 48 h. The trapped particulate matter was rinsed out with Milli-Q water and suction-filtered through 0.45 μm membrane filters, before being oven dried at 60 °C for 48 h and thereafter weighed.

6.2.3 Survey and Sampling of Native Giant Clams

Between May and September 2023, field surveys were conducted to locate giant clams across all study sites. Surveys were carried out with similar units of effort (i.e. number of observers, survey duration and frequency). During surveys, observers conducted random walks with no fixed transects in search of giant clams, specifically the extant species: *T. squamosa* and *T. crocea* (Neo and Todd 2012, 2013; Neo 2024). Giant clams observed were also marked with a GPS and photographed with a scale. For molecular work purposes, tissue biopsies of approximately 1 cm^2 were excised from all amenable individuals (i.e. sufficiently exposed mantle tissues) with ethanol (EtOH) sterilised equipment. Tissues were immediately fixed in 99% EtOH and stored at room temperature before further analyses. Details of sampled individuals in this study are available from Figshare (https://doi.org/10.6084/m9.figshare.c.7686395).

6.2.4 DNA Extraction, Amplification and Sequencing

The extraction of total genomic DNA was performed with a subsample of mantle tissue (ca. 2 mm^2) from each clam using the QIAGEN® DNeasy® Blood and Tissue Kit (Qiagen, Singapore), following the manufacturer's protocol. Derived DNA concentrations and quality were verified via gel electrophoresis (2% agarose) and NanoDrop™ spectrophotometer (Thermo Fisher Scientific, Wilmington, DE, USA). All DNA elutes were later stored in −20 °C prior to PCR amplification and sequencing. For assessing Symbiodiniaceae species diversity, the ITS2 region of the nuclear ribosomal DNA (rDNA) was amplified through PCR with the following primers: 'SYM_VAR_5.8S2′ (5′–GAA TTG CAG AAC TCC GTG AAC C–3′ (forward) and 'SYM_VAR_REV' (5′–CGG GTT CWC TTG TYT GAC TTC ATG C–3′ (reverse) (Hume et al. 2018). Unique 8-base pair (bp) barcodes were generated through Barcode Generator 2.8 (Comai and Howell 2012) and conjugated to the 5′ end of each primer pair for every clam individual.

PCR amplifications were carried out in 25 μL reactions consisting of 12.5 μL of GoTaq master mix (DNA polymerase, dNTP mix, $MgCl_2$ and PCR buffer; Promega Corporation, Singapore), 1 μL each of 10 μM forward and reverse primers, 2 μL of template DNA and 8.5 μL of nuclease-free water. The adopted thermocycling profile, modified from Jain et al. (2020), was as follows: (1) an initial denaturing step at 95 °C for 2 min, (2) 30 cycles of denaturing at 95 °C, annealing at 55 °C and elongation at 72 °C, with each step lasting 30 s and (3) a final extension at 72 °C for 5 min. Negative controls were also prepared for each set of PCRs to confirm the absence of contamination. The presence of PCR amplicons was first verified using gel electrophoresis (2% agarose), followed by purification with the QIAquick PCR Purification Kit (Qiagen, Singapore). The purified products were quantified using the Life Technologies Qubit 3.0 Fluorometer (Thermo Fisher Scientific, Waltham, MA, USA), normalised to ca. 12.6 ng μL^{-1}, and pooled to form a single library of 25 μL volume. The pooled library was sequenced with the Illumina MiSeq sequencing platform for 300-bp paired-end reads (MiSeq Reagent Kit v.3) at Macrogen (Singapore).

6.2.5 Comparing Symbiodiniaceae Diversity Across Reef Taxa in Singapore

The ITS2 sequences for giant clams in the paired demultiplexed fastq.gz files were analysed locally with SymPortal (Hume et al. 2019). Within the framework, the following steps were also taken to minimise non-Symbiodiniaceae and artifact sequences from the dataset (Hume et al. 2019): (1) integrated sequence quality control procedures (MOTHUR 1.39.5) (Schloss et al. 2009), (2) the executables within the BLAST + suite (Camacho et al. 2009) and (3) minimum entropy decomposition (MED) (Eren et al. 2015). The unique validated sequences were pooled and classified to their genus levels accordingly, but groups containing <4 sequence reads per sample were algorithmically omitted from subsequent analyses. Derived ITS2 sequences with >200 sequence reads per sample were identified and assigned their respective defining intragenomic variant (DIV) names according to the SymPortal local database. The identified DIVs were used to characterise the ITS2 type profiles for individual clams present within the dataset. Additional comparisons of the Symbiodiniaceae diversity between giant clams and other reef taxa were carried out using four available studies. These studies primarily focused on four hard coral species, namely *Porites lutea*, *Pocillopora acuta*, *Pachyseris speciosa* and *Diploastrea heliopora* (Tan et al. 2020; Jain et al. 2021; Ong et al. 2022) and one nudibranch species, *Pteraeolidia semperi* (Soon et al. 2023). The original demultiplexed fastq.gz sequencing files were pooled with the giant clam dataset and analysed using SymPortal. The approach for analysing this pooled dataset followed the criteria and methods outlined above. The ITS2 raw sequences for giant clams were deposited at the National Center for Biotechnology Information (NCBI) under the BioProject ID: PRJNA1226285.

6.2.6 Statistical Analyses

As the monthly abiotic data did not fulfil the assumptions of normality (via Shapiro-Wilk test) and homogeneity of variance (via Levene's test) (*car* package; Fox and Weisberg et al. 2019), non-parametric tests were used for subsequent analyses. To compare the variations in the four abiotic parameters characterised across the four sites, we performed mixed-effect models, specifically, linear mixed-effects models (*nlme* package; Pinheiro et al. 2025) for light and temperature data, and generalised linear mixed-effects models (*lme4* package; Bates et al. 2015) for diffusion index and sediment mass values. In all models, 'Site' was fitted as a fixed variable and the 'Month' was nested within 'Site' as a random variable to account for repeated measurements over time. Model fit was evaluated using the Akaike Information Criterion (AIC). *Post-hoc* analyses were conducted using estimated marginal means (*emmeans* package; Lenth 2024) with Tukey's adjusted *p*-values.

Across all taxa, the relative proportions of endosymbiont genera and the relative abundances of the predicted ITS2 type profiles present in individuals were represented in bar plots using values in the ITS2 type profile absolute count table generated by SymPortal. For the giant clams, the Symbiodiniaceae diversity was evaluated by using the absolute abundance of the ITS2 type profiles to compute the Inverse Simpson Diversity Index and Pielou's Evenness Index. As the data did not satisfy the assumptions of normality and homogeneity of variance, the non-parametric Wilcoxon signed-rank and Kruskal-Wallis tests were used to analyse differences in Symbiodiniaceae diversity for *T. crocea* and *T. squamosa*, respectively. All statistical analyses were conducted in R v.4.2.1 (RStudio Team 2015).

6.3 Results

6.3.1 Giant Clam Survey and Sampling Effort

A total of 41 giant clam individuals were observed across all four sites during surveys. Two species were identified: *Tridacna squamosa* ($n = 19$) was present in all four sites and *Tridacna crocea* ($n = 22$) only in two sites, PSE and TR. Among the study sites, TR had the greatest number of giant clams ($n = 24$) compared to the other sites: PSE ($n = 11$), PSW ($n = 3$) and PJ ($n = 3$). For the genetic characterisation of Symbiodiniaceae, mantle biopsies of 19 *T. squamosa* and 14 *T. crocea* were sampled successfully. For the latter species, its burrowing behaviour poses a significant challenge for mantle biopsy, and hence the lower rate of sampling success observed. Furthermore, the observed sample size for each host species was highly variable and this reflected the species' low relative abundances on local reefs (Neo 2024).

6.3.2 Among-Site Variations of Abiotic Parameters

The variation of the average temperatures across all sites was not significantly different ($\chi^2 = 3.86$, $p = 0.275$) (Fig. 6.3a). Minimum and maximum monthly average temperatures were 29.62 °C (±0.05 °C SE) at site TR and 30.04 °C (±0.05 °C SE) at site PJ, respectively. In contrast, light regimes differed significantly across all sites ($\chi^2 = 56.22$, $p < 0.001$), where TR and PSW recorded the highest (325.55 PAR ±20.97 PAR SE) and lowest (137.30 PAR ±7.14 PAR SE) monthly mean PAR values, respectively. *Post-hoc* tests revealed significant pair-wise differences in PAR values, where TR was most different compared to the other three sites: PJ vs. TR (estimate = −116.7, $p = 0.0001$), PSE vs. TR (estimate = −130.4, $p = 0.0002$) and PSW vs. TR (estimate = −143.8, $p < 0.0001$) (Fig. 6.3b).

A wide range of monthly mean hydrodynamic activity (DI) values was recorded, varying between 10.85 (±0.1 DI SE) at site PJ and 3.87 (±0.22 DI SE) at site PSE. These values were also significantly different between sites ($\chi^2 = 113.93$, $p < 0.001$). *Post-hoc* estimated marginal means indicated significant inter-site differences between all pairwise comparisons (PJ vs. PSE: estimate = 0.5, $z = 15.89$, $p < 0.0001$; PJ vs. PSW: estimate = 0.2, $z = 7.190$ $p < 0.0001$; PJ vs. TR: estimate = 0.2, $z = 7.92$, $p < 0.0001$; PSE vs. PSW: estimate = −0.3, $z = -7.28$, $p < 0.0001$; PSE vs. TR: estimate = −0.3, $z = -10.56$, $p < 0.0001$), except for PSW vs. TR (estimate = −0.04, $z = -1.18$, $p = 0.638$) (Fig. 6.3c). Like DI, a broad monthly average of sediment mass was observed, ranging from 1.25 g (±0.36 g SE) at site PJ to 0.039 g (±0.04 g SE) at site PSW (Fig. 6.3d). Significant inter-site differences in sedimentation were also observed ($\chi^2 = 20.95$, $p = 0.0001$) more specifically, for PJ vs. PSE (estimate = −3.6, $z = -3.17$, $p = 0.0080$) and PJ vs. TR (estimate = −3.7, $z = -4.41$, $p = 0.0001$) (Fig. 6.3d).

6.3.3 Endosymbiont Genera and ITS2 Type Profiles Across Giant Clams

Illumina MiSeq sequencing resulted in a total of 10,505,153 assembled paired-end reads from 33 samples, corresponding to ca. 318,337 reads per giant clam sample; of which, 7,970,515 post-QC sequences (75.87%) were retained for analysis. Broadly, three Symbiodiniaceae genera were present across all tridacnine samples: *Symbiodinium*, *Cladocopium* and *Durusdinium* (Fig. 6.4). SymPortal further identified 230 ITS2 sequence types, where 58 of them formed the top 90% of ITS2 sequence types in all giant clam samples. In addition, a collection of 24 DIVs was derived based on the SymPortal database. Among the samples, *Symbiodinium* was the least represented genus (i.e. represented by only one DIV) while *Cladocopium* and *Durusdinium* were relatively well represented with 11 and 12 DIVs, respectively. For both tridacnine host species, the identified DIVs were further classified into 11 predicted Symbiodiniaceae ITS2 type profiles by SymPortal (Fig. 6.5).

6.3.4 Symbiodiniaceae Diversity and Distribution in *Tridacna crocea*

Among the 14 *T. crocea*, the most abundant Symbiodiniaceae genus was *Cladocopium*, followed by *Durusdinium*, while *Symbiodinium* was the least abundant (Fig. 6.4a). In contrast, *Durusdinium* endosymbionts were most common (13 individuals), followed by *Cladocopium* (12 individuals) and *Symbiodinium* (one individual). At the site level, PSE individuals recorded both genera, where *Cladocopium* was more abundant than *Durusdinium* (except individual A9) (Fig. 6.4a). On the other hand, the *Durusdinium* endosymbionts were most common in TR individuals (i.e. eight out of nine individuals), followed by *Cladocopium* (seven individuals). Interestingly, the presence of *Symbiodinium* in *T. crocea* appears to be minor (i.e. a single individual at site TR) (Fig. 6.4a).

Of the three Symbiodiniaceae genera in *T. crocea*, SymPortal predicted the profile *D13* as the most common ITS2 type profile, which was present in half of the sampled *T. crocea* (Fig. 6.4a). In contrast, the other profiles *3845_A*, *C1*, *C91*, *C93a*, *D3* and *D13/D5-D4-D9b* occurred only once in *T. crocea* across the study sites. Notably, the latter type profiles appeared to be unique to specific sites—site SE had profiles *C93a* and *C1,* while site TR had profiles *3845_A*,

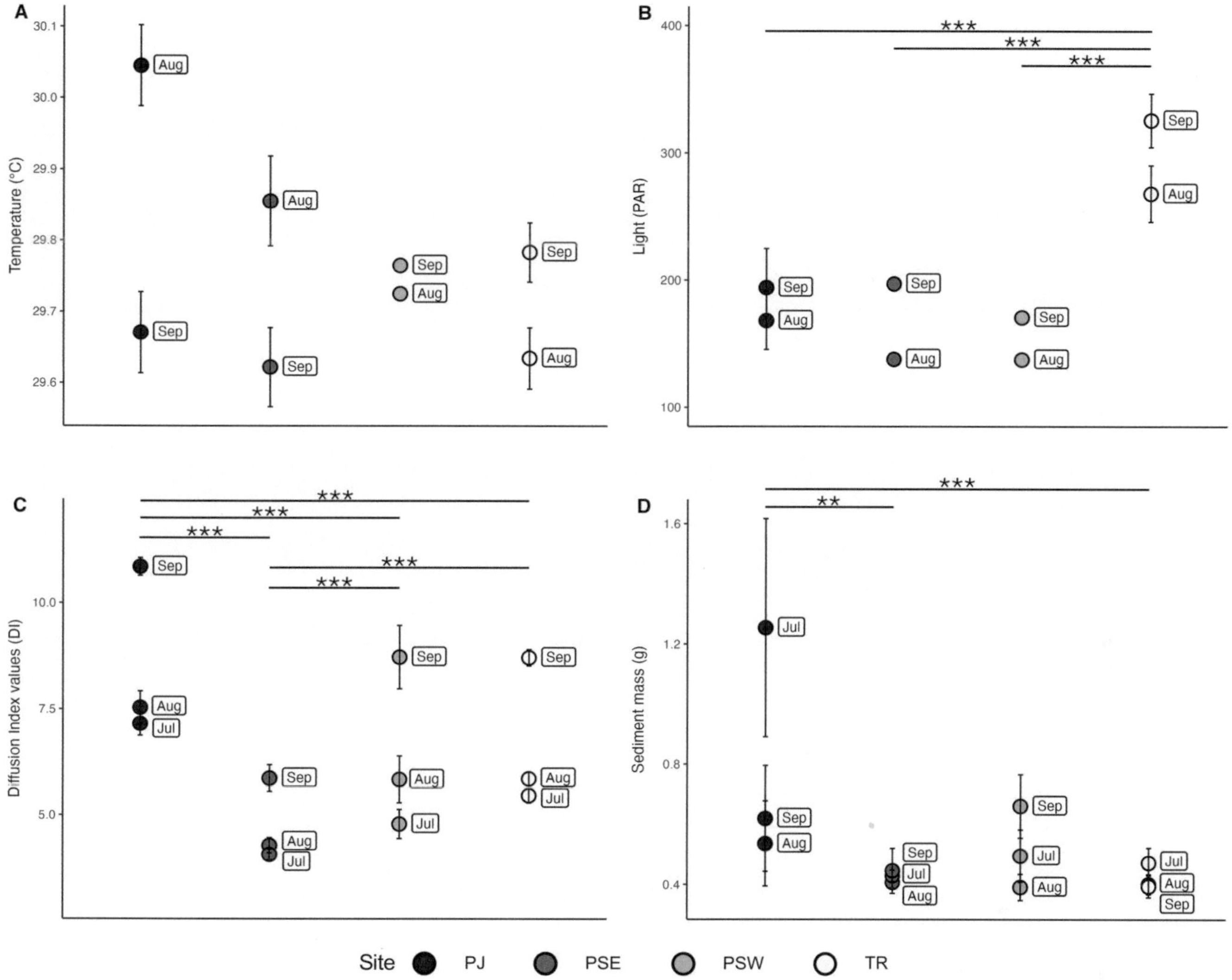

Fig. 6.3 Overview of the mean monthly values for (**a**) temperature, (**b**) light (PAR), (**c**) hydrodynamic activity (diffusion index, DI) and (**d**) sedimentation (sediment mass) across the four study sites. Each data point represents the monthly average value (i.e. alphabet labels). Only 2 months of temperature and light data are presented due to logger failure, while data for all 3 months are presented for DI and sedimentation. Error bars indicate the standard errors. Abiotic parameters that are significantly different across sites are indicated by horizontal lines (*$p < 0.05$, **$p < 0.01$, ***$p < 0.001$). Abbreviations: PJ - Pulau Jong, PSE - Pulau Semakau east, PSW - Pulau Semakau west and TR - Terumbu Raya

C91, *D13/D5-D4-D9b* and *D3*. For the Inverse Simpson index, no significant difference was found for endosymbiont type profile diversity in *T. crocea* between PSE and TR ($p = 0.052$) (Fig. 6.5a). However, TR individuals had significantly higher Pielou's Evenness index than PSE individuals ($p = 0.011$) (Fig. 6.5b).

6.3.5 Symbiodiniaceae Diversity and Distribution in *Tridacna squamosa*

Among the 19 *T. squamosa* individuals, *Durusdinium* was present and dominant in all individuals, except individual B7 (Fig. 6.4b). At the site level, PSE and PSW individuals hosted only *Durusdinium* endosymbionts, while only two individuals (one each from PJ and TR) hosted *Cladocopium* endosymbionts. *Symbiodinium* appears to be absent from *T. squamosa* samples. Of the two Symbiodiniaceae genera in *T. squamosa*, SymPortal predicted the profile *D5/D4-D9b-3143_D-2929_D* as the most common ITS2 profile, which was present in 15 out of the 19 sampled *T. squamosa* across all sites (Fig. 6.4b). In contrast, the remaining profiles *C3/C3v-3960_C-C1q-C1-C1n-3935_C*, *C66*, *D13*, *D3* and *D5* occurred only once in clams among the study sites. Notably, the latter type profiles in *T. squamosa* appeared unique to site, where site PJ had profile *C66*, site TR had profiles *C3/C3v-3960_C-C1q-C1-C1n-3935_C*, *D13* and *D13/*

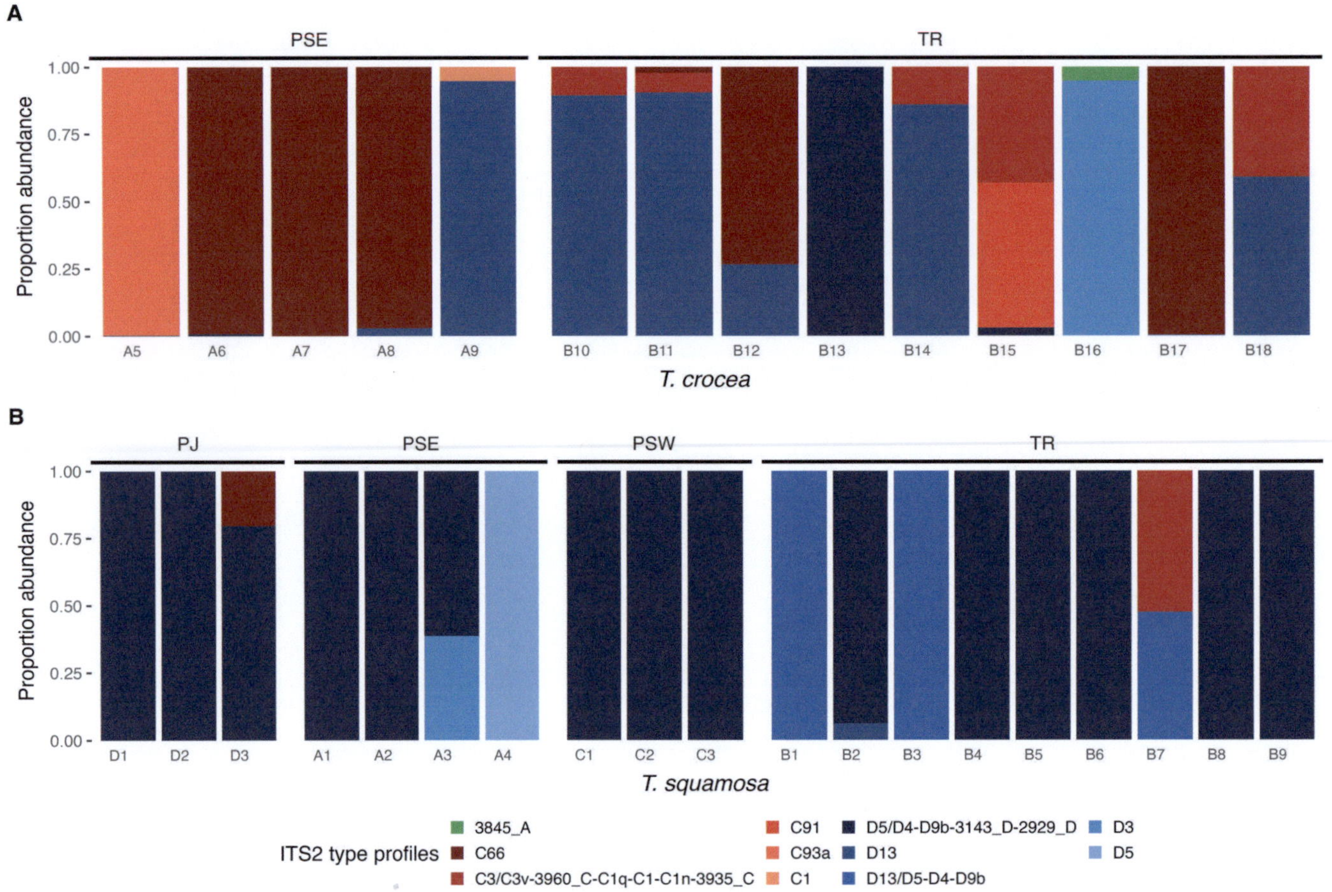

Fig. 6.4 Relative proportions of the predicted ITS2 type profiles identified in the individuals of respective host species (**a**) *Tridacna crocea* and (**b**) *Tridacna squamosa* across the study sites. Abbreviations: PJ - Pulau Jong, PSE - Pulau Semakau east, PSW - Pulau Semakau west, TR - Terumbu Raya

D5-D4-D9b and site PSE had profiles *D3* and *D5*. Overall, Inverse Simpson index (Fig. 6.5c) and Pielou's Evenness index values (Fig. 6.5d) were not significantly different across the study sites ($p > 0.05$).

6.3.6 Comparing Symbiodiniaceae Diversity and Distribution in Endosymbiotic Reef Taxa

Seven host taxa were analysed in this study (Fig. 6.6), specifically *Diploastrea heliopora* (number of individuals, $n = 21$), *Pachyseris speciosa* ($n = 137$), *Pocillopora acuta* ($n = 32$), *Porites lutea* ($n = 155$), *Pteraeolidia semperi* ($n = 49$), *T. crocea* ($n = 14$) and *T. squamosa* ($n = 19$). Of the four Symbiodiniaceae genera identified, the most prevalent genus was *Cladocopium* ($n = 348$), followed by *Durusdinium* ($n = 163$), while the least common was *Symbiodinium* ($n = 8$) and *Gerakladium* ($n = 1$). The distribution of Symbiodiniaceae differed among the seven host taxa, where *T. crocea*, *P. semperi* and *P. lutea* hosted *Symbiodinium*, *Cladocopium* and *Durusdinium*; *T. squamosa*, *D. heliopora* and *P. speciosa* hosted *Cladocopium* and *Durusdinium*; and *P. acuta* hosted only *Durusdinium*. *Gerakladium* was only found in a single individual of host *P. semperi* (Fig. 6.6b). The Symbiodiniaceae community in both *Tridacna* spp. was described above in Sects. 6.3.4 and 6.3.5 (Fig. 6.6a). For the nudibranch *P. semperi*, the majority of individuals hosted *Durusdinium* (75.5%) as the dominant Symbiodiniaceae genera (Fig. 6.6b). On the other hand, the diversity and distribution of Symbiodiniaceae varied across the coral species. Colonies of two species, *D. heliopora* (95%) and *P. acuta* (100%), primarily hosted *Durusdinium* endosymbionts (Fig. 6.6c, d), while *P. speciosa* (97.8%) and *P. lutea* (100%) primarily hosted *Cladocopium* endosymbionts (Fig. 6.6c, d). In contrast, the occurrence of *Symbiodinium* was very low among all host taxa, where it was present in a handful of individuals (i.e. a single *T. crocea*, a single *P. semperi* and six *P. lutea* colonies).

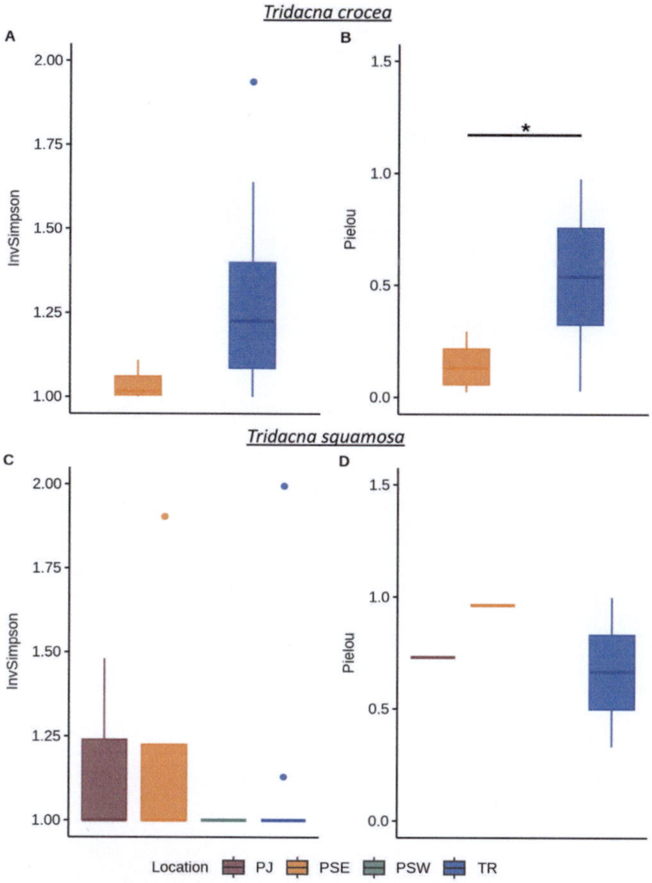

Fig. 6.5 Distribution of Inverse Simpson and Pielou's evenness index values based on the ITS2 type profiles identified for both host species, *Tridacna crocea* (**a**, **b**) and *Tridacna squamosa* (**c**, **d**), occurring across their respective sites. Pielou's evenness was not computed for *T. squamosa* in PSW because only one ITS2 type profile was identified from the clams

Among the seven host species, SymPortal identified 42 unique ITS2 type profiles (Fig. 6.6). The greatest similarities of endosymbiont diversity were between both giant clam species, sharing six ITS2 type profiles (Table 6.1). The nudibranch, *P. semperi* shared five and three ITS2 type profiles with *T. squamosa* and *P. speciosa*, respectively. In addition, the dominant ITS2 type profile was highly variable across all host species (Table 6.1). For both giant clam species, the profiles *D13* and *D5/D4-D9b-3143_D-2929_D* were found in most individuals (Fig. 6.6a), while *D4/D1/D5-D9b-3143_D-2929_D* was common among the *P. semperi* individuals (Fig. 6.6b). Among corals, the dominant ITS2 type profile for each species is as follows: 21 colonies of *D. heliopora* had *D1-D4-D4c*, 79 colonies of *P. speciosa* had *C27-C3-4142-C27a*, 19 colonies of *P. acuta* had *D1-D2d-3856-3886-3895* and 121 colonies of *P. lutea* had *C15* (Fig. 6.6c, d).

6.4 Discussion

6.4.1 Symbiodiniaceae Diversity and Composition Among Singapore's Giant Clams

The endosymbiont communities of giant clams (*T. crocea* and *T. squamosa*) in Singapore's reefs consist of endosymbionts from genera *Symbiodinium*, *Cladocopium* and *Durusdinium*. These genera have been observed in other local photosymbiotic host taxa such as zoanthids (Reimer and Todd 2009), scleractinian corals (Smith et al. 2020; Jain et al. 2021; Quek et al. 2023) and nudibranch (Soon et al. 2023). These findings also align with those of other giant clam populations in the Indo-Pacific region (Lim et al. 2019; Mies 2019; Rossbach et al. 2021).

Between both native species, the composition of Symbiodiniaceae differed, where *T. crocea* mostly harboured *Cladocopium* and *T. squamosa* primarily hosted *Durusdinium*, while *Symbiodinium* was least represented. The prevalance of both *Cladocopium* and *Durusdinium*, in *T. crocea* and *T. squamosa* has also been observed in wild populations from the central Indo-west Pacific and the Western Indian Ocean (Weber 2009; DeBoer et al. 2012). Notably, DeBoer et al. (2012) observed a greater proportion of *Cladocopium* and *Durusdinium* in giant clams inhabiting warmer waters, while those in cooler waters harboured more *Symbiodinium*. An experimental study found that host clams *T. gigas* and *T. derasa* harbouring *Cladocopium* could resist bleaching more than individuals containing *Symbiodinium* (Sison 2003). Furthermore, isolated strains of *Cladocopium* and *Durusdinium* members revealed a lower degree of deterioration in their overall photosynthetic functions and efficiency compared to *Symbiodinium* and *Breviolum* (Sison 2003; DeBoer et al. 2012; Fujise et al. 2018). Endosymbionts of *Cladocopium* and *Durusdinium* are often identified in host species living in disturbed or degraded reefs, such as the marginal reefs in Singapore (Fong and Todd 2021). Instead, the apparent absence of *Symbiodinium* in the local giant clams reflects the influence of the environment on Symbiodiniaceae diversity, as *Symbiodinium* are less suited to environments with low light conditions such as in Singapore (LaJeunesse et al. 2018). The prevalence of stress-tolerant endosymbionts in the native giant clams could therefore provide a buffer against the fluctuating reef conditions, helping them resist major disturbances such as bleaching events.

Despite their co-occurrence in the same reef environments, the differences in the composition of ITS2 type profiles between *T. crocea* and *T. squamosa* may be attributed to their individual physiological and ecological needs. Previous

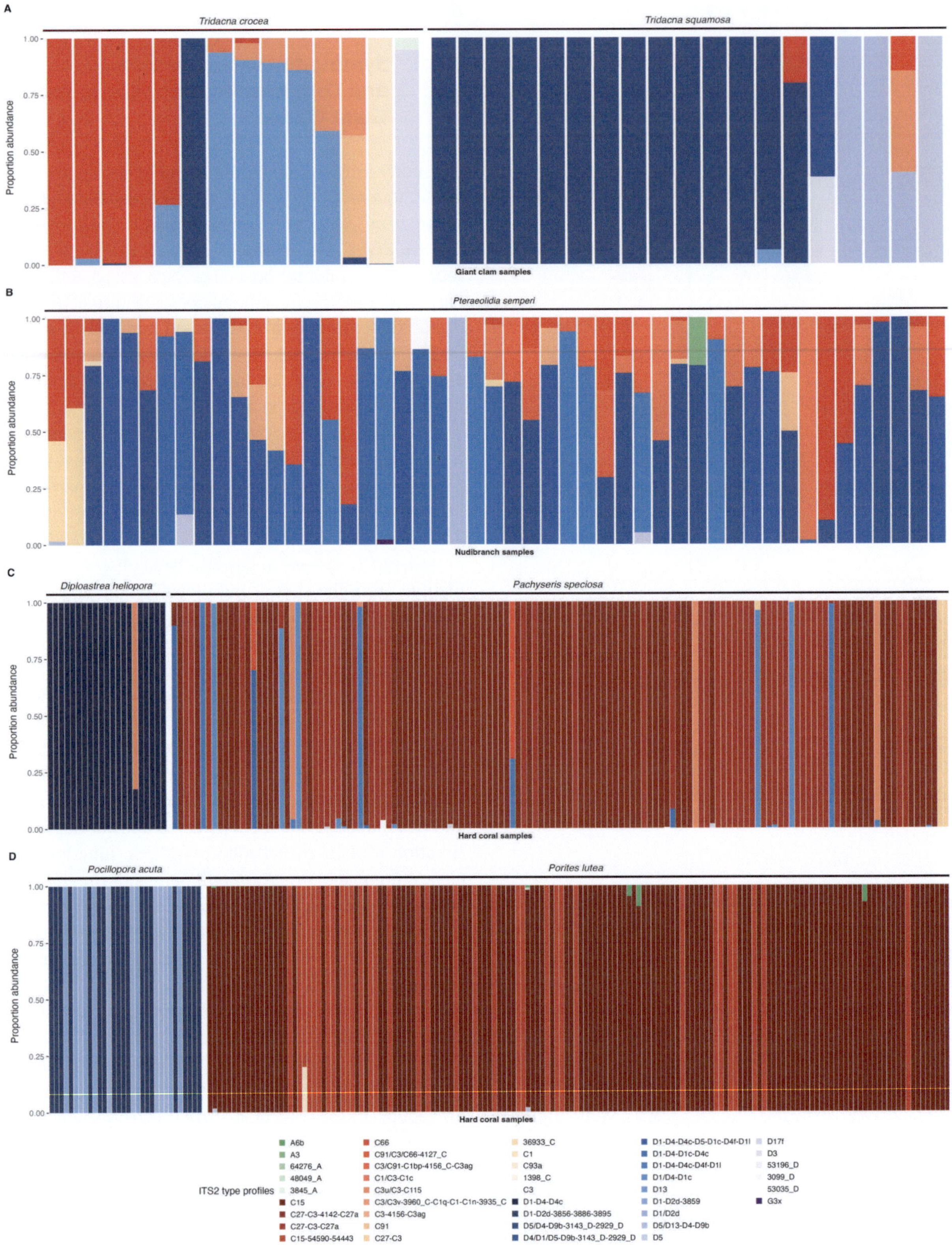

Fig. 6.6 Relative proportions of the various Symbiodiniaceae ITS2 type profiles identified through SymPortal in (**a**) giant clams, (**b**) nudibranchs and (**c**, **d**) corals (n = 427). Each column represents a single individual in each host group, with the respective host species being indicated as bars spanning across multiple columns

Table 6.1 Summary of ITS2 type profiles present in different host species

ITS2 type profile	*Tridacna crocea*	*Tridacna squamosa*	*Pteraeolidia semperi*	*Diploastrea heliopora*	*Pachyseris speciosa*	*Pocillopora acuta*	*Porites lutea*
Symbiodinium							
A6b							×
A3			×				
64276_A							×
48049_A							×
3845_A	×						
Cladocopium							
C15							×
C27-C3-4142-c27a					×		
C27-C3-C27a					×		
C15-54590-54443							×
C66	×	×	×				
C91/C3/C66-4127_C		×	×		×		
C3/C91-C1bp-4156_C-C3ag			×				
C1/C3-C1c	×		×				
C3u/C3-C115				×	×		
C3/ C3v-3960_C-C1q-C1-C1n--3935_C	×	×					
C3-4156-C3ag			×				
C91	×		×				
C27-C3					×		
36933_C			×				
C1			×				
C93A	×						×
1398_C					×		
C3			×				
Durusdinium							
D1-D4-D4c				×			
D1-D2d-3856-3886-3895						×	
D5/D4-D9b-3143_D-2929_D	×	×					
D4/D1/ D5-D9b-3143_D-2929_D		×	×				
D1-D4-D4c-D5-D1c-D4f-D11			×				
D1-D4-D1c-D4c			×		×		
D1-D4-D4c-D4f-D11			×		×		
D1/D4-D1c					×		
D13	×	×					
D1-D2d-3859						×	
D1/D2d						×	
D5/D13-D4-D9b	×	×	×				
D5		×	×				×
D17f					×		
D3	×	×					
53196_D					×		
3099_D							×
53035_D					×		
Gerakladium							
G3x			×				

studies have alluded that the host clam could select and curate its own endosymbiont composition (Ikeda et al. 2017; Lim et al. 2019; Guibert et al. 2024). For instance, Ishikura et al. (2004) found that the presence of giant clam tissue homogenate (specifically, *H. hippopus* and *T. crocea*) significantly improved the growth rates of *ex hospite* cultured *Cladocopium* and *Durusdinium* than *Symbiodinium*. This suggests that growth-related factors are present in respective tridacnine species that regulate the growth of selected Symbiodiniaceae. Although the *Cladocopium* endosymbionts were in greater relative abundance among local *T. crocea* individuals, about half of them also harboured *Durusdinium* (i.e. site Terumbu Raya). The latter confirms the influence of environmental conditions concurrently driving diversity and composition of Symbiodiniaceae between the tridacnine species. Interestingly, local *T. squamosa* exhibited an overwhelming representation of *Durusdinium* type profiles. Other studies, in Japan and Indonesia, reporting the presence of *Durusdinium* in *T. squamosa* did not observe such dominance in hosts (DeBoer et al. 2012; Ikeda et al. 2017). *Durusdinium* members are well-known to be stress-tolerant, but they are also less photosynthetically efficient (Cunning et al. 2015; Matthews et al. 2017). Therefore, the host may not benefit nutritionally by hosting *Durusdinium* in great abundance. In *T. squamosa*'s case, this may be overcome by its strongly heterotrophic lifestyle compared to other giant clam species (Guibert et al. 2024). This unique finding for *T. squamosa* suggests a tradeoff is possible whereby the host clam gains enhanced resilience conferred by *Durusdinium* against stressful conditions in exchange for lower nutritional yields (Cunning et al. 2015).

Numerous studies have confirmed the strong relationship between the *in hospite* Symbiodiniaceae community within hosts and their abiotic environmental conditions (LaJeunesse 2002; Ng et al. 2024). Among the study sites, both giant clam species in Terumbu Raya had the highest species richness in *Symbiodiniaceae* type profiles than the other three sites. While the study sites were geomorphologically similar (i.e. shallow intertidal reef flats), light levels and hydrodynamic currents were significantly greater at Terumbu Raya than at other sites, and there were no inter-site differences in temperature and sedimentation, except between Terumbu Raya and Pulau Jong for sedimentation. Previous studies have observed a greater diversity of Symbiodiniaceae in host organisms occupying environments with high light levels in the Caribbean and the Pacific (Baker and Rowan 1997; LaJeunesse 2002). These high-light environments could provide environmental heterogeneity that enables Symbiodiniaceae to better adapt to fluctuating light regimes (Mies 2019), thereby promoting a wider diversity of Symbiodiniaceae within the *in hospite* community. In addition, the increased hydrodynamic currents and exposure potentially bring more variants of Symbiodiniaceae into an area (Million et al. 2024), which increases the chance of capturing more endosymbionts across the water column and sediments. Hence, this could explain the greater diversity of Symbiodiniaceae in host clams found at Terumbu Raya, which is most exposed compared to the other study sites (Tay et al. 2012). Although the sample sizes across the study sites were uneven, the ITS2 type profiles exhibited by the native giant clams, in particular *T. crocea*, also revealed a greater degree of evenness. Future work could include additional study sites to confirm the effects of the abiotic parameters on the endosymbiont diversity of giant clams.

6.4.2 Symbiodiniaceae Diversity and Composition in Singapore's Coral Reefs

The *in hospite* Symbiodiniaceae diversity in Singapore is mainly composed of *Cladocopium* and *Durusdinium*, with a low relative abundance of *Symbiodinium* and *Gerakladium*. Specifically, *Cladocopium* was most abundant among the individuals of *P. speciosa* and *P. lutea*, while *Durusdinium* was most abundant among the individuals of *T. squamosa*, *P. semperi*, *D. heliopora* and *P. acuta*. Interestingly, both *Cladocopium* and *Durusdinium* were observed to be equally prominent among the *T. crocea* clams. The latter results aligned with a study in Papua New Guinea that also reported an even distribution of both genera among four coral species—*P. speciosa*, *P. lutea*, *P. acuta* and *D. heliopora* (Ng et al. 2024). However, in other locations, the overall representation between the same two Symbiodiniaceae genera varied. For instance, a greater representation of *Cladocopium*, compared to *Durusdinium*, was observed among various coral species in Hong Kong (Saad et al. 2022) and Malaysia (Lee et al. 2022). *Cladocopium* endosymbionts are species-rich and generalists that can adapt to a wide range of temperatures and irradiances across habitats (Ikeda et al. 2017; LaJeunesse et al. 2018). In contrast, *Durusdinium* endosymbionts are usually found in symbiosis with hosts living in stressful environments exhibiting seasonal shifts in temperature or broad fluctuations in water turbidity (Fujise et al. 2018; Jain et al. 2021). For instance, corals exposed to elevated nutrient levels or temperatures were found to harbour a greater diversity and abundance of *Durusdinium* endosymbionts (Gong et al. 2018; Draney 2024).

The non-coral taxa examined in this study, including *T. crocea*, *T. squamosa* and *P. semperi,* take up endosymbionts via horizontal transmission (Kirkendale and Paulay 2017; Soon et al. 2023). Two coral species, *D. heliopora* and *P. speciosa* (Hartmann et al. 2017), also take up endosymbionts via horizontal transmission, while *P. acuta* (Millán-Márquez et al. 2024) and *P. lutea* (LaJeunesse et al. 2004) undergo vertical transmission. The mode of Symbiodiniaceae acquisition is often proposed as a primary determinant driving the *in hospite* endosymbiont diversity. Specifically, vertical transmission usually yields host individuals with a more conserved diversity of Symbiodiniaceae, and the converse is typically the case for horizontal transmission (Baker 2003). The latter mode allows host species to associate with diverse environmental sources of Symbiodiniaceae, explaining the higher diversity of ITS2 type profiles among the horizontal transmitters: giant clams, *P. semperi* and *P. speciosa*. However, an exception is evident in *D. heliopora*, a horizontal transmitter exhibiting relatively low type profile diversity. Although not specific to this coral species, Quigley et al. (2017) identified genetic contributions to the endosymbiont diversity of coral species relying on either vertical or horizontal transmission. Such heritability of the endosymbiont community could explain the lower type profile diversity of *D. heliopora*. These findings underscore the potentially 'plastic' nature of endosymbiont acquisition, regardless of the mode of acquisition (Quigley et al. 2018).

Among the host species, *T. crocea*, *T. squamosa* and *P. semperi* shared two type profiles *C66* and *D5/D13-D4-D9b*, which were absent from all hard coral species. *C66* was previously identified in giant clams from Papua New Guinea and French Polynesia (Weber 2009; Guibert et al. 2020) and in corals occupying shallow waters (LaJeunesse et al. 2008). Terrell et al. (2023) inferred its stress-tolerant potential as there was an increased relative abundance of *C66* in corals exposed to an increase in temperature and pCO_2. The latter *D5/D13-D4-D9b Durusdinium* type profile is associated with *Durusdinium eurythalpos* (LaJeunesse et al. 2018), which was a dominant species in colonies of the coral *Oulastrea crispata*, typically found in marginal habitats (LaJeunesse et al. 2014; Saad et al. 2022). Likewise, *D. eurythalpos* was present in all three non-coral host species living in local marginal reefs but was not necessarily a dominant species. Interestingly, this type profile was absent in the local hard corals, which concurred with the previous studies (LaJeunesse et al. 2014; Saad et al. 2022). They also suggested that *D. eurythalpos* could be specific to the host coral *O. crispata* and not to other coral species. Of the hard coral species, *P. speciosa* harboured ITS2 type profiles also found in the non-coral taxa (i.e. *C91/C3/C66-4127_C*, *D1-D4-D1c-D4c*, *D1-D4-D4c-D4f-D1l*). In contrast, the other three coral species appeared to primarily host type profiles unique to corals such as *C15* and *D1/D2d* (Lee et al. 2022; Saad et al. 2022). The diversity of the endosymbiont composition may be linked to the ecology of the photosymbiotic host. For instance, *P. speciosa* has a generalist life-history strategy that is characterised by its intermediate growth rates and large colony sizes (Darling et al. 2012). Although there is little information on Symbiodiniaceae diversity for the other three coral species, with differing ecologies, a study by Ng et al. (2024) found that *P. speciosa* colonies in Papua New Guinea harboured a wide variety of ITS2 type profiles as they exhibit considerably low endosymbiont specificity. This may also explain the common ITS2 type profiles between *P. speciosa* and the non-coral taxa, as this coral species exhibits flexibility in its endosymbiont composition.

6.5 Conclusions

This study used NGS and SymPortal to characterise the Symbiodiniaceae diversity associated with *T. crocea* and *T. squamosa* living in Singapore's coral reefs. Results indicate that the endosymbiont composition and community structure differed between both giant clam species, which may be explained by their nutritional ecology. Furthermore, abiotic conditions such as light levels and hydrodynamic activity could partially shape the diversity of Symbiodiniaceae. Among reef taxa, the diversity of Symbiodiniaceae appeared to be highly variable, which may be attributed to the host's mode of symbiont transmission and their biology influencing the diversity and distribution of Symbiodiniaceae across host species. High *in hospite* Symbiodiniceae diversity is often linked to greater host resilience and adaptability (Ziegler et al. 2018; Qin et al. 2019). For instance, *P. semperi* may exhibit a greater capacity and flexibility in adapting to changing environments because of its wider diversity of ITS2 type profiles, and vice versa. Nevertheless, the Symbiodiniaceae community represents just one component of the holobiont's adaptation towards sub-optimal environments, alongside other factors including host physiology (Mies et al. 2025). Another notable finding is the prevalence of *Cladocopium* and *Durusdinium* across all reef taxa. These stress-tolerant Symbiodiniaceae genera have often been considered a key factor in the persistence and resilience of local coral reefs amidst a suite of environmental stressors (Tanzil et al. 2016; Tan et al. 2020). These findings serve as baselines to understanding the potential physiological responses of reef taxa to sub-optimal conditions characteristic of a marginal reef system, and furthermore emphasise the relevance of Symbiodiniaceae in the future of coral reefs amidst changing climates.

References

Baker AC (2003) Flexibility and specificity in coral-algal symbiosis: diversity, ecology, and biogeography of *Symbiodinium*. Annu Rev Ecol Evol Syst 34(1):661–689. https://doi.org/10.1146/annurev.ecolsys.34.011802.132417

Baker AC, Rowan R (1997) Diversity of symbiotic dinoflagellates (zooxanthellae) in scleractinian corals of the Caribbean and eastern Pacific. In: Proceedings of the 8th international Coral Reef symposium, Panama City, 24–29 June 1996, vol 2, pp 1301–1306

Bates D et al (2015) Fitting linear mixed-effects models using lme4. J Stat Softw 67:1–48. https://doi.org/10.18637/jss.v067.i01

Baums IB, Devlin-Durante MK, LaJeunesse TC (2014) New insights into the dynamics between reef corals and their associated dinoflagellate endosymbionts from population genetic studies. Mol Ecol 23(17):4203–4215. https://doi.org/10.1111/mec.12788

Bhattacharya D, Stephens TG, Chille EE et al (2024) Facultative lifestyle drives diversity of coral algal symbionts. Trends Ecol Evol 39(3):239–247. https://doi.org/10.1016/j.tree.2023.10.005

Bleuel J, Waechter L, Bender M et al (2024) Taxonomic and functional diversity of zooxanthellate corals and hydrocorals in southwestern Atlantic reefs. Front Ecol Evol 12:1322751. https://doi.org/10.3389/fevo.2024.1322751

Browne NK, Bauman AG (2023) Marginal reef systems: resilience in a rapidly changing world. Diversity 15:703. https://doi.org/10.3390/d15060703

Camacho C, Coulouris G, Avagyan V et al (2009) Blast+: architecture and applications. BMC Bioinform 10:1–9. https://doi.org/10.1186/1471-2105-10-421

Carlos AA, Baillie BK, Kawachi M, Maruyama T (1999) Phylogenetic position of *Symbiodinium* (Dinophyceae) isolates from tridacnids (Bivalvia), cardiids (Bivalvia), a sponge (Porifera), a soft coral (Anthozoa), and a free-living strain. J Phycol 35(5):1054–1062. https://doi.org/10.1046/j.1529-8817.1999.3551054.x

Carlos AA, Baillie BK, Maruyama T (2000) Diversity of dinoflagellate symbionts (zooxanthellae) in a host individual. Mar Ecol Prog Ser 195:93–100. https://doi.org/10.3354/meps195093

Chew SF, Koh CZ, Hiong KC et al (2019) Light-enhanced expression of carbonic anhydrase 4-like supports shell formation in the fluted giant clam *Tridacna squamosa*. Gene 683:101–112. https://doi.org/10.1016/j.gene.2018.10.023

Comai L, Howell T (2012) Barcode Generator. Available at: https://comailab.org/data-and-method/barcode-generator-a-python-program/. Accessed 2 Feb 2025

Courtois de Vicose G, Chou L (1999) Future of giant clam mariculture in Singapore: problems and potential solutions. In: Proceedings of an international workshop on the rehabilitation of degraded coastal systems, Department of Fisheries, Bangkok (Thailand). Phuket Marine Biological Center, Phuket, Thailand, 1998.119–122, 19–24 Jan 1998

Cunning R, Gillette P, Capo T et al (2015) Growth tradeoffs associated with thermotolerant symbionts in the coral *Pocillopora damicornis* are lost in warmer oceans. Coral Reefs 34(1):155–160. https://doi.org/10.1007/s00338-014-1216-4

Darling ES, Alvarez-Filip L, Oliver TA et al (2012) Evaluating life-history strategies of reef corals from species traits. Ecol Lett 15(12):1378–1386. https://doi.org/10.1111/j.1461-0248.2012.01861.x

DeBoer TS, Baker AC, Erdmann MV et al (2012) Patterns of *Symbiodinium* distribution in three giant clam species across the biodiverse Bird's Head region of Indonesia. Mar Ecol Prog Ser 444:117–132. https://doi.org/10.3354/meps09413

Dellaert Z, Vargas PA, La Riviere PJ et al (2022) Uncovering the effects of symbiosis and temperature on coral calcification. Biol Bull 242(1):62–73. https://doi.org/10.1086/716711

Doty MS (1971) Measurement of water movement in reference to benthic algal growth. Bot Mar 14(1):32–35. https://doi.org/10.1515/botm.1971.14.1.32

Draney CL (2024) Altered Symbiodiniaceae community dynamics in *Acropora retusa* corals in Mo'orea, French Polynesia under heat and nutrient stress. Dissertation. Oregon State University

English S, Wilkinson C, Baker V (1997) Survey manual for tropical marine resources. Australian Institute of Marine Science, Townsville

Eren AM, Morrison HG, Lescault PJ et al (2015) Minimum entropy decomposition: unsupervised oligotyping for sensitive partitioning of high-throughput marker gene sequences. ISME J 9(4):968–979. https://doi.org/10.1038/ismej.2014.195

Finney JC, Pettay DT, Sampayo EM et al (2010) The relative significance of host–habitat, depth, and geography on the ecology, endemism, and speciation of coral endosymbionts in the genus *Symbiodinium*. Microb Ecol 60:250–263. https://doi.org/10.1007/s00248-010-9681-y

Fong J, Todd PA (2021) Spatio-temporal dynamics of coral–macroalgal interactions and their impacts on coral growth on urbanised reefs. Mar Pollut Bull 172:112849. https://doi.org/10.1016/j.marpolbul.2021.112849

Fox J, Weisberg S (2019) An R companion to applied regression, 3rd edn. Sage, Thousand Oaks. https://www.john-fox.ca/Companion/

Fujise L, Nitschke MR, Frommlet JC et al (2018) Cell cycle dynamics of cultured coral endosymbiotic microalgae (*Symbiodinium*) across different types (species) under alternate light and temperature conditions. J Eukaryot Microbiol 65(4):505–517. https://doi.org/10.1111/jeu.12497

Gong S, Chai G, Xiao Y et al (2018) Flexible symbiotic associations of *Symbiodinium* with five typical coral species in tropical and subtropical reef regions of the northern South China Sea. Front Microbiol 9:2485. https://doi.org/10.3389/fmicb.2018.02485

Guest JR, Todd PA, Goh E et al (2008) Can giant clam (*Tridacna squamosa*) populations be restored on Singapore's heavily impacted coral reefs? Aquat Conserv: Mar Freshw Ecosyst 18:570–579. https://doi.org/10.1002/aqc.888

Guest J, Low J, Tun K et al (2016) Coral community response to bleaching on a highly disturbed reef. Sci Rep 6:20717. https://doi.org/10.1038/srep20717

Guibert I, Lecellier G, Torda G et al (2020) Metabarcoding reveals distinct microbiotypes in the giant clam *Tridacna maxima*. Microbiome 8:57. https://doi.org/10.1186/s40168-020-00835-8

Guibert I, Conti-Jerpe I, Pons L et al (2024) Trophic niche partitioning in symbiotic marine invertebrates. Preprint at https://www.biorxiv.org/content/10.1101/2024.02.05.578332

Harrison J, Tham A (1973) Chapter 14—the exploitation of animals. In: Animal life and nature in Singapore. Singapore University Press, Singapore, pp 251–259

Hartmann AC, Baird AH, Knowlton N et al (2017) The paradox of environmental symbiont acquisition in obligate mutualisms. Curr Biol 27:3711–3716.e3. https://doi.org/10.1016/j.cub.2017.10.036

Hume BCC, D'Angelo C, Smith EG et al (2015) *Symbiodinium thermophilum* sp. nov., a thermotolerant symbiotic alga prevalent in corals of the world's hottest sea, the Persian/Arabian gulf. Sci Rep 5:8562. https://doi.org/10.1038/srep08562

Hume BC, Ziegler M, Poulain J et al (2018) An improved primer set and amplification protocol with increased specificity and sensitivity targeting the *Symbiodinium* ITS2 region. PeerJ 6:e4816. https://doi.org/10.7717/peerj.4816

Hume BCC, Smith EG, Ziegler M et al (2019) SymPortal: a novel analytical framework and platform for coral algal symbiont next-generation sequencing ITS2 profiling. Mol Ecol Resour 19:1063–1080. https://doi.org/10.1111/1755-0998.13004

Ikeda S, Yamashita H, Kondo S-n et al (2017) Zooxanthellal genetic varieties in giant clams are partially determined by species-intrinsic and growth-related characteristics. PLoS One 12:e0172285. https://doi.org/10.1371/journal.pone.0172285

Ip YK, Chew SF (2021) Light-dependent phenomena and related molecular mechanisms in giant clam-dinoflagellate associations: a review. Front Mar Sci 8:627722. https://doi.org/10.3389/fmars.2021.627722

Tanzil JTI, Ng APK, Tey YQ et al (2016) A preliminary characterisation of *Symbiodinium* diversity in some common corals from Singapore. Cosmos 12:15–27. https://doi.org/10.1142/S0219607716500014

Ishikura M, Hagiwara K, Takishita K et al (2004) Isolation of new *Symbiodinium* strains from tridacnid giant clam (*Tridacna crocea*) and sea slug (*Pteraeolidia ianthina*) using culture medium containing giant clam tissue homogenate. Mar Biotechnol 6:378–385. https://doi.org/10.1007/s10126-004-1800-7

Jain SS, Afiq-Rosli L, Feldman B et al (2020) Homogenization of endosymbiont communities hosted by equatorial corals during the 2016 mass bleaching event. Microorganisms 8:1370. https://doi.org/10.3390/microorganisms8091370

Jain SS, Afiq-Rosli L, Feldman B et al (2021) Endosymbiont communities in *Pachyseris speciosa* highlight geographical and methodological variations. Front Mar Sci 8:666428. https://doi.org/10.3389/fmars.2021.759744

Kirkendale L, Paulay G (2017) Photosymbiosis in Bivalvia. In: Part N, Revised, Volume 1. Treatise Online 89:1–31. https://doi.org/10.17161/to.v0i0.6554

LaJeunesse TC (2001) Investigating the biodiversity, ecology, and phylogeny of endosymbiotic dinoflagellates in the genus *Symbiodinium* using the ITS region: in search of a "species" level marker. J Phycol 37(5):866–880. https://doi.org/10.1046/j.1529-8817.2001.01031.x

LaJeunesse T (2002) Diversity and community structure of symbiotic dinoflagellates from Caribbean coral reefs. Mar Biol 141:387–400. https://doi.org/10.1007/s00227-002-0829-2

LaJeunesse TC, Trench R (2000) Biogeography of two species of *Symbiodinium* (Freudenthal) inhabiting the intertidal sea anemone *Anthopleura elegantissima* (Brandt). Biol Bull 199(2):126–134. https://doi.org/10.2307/1542872

LaJeunesse T, Bhagooli R, Hidaka M et al (2004) Closely related *Symbiodinium* spp. differ in relative dominance in coral reef host communities across environmental, latitudinal and biogeographic gradients. Mar Ecol Prog Ser 284:147–161. https://doi.org/10.3354/meps284147

LaJeunesse T, Bonilla HR, Warner M et al (2008) Specificity and stability in high latitude eastern Pacific coral-algal symbioses. Limnol Oceanogr 53(2):719–727. https://doi.org/10.4319/lo.2008.53.2.0719

LaJeunesse TC, Pettay DT, Sampayo EM et al (2010) Long-standing environmental conditions, geographic isolation and host–symbiont specificity influence the relative ecological dominance and genetic diversification of coral endosymbionts in the genus *Symbiodinium*. J Biogeogr 37(5):785–800. https://doi.org/10.1111/j.1365-2699.2010.02273.x

LaJeunesse TC, Wham DC, Pettay DT et al (2014) Ecologically differentiated stress-tolerant endosymbionts in the dinoflagellate genus *Symbiodinium* (Dinophyceae) Clade D are different species. Phycologia 53(4):305–319. https://doi.org/10.2216/13-186.1

LaJeunesse TC, Parkinson JE, Gabrielson PW et al (2018) Systematic revision of Symbiodiniaceae highlights the antiquity and diversity of coral endosymbionts. Curr Biol 28(16):2570–2580.e6. https://doi.org/10.1016/j.cub.2018.07.008

Lee LK, Leaw CP, Lee LC et al (2022) Molecular diversity and assemblages of coral symbionts (Symbiodiniaceae) in diverse scleractinian coral species. Mar Environ Res 179:105706. https://doi.org/10.1016/j.marenvres.2022.105706

Lenaers G, Scholin C, Bhaud Y et al (1991) A molecular phylogeny of dinoflagellate protists (Pyrrhophyta) inferred from the sequence of 24S rRNA divergent domains D1 and D8. J Mol Evol 32:53–63. https://doi.org/10.1007/BF02099929

Lenth RV (2024) emmeans: Estimated Marginal Means, aka Least-Squares Means. R package version 1.10.6. https://CRAN.R-project.org/package=emmeans

Leveque S, Afiq-Rosli L, Ip YCA et al (2019) Searching for phylogenetic patterns of Symbiodiniaceae community structure among Indo-Pacific Merulinidae corals. PeerJ 7:e7669. https://doi.org/10.7717/peerj.7669

Lim SSQ, Huang D, Soong K et al (2019) Diversity of endosymbiotic Symbiodiniaceae in giant clams at Dongsha Atoll, northern South China Sea. Symbiosis 78:251–262. https://doi.org/10.1007/s13199-019-00615-5

Long C, Zhang Y, Li Y et al (2021) Effects of Symbiodiniaceae phylotypes in clades A–E on progeny performance of two giant clams (*Tridacna squamosa* and *T. crocea*) during early history life stages in the South China Sea. Front Mar Sci 8:633761. https://doi.org/10.3389/fmars.2021.633761

Matthews JL, Crowder CM, Oakley CA et al (2017) Optimal nutrient exchange and immune responses operate in partner specificity in the cnidarian-dinoflagellate symbiosis. Proc Natl Acad Sci USA 114(50):13194–13199. https://doi.org/10.1073/pnas.1710733114

Mies M (2019) Evolution, diversity, distribution and the endangered future of the giant clam–Symbiodiniaceae association. Coral Reefs 38(6):1067–1084. https://doi.org/10.1007/s00338-019-01857-x

Mies M, Braga F, Scozzafave MS et al (2012) Early development, survival and growth rates of the giant clam *Tridacna crocea* (Bivalvia: Tridacnidae). Braz J Oceanogr 60(2):127–133. https://doi.org/10.1590/S1679-87592012000200003

Mies M, Scozzafave MS, Braga F et al (2017) Giant clams. In: Calado R, Olivotto I, Oliver MP et al (eds) Marine Ornamental Species Aquaculture. Wiley-Blackwell, Chichester, pp 510–535. https://doi.org/10.1002/9781119169147.ch23a

Mies M, Garrido AG, Zilberberg C et al (2025) Anatomical complexity allows for heat-stressed giant clams to undergo symbiont shuffling at both organism and organ levels. Mar Biol 172:4. https://doi.org/10.1007/s00227-024-04567-3

Millán-Márquez AM, Velasco-Montoya DA, Terraneo TI et al (2024) Symbiodiniaceae diversity in *Pocillopora* corals in different environments of the Colombian Eastern Pacific: symbiont specificity in spite of coral-host flexibility. Coral Reefs 43:1581–1597. https://doi.org/10.1007/s00338-024-02552-2

Million W, Voolstra CR, Perna G et al (2024) Resolving Symbiodiniaceae diversity across coral microhabitats and reef niches. bioRxiv 2024.09.06:611593. https://doi.org/10.1101/2024.09.06.611593

Mote S, Gupta V, De K et al (2021) Differential Symbiodiniaceae association with coral and coral-eroding sponge in a bleaching impacted marginal coral reef environment. Front Mar Sci 8:666825. https://doi.org/10.3389/fmars.2021.666825

Neo ML (2024) Guardians of the clams: a decadal monitoring effort of endangered giant clams by citizen scientists. J Molluscan Stud 90(4):eyae038. https://doi.org/10.1093/mollus/eyae038

Neo ML, Todd PA (2012) Population density and genetic structure of the giant clams *Tridacna crocea* and *T. squamosa* on Singapore's reefs. Aquat Biol 14(3):265–275. https://doi.org/10.3354/ab00399

Neo ML, Todd PA (2013) Conservation status reassessment of giant clams (Mollusca: Bivalvia: Tridacninae) in Singapore. Nat Singap 6(3):125–133

Neo ML, Eckman W, Vicentuan K et al (2015) The ecological significance of giant clams in coral reef ecosystems. Biol Conserv 181:111–123. https://doi.org/10.1016/j.biocon.2014.11.004

Neo M, Wabnitz C, Braley R et al (2017) Giant clams (Bivalvia: Cardiidae: Tridacninae): a comprehensive update of species and their distribution, current threats and conservation status. In: Hawkins SJ, Evans AJ, Dale AC et al (eds) Oceanography and marine biology: an annual review, vol 55. CRC Press, Boca Raton, pp 87–388. https://doi.org/10.1201/b21944-5

Ng CSL, Huang D, Toh KB et al (2020) Responses of urban reef corals during the 2016 mass bleaching event. Mar Pollut Bull 154:111111. https://doi.org/10.1016/j.marpolbul.2020.111111

Ng MS, Soon N, Afiq-Rosli L et al (2024) Highly diverse Symbiodiniaceae types hosted by corals in a global hotspot of marine biodiversity. Microb Ecol 87(1):92. https://doi.org/10.1007/s00248-023-02351-2

Nitschke MR, Rosset SL, Oakley CA et al (2022) The diversity and ecology of Symbiodiniaceae: a traits-based review. In: Sheppard

C (ed) Advances in marine biology, vol 92. Academic Press, pp 55–127. https://doi.org/10.1016/bs.amb.2022.07.001

Norton J, Shepherd M, Long H et al (1992) The zooxanthellal tubular system in the giant clam. Biol Bull 183(3):503–506. https://doi.org/10.2307/1542028

Ong JH, Wainwright BJ, Jain SS et al (2022) Species and spatio-environmental effects on coral endosymbiont communities in Southeast Asia. Coral Reefs 41(4):1131–1145. https://doi.org/10.1007/s00338-022-02254-7

Ow Yong WL, Todd PA, Ang AC et al (2022) Fluted giant clam (*Tridacna squamosa*) restocking experiment in an urban turbid reef environment. Aquat Conserv 32(4):633–644. https://doi.org/10.1002/aqc.3789

Palacio-Castro AM, Dennison CE, Rosales SM et al (2021) Variation in susceptibility among three Caribbean coral species and their algal symbionts indicates the threatened staghorn coral, *Acropora cervicornis*, is particularly susceptible to elevated nutrients and heat stress. Coral Reefs 40:1601–1613. https://doi.org/10.1007/s00338-021-02184-w

Pappas MK, He S, Hardenstine RS et al (2017) Genetic diversity of giant clams (*Tridacna* spp.) and their associated *Symbiodinium* in the central Red Sea. Mar Biodivers 47:1209–1222. https://doi.org/10.1007/s12526-017-0715-2

Pinheiro J, Bates D, R Core Team (2025) nlme: linear and nonlinear mixed effects models. R package version 3.1-167. Available via CRAN https://CRAN.R-project.org/package=nlme

Pochon X, Gates RD (2010) A new *Symbiodinium* clade (Dinophyceae) from soritid foraminifera in Hawai'i. Mol Phylogenet Evol 56(1):492–497. https://doi.org/10.1016/j.ympev.2010.03.040

Pochon X, Pawlowski J, Zaninetti L et al (2001) High genetic diversity and relative specificity among *Symbiodinium*-like endosymbiotic dinoflagellates in soritid foraminiferans. Mar Biol 139:1069–1078. https://doi.org/10.1007/s002270100674

Pochon X, LaJeunesse T, Pawlowski J (2004) Biogeographic partitioning and host specialization among foraminiferan dinoflagellate symbionts (*Symbiodinium*; Dinophyta). Mar Biol 146:17–27. https://doi.org/10.1007/s00227-004-1427-2

Qin Z, Yu K, Chen B et al (2019) Diversity of Symbiodiniaceae in 15 coral species from the southern South China Sea: potential relationship with coral thermal adaptability. Front Microbiol 10:2343. https://doi.org/10.3389/fmicb.2019.02343

Quek ZR, Tanzil JT, Jain SS et al (2023) Limited influence of seasonality on coral microbiomes and endosymbionts in an equatorial reef. Ecol Indic 146:109878. https://doi.org/10.1016/j.ecolind.2023.109878

Quigley KM, Willis BL, Bay LK (2017) Heritability of the *Symbiodinium* community in vertically-and horizontally-transmitting broadcast spawning corals. Sci Rep 7(1):8219. https://doi.org/10.1038/s41598-017-08179-4

Quigley KM, Warner PA, Bay LK et al (2018) Unexpected mixed-mode transmission and moderate genetic regulation of *Symbiodinium* communities in a brooding coral. Heredity 121(6):524–536. https://doi.org/10.1038/s41437-018-0059-0

Reimer JD, Todd PA (2009) Preliminary molecular examination of zooxanthellate zoanthid (Hexacorallia: Zoantharia) and associated zooxanthellae (*Symbiodinium* spp.) diversity in Singapore. Raffles Bull Zool 22:103–120

Rossbach S, Hume BCC, Cárdenas A et al (2021) Flexibility in Red Sea *Tridacna maxima*-Symbiodiniaceae associations supports environmental niche adaptation. Ecol Evol 11(7):3393–3406. https://doi.org/10.1002/ece3.7299

Rowan R, Powers DA (1991) A molecular genetic classification of zooxanthellae and the evolution of animal-algal symbioses. Science 251(4999):1348–1351. https://doi.org/10.1126/science.251.4999.1348

RStudio Team (2015) RStudio: integrated development for R. RStudio, Inc, Boston, MA. http://www.rstudio.com/

Saad OS, Lin X, Ng TY et al (2022) Species richness and generalists–specialists mosaicism of symbiodiniacean symbionts in corals from Hong Kong revealed by high-throughput ITS sequencing. Coral Reefs 41(1):1–12. https://doi.org/10.1007/s00338-021-02196-6

Sampayo E, Dove S, LaJeunesse T (2009) Cohesive molecular genetic data delineate species diversity in the dinoflagellate genus *Symbiodinium*. Mol Ecol 18(3):500–519. https://doi.org/10.1111/j.1365-294X.2008.04037.x

Schloss PD, Westcott SL, Ryabin T et al (2009) Introducing mothur: open-source, platform-independent, community-supported software for describing and comparing microbial communities. Appl Environ Microbiol 75(23):7537–7541. https://doi.org/10.1128/AEM.01541-09

Silverstein RN, Cunning R, Baker AC (2017) Tenacious D: *Symbiodinium* in clade D remain in reef corals at both high and low temperature extremes despite impairment. J Exp Biol 220(7):1192–1196. https://doi.org/10.1242/jeb.148239

Sison M (2003) The effects of zooxanthellae clades A and C on giant clam host, in particular to its early growth, survival, recruitment and response to thermal stress. MSc Thesis. University of the Philippines, Diliman, Quezon City

Smith EG, Gurskaya A, Hume BC et al (2020) Low Symbiodiniaceae diversity in a turbid marginal reef environment. Coral Reefs 39(3):545–553. https://doi.org/10.1007/s00338-020-01956-0

Soon N, Quek ZBR, Pohl S et al (2023) More than meets the eye: characterizing the cryptic species complex and Symbiodiniaceae communities in the reef-dwelling nudibranch *Pteraeolidia 'semperi'* (Nudibranchia: Aeolidioidea) from Singapore. J Molluscan Stud 89(2):eyad011. https://doi.org/10.1093/mollus/eyad011

Stat M, Gates RD (2011) Clade D *Symbiodinium* in scleractinian corals: a "nugget" of hope, a selfish opportunist, an ominous sign, or all of the above? J Mar Sci 2011(1):730715. https://doi.org/10.1155/2011/730715

Stat M, Carter D, Hoegh-Guldberg O (2006) The evolutionary history of *Symbiodinium* and scleractinian hosts—symbiosis, diversity, and the effect of climate change. Perspect Plant Ecol Evol Syst 8(1):23–43. https://doi.org/10.1016/j.ppees.2006.04.001

Suggett DJ, Warner ME, Leggat W (2017) Symbiotic dinoflagellate functional diversity mediates coral survival under ecological crisis. Trends Ecol Evol 32(10):735–745. https://doi.org/10.1016/j.tree.2017.07.013

Tan YTR, Wainwright BJ, Afiq-Rosli L et al (2020) Endosymbiont diversity and community structure in *Porites lutea* from Southeast Asia are driven by a suite of environmental variables. Symbiosis 80:269–277. https://doi.org/10.1007/s13199-020-00671-2

Tay Y, Todd P, Rosshaug PS, Chou L (2012) Simulating the transport of broadcast coral larvae among the southern islands of Singapore. Aquat Biol 15(3):283–297. https://doi.org/10.3354/ab00433

Terrell AP, Marangon E, Webster NS et al (2023) The promotion of stress tolerant Symbiodiniaceae dominance in juveniles of two coral species under simulated future conditions of ocean warming and acidification. Front Ecol Evol 11:1113357. https://doi.org/10.3389/fevo.2023.1113357

Thomas L, López EH, Morikawa MK, Palumbi SR (2019) Transcriptomic resilience, symbiont shuffling, and vulnerability to recurrent bleaching in reef-building corals. Mol Ecol 28(14):3371–3382. https://doi.org/10.1111/mec.15143

Venn AA, Loram JE, Douglas AE (2008) Photosynthetic symbioses in animals. J Exp Bot 59(5):1069–1080. https://doi.org/10.1093/jxb/erm328

Wall CB, Kaluhiokalani M, Popp BN et al (2020) Divergent symbiont communities determine the physiology and nutrition of a reef coral across a light-availability gradient. ISME J 14(4):945–958. https://doi.org/10.1038/s41396-019-0570-1

Weber M (2009) The biogeography and evolution of *Symbiodinium* in Giant Clams (Tridacnidae). PhD dissertation. University of California, Berkeley

Wilcox TP (1998) Large-subunit ribosomal RNA systematics of symbiotic dinoflagellates: morphology does not recapitulate phylogeny. Mol Phylogenet Evol 10(3):436–448. https://doi.org/10.1006/mpev.1998.0546

Yonge CM (1936) Mode of life, feeding, digestion and symbiosis with zooxanthellae in the Tridacnidae. Sci Rep Gt Barrier Reef Exped 1:283–321

Ziegler M, Roder CM, Büchel C, Voolstra CR (2015) Mesophotic coral depth acclimatization is a function of host-specific symbiont physiology. Front Mar Sci 2:4. https://doi.org/10.3389/fmars.2015.00004

Ziegler M, Eguíluz VM, Duarte CM, Voolstra CR (2018) Rare symbionts may contribute to the resilience of coral–algal assemblages. ISME J 12(1):161–172. https://doi.org/10.1038/ismej.2017.151

7 Microbiomes Associated with Benthic Marine Macroorganisms in Singapore

Lik Tong Tan, Beverly Pi Lee Goh, Deepak Kumar Gupta, Nursheena Parveen Katermeran, and Lindsey Kane Deignan

Abstract

Marine microbiomes associated with coral reef benthic macroorganisms, such as sponges, corals and macroalgae, play crucial roles in maintaining the health and functioning of marine ecosystems. These microbial communities form symbiotic relationships with their hosts, aiding in nutrient cycling, defense against pathogens and the breakdown of organic matter. They also contribute to crucial biogeochemical processes including nitrogen and carbon cycling, that are essential for ocean productivity. Investigating these microbiomes is critical for unlocking their biotechnological potential, including the discovery of novel bioactive compounds with medicinal and industrial applications. This chapter covers research related to microbiomes associated with benthic marine macroorganisms found at coral reefs within highly urbanized and sedimented coastal waters in Singapore. We focus on the diversity and ecological functions of symbiotic microbes associated with benthic organisms, such as scleractinian corals, sponges and macroalgae. In addition, marine microbiome research conducted on Singapore's hard corals, sponges, nudibranchs, macroalgae and seagrasses is reviewed. Furthermore, results from our ongoing research on the exploitation of selected sponge and soft coral-associated microbes as potential sources of unique specialized molecules as anti-infectives are presented. Recent experimental data on the use of symbiotic microbes obtained from relatively resilient hard corals to confer resilience in susceptible hard corals will also be discussed. We conclude with perspectives and recommendations on future research in coastal marine microbiomes in Singapore.

L. T. Tan (✉) · B. P. L. Goh (✉) · D. K. Gupta · N. P. Katermeran
Natural Sciences and Science Education, National Institute of Education, Nanyang Technological University, Singapore, Singapore
e-mail: liktong.tan@nie.edu.sg; beverly.goh@nie.edu.sg

L. K. Deignan
Singapore Centre for Environmental Life Sciences Engineering, Nanyang Technological University, Singapore, Singapore

Keywords

Marine microbiome · Microbial baseline data · Marine biotechnology · Drug discovery · Bacterial manipulation · Probiotics

7.1 Introduction

Marine macroorganisms, such as sponges, corals, macroalgae and seagrasses, coexist with a vast diversity of microorganisms, including protists, bacteria, archaea, fungi and viruses, which make up millions of cells in each millilitre of the ocean's 1.3 billion km^3 of water (Eakins and Sharman 2010). Despite their small size, often just a few micrometres or smaller, these microorganisms play crucial roles in oxygen production, nutrient cycling and organic matter degradation, which are vital to both ocean ecosystems and life on Earth (Hutchins and Fu 2017). For instance, microbes associated with marine animals are part of the animal's microbiome, a community of microorganisms living on or within the host. Some of these microbes originate from the surrounding seawater, while others are passed down through generations of the host (Nussbaumer et al. 2006; Sipkema et al. 2015).

Over the past two decades, genomic and microbiological advancements have significantly improved our understanding of macroorganism-associated microbiomes (Reddy 2019). Although symbiotic relationships between marine animals and microorganisms have long been studied, modern technologies are revealing the incredible diversity of the associated microbial life (Reddy 2019). For instance, reef-building corals are well-known examples of animal-microbial symbiosis, hosting photosynthetic symbionts that provide essential nutrients (Muscatine et al. 1981). Recent studies

L. M. Chou, D. Huang (eds.), *Coral Reefs of Singapore's Urbanised Sea*, Coral Reefs of the World 23,
https://doi.org/10.1007/978-981-95-0903-4_7

have identified a wider array of protists, bacteria, archaea and viruses living with corals, revealing their roles in supporting various functions within the coral host (van Oppen and Blackall 2019). Similarly, research on marine sponges shows that they harbor a complex symbiotic microbial community and they play important roles in the metabolic functions of host sponges, including carbon, nitrogen, sulfur, phosphorus and vitamin cycles (Pita et al. 2018). Moreover, it has been reported that several specialized metabolites are derived from symbiotic bacteria as defensive molecules in protecting the host animals (Tan 2023).

The study of microbiomes in marine macroorganisms is an expanding field within marine science, with a strong focus on identifying consistent or 'core' microbial communities (Shade and Handelsman 2012). After determining 'who is there' using diversity-based surveys of small subunit (SSU) ribosomal RNA (rRNA) genes, such as 16S rRNA, researchers then explore the functions of these microbes, the nature of their associations and how they influence animal health, physiology, ecology and behavior (McFall-Ngai et al. 2013). Moreover, the ocean is undergoing rapid variations due to climate change and human activities (Poloczanska et al. 2016), leading to investigations into the microbiome's role as a potential indicator of the holobiont's response to these shifts (Ainsworth and Gates 2016).

Biotechnology has unlocked the potential of the marine microbiome as a rich source for drug discovery and biocatalysts (de la Calle 2017; de Oliveira et al. 2020). Marine microorganisms, including bacteria, fungi and archaea, can thrive in unique ocean environments, producing bioactive compounds that often exhibit novel structures and functions. These natural products have shown promise in various therapeutic areas, such as antibiotics, antivirals, anticancer agents and anti-inflammatory drugs. Additionally, several pharmaceutically important compounds originally reported from marine animals, including sponges, bryozoans and tunicates, have been conclusively revealed to be produced by their microbial symbionts. These microbial-derived compounds play crucial ecological roles, including chemical defense and antifouling activities for their host organisms. Uncovering the ecological functions of these symbiotic microbial molecules could accelerate the discovery of novel bioactive compounds while also helping to resolve supply challenges for compounds used in drug development. Examples of bioactive compounds produced by microbial symbionts include anticancer bryostatins (Sudek et al. 2007), didemnins (Sings and Rinehart 1996), kahalalides (Zan et al. 2019) and turnercyclamycins (Distel et al. 2002) detected from '*Candidatus* Endobugula sertula', *Synechocystis trididemni*, *Candidatus* Endobryopsis kahalalidefaciens and *Teredinibacter turnerae* T7901, found living within bryozoan, tunicate, marine alga and shipworm, respectively. Advances in genomics, metagenomics and bioinformatics have enabled the exploration of these marine microbiomes, leading to the discovery of new pathways for drug production, thus offering a valuable avenue for developing innovative pharmaceuticals.

The impacts of environmental change and life history events on marine macroorganism-associated microbiomes are increasingly becoming a focal point in research. There is growing interest in understanding the interactions between macroorganisms, their microbiomes and the ocean environment, including how these exchanges take place (Webster et al. 2016). In recent years, there is growing research on the microbiomes of several key coral reef benthic macroorganisms living within Singapore's urban coastal marine environment, which is characterized by high sedimentation. This review provides a comprehensive summary account of research on marine microbiomes found in association with diverse marine macroorganisms in Singapore waters to date. In addition, several findings stemming from our research programme on the biomedical importance of marine microbial symbionts as sources of anti-infectives and their potential application in conferring coral resilience through microbial inoculation experiments will be presented. Due to the scope of this review on marine microbiome research related to reef benthic macroorganisms, microbiome studies in marine vertebrates (e.g. Soh et al. 2024), sediments (Tan et al. 2021), plastic debris (Koh et al. 2023; Bairoliya et al. 2024), coastal water column (Lau et al. 2013; Chénard et al. 2019; Wijaya et al. 2023) and coastal mangroves (Wainwright et al. 2023) will not be included. Lastly, the review will provide perspectives on the future direction of marine microbiome research in Singapore, particularly on the understanding of host–microbiome interactions and functions within a changing ocean scenario.

7.2 Diversity and Ecological Importance of Microbiomes Associated with Benthic Marine Macroorganisms

7.2.1 Coral Microbiome

Of the various benthic marine invertebrates, corals (Cnidaria: Anthozoa) are the most well-known example of animal hosts whose symbiotic relationship with microalgae can shift to a state of dysbiosis, a phenomenon often recognized as coral bleaching (Glynn 1991). Corals serve as a reservoir for a wide range of microorganisms, including bacteria, archaea, viruses, dinoflagellates, ciliates and fungi (Bourne et al. 2016). While some of these microbes are well-studied, others remain elusive and less understood. Microalgae of the dinoflagellate family Symbiodiniaceae were the first and most significant symbionts of corals to be identified (LaJeunesse et al. 2018). Initially thought to be a single species, *Symbiodinium microadriaticum* (Freudenthal 1962), this group is currently classified as family Symbiodiniaceae, comprising seven genera, including *Symbiodinium* (formerly Clade A), *Breviolum* (Clade B), *Cladocopium* (Clade C), *Durusdinium* (Clade D), *Effrenium* (Clade E), *Fugacium* (Clade F) and *Gerakladium* (Clade G) (LaJeunesse et al. 2018). These algae reside specifically in a coral-derived specialized cellular structure called the 'symbiosome', which forms from the coral host's endosome following the phagocytosis of the algal cells (Davy et al. 2012). In addition, the membrane-bound symbiosome protects the Symbiodiniaceae from the host's lysosome degradative action (Mohamed et al. 2016) and facilitates the exchange of nutrients between the coral and the symbionts (Davy et al. 2012).

Corals also host a diverse bacterial microbiome (Blackall et al. 2015), comprising 39 phyla, representing more than one-third of all seawater-derived bacterial phyla (Huggett and Apprill 2019). Some of these bacterial communities are believed to support coral health as well as resilience (Bourne et al. 2016; Meunier et al. 2021). Of the many bacterial phyla associated with corals, the most dominant are Proteobacteria, Bacteroidetes, Cyanobacteria and Firmicutes, as revealed by 16S rRNA gene analysis of over 21,000 sequences (Huggett and Apprill 2019). This is further supported by meta-analysis of 3055 bacterial isolates from more than 50 coral studies, where the majority of cultivable bacteria belong to the Proteobacteria, Firmicutes, Bacteroidetes and Actinobacteria (Sweet et al. 2021). Moreover, the γ-proteobacterium, *Endozoicomonas*, is increasingly recognized as a key component of coral microbiomes, displaying adaptability in its ecological roles in coral health and nutrient regulation (Neave et al. 2017; Tandon et al. 2020).

Disruptions to ocean conditions, such as rising temperatures and ocean acidification, have been shown to negatively impact coral-associated bacteria (Thurber et al. 2009; Meron et al. 2011), including their interactions with *Endozoicomonas* (Morrow et al. 2015). However, some members of the coral microbiome have proven to be resilient across wide environmental gradients (Hernandez-Agreda et al. 2016). Beyond their role in nutrient exchange, the coral microbiome is also critical to coral health and response to stress. For instance, heat and light stress can cause an overproduction of reactive oxygen species (ROS), which is harmful to Symbiodiniaceae and leads to bleaching. However, recent findings suggest that bacteria associated with corals may also release extracellular ROS (Zhang et al. 2016), potentially serving as a signaling mechanism within the microbiome or between coral host and symbionts. Considering the increased frequency of global mass bleaching events (every 5–6 years) on coral reefs (Hughes et al. 2017), corals will likely remain an important model for investigating the dynamics of symbiosis and dysbiosis (Boilard et al. 2020).

7.2.2 Sponge Microbiome

Sponges (Phylum Porifera) are prominent inhabitants of the ocean's diverse benthic ecosystems, and their prevalence, along with their capacity to filter large amounts of seawater, highlight their significant role in shaping both benthic and pelagic processes (Bell 2008). As one of the earliest animal lineages, sponges possess a relatively simple body structure, yet they are frequently found in symbiotic relationships with diverse microbes. Sponges are closely linked to diverse microbial communities, with representatives from more than 60 distinct prokaryotic phyla identified in sponges to date (Moitinho-Silva et al. 2017a). Among these, 13 phyla were consistently found across 81 surveyed sponge species (Thomas et al. 2016).

These sponge-derived microorganisms perform crucial ecological roles in the recycling of nutrients, such as carbon, nitrogen and potentially sulfur and vitamin metabolism (Taylor et al. 2007; Bayer et al. 2008; Hentschel et al. 2012). In addition, these microbes are involved in the production of bioactive secondary metabolites, contributing to the chemical defense of sponges (Wilson et al. 2014). Certain sponge species have also been found to host dense populations of symbiotic microbes within their tissues, while other species are largely free of such microbes (Reiswig 1974). Due to the differences in microbial densities, these sponges are categorized as having either 'high microbial abundance' (HMA) or 'low microbial abundance' (LMA) (Hentschel et al. 2003). Bacterial densities in HMA sponges are two to four orders of magnitude greater than in LMA sponges, with microbial biomass accounting for one-third of the total biomass in HMA sponges (Hentschel et al. 2012). HMA sponges have highly complex microbiomes, whereas LMA sponges are predominantly associated with Proteobacteria and Cyanobacteria (Weisz et al. 2007; Kamke et al. 2010; Giles et al. 2013; Gloeckner et al. 2013). Moreover, the microbial communities within some sponge species have been shown to shift in response to environmental changes, such as rising temperatures (Simister et al. 2012) and ocean acidification (Ribes et al. 2016), as well as combined stressors (Lesser et al. 2016). As such, understanding how these altered interactions between sponges and their microbiomes affect sponge growth and ecology is a key area of future research.

7.2.3 Macroalgal Microbiome

The macroalgal canopy in intertidal zones plays an important role in maintaining the health of intertidal ecosystems. For instance, the removal of certain macroalgal species can change the habitat structure and affect the spatial complexity, influencing the surrounding benthic species (Tait and Schiel 2011). Additionally, the decline of intertidal macroalgal species has been linked to the loss of species that depend on the established algal canopy for survival (Komatsu et al. 2014). Macroalgae also hold significant commercial value due to their richness in minerals and essential trace elements vital for human consumption as well as their use as raw materials in fertilizers, hydrocolloids and biofuel industries (Baghel et al. 2015).

Extensive research has been carried out on the epiphytic and endophytic bacterial communities associated with macroalgae, focusing on their phylogenetic composition and variability (Lachnit et al. 2011; Hollants et al. 2013). Employing a variety of techniques, including morphological characterization, in situ hybridization, electrophoresis, DNA fingerprinting, pyrosequencing and metagenomic approaches, these findings revealed that bacterial communities, particularly those from the phyla Proteobacteria and Firmicutes, are the most commonly associated epiphytic bacteria with macroalgal hosts (Tujula et al. 2010; Lachnit et al. 2011; Bengtsson et al. 2012; Hollants et al. 2013; Bondoso et al. 2014). Research has also highlighted the role of these epiphytic bacteria in the ecophysiology of their macroalgal hosts. Studies have shown that a portion of epiphytic bacterial communities can be classified as 'core microbial species', indicating an obligatory association with their macroalgal hosts (Lu et al. 2023). These bacterial communities have also been found to defend macroalgal surfaces against biofouling organisms by producing biologically active metabolites (Malik et al. 2020). Fundamentally, due to their vital roles in the host's life cycle, the macroalgal host may offer essential organic nutrients to attract and encourage pelagic microbial communities to colonize its surfaces.

7.2.4 Seagrass Microbiome

Seagrasses are widely distributed across coastal waters globally (except Antarctica), thriving at depths ranging from shallow waters to 90 m deep (Duarte 1991). As 'ecosystem engineers', seagrasses can form large meadows that serve as the foundation of complex ecosystems and offer essential ecosystem services (Short et al. 2007), which refer to either natural processes or components that fulfill human needs. In addition, the three-dimensional nature of seagrass meadows provides vital habitats for numerous organisms and acts as a nursery for the juvenile stages of many species, including commercially valuable finfish and shellfish (Jeyabaskaran et al. 2018). Additionally, seagrasses play an important role in stabilizing soft sediments and sequestering significant amounts of carbon dioxide as biomass and dead organic matter, helping to mitigate the effects of anthropogenic impacts (Duarte et al. 2005).

Like terrestrial plants, seagrasses host diverse communities of epi- and endophytic bacteria associated with their leaves and roots (Uku et al. 2007; Garcias-Bonet et al. 2016). The leaves and roots of seagrasses appear to have a core group of epiphytic bacteria distinct from the microbial strains found in surrounding environments, such as the SAR11 cluster, which is predominant in the water column (Pfister et al. 2014; Roth-Schulze et al. 2016). For instance, Jensen et al. (2007) identified Epsilonproteobacteria, particularly sulfur-oxidizing bacteria of the genera *Arcobacter* and *Sulfurimonas*, along with Gammaproteobacteria and Actinobacteria as significant colonizers of *Zostera marina* roots, whereas Deltaproteobacteria were found to dominate the rhizosphere (Jensen et al. 2007). Microorganisms can also exist endophytically within leaf tissues (Garcias-Bonet et al. 2012) and, in roots, between or within the epidermal, exodermal and cortex cells (Nielsen et al. 1999).

The formation of specific microbial communities on seagrasses is influenced by the plant's ability to shape these epiphytic communities through positive interactions, such as nutrient release from the plant surface (Wang et al. 2014), or negative interactions, like the production of antimicrobial compounds (Zidorn 2016). Bacteria can promote plant growth by increasing nutrient availability, such as via nitrogen fixation or the mineralization of organic compounds (Cole and McGlathery 2012). They can also influence shoot and root development by producing or regulating plant hormones (Werner and Schmülling 2009) and protect plants from pathogens by synthesizing antibacterial compounds (Armstrong et al. 2001; Vlamakis et al. 2013). Thus, the seagrass microbial community plays a crucial role in seagrass development, from seed germination and phytohormone production to enhanced nutrient uptake and protection against pathogens and saprophytes.

7.3 Microbiomes Associated with Benthic Marine Macroorganisms in Singapore

7.3.1 Microbiomes Associated with Scleractinian Corals

Due to the importance of symbiotic dinoflagellates, several baseline studies on endosymbionts (Symbiodiniaceae) have been conducted for common hard coral species found at different reef sites in Singapore. These studies include genetic identities and diversity of Symbiodiniaceae in at least six coral species (*Porites lutea*, *Acropora millepora*, *Merulina ampliata*, *Diploastrea heliopora*, *Pachyseris speciosa* and *Pocillopora acuta*) from three reefs (Kusu Island, Pulau Tekukor and Pulau Satumu) (Tanzil et al. 2016), as well as Symbiodiniaceae communities found in *Merulina ampliata*, *Platygyra daedalea*, *Podabacia crustacea* and *Pachyseris speciosa* obtained from six fringing reefs (Smith et al. 2020). The study by Tanzil et al. (2016), using denaturing gradient gel electrophoresis analysis of the nuclear internal transcribed spacer region in 31 coral colonies, revealed the dominance of Symbiodiniaceae clades C (*Cladocopium*) and D (*Durusdinium*). Clade D was the primary symbiont in *Pachyseris speciosa* samples obtained from Pulau Tekukor, while those from Pulau Satumu hosted clade C27, highlighting variability in symbiotic associations within the same species. In addition, the prevalence of the stress-tolerant D clade, particularly in three of the seven coral species studied, suggests that this symbiotic relationship may play a crucial role in supporting the survival of Singapore's impacted reefs, especially under conditions of environmental stress like elevated temperatures and sedimentation. Another study by Smith et al. (2020) revealed low diversity of Symbiodiniaceae communities in several locally abundant coral species, including *Pectinia alcicornis*, *Merulina ampliata*, *Platygyra daedalea*, *Podabacia crustacea* and *Pachyseris speciosa*. Their study suggested that, despite supporting a diverse coral assemblage, Singapore's reefs are subject to strong selective pressures from persistent turbidity, which likely constrains the diversity of the associated symbiont community in hard corals.

A global-scale bleaching event in 2016 presented an opportunity to investigate the relative abundances of *Cladocopium* and *Durusdinium* found within and among the ubiquitous *Pachyseris speciosa* colonies in Singapore (Jain et al. 2020). The relative abundance of *Durusdinium*, known for its thermal tolerance, was revealed to be highest in post-bleaching healthy coral colonies. In contrast, bleached and unbleached tissue samples from bleaching colonies, as well as tissues from healthy coral colonies obtained during the bleaching event, showed reduced amounts of *Durusdinium*. Due to the importance of *Durusdinium* for thermal tolerance and stress response, it is surprising that bleached tissues showed limited change compared to healthy tissues during the bleaching event. Detailed analysis of Symbiodiniaceae in specific coral colonies during stress and recovery could provide valuable insights into this critical symbiosis, with important implications for coral responses during major bleaching events. In a separate study, the role of stress-tolerant endosymbiotic dinoflagellate *Durusdinium* was found to maintain health and function of the coral host *Pocillopora acuta* via differential gene expression related to apoptosis, cell proliferation, cell–extracellular matrix adhesion, DNA damage repair, lipid catabolism and lipid homeostasis (Poquita-Du et al. 2020).

Comparative studies were also conducted on the patterns of Symbiodiniaceae communities in corals within the Indo-Pacific region, including coral samples from Singapore (Leveque et al. 2019; Tan et al. 2020; Ong et al. 2022). For instance, the microalgal (Symbiodiniaceae) symbionts in *Porites lutea*, a massive reef-building coral species found within Southeast Asia, were investigated by Tan et al. (2020). The study employed high-throughput DNA amplicon sequencing based on the nuclear internal transcribed spacer 2 (ITS2) to characterize the diversity, community structure, as well as biogeographic distribution of Symbiodiniaceae in *P. lutea* across Singapore and Peninsular Malaysia. It was revealed that *Cladocopium* was the dominant genus in all samples, with *Cladocopium* C15 being the most prevalent type. The study also revealed previously undetected Symbiodiniaceae forms associated with *P. lutea* in Southeast Asia. Moreover, endosymbiont diversity and community variation were driven by a combination of factors, including site-specific mean monthly cloud cover and variability in monthly sea surface temperature. This study not only provided baseline information for understanding variations in Symbiodiniaceae assemblages hosted by *P. lutea* but also offered insights into how these assemblages might reflect specific environmental conditions and coral responses.

A recent study investigated the endosymbionts associated with samples of common reef corals, including *Pachyseris speciosa*, *Pocillopora acuta* and *Diploastrea heliopora*, collected from ten marine sites along the coasts of the Malay Peninsula as well as three reef sites in Singapore (Ong et al. 2022). Analysis found that Symbiodiniaceae diversity in this region is greater than previously known. Endosymbiont communities are significantly structured by host species, remaining relatively invariant in *D. heliopora*. *Pachyseris speciosa* is associated significantly with *Cladocopium* and *Durusdinium*, while *P. acuta* and *D. heliopora* were dominated by *Durusdinium*. Environmental parameters distinctly affect Symbiodiniaceae communities and diversity across different coral species. In addition, higher sea surface temperatures (SST) positively affect endosymbiont diversity in *P. acuta*, while a higher SST range negatively impacts diversity in *P. speciosa* and *D. heliopora*. This study provided a baseline for subsequent comparative studies on how Southeast Asian reef corals acclimatize and adapt to changing environments by measuring natural changes in endosymbiont communities.

In addition to studies on endosymbiont (Symbiodiniaceae) communities of common hard coral species, other microbial communities associated with scleractinian corals have been investigated locally as well as regionally (Wainwright et al. 2020; Kanisan et al. 2023). Research has investigated the link between the age of coral host, *Porites lutea*, and microbial diversity (Wainwright et al. 2020). The study found no clear, consistent link between the age of the coral and the diversity or richness of its associated microbial communities. While some sites showed a decrease in microbial diversity with age, others displayed an increase, and in many cases, no significant relationship was observed. In addition, in more impacted reefs, there was evidence of a decrease in diversity with age, possibly due to environmental degradation, while in less impacted reefs, an increase in microbial richness and diversity was noted as the coral aged. The study emphasized that location played a significant role in structuring the microbial community, more so than the age of the coral. This suggests that environmental factors and site-specific conditions are critical in shaping the coral microbiome.

A recent study investigated the structure of bacterial communities associated with three scleractinian coral hosts, including *Pachyseris speciosa*, *Porites lutea* and *Diploastrea heliopora*, at three different regions in eastern and western Peninsular Malaysia and Singapore (Kanisan et al. 2023). Their findings revealed that corals from these distinct regions harbor unique bacterial communities. Among the coral species, *Pachyseris speciosa* exhibited the most variable microbiome, followed by *Porites lutea*, while the microbiome of *Diploastrea heliopora* was the least variable across all locations. The microbial communities associated with each region or site are likely influenced by local environmental conditions. In addition, the study identified a core microbiome shared across all three coral species. These core microorganisms may play crucial roles in supporting and maintaining coral health. For example, several of these microbes are known to metabolize dimethylsulfoniopropionate, contributing to sulfur cycling and the suppression of bacterial pathogens.

Several studies have investigated microbiome changes in local hard coral species through transplantation and temporal analyses. For instance, Deignan and McDougald (2022) demonstrated that the microbiome of *Pocillopora acuta* rapidly adapted to new environments following transplantation between the reefs of Pulau Satumu and Kusu Island, two Southern Islands of Singapore separated by approximately 15 km. Specifically, corals transplanted from Kusu to Pulau Satumu exhibited microbiome shifts within 1 day, while those moved from Pulau Satumu to Kusu shifted within 2 days. Earlier research found that *P. acuta* populations at Pulau Satumu and Kusu Island harbor distinct microbial communities despite their geographic proximity (Wainwright et al. 2019a; Fong et al. 2020). These findings underscore the influence of local environmental factors on coral microbiomes and their capacity for rapid adaptation. Despite the environmental differences, a small core microbiome persisted in *P. acuta*, consisting of a few consistent microbial species, including members of the Cyanobiaceae and Rhodobacteraceae families. The composition of these core members varied slightly between the two reefs. Differences in environmental conditions, such as nitrate concentrations, were observed between the two reefs and could potentially influence the microbiome's response to transplantation (Deignan and McDougald 2022). Results from the coral transplantation research suggested that the microbiome of *P. acuta* is highly dynamic and capable of rapid shifts in response to environmental changes. This adaptability may be a key factor in the resilience of corals in Singapore's urbanized reef environments, which are subject to high anthropogenic stress.

Fungal communities and diversity associated with more than 90 colonies of *Pocillopora acuta* from nine islands in Singapore were also investigated (Rabbani et al. 2021a). The mycobiome study revealed no clear spatial structuring of fungal communities based on sampling locations, suggesting that coral-associated fungi play roles beyond pathogenesis within their hosts. In addition, many of these coral-associated fungi have been tentatively identified as saprotrophs, likely involved in nutrient cycling to benefit their animal hosts, while others may function as symbiotrophs. Such study not only provides data on the composition of host-associated fungal communities but also offers valuable insights into the coral holobiont and its functioning.

A temporal study monitored the mucus microbiomes of two coral species, *Pectinia paeonia* and *Platygyra sinensis*, using 16S rRNA gene amplicon sequencing over a period of 1 year (Deignan et al. 2023a). The sampling methods involved the collection of coral colonies on seven occasions across both monsoon seasons at three different reef sites to track the spatial and temporal composition of their microbial communities. It was found that the microbiomes of the corals *Pectinia paeonia* and *Platygyra sinensis* displayed significant site-specific differences, suggesting that local environmental factors heavily influence the composition of coral microbiomes. In addition, the study found temporal shifts in the microbiome composition that correlated with the seasonal monsoons. During the southwest monsoon, an increase in Cyanobacteria and Crenarchaeota was observed, while Proteobacteria, the most abundant phylum, decreased. These shifts reflect the environmental changes brought by the monsoons, such as temperature and nutrient variations. Moreover, *Platygyra sinensis* showed more specific and significant temporal changes in its microbiome, linked to various environmental parameters like temperature and nutrients, especially at certain reef sites like Kusu Island and Pulau Satumu. In contrast, *Pectinia paeonia* exhibited more stability in its microbiome over time. Environmental factors such as nitrate, nitrite, ammonia and phosphate were found to correlate with the temporal changes in the coral microbiomes, particularly during the monsoon seasons. The study suggests that these environmental parameters play a crucial role in shaping the coral microbiome, with specific responses seen in different reef sites. Despite the intense anthropogenic impacts in Singapore's urbanized reefs, corals may have been able to persist at least partly due to the flexibility of their microbiomes. This adaptability could be a key mechanism allowing corals to survive in such challenging environments.

In another seasonal study, the bacterial and Symbiodiniaceae communities in three coral species, *Merulina ampliata*, *Pachyseris speciosa* and *Porites lutea*, were characterized over a two-year period (Quek et al. 2023). Seasonality had a limited influence on the bacterial and Symbiodiniaceae communities associated with these corals. Despite the presence of monsoon seasons, there were no consistent patterns of diversity changes linked to these seasonal variations. Bacterial communities exhibited significant temporal variability, but this variability was not consistently associated with environmental factors like temperature and rainfall. Instead, the bacterial communities were highly dynamic over time, showing changes that were independent of the monsoon seasons. In contrast to the bacterial communities, the Symbiodiniaceae communities remained stable across time and space, showing little variation in response to the monsoons or other environmental changes. This suggests that the coral endosymbionts are more resilient to seasonal changes than the associated bacterial communities. In addition, the study identified multiple potential biogeochemical processes facilitated by the interactions between bacteria and Symbiodiniaceae within the coral holobiont. The findings underscore the importance of long-term monitoring to understand the temporal dynamics of coral-associated microbial communities, particularly in seemingly stable equatorial environments where seasonal changes may have less impact.

The effects of macroalgae on the physiology and microbiome of several local hard coral species have been investigated. One study investigated the effects of the macroalgae *Lobophora* sp. and *Hypnea pannosa* on the physiology and microbiome of three Indo-Pacific coral species, *Merulina ampliata*, *Montipora stellata* and *Pocillopora acuta*, obtained from two reefs, Pulau Satumu and Kusu Island (Fong et al. 2020). Direct contact with the macroalga *Lobophora* sp. led to significant tissue bleaching in corals, with an average of 12.26% of the coral tissue affected. This contact also led to a notable suppression of the maximum quantum yield (Fv/Fm), decreasing to 0.529 as compared to 0.760 in control corals. In contrast, direct contact with *Hypnea pannosa* resulted in only slight but significant suppression of Fv/Fm. While direct contact with macroalgae had detrimental effects on coral physiology, no significant negative impacts were observed when corals were near the macroalgae or placed in contact with algal mimics. This suggests that water-mediated interactions did not compromise coral health in the same way as direct contact did. Both direct contact and water-mediated interactions with *Lobophora* sp. and *Hypnea pannosa* changed the prokaryotic community structures in the coral species *Montipora stellata*. However, the changes in microbiomes for *Merulina ampliata* and *Pocillopora acuta* were relatively more impacted by the source reefs from which the coral colonies were obtained, indicating that environmental factors also play a significant role. This research highlights the critical impact of macroalgal interactions on coral health, particularly through direct contact, while also emphasizing the complexity of these interactions in relation to the coral microbiome within a heavily urbanized reef system.

A recent field-based study further extended the investigation by using organic extracts of four macroalgal species, including *Bryopsis* sp., *Endosiphonia horrida*, *Hypnea pannosa* and *Lobophora challengeriae*, on coral health and microbiome of the same three hard coral species (Fong et al. 2023). After 24 h of exposure, crude extracts from all four macroalgal species significantly caused coral tissue bleaching and reduced effective quantum yield. The corals subsequently recovered within 3 days after the extracts were removed, except for corals exposed to *L. challengeriae*. The genus *Lobophora* is known to be chemically rich and it would be interesting to determine the causative agents leading to tissue bleaching in these corals (Vieira et al. 2017). While some macroalgal extracts led to an increase in the alpha diversity of coral microbiomes, no significant differences were observed in the composition and variability of coral microbiomes between the controls and the macroalgal extracts at each sampling time point. However, DESeq2 differential abundance analyses revealed species-specific responses in coral microbiomes. Overall, their findings suggested that chemically mediated interactions with macroalgae have a limited effect on coral microbiomes and highlighted the corals' ability to recover quickly from exposure to macroalgal-derived chemicals.

7.3.2 Microbiomes Associated with Marine Sponges

Within the Singapore marine environment, the diversity of bacterial communities in sponges is largely underexplored. This is despite numerous sponge species being documented in local waters (Lim et al. 2012a, 2016, 2023). To our knowledge, only two sponge species have been investigated for their microbial communities, *Xestospongia testudinaria* and *Cliona patera*.

The occurrence of the giant barrel sponges, *Xestospongia*, is ubiquitous in tropical reefs in Indo-Pacific and Atlantic (McGrath et al. 2018). These sponges are long-lived, able to grow beyond a metre and capable of processing large volumes of seawater, as well as participating in nutrient cycling pathways (Fiore et al. 2015). Some *Xestospongia* species can measure up to a base diameter of greater than 2.5 m, with coverage as high as 9% in some reef systems. Currently, more than 30 *Xestospongia* species, including *X. testudinaria* , *X. bergquistia*, *X. muta* and *X. exigua*, are recognized. Within the Southeast Asian region, these species are typically found in shallow tropical reefs as well as deeper waters.

Recent molecular-based research revealed that the giant barrel sponges belong to a pan-global species complex with 17 genetic clusters, consisting of up to nine cryptic species (Swierts et al. 2017; Evans et al. 2021). Morphologically similar species can occur sympatrically and could exhibit differences in their physiological roles. Members of the *Xestospongia* genus have also been well studied for their rich source of secondary metabolites, including steroids, terpenoids and alkaloids, with associated biological activities ranging from anti-inflammatory, antimicrobial to anticancer properties (Zhou et al. 2010).

Among the *Xestospongia* species, *X. testudinaria* is a well-known high microbial abundance sponge and has been studied due to its ecological importance as a habitat engineer and source of therapeutic drug agents (McGrath et al. 2018). This species is also one of the most common sponge species in Singapore waters (Lim et al. 2012a). Several microbiome studies have reported on the microbial community of *X. testudinaria* collected from different locations, including the Red Sea (Moitinho-Silva et al. 2014), Malaysia (Hayami et al. 2023), north and south Sulawesi (Cleary et al. 2015), Northwest Java (de Voogd et al. 2015), East Kalimantan (Polónia et al. 2017), West Papua (Cleary et al. 2018), Vietnam (Dat et al. 2018), South China Sea (Jin et al. 2014), as well as Singapore (Swierts et al. 2018; Pires et al. 2019). Using culture-independent methods, these studies revealed that *X. testudinaria* tends to have consistent microbial communities dominated by phyla Acidobacteria, Actinobacteria, Chloroflexi and Proteobacteria (Pires et al. 2019).

Several functional roles have been attributed to the sponge-associated microbes that could contribute to the resilience of the host animal to environmental perturbations. For instance, microbial genes involved in the nitrogen cycle, including nosZ, nifD, nirK, norB and nrfA genes, were found to be enriched in *X. testudinaria* sponge samples (de Voogd et al. 2015). Another study found *X. testudinaria* to be enriched for KEGG orthologs related to the toxins-antitoxin system, nutrient starvation and export of heavy metals (Cleary et al. 2015). Moreover, studies involving the isolation of associated microbes from the host samples revealed that their microbial extracts exhibited potent antibacterial and cytotoxic activities, suggesting that these microbial symbionts provided sponge chemical defense (Cita et al. 2017).

The microbiome of *Xestospongia testudinaria* occurring within Singapore waters was previously investigated using culture-independent methods, such as pyrosequencing/amplicon sequencing (Swierts et al. 2018; Pires et al. 2019). These studies consistently revealed that phylum Chloroflexi was the most abundant microbial group in the host sponge. Previous studies showed that Chloroflexi is one of the most common and diverse bacterial group in sponges and consists of many sponge-specific lineages (Schmitt et al. 2011). Metagenomic and single-cell genomic methods to characterize the functional gene repertoire of Chloroflexi symbionts in marine sponges suggested clade-specific metabolic specialization, and that these unique symbionts have the genomic potential for dissolved organic matter degradation from seawater (Bayer et al. 2018). Moreover, members of the Chloroflexi phylum have the capacity to synthesize bioactive polyketides in marine sponges, suggesting that they could play a role in sponge defense against predators (Bayer et al. 2018).

Other abundant microbial phyla found in local *X. testudinaria* include Crenarchaeota, Acidobacteriota, Proteobacteria and Actinobacteriota. High diversity was also found in a pyrosequencing-based analysis of *X. testudinaria* from the Red Sea, where Chloroflexi was also dominant but followed by Proteobacteria and Firmicutes. The microbiome composition of *X. testudinaria* did not differ across reef sites at Pulau Hantu, Sisters' Islands Marine Park and Kusu Island (Deignan et al. 2023b). Interestingly, the study found significant differentiation in microbiome composition by host sponge *X. testudinaria* genetics in a manner that was aligned with the proposed species boundaries. In another study by Swierts et al. (2018), samples of *X. testudinaria* were collected from eight areas across the Indo-Pacific region, including Red Sea, Mayotte, Phuket, Koh Tao, Pattaya, Vietnam, Lembeh Makassar, Taiwan and Singapore. It was revealed that variation in the prokaryotic community of the sponge is primarily associated with geography as opposed to phylogenetic relatedness (Swierts et al. 2018).

In view of the ecological importance of *X. testudinaria* as well as its common occurrence in Singapore waters, a pilot-scale study using culture-dependent methods was initiated to investigate the nature of bacteria associated with samples of *X. testudinaria* collected from three locations, Pulau Hantu, Kusu Island and Pulau Semakau (unpublished). The study used three different marine media, as well as phylogenetic analysis, to uncover associated bacteria from *X. testudinaria* at the three reefs. Previous microbiome studies conducted on local *X. testudinaria* used culture-independent approaches, such as the 16S rRNA gene barcoded pyrosequencing method, to assess the diversity of associated microbes (Deignan et al. 2023b). The culture-dependent approach can complement previous culture-independent studies and has implications for further studies, such as uncovering the biotechnological/therapeutic potential and ecological roles of isolated marine bacterial strains.

Samples of *X. testudinaria* were collected from three locations: Pulau Hantu, Kusu Island and Pulau Semakau, at a depth of about 4.5 m. Nine sponge specimens were collected using SCUBA and transported to a laboratory within 2 h for further processing. To isolate and culture bacteria associated with the sponge, three types of seawater media (M1, M2 and M3) were used. The growth of bacterial colonies was monitored over 5 weeks. Medium M2, which was

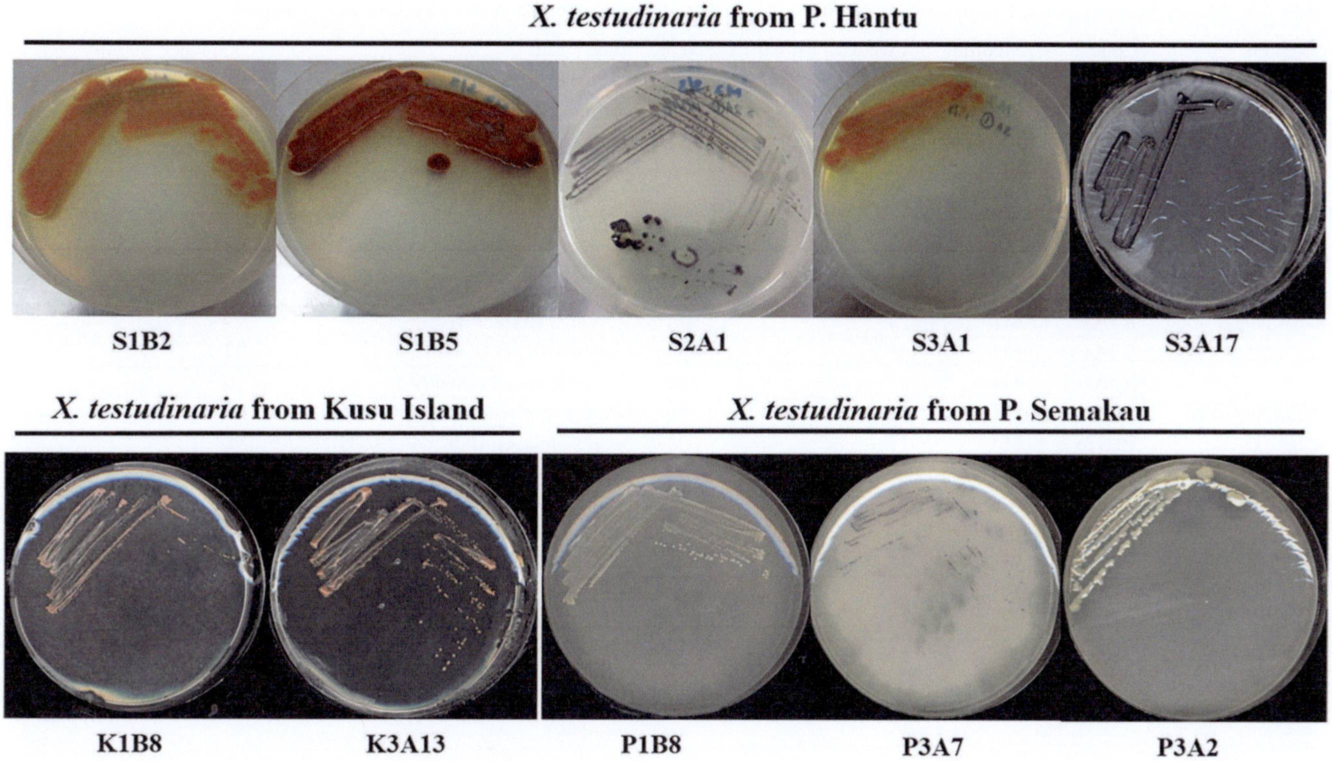

Fig. 7.1 A selection of 10 isolated marine bacterial colonies plated on marine agar petri dishes (100 mm × 15 mm) from samples of *Xestospongia testudinaria* obtained from Pulau Hantu, Kusu Island and Pulau Semakau

the least selective, showed rapid microbial growth, with many bacterial colonies appearing in the first week. In contrast, media M1 and M3 displayed slower bacterial growth with fewer colonies. After weeks 3–5, new bacterial colonies were still observed on M1 and M3, but M2 plates were mostly covered by the second week, with little new growth later.

Different microbial growth patterns were observed using these marine media. The differences in the growth rates and quantity of bacteria found on the different media plates could be due to the different amounts of nutrients present in the media. For instance, M2 plates contained 10 g/L of soluble starch as a carbon source, which is much higher in concentration as compared to M1 and M3 media. The higher concentration of carbon source in medium M2 could likely facilitate the growth of fast-growing bacteria, which tended to cover the plate over a short period of 2–3 weeks. The smaller amounts of carbon source in media M1 and M3 most likely explained the smaller number of bacterial colonies and slower growth rates observed. Regardless of the differences in media compositions, the recovery of associated marine bacteria from sponge samples using media M1 to M3 in this study was consistent with reports by other researchers for the isolation of heterotrophic microbes from environmental marine samples (Jensen et al. 1991).

Following Gram-staining for easier visualization under a microscope, the bacterial colonies displayed various pigments, including dark red, orange, pink, purplish-black and black, with some colonies being white or colorless (Fig. 7.1). Of the 84 marine bacterial strains isolated, slightly less than 50% had their genomic DNA successfully extracted based on the DNA extraction protocol by Lee et al. (2003) as compared to only 10% success rate using commercial DNA extraction kits. The 16S rRNA gene fragment (~1500 bp in length) was then PCR-amplified using the universal primers 27F and 1500R (Woese et al. 1983). Out of 40 bacterial genomic DNA samples that underwent PCR, 29 of the samples showed PCR success.

Among the 29 amplified PCR products, only 20 bacterial strains had DNA sequences that were of high enough quality to be aligned using BLAST with known 16S rRNA gene sequences in the GenBank database (Table 7.1). Bacterial PCR samples that could be aligned using BLAST were derived from eight, three and nine marine bacterial strains obtained from sponge samples at Pulau Hantu, Kusu Island and Pulau Semakau, respectively (Table 7.1). Moreover, among the isolated bacterial strains that were successfully sequenced, most of these strains were isolated from medium M1, with 55% of the sequenced bacterial strains from media M1, 15% from M2 and 30% from M3.

Table 7.1 Marine bacteria isolated from samples of *Xestospongia testudinaria* collected at three locations in Singapore

Location	Code	Medium	Week of growth	Nearest type strain	Sequence identity (%)
Pulau Hantu	S1A9	M1	1	*Bacillus altitudinis* (KM369867.1)	100
	S1B10	M3	1	*Bacillus altitudinis* (KM369867.1)	99
	S1B11	M1	2	*Labrenzia alba* (FR750958.1)	63
	S1A13	M3	3	*Staphylococcus warneri* (KM660697.1)	100
	S1B16	M1	4	*Bacillus altitudinis* (KM369867.1)	100
	S1A17	M3	4	*Vibrio tubiashii* (CP009354.1)	99
	S2A10	M3	1	*Bacillus vietnamensis* (KF933662.1)	79
	S3A22	M1	4	*Pseudoalteromonas lipolytica* (KJ917337.1)	100
Kusu Island	K3A3	M1	1	*Vibrio neptunius* (EU834014.1)	99
	K3A13	M3	3	*Bacillus subtilis* (KF009681.1)	97
	K3A14	M1	3	*Vibrio alginolyticus* (JN188409.1)	100
Pulau Semakau	P1B8	M1	2	*Phaeobacter daeponensis* (NR_044026.1)	99
	P1B15	M3	2	*Marinobacter lacisalsi* (NR_044392.1)	98
	P2A1	M1	1	*Vibrio sinaloensis* (JN871696.1)	99
	P2A3	M2	1	*Vibrio fortis* (KJ577028.1)	100
	P2A6	M1	4	*Vibrio fortis* (KJ577028.1)	99
	P3A2	M2	2	*Bacillus altitudinis* (KM253188.1)	100
	P3A9	M1	2	*Pseudomonas pachastrellae* (AY880300.1)	100
	P3B12	M2	2	*Bacillus altitudinis* (KM369867.1)	99
	P3B16	M1	5	*Halomonas cupida* (AB681327.1)	99

The BLAST analysis carried out via search through GenBank revealed that the 20 bacterial isolates belonged to nine genera: *Labrenzia*, *Halomonas*, *Pseudomonas*, *Marinobacter*, *Phaeobacter*, *Pseudoalteromonas*, *Staphylococcus*, *Bacillus* and *Vibrio*. Most of the marine bacterial strains were from the genera *Bacillus* and *Vibrio*, comprising 35% and 30% of all isolated strains, respectively. Among the 20 bacterial strains, 60% of the strains isolated were phylum Proteobacteria, while the remaining 40% were phylum Firmicutes. In general, both *Vibrio* and *Bacillus* species were found associated with sponge samples from all three locations (Table 7.1). In addition, *Pseudoalteromonas*, *Staphylococcus* and *Labrenzia* species were only identified in bacterial samples associated with sponge samples from Pulau Hantu, while *Pseudomonas*, *Halomonas*, *Marinobacter* and *Phaebacter* species were found in bacterial strains obtained from sponge samples derived from Pulau Semakau.

A majority of the 20 isolated marine bacterial strains had high percentage identity (more than 97%) with known bacterial species, including *Bacillus altitudinis*, *B. subtilis*, *Staphylococcus warneri*, *Vibrio alginolyticus*, *V. tubiashii*, *V. neptunius*, *V. sinaloensis*, *V. fortis*, *Pseudoalteromonas lipolytica*, *Phaeobacter daeponensis*, *Marinobacter lacisalsi*, *Pseudomonas pachastrellae* and *Halomonas cupida* (Table 7.1). Several of the marine bacterial strains with high percentage identity with known bacterial strains belonged to bacterial genera known to be prolific producers of bioactive compounds and useful extracellular agents. These bacterial genera include *Bacillus* (Mondol et al. 2013), *Vibrio* (Mansson et al. 2011), *Pseudoalteromonas* (Holmström and Kjelleberg 1999), *Pseudomonas* (Isnansetyo and Kamei 2009) and *Halomonas* (Biswas et al. 2023). In addition, two marine bacterial strains, S1B11 and S2A10, isolated from sponge samples from Pulau Hantu, could potentially be novel strains due to their low percentage identity with their nearest bacterial match with *Labrenzia alba* at 63% and *Bacillus vietnamensis* at 79%, respectively.

Several known marine bacterial species found closely clustered with the isolated bacterial strains in this preliminary study were also reported from various marine sources and are known to produce bioactive molecules. For instance, the heterotrophic *B. altitudinis* has been found associated with marine algae and the sponge, *Haliclona simulans*, and its bacterial extract is reported to possess anti-proliferative activity against human hepatocellular adenocarcinoma (Phelan et al. 2012). Previous research has shown that a strain of *B. altitudinis*, isolated from marine sediment, can produce antifungal compounds against *Magnaporthe oryzae*, a fungal strain that causes rice blast (Jin et al. 2012). Another strain, *B. vietnamensis*, an endospore-forming halotolerant bacterium, was previously reported from Vietnamese fish sauce, sponge and marine sediments (Noguchi et al. 2004).

An isolated bacterial strain in this study, S3A22 (Table 7.1), is most closely related to *Pseudoalteromonas lipolytica*, known for synthesizing bioactive compounds. For instance, a brown pigment, pyomelanin, has been isolated from *P. lipolytica*, whereby its biofilms have displayed anti-larval activity against the settlement and metamorphosis of the mussel *Mytilus coruscus* (Zeng et al. 2015, 2017). Strains K3A3 and K3A14, whose closest relatives are *V. neptunius* and *V. alginolyticus*, respectively, are known to produce secondary metabolites (Table 7.1) (Mansson et al. 2011). Lastly, two known marine bacterial strains, *Pseudomonas pachastrellae* and *Phaeobacter daeponensis*, are closely related to strains P3A9 and P1B8, respectively (Table 7.1). The former marine bacterium was previously isolated from a deep-sea marine sponge, *Pachastrella* sp. (Romanenko et al. 2005), while *P. daeponensis*, a facultatively anaerobic bacterium, was cultivated from marine sediment (Dogs et al. 2013). Taken together, the isolation of strains belonging to bacterial species/genera known to be prolific producers of bioactive metabolites could suggest their defensive roles in the host sponge, *X. testudinaria*, and contribute to host resilience and its widespread occurrence in the Indo-Pacific reef systems. Further ecology-based research is needed to verify the defensive roles of these associated marine bacteria.

The marine bacterial strains isolated here did not belong to the typically dominant bacterial phyla, such as Chloroflexi and Acidobacteria, which were detected in previous studies using culture-independent methods (Deignan et al. 2023b). Of the 84 strains initially isolated, 20 were identified as belonging to Proteobacteria or Firmicutes. The remaining unidentified strains may belong to Chloroflexi or Acidobacteria, but challenges in obtaining high-quality DNA hindered their phylogenetic analysis. These phyla may also be difficult to isolate due to the slow-growing nature of certain microorganisms. Future studies could include longer incubation times (e.g. months) in combination with increasing agar media types and detection of minicolonies (Davis et al. 2011). These techniques may prove useful in the isolation of hard to culture microbes from marine sponges to provide further insights into the roles these bacterial groups play in sponge-microbe interactions as well as in biodiscovery efforts.

7.3.3 Microbiomes Associated with Nudibranchs

DNA metabarcoding techniques have been employed to investigate the microbial (e.g. bacterial and fungal) communities associated with the 'blue dragon' nudibranch *Pteraeolidia semperi* (Ng et al. 2022). This species is widely distributed across its native Indo-Pacific range and is commonly found in Singapore reefs at shallow water depths. The bacterial *Mycoplasma* and *Endozoicomonas* genera were consistently detected across all samples and locations. In addition, the fungal genus *Leucoagaricus* had the highest occurrence, marking the first report of its presence in marine environments. Analysis of the co-occurrence network indicated limited interactions between bacterial and fungal communities. However, the bacterial family Colwelliaceae emerged as a potential keystone taxon due to its disproportionately high number of network connections. Colwelliaceae was also found to cluster with other bacterial families, such as Pseudoalteromonadaceae and Alteromonadaceae, all of which are likely involved in food digestion of the host animal.

A recent study based on sequencing and phylogenetic analysis of the cytochrome c oxidase subunit I marker from 48 *P. 'semperi'* samples obtained in Singapore's southern waters, along with morphological assessments using scanning electron microscopy of the radula, revealed the presence of several distinct cryptic lineages within *P. 'semperi'* (Soon et al. 2023). Based on metabarcoding of the internal transcribed spacer 2 gene, it was shown that the Symbiodiniaceae communities associated with *P. 'semperi'* in Singapore predominantly host *Cladocopium* and *Durusdinium* symbionts and is significantly influenced by biogeography.

7.3.4 Microbiomes Associated with Macroalgae

Large, canopy-forming macroalgae (seaweed) are important primary producers that naturally inhabit many tropical and temperate reef ecosystems. They create complex habitats and provide food for a wide variety of reef-associated species (Evans et al. 2014), but also compete with other benthic organisms for space (Hughes et al. 2007). On coral reefs, significant shifts from coral-dominated to macroalgal-dominated systems after disturbances have underscored the competitive dynamics of these interactions (Graham et al. 2006) and contributed to the prevailing view that large macroalgae pose a threat to coral reef resilience (Hughes et al. 2007). *Sargassum* (Phaeophyceae), a brown leathery macroalga, is the dominant canopy-forming species on many Indo-Pacific coral reefs (Hughes et al. 2007). Numerous *Sargassum* species display distinct seasonal growth patterns as they maintain their holdfasts year-round while shedding lateral branches following reproduction or in response to heightened stress (Ang Jr 1986). The temporal and spatial variations in the density and length of *Sargassum ilicifolium* thalli were previously studied on coral reefs in Singapore, with thallus length exhibiting strong seasonal patterns and temperature identified as a key influencing factor. In addition, significant spatial variations in both thallus density and length were also observed across different reefs in Singapore (Low et al. 2019; Kwan et al. 2022).

An investigation on the microbial communities found within the macroalga *Sargassum ilicifolium* in a highly urbanized marine environment was conducted (Oh et al. 2021). The study employed several methods to analyze the microbiome associated with the macroalga *S. ilicifolium*. Eight complete thalli of *S. ilicifolium* from eight island locations in Singapore along a 14 km east-to-west transect were analyzed, revealing that microbial communities associated with different parts of *S. ilicifolium* (holdfast, vesicles and leaves) are significantly different from one another. The holdfasts exhibited the most distinct microbial community composition compared to the other structures. Despite the short distances between sampling locations, significant variation in microbial community composition was revealed. This suggests that even small spatial scales can lead to microbial community divergence, influenced by local environmental conditions. Metagenome functional predictions indicated that leaf-associated microbial communities had a higher representation of taxa involved in nitrogen cycling, while holdfast-associated communities were enriched with taxa related to sulfur cycling. This suggests that different parts of the macroalga may play distinct roles in nutrient cycling within the marine ecosystem. The research provides useful baseline data that can be employed for monitoring changes in microbial communities over time and lays the groundwork for understanding the complexities of reef-associated microbial communities and their ecological roles in the functioning of marine ecosystems.

A separate study was conducted to characterize the mycobiome of *S. ilicifolium* at eight sites in Singapore (Wainwright et al. 2019b). It was discovered that fungal communities exhibited significant differences among sampling localities, even at relatively small spatial scales (≤12 km). This indicates that geographic locality plays a crucial role in shaping fungal diversity. The Aspergillaceae family was the most encountered across all islands, with the genus *Aspergillus* being particularly prevalent. In addition, a large proportion of the fungal taxa identified were unclassified, highlighting the potential for discovering many novel fungal taxa in marine environments. For instance, between 40% and 75% of the recovered taxa was unable to be assigned to a specific class, indicating a significant gap in our current understanding of marine fungal biodiversity. These findings underscore the complexity and variability of fungal communities in marine ecosystems and the need for further research to fully characterize this biodiversity.

7.3.5 Microbiomes Associated with Seagrasses

Halophila ovalis, a fast-growing seagrass species found in both the tropical Atlantic and Pacific bioregions, is known for its high potential for long-distance dispersal (Short et al. 2007; Wu et al. 2016). This seagrass is a preferred food source for the charismatic and threatened megaherbivore *Dugong dugon* (Preen 1995; Nakaoka and Aioi 1999). Due to the contribution of mycobiome in promoting host health and limited studies on seagrass-associated fungi, an investigation on the mycotic communities associated with *H. ovalis* across the Malay Peninsula and Singapore was conducted (Quek et al. 2021). The study revealed that geographic location significantly influences the structure of fungal communities associated with *H. ovalis*, suggesting that location is a more critical factor than the specific plant structure, such as leaf, rhizome or root, in shaping fungal communities. Given that location plays a significant role in determining fungal community structure, the study recommends that seagrass transplantation efforts should consider the origin of the seagrass and its associated fungal communities to ensure the success of restoration projects.

Bacterial communities found on different parts of *Halophila ovalis* were also investigated using high-throughput 16S rRNA amplicon sequencing methods (Yan et al. 2021). It was found that the bacterial assemblages associated with *H. ovalis* varied significantly between locations, with distinct bacterial communities found in different plant parts. In addition, it was revealed that a distance–decay relationship was observed, indicating that dispersal limitations may account for the structuring of bacterial communities. Moreover, bacterial indicator amplicon sequence variants (ASVs) associated with either degraded or healthy seagrass meadows were identified in the study. These indicator ASVs, which may signal anthropogenic stress or environmental decline in seagrass habitats, are useful in developing proactive conservation and management strategies for seagrass ecosystems.

A similar microbiome study was conducted on another seagrass species, *Enhalus acoroides*, collected from Singapore and Peninsular Malaysia. Significant differences in fungal communities associated with the seagrass were found across different locations (Wainwright et al. 2019c). In a related study, the bacterial communities associated with each plant part and the corresponding sediment samples were analyzed based on high-throughput 16S rRNA gene amplicon sequencing (Rabbani et al. 2021b). The results indicate that bacterial communities exhibit geographic structuring, with a significant distance–decay pattern, suggesting that dispersal limitations contribute to the observed differences in bacterial community structure. The bacterial communities also varied based on the specific part of the collected seagrass, such as leaf, root and rhizome, and several microbial indicator species associated with different plant parts. Further analysis uncovered the presence of microbial taxa, recognized in previous studies, as indicators of 'unhealthy' or 'stressed' seagrass meadows.

7.4 Biomedical and Biotechnological Applications of Marine Microbiomes

Marine microbiomes associated with marine invertebrates are a prolific source of natural products with immense biotechnological and pharmaceutical potential (Tan 2023). These microbial communities often form symbiotic relationships with their host invertebrates, such as sponges, corals and molluscs, contributing to the production of bioactive compounds that can protect the host from predators, diseases and environmental stressors. These compounds, which include anti-infectives, antifungals, antivirals and anticancer agents, have attracted significant attention due to their potential applications in medicine and industry. The unique conditions of their native environments, such as high pressure, salinity, and temperature variations, drive the evolution of novel chemical structures within these microbial communities, making them a valuable reservoir of new and unexplored natural products (de la Calle 2017).

The study of marine microbiomes associated with invertebrates offers insights into the complex ecological interactions and evolutionary processes that give rise to these bioactive compounds. Understanding these interactions can lead to the discovery of new molecules with unique mechanisms of action, which are crucial in the development of new drugs, especially in an era where antibiotic resistance is a growing concern. Additionally, exploring the marine microbiome opens avenues for sustainable bioprospecting, as many marine invertebrates are difficult to cultivate or harvest in large quantities. By focusing on their microbial symbionts, researchers can potentially harness these natural products without overexploiting marine resources, thereby contributing to the conservation of marine biodiversity (Tan 2023).

Symbiotic microbes associated with several species of common benthic marine invertebrates, including sponges and soft corals, found in Singapore waters, have been explored for their bioactive natural products. Using an integrated approach, including culture-dependent method, MS-molecular networking, bioinformatics and bioassay screening, several microbial strains have been shown to produce bioactive natural products, as well as the presence of unique biosynthetic gene clusters. The following sections highlight some of these research findings.

7.4.1 Biomedical Potential of Marine Bacteria Associated with Common Benthic Invertebrates

Marine bacteria associated with invertebrates, such as sponges and soft corals, are important sources of new bioactive natural products as they can comprise diverse complex microbial communities in different physiological conditions (Blockley et al. 2017). These can produce a wide variety of chemical compounds with unique structures and ecological functions (e.g. chemical defense), and many of these compounds can be used for the development of novel drugs. The growing recognition that marine bacteria associated with invertebrate hosts are involved in the biosynthesis of secondary metabolites offers new alternatives for the discovery and development of bioactive marine natural products (Simmons et al. 2008). However, there are bottlenecks when developing potential drugs from marine invertebrate-associated bacteria responsible for natural product biosynthesis, including isolation and cultivation of bacterial symbionts. These issues present significant challenges due to microbial symbionts' dependence on host-supplied factors for growth and secondary metabolite production (Haygood et al. 1999). Moreover, less than 1% of microorganisms can be cultured, and these are not representative of total microbial diversity (Rodrigues and de Carvalho 2022).

The significance of microbial associates in sponges as sources of biologically active secondary metabolites became evident when a striking structural similarity was discovered between sponge-derived compounds and those identified in terrestrial organisms from entirely different taxa (McCauley et al. 2020). For example, similarities have been found between the chemical structures of mycalamides A and B, derived from the marine sponge *Mycale hentscheli* and pederin, a toxin previously isolated from the South American *Paederus* beetles (Perry et al. 1988; Tammam and El-Demerdash 2023). This suggests that certain bioactive sponge-derived molecules are possibly produced by biosynthetic enzyme clusters originating from the sponges and/or associated microorganisms. Other examples are the polybrominated biphenyl ether antibiotics originally isolated from the sponge *Dysidea herbacea* that are found to be produced by the endosymbiotic cyanobacterium, *Oscillatoria spongeliae* (Schmitt et al. 2021).

Octocorals, or soft corals, are known to host diverse microbial communities (van de Water et al. 2018). Studies have demonstrated that the coral microbiome plays a crucial role in protecting the coral host through the production of antimicrobial compounds, inhibition of pathogenic enzymes, interference of pathogen communication and competitively excluding pathogens from host surfaces (van Oppen and Blackall 2019). In addition, coral-associated bacteria are capable of producing a wide array of structurally diverse compounds with various bioactivities, including antimicrobial agents effective against a broad range of coral pathogens (Sang et al. 2019).

In view of the importance of the marine microbiome associated with benthic macroorganisms as sources of bioactive compounds, a pilot-scale study was initiated to explore the biomedical potential of these microbes isolated from common benthic invertebrates in Singapore (unpublished). A total of six common marine invertebrates were selected, including three sponges, *Gelliodes fibulata*, *Clathria reinwardti* and *Xestospongia testudinaria,* as well as three soft corals *Sacrophyton tenuispiculatum*, *Sinularia* sp. and *Lobophytum crassum*. The culturable marine bacteria were isolated in three different marine media, such as M1, M2 and M3 marine media. Small-scale fermentations of these isolated bacterial strains were performed, and their organic extracts were prepared for biological assessment in a panel of assays, including the brine shrimp (*Artemia salina*) toxicity, cytotoxicity (MOLT-4, human leukaemic cell line) and antibacterial assays. Antibacterial activity was tested against three gram-positive, namely, *Bacillus cereus*, Methicillin-resistant *Staphylococcus aureus* (MRSA) and *Enterococcus faecalis* V583 as well as five gram-negative bacteria, namely, *Escherichia coli*, *Escherichia coli* EC 598, *Pseudomonas aeruginosa*, *Burkholderia cenocepacia* and *Acinetobacter baumannii*. In total, 57 bacterial isolates were isolated from six benthic marine invertebrates using three different marine media.

Small-scale fermentation (1L) was performed on all 57 marine bacterial isolates using the marine medium from which they were originally isolated. Fermentation flasks were then incubated at 28 °C on a rotary shaker at 150 rpm for a period of 21 days. Organic solvents, including ethyl acetate and methanol, were used for the extraction of natural products for further testing in the brine shrimp (*Artemia salina*) toxicity, antibacterial and cytotoxic (MOLT-4) assays. Twelve (21%) of the crude bacterial extracts displayed activity (>25%) in the brine shrimp toxicity assay when tested at 1000 ppm. These bacterial isolates were isolated from sponges *Gelliodes fibulata* (S1 and S4), *Clathria reinwardti* (S6) and *Xestospongia testudinaria* (S16), as well as from soft corals *Sacrophyton tenuispiculatum* (C3, C4, C5 and C8) and *Sinularia* sp. (C9, C10, C11 and C13). In addition, eight bacterial extracts (S1, S6, C3, C4, C8, C9, C11 and C13) exhibited significant toxicity of 100% at 1000 ppm.

In vitro cytotoxicity assay based on MOLT-4 cancer cells line was used to investigate anticancer potential of these bacterial isolates. Of the 57 crude bacterial extracts tested, 26 and 6 of the bacterial extracts displayed cytotoxicity of >25% when tested at 1000 ppm and 100 ppm, respectively. Ten (S6, S10, S22, S24, C6, C8, C9, C10 and C11) and two (S6 and C8) bacterial extracts exhibited 100% activity at 1000 ppm and 100 ppm, respectively. Of the 26 (~46%) bacterial extracts, which exhibited activity (>25%) at the 1000 ppm, 14 and 12 bacterial extracts came from bacteria isolated from sponges and soft corals, respectively.

Hospital-acquired infection (HAI) is one of the major threats to public health and is a substantial cause of morbidity and mortality (Sunarti 2024). The majority of HAIs are caused by a few groups of bacterial species, including *Escherichia coli*, *Pseudomonas aeruginosa* and *Acinetobacter baumannii*, which are commonly found in hospital environments (Ventola 2015). In addition, with the increasing antibiotic usage, as well as misapplication of these drugs, there is a significant increase in antibiotic-resistant bacterial strains for all the dominant species causing HAIs, which further complicates the therapeutic strategies. Thus, the development of new drugs or control strategies as alternative therapies is clearly a matter of urgency. At present, the vast majority of currently available antibiotics have been isolated from terrestrial sources, with recent research strongly suggesting that marine microbes represent an untapped source of new antibacterial compounds (Monciardini et al. 2014).

Here, crude organic extracts from the 57 marine bacterial isolates (from sponges and soft corals) were screened for antibacterial activity using well diffusion on Mueller–Hinton agar plates (Table 7.2). In primary screening against *Escherichia coli* and *Bacillus cereus* at 1000 ppm, 14 extracts (24%) showed zones of inhibition (ZOI) ≥ 5 mm. Specifically, eight extracts from sponges and six from soft corals were effective. Against *E. coli*, 12 extracts produced ZOI of 9.5–29 mm, while against *B. cereus*, four extracts produced ZOI of 8–16 mm. At 100 ppm, nine extracts were active against *E. coli* (ZOI 5–18.5 mm), but none against *B. cereus*. Notably, extracts S8 and S20, derived from sponges *Clathria reinwardti* and *X. testudinaria*, inhibited both bacteria at 1000 ppm. These promising results prompted further secondary screening against pathogenic bacterial strains.

Table 7.2 Bacterial strains used for antibiotic screening of crude organic marine-derived bacterial organic extracts

Screening	Gram Staining	Bacterial strain	Disease
Primary	Negative	*Escherichia coli* HB 101	Gastrointestinal infection
	Positive	*Bacillus cereus* 111	Nausea, vomiting and diarrhea (foodborne illness)
Secondary (Pathogen)	Negative	*Escherichia coli* EC598	Urinary tract infection
		Pseudomonas. aeroginosa PA01	Meningitis, endocarditis and microbial keratitis
		Burkholderia cenocepacia WT-H111	Immunocompromised and cystic fibrosis
		Acinetobacter baumannii LY1	Urinary tract infection
	Positive	Methicillin-resistant *Staphylococcus aureus* (MRSA) BAA-40 ATCC	Skin infection
		Enterococcus faecalis V583	Gastrointestinal and genital tracts of humans

Fourteen bacterial extracts that displayed activity in the primary screening at 1000 ppm were selected for secondary screening against six antibiotic-resistant bacteria, including Methicillin-Resistant *Staphylococcus aureus* (MRSA), *Enterococcus faecalis* V583, *Escherichia coli* EC 598, *Pseudomonas aeruginosa*, *Burkholderia cenocepacia* and *Acinetobacter baumannii*. Nine bacterial extracts displayed activity (ZOI ≈ 17 mm to 19 mm) against resistant pathogens, *B. cenocepacia*, *A. baumannii* and/or *P. aeruginosa* at 1000 ppm, which is either comparable or greater than positive control standard antibiotics, kanamycin sulfate, tetracycline, chloramphenicol and ampicillin (ZOI ≈ 11 mm to 22 mm). At a lower concentration of 100 ppm, none of these bacterial extracts displayed any activity. However, bacterial extracts of S14 (ZOI ≈ 18 mm) and C22 (ZOI ≈ 19 mm) displayed activity only against *B. cenocepacia* and *P. aeruginosa*, respectively, while S8, S18. S20, C14, C19, C20 and C26 (ZOI ≈ 17 mm to 19 mm) displayed activity only against *A. baumannii* at 1000 ppm. Of the nine bacterial extracts which displayed activity ZOI ≥ 5 mm at 1000 ppm, four (S8, S14, S18 and S20) and five (C14, C19, C20, C22 and C26) were obtained from sponges (namely *Clathria reinwardti* and *Xestospongia testudinaria*) and soft corals (i.e. *Sinularia* sp. and *Lobophytum crassum*), respectively.

Of the 57 bacterial isolates, only eight bacterial isolates had successful pure DNA extractions using the liquid nitrogen method for DNA extractions. Phylogenetic analysis based on 16S rDNA gene sequences analysis of these bacterial isolates revealed that S21 and S16 represent 88% homology between each other, whereas bacterial isolate S4 is closely related to *Arthrobacter* genus with 55% homology. The genus *Arthrobacter* is identified as a member of the high G+C content branch of Gram-positive bacteria 'actinomycete', phylogenetically closely related to other *Coryneform* genera such as *Aureobacterium*, *Cellulomonas*, *Curtobacterium* and *Microbacterium*. Actinomycetes are virtually inexhaustible sources of novel compounds with many therapeutic applications and hold an outstanding position due to their diversity and demonstrated ability to produce novel bioactive compounds. Phylogenetic analysis of five bacterial isolates, C14, C16, C20, C22 and C26, revealed that soft coral samples *Sinularia* sp. and *Lobophytum crassum* have bacterial communities related with *Staphylococcus* genus of Firmicutes phylum.

This study presents a pilot-scale analysis of the biomedical potential of marine bacteria isolated from marine invertebrates (sponges and soft corals) from reefs in Singapore. About 65% (37 of 57) of bacterial isolates displayed activities, in which 61% (19 of 31) and 69% (18 of 26) of bacterial strains were obtained from sponges and soft coral, respectively. Significantly, 21%, 46% and 24% of the bacterial extracts displayed activity in brine shrimp toxicity (>25%), cytotoxicity (anticancer) (>25%), and in antibacterial assay (ZOI ≥ 5 mm), respectively. Interestingly, one (C3) of the bacterial extracts obtained from soft coral *Sarcophyton tenuispiculatum* displayed activities in all three bioassays. These results underscore the potential of marine bacteria isolated from marine invertebrates collected from Singapore for drug discovery, as they are still widely unexplored and might provide a valuable resource for new pharmaceutical lead compounds. The combination of different bioactivity assays with chemical analysis is a useful strategy to prioritize bacterial isolates for further bioassay-guided isolation work.

7.4.2 Integrated Omics Approach to Drug Discovery from Microbes Associated with Biotic and Abiotic Sources

To facilitate the discovery of anti-quorum-sensing molecules from microbes associated with marine samples, an integrated omics approach, including metabolomics and genomics, has been applied (Ong et al. 2019). Based on a culture-dependent method, marine bacteria were obtained from 13 subtidal marine samples, including sponges and sediments, procured using a rectangular dredge in the Singapore Strait. Preliminary crude organic extracts, obtained through small-scale fermentation of isolated marine bacterial colonies, were screened for anti-quorum-sensing property using the *Pseudomonas aeruginosa* bioreporter strain. Selected bacterial extracts that showed quorum-sensing inhibitory (QSI) activity were subsequently analyzed using a mass-spectrometric-based metabolomics Global Natural Products Social Molecular Networking platform (GNPS; https://gnps.ucsd.edu/) for compound dereplication (Allard et al. 2016). Furthermore, marine bacterial strains that showed significant biological activity were subjected to whole genome sequencing for annotation of biosynthetic gene clusters based on antiSMASH bioinformatics tool (Blin et al. 2019). The integration of metabolomics and genomic methods used in the study is effective in the selection of marine bacterial strains with a high probability of novel bioactive compound discovery.

Homogenates from 13 marine samples were plated onto eight types of marine media based on previous publications (Magarvey et al. 2004; Choi et al. 2015; Duncan et al. 2015; Maansson et al. 2016). Colonies that showed interesting morphologies, such as bright colors, matte textures or unique colony shapes, were selected as colonies of interest. Certain common microbial colonies appearing across different isolation agar plates were also isolated as colonies of interest to ensure non-biased colony selection. As a result, a total of 102 bacterial colonies of interest were obtained over a two-months' incubation period. Sponge samples from *Geodia* sp.

yielded the highest number of bacterial isolates (20), followed by *Stelletta* sp. (13), *Coelocarteria singaporensis* (11), *Haliclona* sp. (9), *Dysidea* sp. (8), *Xestospongia testudinaria* (7), *Rhabdastrella globostellata* and cf. *Leiodermatium* sp. (2), *Halichondria* sp. and *Ircinia* sp. (1).

Studies have shown that less than 1% of the bacteria associated with marine samples can be isolated and cultured successfully in the laboratory (Rodrigues and de Carvalho 2022). As such, the use of appropriate marine media for bacterial isolation is important at the onset of the study. The eight marine media employed in the study consisted of both nutrient-rich and minimal nutrient media types (Magarvey et al. 2004; Choi et al. 2015; Duncan et al. 2015; Maansson et al. 2016). Media plates based on A4HT, a minimal nutrient medium, produced the highest number of colonies of interest (29 isolates), while plates with A1, a nutrient-rich medium, produced only four bacterial colonies. A possible explanation could be that the nutrient-rich media promoted fast-growing bacterial types to outgrow the slower growing colonies in the isolation agar plates. Although the nutrient-rich media provided numerous bacterial colonies compared to the plates with minimal nutrient media, a majority of these fast-growing bacteria were not of interest for isolation. The focus was on the slow-growing bacterial types, which are usually rare and challenging to isolate.

Subsequently, the organic extracts were obtained from 102 colonies of interest, derived from 13 marine samples, via small-scale culturing and screened, at 50 mg/mL, in the bioreporter strain, *Pseudomonas aeruginosa* PAO1 *lasB-gfp*, (Rasmussen et al. 2005) for QSI property. Bacterial extracts that indicated a reduction of the green fluorescent protein (GFP) production in relation to the controls were considered as positive hits. The induction of the *lasB* gene, encoding elastase, is a proxy to detect LasR activity, and any decrease in GFP would indicate the presence of possible antagonist(s) of the autoinducer, 3-oxo-C12-HSL, which results in the inhibition of lasB expression. Many bacterial types, including pathogenic strains, employ a quorum-sensing system to coordinate virulence and biofilm development (Haque et al. 2018). The quorum-sensing systems in the pathogenic strain, *Pseudomonas aeruginosa*, comprise two circuits that utilize acyl-homoserine lactone signals, along with a third system that employs quinolone signals. Collectively, these three interconnected QS systems control the expression of hundreds of genes, including virulence factors.

From the preliminary screening efforts, EtOAc extracts derived from five marine bacterial strains, including #24, #27, #33 and #34 from *Geodia* sp. and #52 from *Coelocarteria singaporensis,* inhibited the GFP expression in the bioreporter strain. Extracts were subsequently assayed in a dose-dependent manner to confirm the reduction of the GFP expression without affecting the bioreporter bacterial growth rates. In addition, metabolomic analysis of the bacterial extracts using the MS-based molecular networking platform was carried out (Wang et al. 2016). Unfortunately, marine bacterial extracts #27 and #33 did not achieve optimal ionization in the tandem mass spectrometry based on their respective total ion chromatogram (TIC) profiles. As such, only marine extracts #24, #34 and #52 were analyzed for compound dereplication and comparative analyses. The compound dereplication process is a rapid analytical tool to shortlist crude marine bacterial extracts for downstream chemical investigations. The analysis provided an informed decision in prioritizing certain bacterial strains for eventual large-scale culturing and isolation work for potential new molecules. Based on the molecular networking analysis, bacterial strains #24 and #52 indicated unique clusters of ionized compounds. For instance, the molecular clusters from strain #24 revealed relatively higher molecular-weight compounds than in strain #52. The use of MS-based metabolomic technique coupled with relevant bioassay resulted in the selection of the three bacterial strains for the downstream large-scale fermentation and chemical investigations.

Isolation of genomic material from these selected marine bacterial strains were carried out for phylogenetic analysis. Several marine bacterial strains derived from the sponge, *Geodia* sp., that did not show clear QSI activity were included for the phylogenetic analysis due to the high number of bacterial strains isolated from this sponge species. By including these other bacterial strains obtained from this sponge, we hope to gain insights on the sponge-associated bacterial communities. Only 29 marine bacterial strains were eventually identified based on their 16S rRNA gene sequencing. Based on sequences in the NCBI GenBank database, these bacterial strains were affiliated with 14 different bacterial genera, of which one belonged to the uncultured group. Two out of the three priority strains that were of interest, due to their positive QSI activity and molecular networking analysis, exhibited more than 98% confirmation to uncultured bacterial clones (Garza and Dutilh 2015). The findings suggested that the two marine bacterial strains probably belong to a putatively novel bacterial genus class or that these marine bacterial strains were not previously studied in detail.

The highest number of colonies of interest isolated were found to be affiliated with the genus *Kocuria* (nine colonies), followed by *Micrococcus* and uncultured (three colonies each), *Micromonospora*, *Gordonia* and *Bacillus* (two colonies each). The majority of the bacterial colonies identified belong to the Actinomycetales phylum (e.g. *Kocuria*, *Micrococcus*, *Micromonospora*, *Streptomonospora* and *Gordonia*), with others classified under Proteobacteria (e.g. *Kiloniella*, *Pseudomonas*, *Alcanivorax* and *Vibrio*) and Firmicutes (e.g. *Bacillus*, *Staphylococcus* and *Paenibacillus*). Phylogenetic analysis showed that three uncultured bacterial clones (#24, #25 and #52) form a distinct clade, potentially representing a novel genus closely related to *Kiloniella* and *Bacillus*. Further phenotypic and genotypic studies are needed to confirm their taxonomic classification.

Whole genome sequencing was performed on two priority bacterial strains (#24 and #52) to identify natural product biosynthetic gene clusters (BGCs) using antiSMASH 4.0.2 (Blin et al. 2017). These strains, which exhibit anti-quorum-sensing activity, are potential novel marine bacterial taxa. Strain #24 has a genome size of 3.8 Mb and a GC content of 41.3%, while strain #52 has a 5.6 Mb genome with 32.5% GC content. Six putative BGCs were identified in #24, including clusters for siderophore, bacteriocin, terpene, a type III polyketide/saccharide hybrid and two non-ribosomal peptide synthetase (NRPS) clusters, with similarities to lichenysin and bacilysin BGCs. In strain #52, 13 putative BGCs were found, including clusters for siderophore (petrobactin), NRPS, terpene, bacteriocins, sactipeptide (thurincin H), arylpolyene and others. This genomic analysis lays the groundwork for future studies on their biosynthetic and bioactive potential.

This study is the first in Singapore to combine biological, genomic and metabolomic methods to discover anti-quorum-sensing compounds from marine samples. Five isolated marine bacteria produced extracts with anti-quorum-sensing activity, with three from sponges (*Geodia* sp. and *Coelocarteria singaporensis*) likely producing novel or new analogue compounds, as no matches were found in the GNPS library. The findings emphasize the value of a targeted, integrated approach in bioprospecting and demonstrate the usefulness of molecular networking-based dereplication for prioritizing bacterial isolates for further chemical investigation and large-scale compound purification (Purves et al. 2016).

7.4.3 Microbiome of *Cliona patera* and Its Biomedical Importance

This study investigates the diversity and biomedical potential of marine bacteria associated with the Neptune's Cup sponge, *Cliona patera*, in Singapore using both culture-dependent and culture-independent methods (Ho et al. 2021). *C. patera*, known for its distinctive wine glass shape and large size, was first described in Singapore in the nineteenth century but was presumed extinct by the early twentieth century due to overcollection (Hardwicke 1822; Low 2012; Lim et al. 2012b). Remarkably, it was rediscovered in Singapore in 2011 and later found in Cambodia in 2018. It was revealed that the rediscovered population of *C. patera* in Singapore exhibited low genetic diversity and faces challenges to recovery due to environmental pressures, anthropogenic impacts and severe predation reducing sponge sizes by up to 5.6% per month (Quek et al. 2022).

The rediscovery of *C. patera* in Singapore provides a rare opportunity to study its natural microbiome. Researchers analyzed the prokaryotic diversity in different parts of the sponge using 16S rRNA amplicon sequencing and screened cultured bacteria from two sponge colonies for QSI activity using *Pseudomonas aeruginosa* PAO1 biosensor strains. Since disrupting bacterial quorum sensing offers a novel strategy against infections (Saurav et al. 2016), selected QSI-active strains were further analyzed through mass spectrometry-based metabolomics using the GNPS platform to annotate their metabolomes (Nguyen et al. 2013).

The study revealed that *Cliona patera* harbors a complex and diverse prokaryotic microbiome, with over twice the number of distinct OTUs compared to surrounding seawater (Ho et al. 2021). The microbiome was consistent across all six sponge individuals and clearly distinct from the seawater community, indicating a host-specific microbial assemblage, consistent with findings from other sponge species (Schmitt et al. 2007; Sharp et al. 2007; Turon et al. 2018). No significant differences were observed between samples collected in January and February, though further sampling is needed to assess the microbiome's long-term stability.

The microbial community of *C. patera* is dominated by Proteobacteria, particularly Gammaproteobacteria and Alphaproteobacteria, which together make up 70% of its microbiome, consistent with other studies on Clionaidae sponges (Webster and Thomas 2016; Pita et al. 2018; Sacristán-Soriano et al. 2020). This dominance supports *C. patera*'s classification as a low microbial abundance (LMA) sponge (Moitinho-Silva et al. 2017b). Other notable taxa include Bacteroidetes, Actinobacteria, Acidobacteria, Archaea and several minor groups, with 10.5% of the community unclassified. Unlike many sponges, *C. patera* lacks a significant presence of Chloroflexi and Poribacteria (Thomas et al. 2016). The genus *Ruegeria*, part of the metabolically versatile *Roseobacter* clade (Karimi et al. 2019), was found in all sponge samples but not in the surrounding seawater, highlighting its potential ecological and biomedical importance.

Using culture-dependent methods, 110 bacterial colonies were isolated from *C. patera* sponge samples, with diverse media contributing to the range of isolates. About 30.9% of the bacterial extracts showed moderate to strong QSI activity against *Pseudomonas aeruginosa* PAO1 *lasB-gfp* tested at 100 μg/mL. These QSI-active strains were found in both the cup and stem regions of the sponge, marking the first report of such activity from different sponge regions. The active strains belonged to Firmicutes (class Bacilli) and Proteobacteria (class Alphaproteobacteria), both known producers of diverse bioactive compounds.

Mass spectrometry-based molecular networking of organic extracts from QSI active marine bacterial strains (Bacilli and Alphaproteobacteria) revealed a chemically diverse profile. Initial analysis of five highly active extracts identified 591 precursor ions, corresponding to various natural product families, including siderophores (e.g. dehydroxynocardamine), pyrrolidine derivatives (e.g. anisomycin), biosurfactants (e.g. surfactin D), diketopiperazines, indole alkaloids (e.g. harmaline), aromatic polyketides and pyrone derivatives. Further analysis of extracts from 11 QSI-active strains uncovered additional surfactin derivatives, highlighting a broad and rich chemical diversity.

The study by Ho et al. (2021) provides the first comprehensive characterization of the microbiome of the Neptune's Cup sponge, *C. patera*, establishing important baseline data for future reef health monitoring in Singapore. The sponge hosted 5222 distinct OTUs, mainly from Proteobacteria, especially Gammaproteobacteria and Alphaproteobacteria. Several associated marine bacteria, particularly from the classes Bacilli and Alphaproteobacteria, showed strong potential for producing novel QSI compounds based on mass spectrometry-based molecular networking analysis.

7.4.4 Coral Microbiome Manipulation Towards Long-Term Climate Resilience

Microbiome manipulation is a well-established technique in medicine, such as using fecal microbiota transplantation to treat *Clostridium* infections (Harkins et al. 2020). However, its application in non-humans is still in its early stages. Research has demonstrated the feasibility of altering coral-associated bacterial communities through bacterial inoculation (Damjanovic et al. 2019). Initial studies have focused on treating coral diseases, e.g. white pox disease (Alagely et al. 2011), or on bioremediation of oil pollution (Fragoso Ados Santos et al. 2015; Silva et al. 2021). More recent studies exploring microbiome manipulation to improve coral bleaching resilience are promising, though no direct link between inoculated bacteria and increased bleaching tolerance has been definitively established. For instance, inoculation of coral species, *Pocillopora damicornis*, with an associated bacterial consortium isolated from corals and surrounding seawater had potential benefits, such as antagonistic activity against a common coral pathogen, ROS-scavenging enzyme catalase activity, as well as possible contributions to sulfur and nitrogen cycling (Rosado et al. 2019). A similar study on *Mussismilia hispida* demonstrated that a probiotic mix reduced bleaching and enhanced recovery after thermal stress (Santoro et al. 2021). Another study showed short-term bleaching mitigation in previously heat-sensitive corals after inoculation with microbiomes from heat-tolerant conspecifics (Doering et al. 2021).

Overall, these studies showed limited, short-term uptake of inoculated bacteria or changes in the coral's bacterial community composition after inoculation. Long-term stability of inoculated bacteria has not been demonstrated, though one study observed divergent microbiome communities in coral juveniles 4 months after a single transplant from different adult coral species. For microbiome manipulation to provide lasting benefits and enhance coral bleaching resilience, it is crucial that beneficial bacteria establish a stable, long-term association with the coral host, reducing the need for repeated inoculations.

A pilot study was conducted in Singapore in 2020 to examine if inoculation of bacterial communities from a coral species resilient to bleaching to a susceptible coral species would improve recovery from bleaching under heat stress. In this (unpublished) study, mucus containing microbiome extracted from *Acropora millepora* (bleaching resilient) was inoculated in *Pachyseris speciosa* (bleaching susceptible) and experimental coral fragments were subjected to temperature stress conditions (31 °C). Treatments consisted of un-inoculated and inoculated *P. speciosa* and *A. millepora* fragments (Controls) subjected to ambient, low temperature (28 °C) and high temperature (31 °C) seawater conditions over 10 days. Corals were then examined for recovery using pulse amplitude-modulated (PAM) fluorometry.

The photosynthetic efficiency of the coral fragments was assessed as the maximum quantum yield after dark adaptation and effective quantum yield after light adaptation using a PAM fluorometer (Diving-PAM, Walz, Germany). Measurements for maximum quantum yield were conducted after coral fragments were dark adapted for 15 h at the start and the end of the experiment. Additionally, measurements for effective quantum yield were conducted after coral fragments were light adapted for 5 h on days 1, 3, 4, 5, 6, 7, 8 and 10. Readings were taken in triplicate per treatment tank (3 fragments per tank). The main PAM settings were Measuring Light Intensity (MI) = 3 and Gain (G) = 1.

Results showed significant interaction effects between exposure time, treatment (inoculation) and temperature regimes in the maximum quantum yields measured after dark adaptation. Examination of the interaction plots of maximum quantum yield at Days 1 and 10 of the experiment for each combination of treatments showed a general increase in maximum quantum yield after 10 days in most treatment conditions, except for *A. millepora* controls in low temperature regimes and inoculated *P. speciosa* in high temperature regimes where the Day 10 maximum quantum yields were lower than at the start of the experiment (Fig. 7.2). This may suggest that the photosynthetic efficiency of inoculated *P. speciosa* after dark adaptation (maximum yield) improved over time only at 28 °C temperature regimes, and not at 31 °C temperature regimes. These observations, together with the upward trends in photosynthetic efficiency observed for inoc-

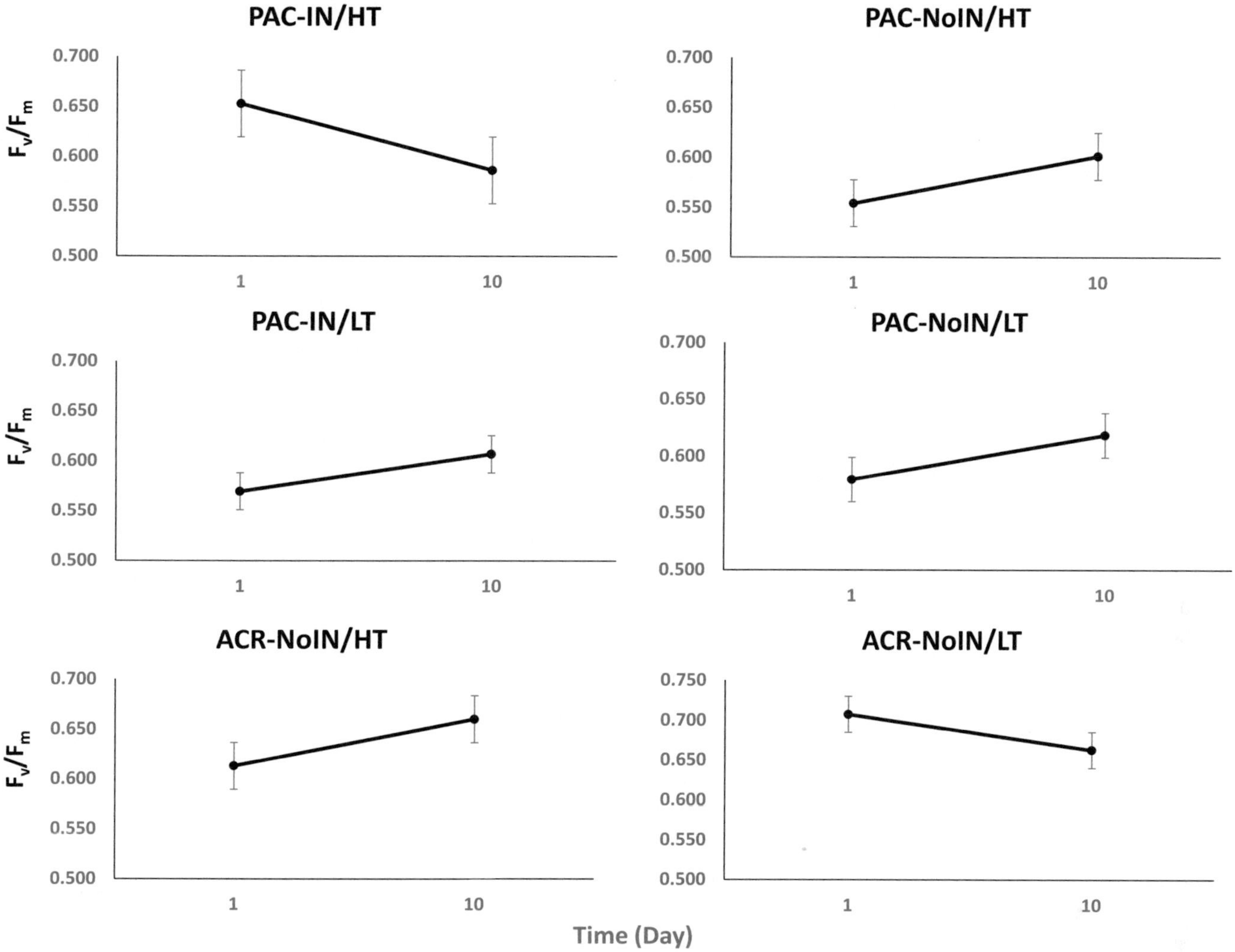

Fig. 7.2 Interaction plots of mean maximum quantum yield of *Pachyseris speciosa* and *Acropora millepora* under different treatment and temperature conditions after dark adaptation from Day 1 and Day 10. Each data point represents the mean ± 1 standard error, $n = 3$. PAC—*Pachyseris speciosa*; ACR—*Acropora millepora*; IN—Inoculated; NoIN—Non-inoculated; LT—28 °C; HT—31 °C

ulated *P. speciosa* at 28 °C, non-inoculated *P. speciosa* at 28 °C and 31 °C temperature regimes, suggest that the inoculation of resilient coral mucus did not significantly increase photosynthetic efficiency at the high temperature regimes.

Effective quantum yield measurements (representing photosynthetic efficiency) on coral fragments after light adaptation also showed similar general trends but did not yield statistical differences. Results for effective quantum yields in *P. speciosa* did not show significant differences from Day 3 to Day 10 after light adaptation. There was an overall decrease in photosynthetic efficiency between Day 3 and Day 10 in both inoculated and non-inoculated *P. speciosa* fragments at the high temperature (31 °C) treatments (Fig. 7.3), while a slight increase in photosynthetic efficiency in inoculated and non-inoculated *P. speciosa* fragments was observed after 10 days in 28 °C treatments, similar to the results from the measurement of maximum quantum yield (Fig. 7.2). Hence, both photosynthetic efficiency measurements (after dark and light adaptations) seem to indicate that the inoculation of resilient coral mucus only improved the photosynthetic efficiency in *P. speciosa* at lower temperatures, although the results were not statistically significant. Measurements of photosynthetic efficiency from the control *A. millepora* fragments at both temperature regimes also seem to indicate that this species is more resilient to temperature stress.

In terms of species composition, the microbiomes of all coral samples analysed in this study were dominated by Proteobacteria (61.9% ± 17.9), the most abundant bacterial phylum in all sample groups (Fig. 7.4). In contrast, the field control *A. millepora* fragments yielded a higher abundance of Cyanobacteria (25%). Desulfobacterota was the second most abundant phylum (8.9% ± 8.2) present in all samples, followed by Firmicutes (7.2% ± 5.3). The high abundance of Desulfobacterota (mainly Desulfobacteraceae and

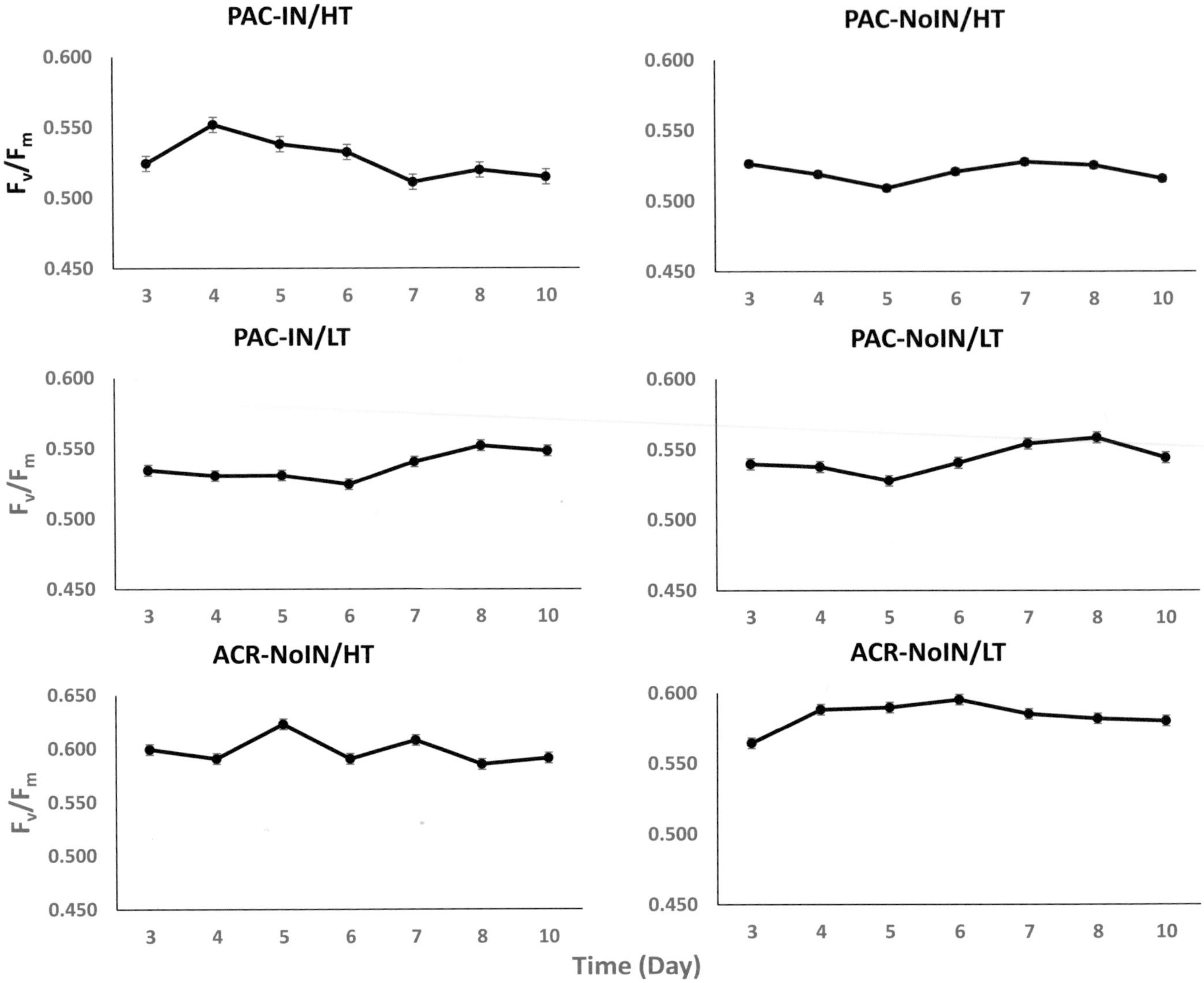

Fig. 7.3 Mean effective quantum yield of *Pachyseris speciosa* and *Acropora millepora* under different treatment and temperature conditions after light adaptation from Day 3 to Day 10. Each data point represents the mean ± 1 standard error. PAC—*Pachyseris speciosa*; ACR—*Acropora millepora*; IN—Inoculated; NoIN—Non-inoculated; LT—28 °C; HT—31 °C

Desulfovibrionaceae) was largely represented in the Experimental tank fragments (10.8%) and less so by the Field control and Acclimated samples (0.0%).

Analysis of microbiome communities showed that *A. millepora* field control samples had 23 indicator ASVs that were significantly different from treatment coral fragments ($p < 0.01$). These indicator ASVs comprised 40% cyanobacteria Synechococcales and 40% halophilic archaea bacteria Halobacterales. The remaining 20% comprised two ASVs of Sphingomonadales, two ASVs of Staphylococcales and one unidentified Gammaproteobacteria. Fragments of *P. speciosa* yielded two indicator ASVs each for Field control and Acclimated samples, which belonged to Synechoccocales and Oceanospirillales (genus *Endozoicomonas*), respectively. All other location treatments for both species did not have indicator species above an indicator value of 0.7. Several ASVs belonging to Stenotrophomonas (2 ASVs) and Rhodobacteraceae (1 ASV) were found to drive the dissimilarities in microbiome communities of *A. millepora* between the acclimation and experimental tanks ($p < 0.01$). In addition, ASV76 and ASV98, belonging to the family Vibrionaceae, were found exclusively on *A. millepora* from the acclimation tank. This was the species that was subsequently used as the inoculant. The two ASVs were in high average abundance of 28% and 24%, respectively, and contributed 30% of the total abundance. Field control samples of both *A. millepora* and *P. speciosa* had a larger abundance of Cyanobacteria (Fig. 7.4).

Microbial communities of *A. millepora* and *P. speciosa* were significantly different between coral species

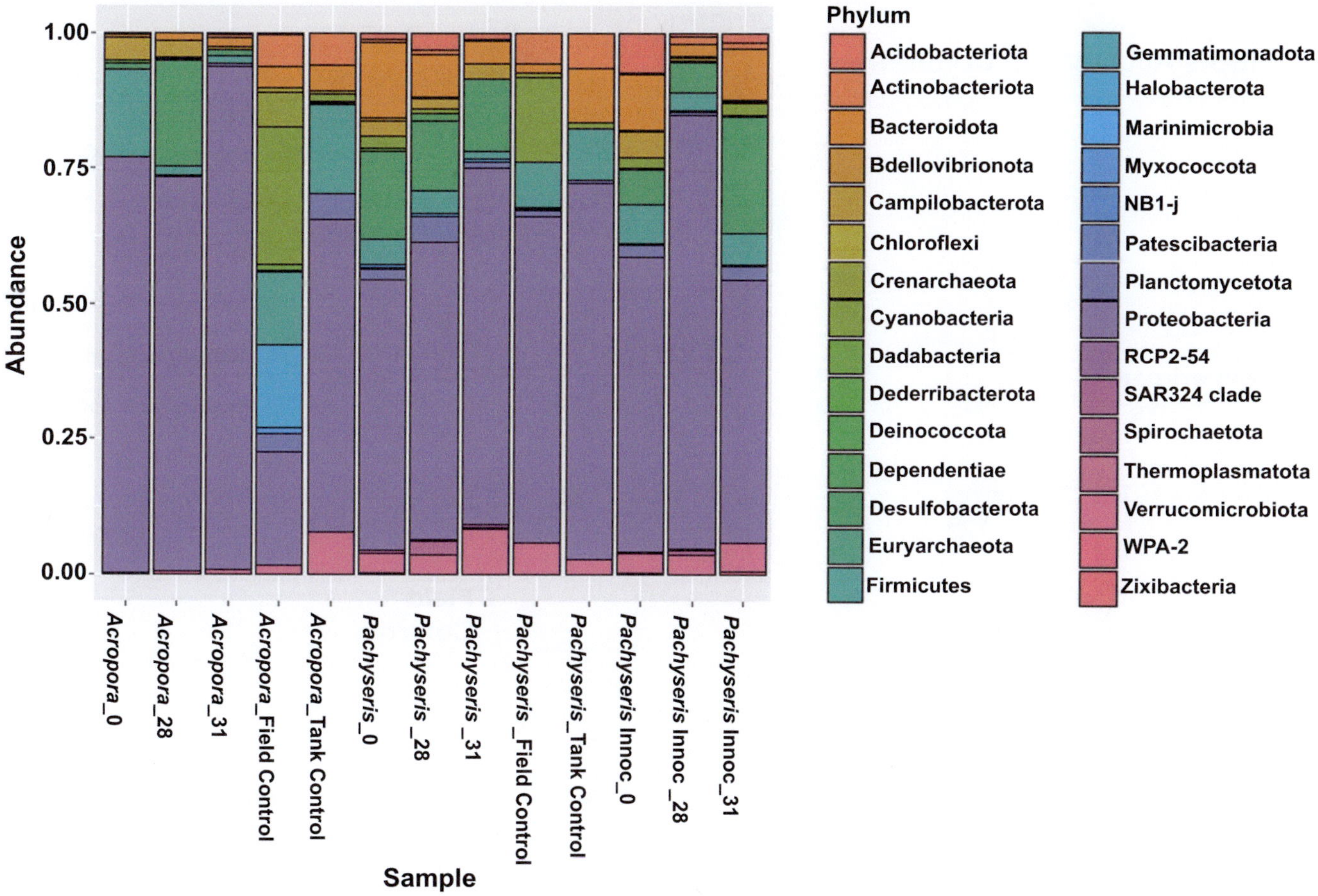

Fig. 7.4 Mean relative abundance of all phyla separated by treatment groups

(PERMANOVA, pseudo-F = 2.4827, P = 0.001). Coral microbial communities of both coral species were highly dynamic, shifting significantly with each translocation (PERMANOVA, pseudo-F = 3.1559, P = 0.001), specifically between the experimental conditions (tanks) and acclimated and field control samples. Lower dissimilarity was observed between the acclimated and field control samples (Fig. 7.5). Distinct clusters of microbiome communities were observed for *A. millepora* and *P. speciosa* in both the Experimental tank and Field control samples on the nMDS plot (Anderson 2008), in line with the PERMANOVA results. The microbiome structure of *A. millepora* field control fragments was more distinctly dissimilar to all other samples (Fig. 7.5).

Analysis of microbiome communities under different treatments in Day 10 fragments using PERMANOVA showed that temperature was not a significant factor in determining dissimilarities in microbial communities within the Experimental tank samples, as well as within treatment groups for *A. millepora,* inoculated *P. speciosa* and uninoculated *P. speciosa*. The microbial communities of inoculated and uninoculated *P. speciosa* were not significantly dissimilar (Fig. 7.6). The microbiome from inoculated *P. speciosa* was also not dissimilar to the original *A. millepora* inoculant. The only significant dissimilarity in microbiomes was between uninoculated *P. speciosa* and the *A. millepora* inoculant (Fig. 7.6). The differences observed between the inoculant and uninoculated *P. speciosa* were driven by eight ASVs belonging to the phyla Desulfobacterota (4 ASVs), Verrucomicrobiota (2 ASVs), Firmicutes (1 ASV) and Proteobacteria (1 ASV).

Coral microbial communities are known to be species specific but temporally and spatially dynamic (Dunphy et al. 2019; van Oppen and Blackall 2019), and environmental stressors can increase both alpha and beta diversity (McDevitt-Irwin et al. 2017). Increased dispersion in the translocated coral samples of the Experimental and Acclimation tanks indicates increased beta-diversity, which may be indicative of stress. Experimental tank samples also had relatively high abundances of Desulfobacteraceae and Desulfovibrionaceae, which are stress-associated bacteria capable of degrading dimethylsulfoniopropionate (DMSP) (Rice et al. 2019), and stressed corals have been reported to increase production of DMSP, which, when degraded, produces antioxidative compounds functionally important in the coral stress response (Raina et al. 2013).

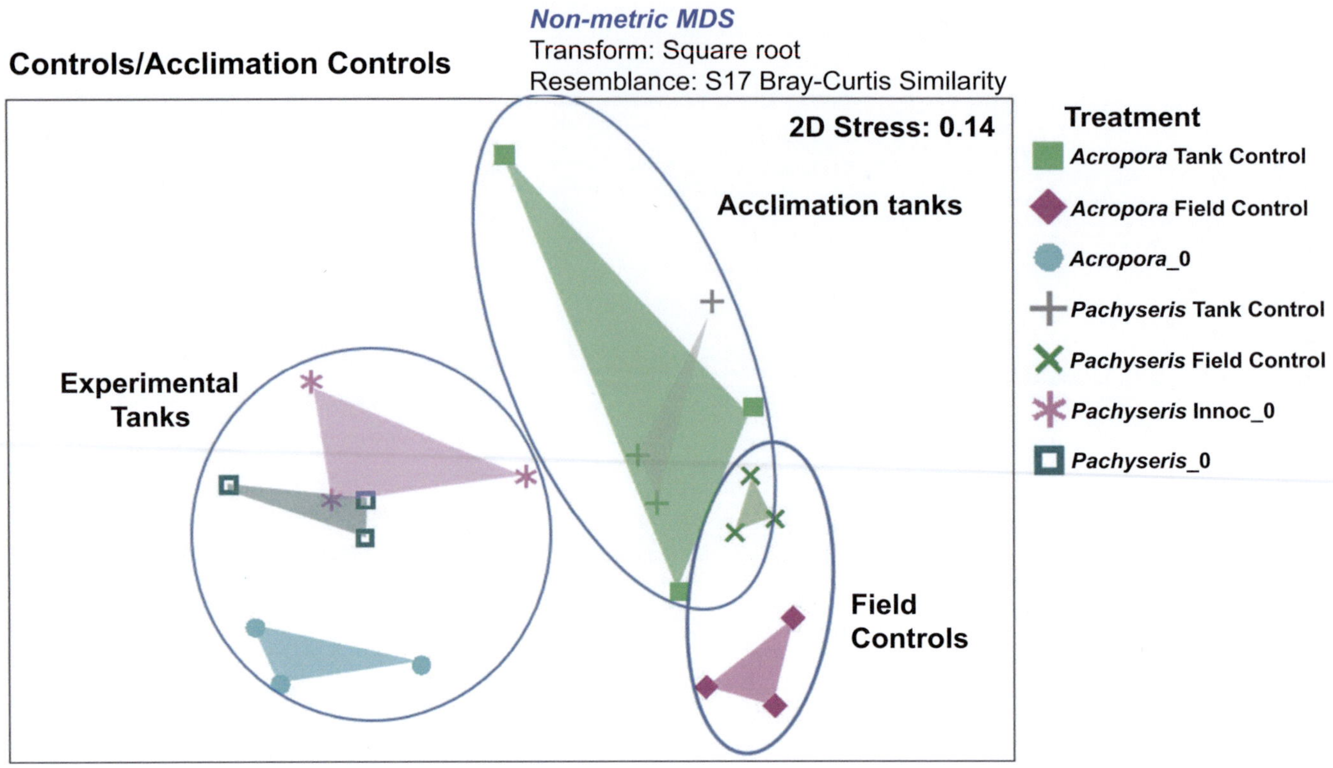

Fig. 7.5 nMDS plot of Field control, Acclimated and Experimental tank (Day 0) coral samples over the first 19 days of the experiment. Experimental tank (Day 10) samples are not included in this plot

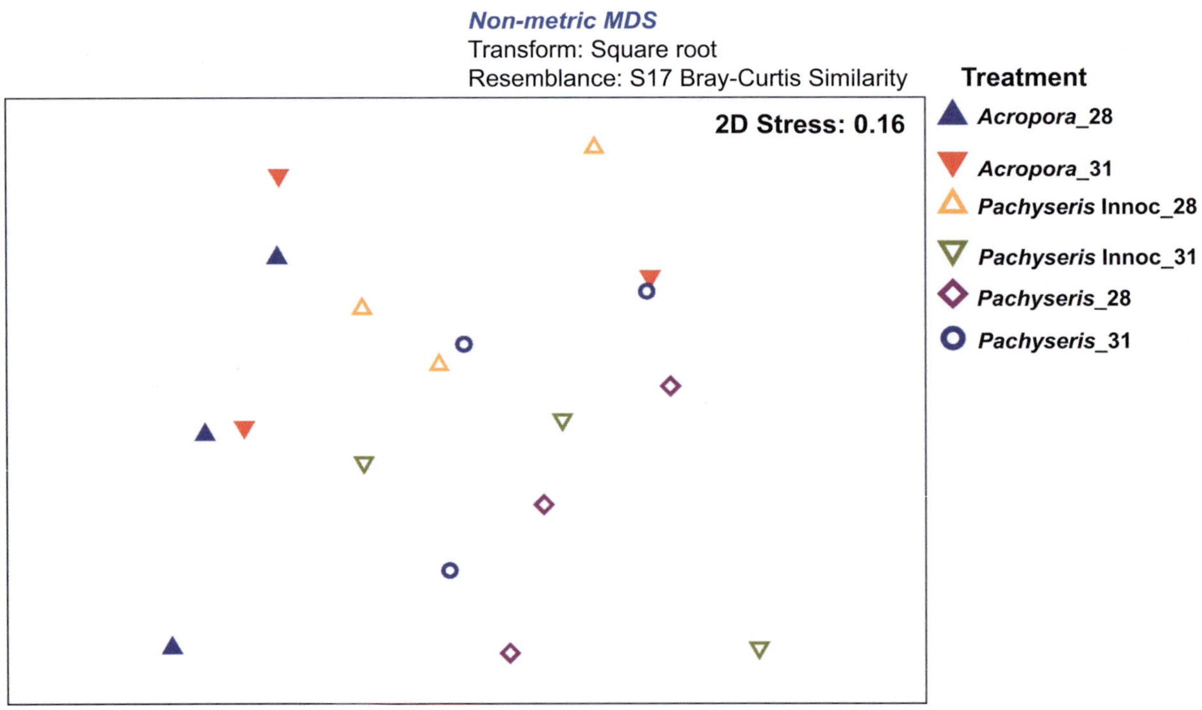

Fig. 7.6 nMDS plot of microbial communities found on temperature-treated *Acropora millepora* and *Pachyseris speciosa*

During stress events, the coral microbiome is susceptible to opportunistic bacteria colonization, which could have resulted in the high number of unique ASVs observed in the Experimental tank samples of *P. speciosa*. The nMDS plots (Fig. 7.5 and 7.6) suggest that the Experimental tank and Field control samples had more distinguishable species-specific clusters than the Acclimated samples, which were held in a crowded holding tank with several coral species and bacteria in the water column that may have influenced the microbiome of the mucus layer (Guppy and Bythell 2006). Further, pathogenic Vibrionaceae (ASV76 and 98) were found in high abundance (30% of total abundance) exclusively in samples of the Acclimation tank *A. millepora*. The shift in cyanobacteria dominance of the Field control *A. millepora* may have exposed a niche for opportunistic species (Kvennefors et al. 2012). Given that these corals were used for inoculation, it is possible that harmful pathogenic bacteria were inoculated into the recipient corals. Despite this, no pathogenic bacteria were retained by the inoculated *P. speciosa*.

The phyla Proteobacteria and Cyanobacteria, both dominant taxa in this study, and their presence concurs with previous studies on four local coral species, including *Merulina ampliata*, *Montipora stellata*, *Pocillopora acuta* and *Porites lutea* (Fong et al. 2020; Wainwright et al. 2020). Field samples contained more Cyanobacteria, which were then lost in acclimation and experimental tank samples. The core microbial taxa in both *A. millepora* and *P. speciosa* were largely similar, with the only difference being *Pseudomonas* and *Endozoicomonas*, associated with the core taxa of *A. millepora* and *P. speciosa*, respectively. *Endozoicomonas* (ASV17) was the single core ASV present in 100% of the *P. speciosa* samples, while no ubiquitous ASV was found in *A. millepora*. A decrease in *Endozoicomonas* abundance was observed within *P. speciosa* fragments translocated from the Acclimation tank to the Experimental tanks. Interestingly, Endozoicomonas abundance was also comparatively lower in the uninoculated *P. speciosa*, as well as in the heat stress treatment at the end of the experiment. *Endozoicomonas* has been suggested as a beneficial taxon that confers thermal or bleaching protection in corals (Pantos et al. 2015; Epstein et al. 2019), and its abundance can be reduced in stressed corals (McDevitt-Irwin et al. 2017), although reduction can also be due to natural variation within species due to abiotic factors (Hernandez-Agreda et al. 2017). Its prevalence in *P. speciosa*, especially in the Acclimation tank samples where the water temperature was generally higher than the Experimental tanks may be indicative of its protective benefits for the more heat stress susceptible *P. speciosa* (NParks field observation). However, more studies are needed to tease apart the true nature of *Endozoicomonas* in *P. speciosa*.

The difference in microbiomes between Acclimated and Experimental tank *A. millepora* was driven by phylum Proteobacteria (2 ASVs of family Xanthomonadaceae and 1 Rhodobacteraceae) and Verrucomicrobiota (1 ASV of class Lentisphaeria). All four of these ASVs were found exclusively in the Experimental tank samples of both *A. millepora* and *P. speciosa*. Xanthomonadaceae (Polson 2007) and Lentisphaeria (De Castro et al. 2010) were previously found in healthy corals, whereas Rhodobacteraceae was identified in previous studies as opportunistic pathogens and stress-associated taxa (Thurber et al. 2009; Meyer et al. 2014; Pootakham et al. 2019). It is unclear what caused the shift in microbial community structure in this study.

Inoculation of *P. speciosa* with microbiome collected from *A. millepora* did not lead to a significant shift from its original microbiome. However, despite *P. speciosa* being a different host species, the microbiome of the inoculated fragments was not significantly different from *A. millepora*. This signifies a shift in the microbiome of the inoculated *P. speciosa* to be closer to that of the inoculant, while simultaneously resembling the original microbial community.

The low-temperature treatment at 28 °C and the heat stress treatment at 31 °C did not significantly affect overall and within-treatment-group microbiome differences. Given that Singapore's mean monthly seawater temperature is 29.1 °C ± 0.2 °C and can reach over 30 °C (Low et al. 2019), the thermal tolerance of local corals may be higher than 31 °C, a degree above the upper limit in situ. Since there was no heat stress effect as evident by the treatment groups, it is not surprising that the inoculated *P. speciosa* was not driven by heat stress to retain significantly different ASVs than the control temperature treatment. However, there were observed differences in the ASVs retained by the two temperature treatment groups, where one ASV of order Enterobacterales and two ASVs of Burkholderiales (genus *Ralstonia*) were retained by both treatment groups, and two ASVs of Burkholderiales (family Comamonadaceae) and one unidentified Gammaproteobacteria were retained only in the corals under heat stress treatment. However, the differences observed were not statistically significant.

In conclusion, this (unpublished) pilot study examined the potential beneficial effects of inoculating *P. speciosa*, a species susceptible to bleaching, with the microbiome of the thermally resilient *A. millepora*. *Acropora millepora* has been observed to retain its photosynthetic zooxanthellae despite periods of heat stress observed during bleaching events. No significant differences were observed in both the photosynthetic efficiency and microbiome communities of the inoculated and uninoculated *P. speciosa* subjected to heat-treatment. It is likely that the heat treatments were not sufficiently high for an acute short-term stress exposure on Singapore coral species that may already have been previously exposed and were conditioned to overcoming bleaching events. Results from the inoculations showed some shift in microbiome composition, where inoculated *P. speciosa* microbiomes became indistinct from the inoculant *A. millepora* microbiome. While many coral probiotics studies focus on the selection of known and culturable beneficial microbes (Damjanovic et al. 2019; Epstein et al. 2019; Rosado et al. 2019) to counteract effects of coral stressors, this pilot study was one of a few to focus on inoculation with naturally produced mucus of a heat-tolerant phenotype, *A. millepora* (Damjanovic et al. 2017). Although the ability of coral mucus to confer thermal resistance via microbiome inoculation remains to be proven, the information from this study could assist in future studies that seek to optimize the use of entire microbiomes from coral mucus in understanding tolerance to heat stress.

7.5 The Future of Marine Microbiome Research in Singapore

7.5.1 Local Marine Microbiome Research

Marine microbiome research in Singapore offers a unique perspective due to its urbanized coastal environment, characterized by high sedimentation and anthropogenic stress. Singapore's marine ecosystems, including coral reefs, seagrasses, soft corals and sponges, host diverse microbiomes that play crucial roles in nutrient cycling, resilience to stress and the production of bioactive compounds. For example, research has identified stress-tolerant symbionts in corals, such as *Durusdinium*, which help mitigate bleaching effects under thermal stress, underscoring the potential of these symbionts in enhancing coral resilience. The microbiomes of benthic macroorganisms like sponges and coral corals also present a rich source for biotechnological applications, including drug discovery. A significant aspect of this research is its focus on understanding the impact of local environmental conditions on microbiome diversity and function. For instance, studies have revealed the variability in microbial communities in coral species across different reefs in Singapore, shaped by factors like sedimentation and turbidity. These insights not only advance our understanding of marine microbiomes but also have broader implications for ecological conservation and sustainable use of marine resources in urbanized settings like Singapore.

7.5.2 Future Marine Microbiome Research

Research on marine microbiomes associated with benthic marine macroorganisms within urban coastal marine environment having relatively high sedimentation rates is critical for understanding the role of microbial communities in ecosystem health and resilience (Bourne et al. 2016; Stévenne et al. 2021; Ramljak et al. 2024). Some priority areas of research in this field within Singapore's coastal waters include:

- Microbial composition and diversity: Study the core microbiomes of benthic invertebrates in high-sedimentation environments, including bacterial, archaeal, fungal and protist taxa; utilize this knowledge in restoration efforts by identifying keystone microbial species that promote ecosystem health.
- Ecosystem function and nutrient cycling: Analyze microbiomes' roles in carbon, nitrogen and sulfur cycles, particularly in sediment-rich ecosystems; examine their capacity to detoxify pollutants and degrade organic matter, supporting ecosystem stability; integrate microbiome insights into conservation strategies to manage coastal habitats effectively.
- Impacts of sedimentation on microbial communities and holobiont health: Study how sedimentation alters microbial diversity, structure and function, and its effects on host-microbiome symbiosis; identification of certain microbial taxonomic group(s) as promising disturbance indicators, explore sediment-associated pathogen dynamics and their role in disease outbreaks and use of this knowledge to develop restoration strategies that enhance resilience in sediment-stressed habitats.
- Impacts of climate change: Investigate how climate change exacerbates sedimentation and impacts marine microbiomes and host species; focus on resilience mechanisms provided by microbiomes against climate-induced stresses like temperature increases, acidification and hypoxia; apply findings to inform conservation practices addressing the combined pressures of sedimentation and climate change.

These recommended research priorities can contribute to better understanding how benthic macroorganisms and their associated microbiomes respond to sedimentation and other environmental stressors, with important implications for conservation and ecosystem management in Singapore.

Acknowledgements We would like to acknowledge funding from the NParks-Marine Invert grant and the National Research Foundation, Prime Minister's Office, Singapore, under its Marine Science Research and Development Programme (award nos. MSRDP-P15 and MSRDP-P34).

References

Ainsworth TD, Gates RD (2016) Corals' microbial sentinels. Science 352:1518–1519

Alagely A, Krediet CJ, Ritchie KB, Teplitski M (2011) Signaling-mediated cross-talk modulates swarming and biofilm formation in a coral pathogen *Serratia marcescens*. ISME J 5:1609–1620

Allard P-M, Péresse T, Bisson J, Gindro K, Marcourt L, Pham VC, Roussi F, Litaudon M, Wolfender J-L (2016) Integration of molecular networking and *In-silico* MS/MS fragmentation for natural products dereplication. Anal Chem 88:3317–3323

Anderson M (2008) PERMANOVA+ for PRIMER: guide to software and statistical methods. PRIMER-E, Plymouth Marine Laboratory, p 214

Ang P Jr (1986) Analysis of the vegetation structure of a *Sargassum* community in The Philippines. Mar Ecol Prog Ser 28:9–19

Armstrong E, Yan L, Boyd KG, Wright PC, Burgess JG (2001) The symbiotic role of marine microbes on living surfaces. Hydrobiologia 461:37–40

Baghel RS, Trivedi N, Gupta V, Neori A, Reddy C, Lali A, Jha B (2015) Biorefining of marine macroalgal biomass for production of biofuel and commodity chemicals. Green Chem 17:2436–2443

Bairoliya S, Koh J, Cho ZT, Cao B (2024) Phototrophs as the central components of the plastisphere microbiome in coastal environments. Environ Int 190:108901

Bayer K, Schmitt S, Hentschel U (2008) Physiology, phylogeny and in situ evidence for bacterial and archaeal nitrifiers in the marine sponge *Aplysina aerophoba*. Environ Microbiol 10:2942–2955

Bayer K, Jahn MT, Slaby BM, Moitinho-Silva L, Hentschel U (2018) Marine sponges as *Chloroflexi* hot spots: genomic insights and high-resolution visualization of an abundant and diverse symbiotic clade. mSystems 3:e00150-18

Bell JJ (2008) The functional roles of marine sponges. Estuar Coast Shelf Sci 79:341–353

Bengtsson MM, Sjøtun K, Lanzén A, Øvreås L (2012) Bacterial diversity in relation to secondary production and succession on surfaces of the kelp *Laminaria hyperborea*. ISME J 6:2188–2198

Biswas J, Jana SK, Mandal S (2023) Biotechnological impacts of *Halomonas*: a promising cell factory for industrially relevant biomolecules. Biotechnol Genet Eng Rev 39:348–377

Blackall LL, Wilson B, Van Oppen MJ (2015) Coral - the world's most diverse symbiotic ecosystem. Mol Ecol 24:5330–5347

Blin K, Wolf T, Chevrette MG, Lu X, Schwalen CJ, Kautsar SA, Suarez Duran HG, De Los Santos EL, Kim HU, Nave M (2017) AntiSMASH 4.0—improvements in chemistry prediction and gene cluster boundary identification. Nucleic Acids Res 45:W36–W41

Blin K, Kim HU, Medema MH, Weber T (2019) Recent development of antiSMASH and other computational approaches to mine secondary metabolite biosynthetic gene clusters. Brief Bioinform 20:1103–1113

Blockley A, Elliott DR, Roberts AP, Sweet M (2017) Symbiotic microbes from marine invertebrates: driving a new era of natural product drug discovery. Diversity 9:49

Boilard A, Dubé CE, Gruet C, Mercière A, Hernandez-Agreda A, Derome N (2020) Defining coral bleaching as a microbial dysbiosis within the coral holobiont. Microorganisms 8:1682

Bondoso J, Balague V, Gasol JM, Lage OM (2014) Community composition of the Planctomycetes associated with different macroalgae. FEMS Microbiol Ecol 88:445–456

Bourne DG, Morrow KM, Webster NS (2016) Insights into the coral microbiome: underpinning the health and resilience of reef ecosystems. Ann Rev Microbiol 70:317–340

Chénard C, Wijaya W, Vaulot D, Lopes dos Santos A, Martin P, Kaur A, Lauro FM (2019) Temporal and spatial dynamics of Bacteria, Archaea and protists in equatorial coastal waters. Sci Rep 9:16390

Choi EJ, Nam S-J, Paul L, Beatty D, Kauffman CA, Jensen PR, Fenical W (2015) Previously uncultured marine bacteria linked to novel alkaloid production. Chem Biol 22:1270–1279

Cita YP, Suhermanto A, Radjasa OK, Sudharmono P (2017) Antibacterial activity of marine bacteria isolated from sponge *Xestospongia testudinaria* from Sorong, Papua. Asian Pac J Trop Biomed 7:450–454

Cleary DF, de Voogd NJ, Polónia AR, Freitas R, Gomes NC (2015) Composition and predictive functional analysis of bacterial communities in seawater, sediment and sponges in the Spermonde Archipelago, Indonesia. Microb Ecol 70:889–903

Cleary DFR, Polónia ARM, Becking LE, de Voogd NJ, Purwanto GH, Gomes NCM (2018) Compositional analysis of bacterial communities in seawater, sediment, and sponges in the Misool coral reef system, Indonesia. Mar Biodivers 48:1889–1901

Cole LW, McGlathery KJ (2012) Nitrogen fixation in restored eelgrass meadows. Mar Ecol Prog Ser 448:235–246

Damjanovic K, Blackall LL, Webster NS, van Oppen MJ (2017) The contribution of microbial biotechnology to mitigating coral reef degradation. Microb Biotechnol 10:1236–1243

Damjanovic K, van Oppen MJ, Menéndez P, Blackall LL (2019) Experimental inoculation of coral recruits with marine bacteria indicates scope for microbiome manipulation in *Acropora tenuis* and *Platygyra daedalea*. Front Microbiol 10:1702

Dat TTH, Steinert G, Cuc NTK, Smidt H, Sipkema D (2018) Archaeal and bacterial diversity and community composition from 18 phylogenetically divergent sponge species in Vietnam. PeerJ 6:e4970

Davis KE, Sangwan P, Janssen PH (2011) Acidobacteria, Rubrobacteridae and Chloroflexi are abundant among very slow-growing and mini-colony-forming soil bacteria. Environ Microbiol 13:798–805

Davy SK, Allemand D, Weis VM (2012) Cell biology of cnidarian-dinoflagellate symbiosis. Microbiol Mol Biol Rev 76:229–261

De Castro AP, Araújo SD, Reis AM, Moura RL, Francini-Filho RB, Pappas G, Rodrigues TB, Thompson FL, Krüger RH (2010) Bacterial community associated with healthy and diseased reef coral *Mussismilia hispida* from eastern Brazil. Microb Ecol 59:658–667

de la Calle F (2017) Marine microbiome as source of natural products. Microb Biotechnol 10:1293–1296

de Oliveira BFR, Carr CM, Dobson AD, Laport MS (2020) Harnessing the sponge microbiome for industrial biocatalysts. Appl Microbiol Biotechnol 104:8131–8154

de Voogd NJ, Cleary DF, Polonia AR, Gomes NC (2015) Bacterial community composition and predicted functional ecology of sponges, sediment and seawater from the thousand islands reef complex, West Java, Indonesia. FEMS Microbiol Ecol 91:fiv019

Deignan LK, McDougald D (2022) Differential response of the microbiome of *Pocillopora acuta* to reciprocal transplantation within Singapore. Microb Ecol 83:608–618

Deignan LK, Pwa KH, Loh AAR, Rice SA, McDougald D (2023a) The microbiomes of two Singaporean corals show site-specific differentiation and variability that correlates with the seasonal monsoons. Coral Reefs 42:677–691

Deignan LK, Dansson R, Loh AAR, Pwa KH (2023b) Examining the giant barrel sponge species complex: molecular and microbial differentiation of *Xestospongia testudinaria* in Singapore. Mar Biol 170:166

Distel DL, Morrill W, MacLaren-Toussaint N, Franks D, Waterbury J (2002) *Teredinibacter turnerae* gen. nov., sp. nov., a dinitrogen-fixing, cellulolytic, endosymbiotic gamma-proteobacterium isolated from the gills of wood-boring molluscs (Bivalvia: Teredinidae). Int J Syst Evol Microbiol 52:2261–2269

Doering T, Wall M, Putchim L, Rattanawongwan T, Schroeder R, Hentschel U, Roik A (2021) Towards enhancing coral heat tolerance: a "microbiome transplantation" treatment using inoculations of homogenized coral tissues. Microbiome 9:102

Dogs M, Teshima H, Petersen J, Fiebig A, Chertkov O, Dalingault H, Chen A, Pati A, Goodwin LA, Chain P (2013) Genome sequence of *Phaeobacter daeponensis* type strain (DSM 23529 T), a facultatively anaerobic bacterium isolated from marine sediment, and emendation of *Phaeobacter daeponensis*. Stand Genomic Sci 9:142–159

Duarte CM (1991) Seagrass depth limits. Aquat Bot 40:363–377

Duarte CM, Middelburg JJ, Caraco N (2005) Major role of marine vegetation on the oceanic carbon cycle. Biogeosciences 2:1–8

Duncan KR, Crüsemann M, Lechner A, Sarkar A, Li J, Ziemert N, Wang M, Bandeira N, Moore BS, Dorrestein PC (2015) Molecular networking and pattern-based genome mining improves discovery of biosynthetic gene clusters and their products from Salinispora species. Chem Biol 22:460–471

Dunphy CM, Gouhier TC, Chu ND, Vollmer SV (2019) Structure and stability of the coral microbiome in space and time. Sci Rep 9:6785

Eakins BW, Sharman G (2010) Volumes of the world's oceans from ETOPO1. NOAA National Geophysical Data Center, Boulder, p 7

Epstein HE, Smith HA, Torda G, van Oppen MJ (2019) Microbiome engineering: enhancing climate resilience in corals. Front Ecol Environ 17:100–108

Evans R, Wilson S, Field S, Moore J (2014) Importance of macroalgal fields as coral reef fish nursery habitat in north-west Australia. Mar Biol 161:599–607

Evans JS, López-Legentil S, Pawlik JR, Turnbull IG, Erwin PM (2021) Molecular detection and microbiome differentiation of two cryptic lineages of giant barrel sponges from Conch Reef, Florida Keys. Coral Reefs 40:853–865

Fiore CL, Labrie M, Jarett JK, Lesser MP (2015) Transcriptional activity of the giant barrel sponge, *Xestospongia muta* holobiont: molecular evidence for metabolic interchange. Front Microbiol 6:364

Fong J, Deignan LK, Bauman AG, Steinberg PD, McDougald D, Todd PA (2020) Contact-and water-mediated effects of macroalgae on the physiology and microbiome of three Indo-Pacific coral species. Front Mar Sci 6:831

Fong J, Tang PP, Deignan LK, Seah JC, McDougald D, Rice SA, Todd PA (2023) Chemically mediated interactions with macroalgae negatively affect coral health but induce limited changes in coral microbiomes. Microorganisms 11:2261

Fragoso Ados Santos H, Duarte GAS, Rachid CTC, Chaloub RM, Calderon EN, Marangoni LFB, Bianchini A, Nudi AH, Do Carmo FL, Van Elsas JD (2015) Impact of oil spills on coral reefs can be reduced by bioremediation using probiotic microbiota. Sci Rep 5:18268

Freudenthal HD (1962) *Symbiodinium* gen. nov. and *Symbiodinium microadriaticum* sp. nov., a zooxanthella: taxonomy, life cycle, and morphology. J Protozool 9:45–52

Garcias-Bonet N, Arrieta JM, de Santana CN, Duarte CM, Marbà N (2012) Endophytic bacterial community of a Mediterranean marine angiosperm (*Posidonia oceanica*). Front Microbiol 3:342

Garcias-Bonet N, Arrieta JM, Duarte CM, Marbà N (2016) Nitrogen-fixing bacteria in Mediterranean seagrass (*Posidonia oceanica*) roots. Aquat Bot 131:57–60

Garza DR, Dutilh BE (2015) From cultured to uncultured genome sequences: metagenomics and modeling microbial ecosystems. Cell Mol Life Sci 72:4287–4308

Giles EC, Kamke J, Moitinho-Silva L, Taylor MW, Hentschel U, Ravasi T, Schmitt S (2013) Bacterial community profiles in low microbial abundance sponges. FEMS Microbiol Ecol 83:232–241

Gloeckner V, Hentschel U, Ereskovsky AV, Schmitt S (2013) Unique and species-specific microbial communities in *Oscarella lobularis* and other Mediterranean *Oscarella* species (Porifera: Homoscleromorpha). Mar Biol 160:781–791

Glynn PW (1991) Coral reef bleaching in the 1980s and possible connections with global warming. Trends Ecol Evol 6:175–179

Graham NA, Wilson SK, Jennings S, Polunin NV, Bijoux JP, Robinson J (2006) Dynamic fragility of oceanic coral reef ecosystems. Proc Natl Acad Sci 103:8425–8429

Guppy R, Bythell JC (2006) Environmental effects on bacterial diversity in the surface mucus layer of the reef coral *Montastraea faveolata*. Mar Ecol Prog Ser 328:133–142

Haque S, Ahmad F, Dar SA, Jawed A, Mandal RK, Wahid M, Lohani M, Khan S, Singh V, Akhter N (2018) Developments in strategies for quorum sensing virulence factor inhibition to combat bacterial drug resistance. Microb Pathog 121:293–302

Hardwicke T (1822) Description of a zoophyte, commonly found about the coasts of Singapore Island, with a plate. Asiatic Res 14:180–181

Harkins CP, Kong HH, Segre JA (2020) Manipulating the human microbiome to manage disease. JAMA 323:303–304

Hayami Y, Ambalavanan L, Zainathan SC, Danish-Daniel M, Sharifah NE, Iehata S (2023) Diversity and functional roles of the symbiotic microbiome associated to marine sponges off Karah Island, Terengganu, Malaysia. Reg Stud Mar Sci 62:102967

Haygood MG, Schmidt EW, Davidson SK, Faulkner DJ (1999) Microbial symbionts of marine invertebrates: opportunities for microbial biotechnology. J Mol Microbiol Biotechnol 1:33–43

Hentschel U, Fieseler L, Wehrl M, Gernert C, Steinert M, Hacker J, Horn M (2003) Microbial diversity of marine sponges. Prog Mol Subcell Biol 37:59–88

Hentschel U, Piel J, Degnan SM, Taylor MW (2012) Genomic insights into the marine sponge microbiome. Nat Rev Microbiol 10:641–654

Hernandez-Agreda A, Leggat W, Bongaerts P, Ainsworth T, Rosenberg E, Azam F (2016) The microbial signature provides insight into the mechanistic basis of coral success across reef habitats. mBio 7:e00560-16

Hernandez-Agreda A, Gates RD, Ainsworth TD (2017) Defining the core microbiome in corals' microbial soup. Trends Microbiol 25:125–140

Ho XY, Katermeran NP, Deignan LK, Phyo MY, Ong JFM, Goh JX, Ng JY, Tun K, Tan LT (2021) Assessing the diversity and biomedical potential of microbes associated with the Neptune's cup sponge, *Cliona patera*. Front Microbiol 12:631445

Hollants J, Leliaert F, Verbruggen H, Willems A, De Clerck O (2013) Permanent residents or temporary lodgers: characterizing intracellular bacterial communities in the siphonous green alga *Bryopsis*. Proc Biol Sci 280:20122659

Holmström C, Kjelleberg S (1999) Marine *Pseudoalteromonas* species are associated with higher organisms and produce biologically active extracellular agents. FEMS Microbiol Ecol 30:285–293

Huggett MJ, Apprill A (2019) Coral microbiome database: integration of sequences reveals high diversity and relatedness of coral-associated microbes. Environ Microbiol Rep 11:372–385

Hughes TP, Rodrigues MJ, Bellwood DR, Ceccarelli D, Hoegh-Guldberg O, McCook L, Moltschaniwskyj N, Pratchett MS, Steneck RS, Willis B (2007) Phase shifts, herbivory, and the resilience of coral reefs to climate change. Curr Biol 17:360–365

Hughes TP, Kerry JT, Álvarez-Noriega M, Álvarez-Romero JG, Anderson KD, Baird AH, Babcock RC, Beger M, Bellwood DR, Berkelmans R (2017) Global warming and recurrent mass bleaching of corals. Nature 543:373–377

Hutchins DA, Fu F (2017) Microorganisms and ocean global change. Nat Microbiol 2:17058

Tanzil JT, Ng APK, Tey YQ, Tan BHY, Yun EY, Huang D (2016) A preliminary characterisation of *Symbiodinium* diversity in some common corals from Singapore. Cosmos 12:15–27

Isnansetyo A, Kamei Y (2009) Bioactive substances produced by marine isolates of *Pseudomonas*. J Ind Microbiol Biotechnol 36:1239–1248

Jain SS, Afiq-Rosli L, Feldman B, Levy O, Phua JW, Wainwright BJ, Huang D (2020) Homogenization of endosymbiont communities hosted by equatorial corals during the 2016 mass bleaching event. Microorganisms 8:1370

Jensen PR, Dwight R, Fenical W (1991) Distribution of actinomycetes in near-shore tropical marine sediments. Appl Environ Microbiol 57:1102–1108

Jensen SI, Kühl M, Prieme A (2007) Different bacterial communities associated with the roots and bulk sediment of the seagrass *Zostera marina*. FEMS Microbiol Ecol 62:108–117

Jeyabaskaran R, Jayasankar J, Ambrose T, Vineetha Valsalan K, Divya N, Raji N, Vysakhan P, John S, Prema D, Kaladharan P (2018) Conservation of seagrass beds with special reference to associated species and fishery resources. J Mar Biol Assoc India 60:62–70

Jin X-B, Sun R-J, Zhu J-Q, Xu Z-J, Liu Z, Wang Q, Ye X-Y (2012) Isolation and identification of *Bacillus altitudinis* ZJ 186 from marine soil samples and its antifungal activity against *Magnaporthe oryzae*. Curr Res Bacteriol 5:13–23

Jin L, Liu F, Sun W, Zhang F, Karuppiah V, Li Z (2014) Pezizomycotina dominates the fungal communities of South China sea sponges *Theonella swinhoei* and *Xestospongia testudinaria*. FEMS Microbiol Ecol 90:935–945

Kamke J, Taylor MW, Schmitt S (2010) Activity profiles for marine sponge-associated bacteria obtained by 16S rRNA vs 16S rRNA gene comparisons. ISME J 4:498–508

Kanisan DP, Quek ZR, Oh RM, Afiq-Rosli L, Lee JN, Huang D, Wainwright BJ (2023) Diversity and distribution of microbial communities associated with reef corals of the Malay Peninsula. Microb Ecol 85:37–48

Karimi E, Keller-Costa T, Slaby BM, Cox CJ, da Rocha UN, Hentschel U, Costa R (2019) Genomic blueprints of sponge-prokaryote symbiosis are shared by low abundant and cultivatable Alphaproteobacteria. Sci Rep 9:1999

Koh J, Bairoliya S, Salta M, Cho ZT, Fong J, Neo ML, Cragg S, Cao B (2023) Sediment-driven plastisphere community assembly on plastic debris in tropical coastal and marine environments. Environ Int 179:108153

Komatsu T, Fukuda M, Mikami A, Mizuno S, Kantachumpoo A, Tanoue H, Kawamiya M (2014) Possible change in distribution of seaweed, *Sargassum horneri*, in northeast Asia under A2 scenario of global warming and consequent effect on some fish. Mar Pollut Bull 85:317–324

Kvennefors ECE, Sampayo E, Kerr C, Vieira G, Roff G, Barnes AC (2012) Regulation of bacterial communities through antimicrobial activity by the coral holobiont. Microb Ecol 63:605–618

Kwan V, Fong J, Ng CSL, Huang D (2022) Temporal and spatial dynamics of tropical macroalgal contributions to blue carbon. Science of the Total Environment 828:154369

Lachnit T, Meske D, Wahl M, Harder T, Schmitz R (2011) Epibacterial community patterns on marine macroalgae are host-specific but temporally variable. Environ Microbiol 13:655–665

LaJeunesse TC, Parkinson JE, Gabrielson PW, Jeong HJ, Reimer JD, Voolstra CR, Santos SR (2018) Systematic revision of Symbiodiniaceae highlights the antiquity and diversity of coral endosymbionts. Curr Biol 28:2570–2580

Lau SC, Zhang R, Brodie EL, Piceno YM, Andersen G, Liu W-T (2013) Biogeography of bacterioplankton in the tropical seawaters of Singapore. FEMS Microbiol Ecol 84:259–269

Lee Y, Kim H, Liu C, Lee H (2003) A simple method for DNA extraction from marine bacteria that produce extracellular materials. J Microbiol Methods 52:245–250

Lesser MP, Fiore C, Slattery M, Zaneveld J (2016) Climate change stressors destabilize the microbiome of the Caribbean barrel sponge, *Xestospongia muta*. J Exp Mar Biol Ecol 475:11–18

Leveque S, Afiq-Rosli L, Ip YCA, Jain SS, Huang D (2019) Searching for phylogenetic patterns of Symbiodiniaceae community structure among Indo-Pacific Merulinidae corals. PeerJ 7:e7669

Lim SC, de Voogd NJ, Tan K-S (2012a) Biodiversity of shallow-water sponges (Porifera) in Singapore and description of a new species of *Forcepia* (Poecilosclerida): Coelosphaeridae. Contrib Zool 81:55–71

Lim SC, Tun K, Goh E (2012b) Rediscovery of the Neptune's cup sponge in Singapore: *Cliona* or *Poterion*. Contrib Mar Sci 2012:49–56

Lim SC, Putchakarn S, Thai M-Q, Wang D, Huang YM (2016) Inventory of sponge fauna from the Singapore Strait to Taiwan Strait along the western coastline of the South China Sea. Raffles Bullet Zool 34:104

Lim SF, Tan LT, Goh BPL (2023) Distribution and abundance of intertidal Sponge (Porifera) communities in coral reefs of Singapore. Ocean Sci J 58:30

Low ME (2012) The date of publication of *Cliona patera* (Hardwicke), the 'sponge plant from the shores of Singapore'(Porifera: Hadromerida: Clionaidae). Nat Singapore 5:223–227

Low JK, Fong J, Todd PA, Chou LM, Bauman AG (2019) Seasonal variation of *Sargassum ilicifolium* (Phaeophyceae) growth on equatorial coral reefs. J Phycol 55:289–296

Lu D-C, Wang F-Q, Amann RI, Teeling H, Du Z-J (2023) Epiphytic common core bacteria in the microbiomes of co-located green (*Ulva*), brown (*Saccharina*) and red (*Grateloupia*, *Gelidium*) macroalgae. Microbiome 11:126

Maansson M, Vynne N, Klitgaard A, Nybo J, Melchiorsen J, Nguyen D, Sanchez L, Ziemert N, Dorrestein P, Andersen M (2016) An integrated metabolomic and genomic mining workflow to uncover the biosynthetic potential of bacteria. mSystems 1:e00028-15

Magarvey NA, Keller JM, Bernan V, Dworkin M, Sherman DH (2004) Isolation and characterization of novel marine-derived actinomycete taxa rich in bioactive metabolites. Appl Environ Microbiol 70:7520–7529

Malik SAA, Bedoux G, Maldonado JQG, Freile-Pelegrín Y, Robledo D, Bourgougnon N (2020) Defence on surface: macroalgae and their surface-associated microbiome Advances in Botanical Research. Elsevier, pp 327–368

Mansson M, Gram L, Larsen TO (2011) Production of bioactive secondary metabolites by marine Vibrionaceae. Mar Drugs 9:1440–1468

McCauley EP, Piña IC, Thompson AD, Bashir K, Weinberg M, Kurz SL, Crews P (2020) Highlights of marine natural products having parallel scaffolds found from marine-derived bacteria, sponges, and tunicates. J Antibiot 73:504–525

McDevitt-Irwin JM, Baum JK, Garren M, Vega Thurber RL (2017) Responses of coral-associated bacterial communities to local and global stressors. Front Mar Sci 4:262

McFall-Ngai M, Hadfield MG, Bosch TC, Carey HV, Domazet-Lošo T, Douglas AE, Dubilier N, Eberl G, Fukami T, Gilbert SF (2013) Animals in a bacterial world, a new imperative for the life sciences. Proc Natl Acad Sci 110:3229–3236

McGrath EC, Woods L, Jompa J, Haris A, Bell JJ (2018) Growth and longevity in giant barrel sponges: redwoods of the reef or Pines in the Indo-Pacific? Sci Rep 8:15317

Meron D, Atias E, Iasur Kruh L, Elifantz H, Minz D, Fine M, Banin E (2011) The impact of reduced pH on the microbial community of the coral *Acropora eurystoma*. ISME J 5:51–60

Meunier V, Geissler L, Bonnet S, Rädecker N, Perna G, Grosso O, Lambert C, Rodolfo-Metalpa R, Voolstra CR, Houlbrèque F (2021) Microbes support enhanced nitrogen requirements of coral holobionts in a high CO_2 environment. Mol Ecol 30:5888–5899

Meyer JL, Paul VJ, Teplitski M (2014) Community shifts in the surface microbiomes of the coral *Porites astreoides* with unusual lesions. PLoS One 9:e100316

Mohamed A, Cumbo V, Harii S, Shinzato C, Chan C, Ragan M, Bourne D, Willis B, Ball E, Satoh N (2016) The transcriptomic response of the coral *Acropora digitifera* to a competent *Symbiodinium* strain: the symbiosome as an arrested early phagosome. Mol Ecol 25:3127–3141

Moitinho-Silva L, Bayer K, Cannistraci CV, Giles EC, Ryu T, Seridi L, Ravasi T, Hentschel U (2014) Specificity and transcriptional activity of microbiota associated with low and high microbial abundance sponges from the Red Sea. Mol Ecol 23:1348–1363

Moitinho-Silva L, Nielsen S, Amir A, Gonzalez A, Ackermann GL, Cerrano C, Astudillo-Garcia C, Easson C, Sipkema D, Liu F (2017a) The sponge microbiome project. Gigascience 6:gix077

Moitinho-Silva L, Steinert G, Nielsen S, Hardoim CC, Wu Y-C, McCormack GP, López-Legentil S, Marchant R, Webster N, Thomas T (2017b) Predicting the HMA-LMA status in marine sponges by machine learning. Front Microbiol 8:752

Monciardini P, Iorio M, Maffioli S, Sosio M, Donadio S (2014) Discovering new bioactive molecules from microbial sources. Microb Biotechnol 7:209–220

Mondol MAM, Shin HJ, Islam MT (2013) Diversity of secondary metabolites from marine *Bacillus* species: chemistry and biological activity. Mar Drugs 11:2846–2872

Morrow KM, Bourne DG, Humphrey C, Botté ES, Laffy P, Zaneveld J, Uthicke S, Fabricius KE, Webster NS (2015) Natural volcanic CO_2 seeps reveal future trajectories for host–microbial associations in corals and sponges. ISME J 9:894–908

Muscatine L, Porter JW (1977) Reef corals: mutualistic symbioses adapted to nutrient-poor environments. Biosciences 27:454–460

Muscatine LR, McCloskey LE, Marian R (1981) Estimating the daily contribution of carbon from zooxanthellae to coral animal respiration. Limnol Oceanogr 26:601–611

Nakaoka M, Aioi K (1999) Growth of seagrass *Halophila ovalis* at dugong trails compared to existing within-patch variation in a Thailand intertidal flat. Mar Ecol Prog Ser 184:97–103

Neave MJ, Michell CT, Apprill A, Voolstra CR (2017) *Endozoicomonas* genomes reveal functional adaptation and plasticity in bacterial strains symbiotically associated with diverse marine hosts. Sci Rep 7:40579

Ng MS, Soon N, Chang Y, Wainwright BJ (2022) Bacterial and fungal co-occurrence in the nudibranch, *Pteraeolidia semperi*. Life 12:1988

Nguyen DD, Wu C-H, Moree WJ, Lamsa A, Medema MH, Zhao X, Gavilan RG, Aparicio M, Atencio L, Jackson C (2013) MS/MS networking guided analysis of molecule and gene cluster families. Proc Natl Acad Sci 110:E2611–E2620

Nielsen JT, Liesack W, Finster K (1999) *Desulfovibrio zosterae* sp. nov., a new sulfate reducer isolated from surface-sterilized roots of the seagrass *Zostera marina*. Int J Syst Evol Microbiol 49:859–865

Noguchi H, Uchino M, Shida O, Takano K, Nakamura LK, Komagata K (2004) *Bacillus vietnamensis* sp. nov., a moderately halotolerant, aerobic, endospore-forming bacterium isolated from Vietnamese fish sauce. Int J Syst Evol Microbiol 54:2117–2120

Nussbaumer AD, Fisher CR, Bright M (2006) Horizontal endosymbiont transmission in hydrothermal vent tubeworms. Nature 441:345–348

Oh RM, Bollati E, Maithani P, Huang D, Wainwright BJ (2021) The microbiome of the reef macroalga *Sargassum ilicifolium* in Singapore. Microorganisms 9:898

Ong JFM, Goh HC, Lim SC, Pang LM, Chin JSF, Tan KS, Liang Z-X, Yang L, Glukhov E, Gerwick WH (2019) Integrated genomic and metabolomic approach to the discovery of potential anti-quorum sensing natural products from microbes associated with marine samples from Singapore. Mar Drugs 17:72

Ong JH, Wainwright BJ, Jain SS, Afiq-Rosli L, Lee JN, Huang D (2022) Species and spatio-environmental effects on coral endosymbiont communities in Southeast Asia. Coral Reefs 41:1131–1145

Pantos O, Bongaerts P, Dennis PG, Tyson GW, Hoegh-Guldberg O (2015) Habitat-specific environmental conditions primarily control the microbiomes of the coral *Seriatopora hystrix*. ISME J 9:1916–1927

Perry NB, Blunt JW, Munro MH, Pannell LK (1988) Mycalamide A, an antiviral compound from a New Zealand sponge of the genus *Mycale*. J Am Chem Soc 110:4850–4851

Pfister CA, Gilbert JA, Gibbons SM (2014) The role of macrobiota in structuring microbial communities along rocky shores. PeerJ 2:e631

Phelan R, O'Halloran J, Kennedy J, Morrissey J, Dobson A, O'Gara F, Barbosa T (2012) Diversity and bioactive potential of endospore-forming bacteria cultured from the marine sponge *Haliclona simulans*. J Appl Microbiol 112:65–78

Pires A, Cleary D, Polónia A, Lim SC, De Voogd N, Oliveira V, Gomes N (2019) Comparison of bacterial communities associated with *Xestospongia testudinaria*, sediment and seawater in a Singaporean coral reef ecosystem. J Mar Biol Assoc UK 99:331–342

Pita L, Rix L, Slaby BM, Franke A, Hentschel U (2018) The sponge holobiont in a changing ocean: from microbes to ecosystems. Microbiome 6:46

Poloczanska ES, Burrows MT, Brown CJ, García Molinos J, Halpern BS, Hoegh-Guldberg O, Kappel CV, Moore PJ, Richardson AJ, Schoeman DS (2016) Responses of marine organisms to climate change across oceans. Front Mar Sci 3:62

Polónia ARM, Cleary DFR, Freitas R, Gomes NCM, de Voogd NJ (2017) Archaeal and bacterial communities of *Xestospongia testudinaria* and sediment differ in diversity, composition and predicted function in an Indonesian coral reef environment. J Sea Res 119:37–53

Polson SW (2007) Comparative analysis of microbial community structure associated with acroporid corals during a disease outbreak in the Florida Reef Tract. Medical University of South Carolina

Pootakham W, Mhuantong W, Yoocha T, Putchim L, Jomchai N, Sonthirod C, Naktang C, Kongkachana W, Tangphatsornruang S (2019) Heat-induced shift in coral microbiome reveals several members of the Rhodobacteraceae family as indicator species for thermal stress in *Porites lutea*. Microbiology 8:e935

Poquita-Du RC, Huang D, Chou LM, Todd PA (2020) The contribution of stress-tolerant endosymbiotic dinoflagellate *Durusdinium* to *Pocillopora acuta* survival in a highly urbanized reef system. Coral Reefs 39:745–755

Preen A (1995) Impacts of dugong foraging on seagrass habitats: observational and experimental evidence for cultivation grazing. Mar Ecol Prog Ser 124:201–213

Purves K, Macintyre L, Brennan D, Hreggviðsson GÓ, Kuttner E, Ásgeirsdóttir ME, Young LC, Green DH, Edrada-Ebel R, Duncan KR (2016) Using molecular networking for microbial secondary metabolite bioprospecting. Meta 6:2

Quek ZR, Zahn G, Lee NLY, Ooi JLS, Lee JN, Huang D, Wainwright BJ (2021) Biogeographic structure of fungal communities in seagrass *Halophilia ovalis* across the Malay Peninsula. Environ Microbiol Rep 13:871–877

Quek ZR, Ng JY, Jain SS, Long JXS, Lim SC, Tun K, Huang D (2022) Low genetic diversity and predation threaten a rediscovered marine sponge. Sci Rep 12:22499

Quek ZR, Tanzil JT, Jain SS, Yong WLO, Yu DCY, Soh Z, Ow YX, Tun K, Huang D, Wainwright BJ (2023) Limited influence of seasonality on coral microbiomes and endosymbionts in an equatorial reef. Ecol Indic 146:109878

Rabbani G, Huang D, Wainwright BJ (2021a) The mycobiome of *Pocillopora acuta* in Singapore. Coral Reefs 40:1419–1427

Rabbani G, Yan BC, Lee NLY, Ooi JLS, Lee JN, Huang D, Wainwright BJ (2021b) Spatial and structural factors shape seagrass-associated bacterial communities in Singapore and Peninsular Malaysia. Front Mar Sci 8:659180

Raina J-B, Tapiolas DM, Forêt S, Lutz A, Abrego D, Ceh J, Seneca FO, Clode PL, Bourne DG, Willis BL (2013) DMSP biosynthesis by an animal and its role in coral thermal stress response. Nature 502:677–680

Ramljak A, Žučko J, Lučić M, Babić I, Morić Z, Fafanđel M, Furdek Turk M, Matijević S, Karpouzas D, Udiković-Kolić N, Petrić I (2024) Microbial communities as indicators of marine ecosystem health: insights from coastal sediments in the eastern Adriatic Sea. Mar Pollut Bull 205:116649

Rasmussen TB, Bjarnsholt T, Skindersoe ME, Hentzer M, Kristoffersen P, Kote M, Nielsen J, Eberl L, Givskov M (2005) Screening for quorum-sensing inhibitors (QSI) by use of a novel genetic system, the QSI selector. J Bacteriol 187:1799–1814

Reddy B (2019) Advancements in microbial genome sequencing and microbial community characterization. In: Tripathi V, Kumar P, Tripathi P, Kishore A (eds) Microbial genomics in sustainable agroecosystems: Volume 1. Springer, Singapore, pp 87–113

Reiswig HM (1974) Water transport, respiration and energetics of three tropical marine sponges. J Exp Mar Biol Ecol 14:231–249

Ribes M, Calvo E, Movilla J, Logares R, Coma R, Pelejero C (2016) Restructuring of the sponge microbiome favors tolerance to ocean acidification. Environ Microbiol Rep 8:536–544

Rice MM, Maher RL, Thurber RV, Burkepile DE (2019) Different nitrogen sources speed recovery from corallivory and uniquely alter the microbiome of a reef-building coral. PeerJ 7:e8056

Rodrigues CJ, de Carvalho CC (2022) Cultivating marine bacteria under laboratory conditions: overcoming the "unculturable" dogma. Front Bioeng Biotechnol 10:964589

Romanenko LA, Uchino M, Falsen E, Frolova GM, Zhukova NV, Mikhailov VV (2005) *Pseudomonas pachastrellae* sp. nov., isolated from a marine sponge. Int J Syst Evol Microbiol 55:919–924

Rosado PM, Leite DC, Duarte GA, Chaloub RM, Jospin G, Nunes da Rocha U, Saraiva JP, Dini-Andreote F, Eisen JA, Bourne DG (2019) Marine probiotics: increasing coral resistance to bleaching through microbiome manipulation. ISME J 13:921–936

Roth-Schulze AJ, Zozaya-Valdés E, Steinberg PD, Thomas T (2016) Partitioning of functional and taxonomic diversity in surface-associated microbial communities. Environ Microbiol 18:4391–4402

Sacristán-Soriano O, Turon X, Hill M (2020) Microbiome structure of ecologically important bioeroding sponges (family Clionaidae): the role of host phylogeny and environmental plasticity. Coral Reefs 39:1285–1298

Sang VT, Dat TTH, Vinh LB, Cuong LCV, Oanh PTT, Ha H, Kim YH, Anh HLT, Yang SY (2019) Coral and coral-associated microorganisms: a prolific source of potential bioactive natural products. Mar Drugs 17:468

Santoro EP, Borges RM, Espinoza JL, Freire M, Messias CS, Villela HD, Pereira LM, Vilela CL, Rosado JG, Cardoso PM (2021) Coral microbiome manipulation elicits metabolic and genetic restructuring to mitigate heat stress and evade mortality. Sci Adv 7:eabg3088

Saurav K, Bar-Shalom R, Haber M, Burgsdorf I, Oliviero G, Costantino V, Morgenstern D, Steindler L (2016) In search of alternative antibiotic drugs: quorum-quenching activity in sponges and their bacterial isolates. Front Microbiol 7:416

Schmitt S, Weisz JB, Lindquist N, Hentschel U (2007) Vertical transmission of a phylogenetically complex microbial consortium in the viviparous sponge *Ircinia felix*. Appl Environ Microbiol 73:2067–2078

Schmitt S, Deines P, Behnam F, Wagner M, Taylor MW (2011) Chloroflexi bacteria are more diverse, abundant, and similar in high than in low microbial abundance sponges. FEMS Microbiol Ecol 78:497–510

Schmitt L, Hinxlage I, Cea PA, Gohlke H, Wesselborg S (2021) 40 years of research on polybrominated diphenyl ethers (PBDEs)—a historical overview and newest data of a promising anticancer drug. Molecules 26:995

Shade A, Handelsman J (2012) Beyond the Venn diagram: the hunt for a core microbiome. Environ Microbiol 14:4–12

Sharp KH, Eam B, Faulkner DJ, Haygood MG (2007) Vertical transmission of diverse microbes in the tropical sponge *Corticium* sp. Appl Environ Microbiol 73:622–629

Short F, Carruthers T, Dennison W, Waycott M (2007) Global seagrass distribution and diversity: a bioregional model. J Exp Mar Biol Ecol 350:3–20

Silva DP, Villela HD, Santos HF, Duarte GA, Ribeiro JR, Ghizelini AM, Vilela CL, Rosado PM, Fazolato CS, Santoro EP (2021) Multidomain probiotic consortium as an alternative to chemical remediation of oil spills at coral reefs and adjacent sites. Microbiome 9:118

Simister R, Taylor MW, Tsai P, Fan L, Bruxner TJ, Crowe ML, Webster N (2012) Thermal stress responses in the bacterial biosphere of the Great Barrier Reef sponge, *Rhopaloeides odorabile*. Environ Microbiol 14:3232–3246

Simmons TL, Coates RC, Clark BR, Engene N, Gonzalez D, Esquenazi E, Dorrestein PC, Gerwick WH (2008) Biosynthetic origin of natural products isolated from marine microorganism–invertebrate assemblages. Proc Natl Acad Sci 105:4587–4594

Sings H, Rinehart K (1996) Compounds produced from potential tunicate-blue-green algal symbiosis: a review. J Ind Microbiol Biotechnol 17:385–396

Sipkema D, de Caralt S, Morillo JA, Al-Soud WA, Sørensen SJ, Smidt H, Uriz MJ (2015) Similar sponge-associated bacteria can be acquired via both vertical and horizontal transmission. Environ Microbiol 17:3807–3821

Smith EG, Gurskaya A, Hume BC, Voolstra CR, Todd PA, Bauman AG, Burt JA (2020) Low Symbiodiniaceae diversity in a turbid marginal reef environment. Coral Reefs 39:545–553

Soh M, Tay YC, Lee CS, Low A, Orban L, Jaafar Z, Seedorf H (2024) The intestinal digesta microbiota of tropical marine fish is largely uncultured and distinct from surrounding water microbiota. npj Biofilms Microbiomes 10:11

Soon N, Quek ZR, Pohl S, Wainwright BJ (2023) More than meets the eye: characterizing the cryptic species complex and Symbiodiniaceae communities in the reef-dwelling nudibranch *Pteraeolidia 'semperi'*(Nudibranchia: Aeolidioidea) from Singapore. J Molluscan Stud 89:eyad011

Stévenne C, Micha M, Plumier J-C, Roberty S (2021) Corals and sponges under the light of the holobiont concept: how microbiomes underpin our understanding of marine ecosystems. Front Mar Sci 8:698853

Sudek S, Lopanik NB, Waggoner LE, Hildebrand M, Anderson C, Liu H, Patel A, Sherman DH, Haygood MG (2007) Identification of the putative bryostatin polyketide synthase gene cluster from "Candidatus *Endobugula sertula*", the uncultivated microbial symbiont of the marine bryozoan *Bugula neritina*. J Nat Prod 70:67–74

Sunarti LS (2024) Hospital acquired infection revisited: route of transmission. Int J Pathog Res 13:24–35

Sweet M, Villela H, Keller-Costa T, Costa R, Romano S, Bourne DG, Cárdenas A, Huggett MJ, Kerwin AH, Kuek F (2021) Insights into the cultured bacterial fraction of corals. mSystems 6:e01249-20

Swierts T, Peijnenburg KT, de Leeuw CA, Breeuwer JA, Cleary DF, de Voogd NJ (2017) Globally intertwined evolutionary history of giant barrel sponges. Coral Reefs 36:933–945

Swierts T, Cleary D, De Voogd N (2018) Prokaryotic communities of Indo-Pacific giant barrel sponges are more strongly influenced by geography than host phylogeny. FEMS Microbiol Ecol 94:fiy194

Tait LW, Schiel DR (2011) Legacy effects of canopy disturbance on ecosystem functioning in macroalgal assemblages. PLoS One 6:e26986

Tammam MA, El-Demerdash A (2023) Pederins, mycalamides, onnamides and theopederins: distinctive polyketide families with intriguing therapeutic potentialities. Curr Res Biotechnol 6:100145

Tan LT (2023) Impact of marine chemical ecology research on the discovery and development of new pharmaceuticals. Mar Drugs 21:174

Tan YTR, Wainwright BJ, Afiq-Rosli L, Ip YCA, Lee JN, Nguyen NTH, Pointing SB, Huang D (2020) Endosymbiont diversity and community structure in *Porites lutea* from Southeast Asia are driven by a suite of environmental variables. Symbiosis 80:269–277

Tan SM, Ismail MH, Cao B (2021) Biodiversity of magnetotactic bacteria in the tropical marine environment of Singapore revealed by metagenomic analysis. Environ Res 194:110714

Tandon K, Lu C-Y, Chiang P-W, Wada N, Yang S-H, Chan Y-F, Chen P-Y, Chang H-Y, Chiou Y-J, Chou M-S (2020) Comparative genomics: dominant coral-bacterium *Endozoicomonas acroporae* metabolizes dimethylsulfoniopropionate (DMSP). ISME J 14:1290–1303

Taylor MW, Radax R, Steger D, Wagner M (2007) Sponge-associated microorganisms: evolution, ecology, and biotechnological potential. Microbiol Mol Biol Rev 71:295–347

Thomas T, Moitinho-Silva L, Lurgi M, Björk JR, Easson C, Astudillo-García C, Olson JB, Erwin PM, López-Legentil S, Luter H (2016) Diversity, structure and convergent evolution of the global sponge microbiome. Nat Commun 7:11870

Thurber RV, Willner-Hall D, Rodriguez-Mueller B, Desnues C, Edwards RA, Angly F, Dinsdale E, Kelly L, Rohwer F (2009) Metagenomic analysis of stressed coral holobionts. Environ Microbiol 11:2148–2163

Tujula NA, Crocetti GR, Burke C, Thomas T, Holmström C, Kjelleberg S (2010) Variability and abundance of the epiphytic bacterial community associated with a green marine *Ulvacean* alga. ISME J 4:301–311

Turon M, Cáliz J, Garate L, Casamayor EO, Uriz MJ (2018) Showcasing the role of seawater in bacteria recruitment and microbiome stability in sponges. Sci Rep 8:15201

Uku J, Björk M, Bergman B, Díez B (2007) Characterization and comparison of prokaryotic epiphytes associated with three East African seagrasses. J Phycol 43:768–779

van de Water JA, Allemand D, Ferrier-Pagès C (2018) Host-microbe interactions in octocoral holobionts-recent advances and perspectives. Microbiome 6:1–28

van Oppen MJ, Blackall LL (2019) Coral microbiome dynamics, functions and design in a changing world. Nat Rev Microbiol 17:557–567

Ventola CL (2015) The antibiotic resistance crisis: part 1: causes and threats. Pharm Ther 40:277–283

Vieira C, Gaubert J, De Clerck O, Payri C, Culioli G, Thomas OP (2017) Biological activities associated to the chemodiversity of the brown algae belonging to genus *Lobophora* (Dictyotales, Phaeophyceae). Phytochem Rev 16:1–17

Vlamakis H, Chai Y, Beauregard P, Losick R, Kolter R (2013) Sticking together: building a biofilm the *Bacillus subtilis* way. Nat Rev Microbiol 11:157–168

Wainwright BJ, Afiq-Rosli L, Zahn GL, Huang D (2019a) Characterisation of coral-associated bacterial communities in an urbanised marine environment shows strong divergence over small geographic scales. Coral Reefs 38:1097–1106

Wainwright BJ, Bauman AG, Zahn GL, Todd PA, Huang D (2019b) Characterization of fungal biodiversity and communities associated with the reef macroalga *Sargassum ilicifolium* reveals fungal community differentiation according to geographic locality and algal structure. Mar Biodivers 49:2601–2608

Wainwright BJ, Zahn GL, Zushi J, Lee NLY, Ooi JLS, Lee JN, Huang D (2019c) Seagrass-associated fungal communities show distance decay of similarity that has implications for seagrass management and restoration. Ecol Evol 9:11288–11297

Wainwright BJ, Zahn GL, Afiq-Rosli L, Tanzil JT, Huang D (2020) Host age is not a consistent predictor of microbial diversity in the coral *Porites lutea*. Sci Rep 10:14376

Wainwright BJ, Millar T, Bowen L, Semon L, Hickman K, Lee JN, Yeo ZY, Zahn G (2023) The core mangrove microbiome reveals shared taxa potentially involved in nutrient cycling and promoting host survival. Environ Microb 18:47

Wang X, Chen RF, Cable JE, Cherrier J (2014) Leaching and microbial degradation of dissolved organic matter from salt marsh plants and seagrasses. Aquat Sci 76:595–609

Wang M, Carver JJ, Phelan VV, Sanchez LM, Garg N, Peng Y, Nguyen DD, Watrous J, Kapono CA, Luzzatto-Knaan T, Porto C, Bouslimani A, Melnik AV, Meehan MJ, Liu W, Crüsemann M, Boudreau PD, Esquenazi E, Sandoval-Calderón M, Kersten RD, Pace LA, Quinn RA, Duncan KR, Hsu C, Floros DJ, Gavilan RG, Kleigrewe K, Northen T, Dutton RJ, Parrot D, Carlson EE, Aigle B, Michelsen CF, Jelsbak L, Sohlenkamp C, Pevzner P, Edlund A,McLean J, Piel J, Murphy BT, Gerwick L, Liaw C, Yang Y, Humpf H, Maansson M, Keyzers RA, Sims AC, Johnson AR, Sidebottom AM, Sedio BE, Klitgaard A, Larson CB, Boya P CA, Torres-Mendoza D, Gonzalez DJ, Silva DB, Marques LM, Demarque DP, Pociute E, O'Neill EC, Briand E, Helfrich EJN, Granatosky EA, Glukhov E, Ryffel F, Houson H, Mohimani H, Kharbush JJ, Zeng Y, Vorholt JA, Kurita KL, Charusanti P, McPhail KL, Nielsen KF, Vuong L, Elfeki M, Traxler MF, Engene N, Koyama N, Vining OB, Baric R, Silva RR, Mascuch SJ, Tomasi S, Jenkins S, Macherla V, Hoffman T, Agarwal V, Williams PG, Dai J, Neupane R, Gurr J, Rodríguez AMC, Lamsa A, Zhang C, Dorrestein K, Duggan BM, Almaliti J, Allard PM, Phapale P, Nothias LF, Alexandrov T, Litaudon M, Wolfender JL, Kyle JE, Metz TO, Peryea T, Nguyen DT, VanLeer D, Shinn P, Jadhav A, Müller R, Waters KM, Shi W, Liu X, Zhang L, Knight R, Jensen PR, Palsson BØ, Pogliano K, Linington RG, Gutiérrez M, Lopes NP, Gerwick WH, Moore BS, Dorrestein PC, Bandeira N (2016) Sharing and community curation of mass spectrometry data with Global Natural Products Social Molecular Networking. Nat Biotechnol 34:828–837

Webster NS, Thomas T (2016) The sponge hologenome. mBio 7:e00135-16

Webster N, Negri A, Botté E, Laffy P, Flores F, Noonan S, Schmidt C, Uthicke S (2016) Host-associated coral reef microbes respond to the cumulative pressures of ocean warming and ocean acidification. Sci Rep 6:19324

Weisz JB, Hentschel U, Lindquist N, Martens CS (2007) Linking abundance and diversity of sponge-associated microbial communities to metabolic differences in host sponges. Mar Biol 152:475–483

Werner T, Schmülling T (2009) Cytokinin action in plant development. Curr Opin Plant Biol 12:527–538

Wijaya W, Suhaimi Z, Chua CXE, Sunil RS, Kolundžija S, Bin Rohaizat AM, Md Azmi NB, Hazrin-Chong NH, Lauro FM (2023) Frequent pulse disturbances shape resistance and resilience in tropical marine microbial communities. ISME Commun 3:55

Wilson MC, Mori T, Rückert C, Uria AR, Helf MJ, Takada K, Gernert C, Steffens UA, Heycke N, Schmitt S (2014) An environmental bacterial taxon with a large and distinct metabolic repertoire. Nature 506:58–62

Woese CR, Gutell R, Gupta R, Noller H (1983) Detailed analysis of the higher-order structure of 16S-like ribosomal ribonucleic acids. Microbiol Rev 47:621–669

Wu K, Chen C-NN, Soong K (2016) Long distance dispersal potential of two seagrasses *Thalassia hemprichii* and *Halophila ovalis*. PLoS One 11:e0156585

Yan BC, Rabbani G, Lee NLY, Ooi JLS, Lee JN, Huang D, Wainwright BJ (2021) The microbiome of the seagrass *Halophila ovalis*: community structuring from plant parts to regional scales. Aquat Microb Ecol 87:139–150

Zan J, Li Z, Tianero MD, Davis J, Hill RT, Donia MS (2019) A microbial factory for defensive kahalalides in a tripartite marine symbiosis. Science 364:eaaw6732

Zeng Z, Guo X-P, Li B, Wang P, Cai X, Tian X, Zhang S, Yang J-L, Wang X (2015) Characterization of self-generated variants in *Pseudoalteromonas lipolytica* biofilm with increased antifouling activities. Appl Microbiol Biotechnol 99:10127–10139

Zeng Z, Guo XP, Cai X, Wang P, Li B, Yang JL, Wang X (2017) Pyomelanin from *Pseudoalteromonas lipolytica* reduces biofouling. Microb Biotechnol 10:1718–1731

Zhang T, Diaz JM, Brighi C, Parsons RJ, McNally S, Apprill A, Hansel CM (2016) Dark production of extracellular superoxide by the coral *Porites astreoides* and representative symbionts. Front Mar Sci 3:232

Zhou X, Xu T, Yang XW, Huang R, Yang B, Tang L, Liu Y (2010) Chemical and biological aspects of marine sponges of the genus *Xestospongia*. Chem Biodivers 7:2201–2227

Zidorn C (2016) Secondary metabolites of seagrasses (Alismatales and Potamogetonales; Alismatidae): chemical diversity, bioactivity, and ecological function. Phytochemistry 124:5–28

Coral Reef Fishes Thriving in the Urban Reef System of Singapore

8

Daisuke Taira and Zeehan Jaafar

Abstract

Coral reefs host thousands of fish species, with their distribution and diversity being strongly influenced by biotic and abiotic characteristics of the reef habitat. However, the global decline of coral reefs—driven by human activities and mass bleaching events—is altering assemblages and behaviours of reef-associated fishes. Despite the heavily disturbed and modified environment driven by modernisation as a major shipping hub and shoreline hardening, Singapore's coral reefs continue to flourish, supporting high fish diversity. Ecological studies on reef fishes in Singapore began to gain momentum in the late 1980s, but there has yet to be a comprehensive effort to synthesise decades of studies and provide an overview of our current understanding. Here, we outline the history of marine resource exploitation, the development of formal studies on reef fishes in Singapore and describe current reef fish assemblage patterns. We also discuss studies investigating how human activities and climate change impact coral reef fishes, while also identifying key research gaps that need to be addressed to better understand these impacts. Finally, we highlight ongoing efforts to preserve and conserve coral reef fishes in Singapore, emphasising the crucial role of stakeholder engagement.

Keywords

Fish ecology · Fish diversity · Coastal urbanisation · Shoreline armouring · Climate change impact · Habitat enhancement · Marine conservation · Stakeholder engagement

D. Taira (✉)
Centre for Nature-based Climate Solutions, National University of Singapore, Singapore, Singapore

Department of Biological Sciences, National University of Singapore, Singapore, Singapore
e-mail: d.taira@nus.edu.sg

Z. Jaafar
Department of Biological Sciences, National University of Singapore, Singapore, Singapore

8.1 Introduction

Fishes are the most speciose and abundant vertebrate group in coral reef ecosystems and play important ecological roles in shaping benthic community structures (Ceccarelli 2007; Doropoulos et al. 2013) as well as nutrient recycling through multiple trophic pathways (Meyer et al. 1983; Morais and Bellwood 2019). They are also socio-economically important, substantially contributing to food supply (Roberts et al. 2001) and recreational activities such as SCUBA diving and angling (Brander et al. 2007). However, the ongoing global coral reef crisis, driven by anthropogenic activities (e.g. overfishing, coastal development) and climate change (e.g. mass coral bleaching), is drastically impacting coral reef ecosystem services (Eddy et al. 2021) and biodiversity (Bellwood et al. 2004; Sully et al. 2022), including reef fishes (Strona et al. 2021).

Singapore, a highly urbanised city-state, has seen its territorial waters historically impacted by commercial fishing (Ng et al. 2025) and, more recently, significantly modified through intensified coastal development (Lai et al. 2015). Yet, coral reefs persist near its urban centre (Heery et al. 2018) and continue to support a remarkable diversity of fish species (Low and Chou 1992; Taira et al. 2018). Despite decades of coral reef fish research, to date, there is no comprehensive study that summarises ecological studies of reef fishes in Singapore. Here, we aim to provide an overview of our current understanding of ecological research on fish in this unique urban reef system. We synthesise existing reef fish literature in Singapore to identify research gaps by highlighting the history of fisheries and coastal development, the development of fish ecology research, local fish assemblage patterns, ecological impacts of human activities and mass bleaching events, as well as ongoing biodiversity conservation efforts.

L. M. Chou, D. Huang (eds.), *Coral Reefs of Singapore's Urbanised Sea*, Coral Reefs of the World 23,
https://doi.org/10.1007/978-981-95-0903-4_8

8.2 Coral Reef Fishes as Natural Resources

In Singapore, the commercial fisheries industry grew from the late nineteenth century with the increasing human population and demand for seafood (Boomgaard et al. 2005). Many satellite islands were once inhabited by native communities such as the Orang Laut (Sea People in Malay) and Orang Pulau (Island People in Malay). These island communities lived off coastal and marine resources—for food and building material—that were once abundant (Tan 1966; Chew 1982). There have been numerous records of fishing activities at and around coral reefs. Native islanders practised small-scale fishing using traditional methods such as hook and line ('panching'), fish traps ('bubu'), drift nets ('jaring hanyut'), reef nets ('jaring karang') and palisade fish traps ('kelong') (Tan 1966). A wide range of fish species were targeted, including some common coral reef fishes such as groupers, snappers and rabbitfishes (Tan 1966; Chuang 1973). There was also an influx of migrant fisherfolks from Japan in the early twentieth century who made significant contributions to local fish catches using the 'muro ami' method to target fusiliers (*Caesio* spp.) at the relatively unexploited offshore coral reefs (Shimizu 1997). There are further records of fish harvesting (e.g. seahorses, anemonefishes) from shallow reefs for the aquarium trade by residents of the Malay kampong ('village') at Pulau Sudong (Chew 1982). Nevertheless, the local subsistence fisheries did not seem to have significantly impacted reef fish populations, partly because Singapore's fish populations are believed to be replenished by migration from the surrounding waters (Tham 1976).

From the 1960s to the 1980s, commercial fisheries in Singapore's local waters drastically reduced in scale (Chia et al. 1988). This is due to the increasing cost of living and unreliable income from local fisheries, as well as the nation's shift to the shipping industry and maritime operations to boost economic growth (Reeves and Reeves 2021). Since the 1960s, major waterfront development and shipping activities have resulted in a substantial loss of coral reef areas (Hilton and Manning 1995; Lai et al. 2015) and long-lasting water quality deterioration (Chou 1996; Fabricius 2005). Such chronic changes influenced fisher behaviour as well as fish assemblage patterns (Reeves and Reeves 2021). Consequently, present-day commercial capture fisheries remain, albeit at levels that are negligible; most activities at present are artisanal, with few old-timer fishers setting drift nets and fish traps.

8.3 Advancement in Coral Reef Fish Research in Singapore

8.3.1 Natural History Collections

The systematic study of marine fishes in Singapore was first formalised by European naturalists in the early 1800s. The first ever fish species to be described from Singapore was the Bearded leatherjacket, *Anacanthus barbatus* (Gray 1830), which typically occurs in estuaries and coastal bays (Fig. 8.1a). This species was described by John Edward Gray in 1830 from the natural history collections of Major-General Thomas Hardwicke (Gray 1830). The latter gentleman, a soldier, was posted to India from 1777 to 1823, during which time he travelled the region extensively, and made natural history collections and paintings, spanning many animal groups from fishes to mammals (Dawson 1946). One example of such natural history paintings was the Blue-spotted ribbontail ray *Taeniura lymma*, then described as *Trygon ornatus* Fabricius, 1775, based on specimens from Singapore (Fig. 8.1b).

Over the next century, discoveries of fishes burgeoned in Singapore and the region. Notably, in 1842, Pieter Bleeker, a Dutch medical doctor posted to Batavia in the Dutch East Indies (present day Jakarta, Indonesia), where he stayed until his return to the Netherlands in 1860 (Carpenter 2007), was an avid natural historian with a keen interest in fishes, amphibians and reptiles. Now considered the 'Father of Indo-Pacific Ichthyology', Bleeker's prolific works included three articles specifically on fishes in Singapore, and another 14 in which new fish species were named after specimens obtained from Singapore. Efforts of ichthyologists like Gray, Bleeker, and many others have resulted in approximately 117 marine fish species described from Singapore; 58 of these names are still valid. One such fish that bears the country name, the Truncate Cardinalfish *Cheilodipterus singapurensis* Bleeker, 1840, is still common in its coral reefs. To date, over 700 fish species are reported to occur in Singapore's marine territorial areas, including coral reefs (Jaafar et al. 2024). The type specimens of these species are deposited at many natural history museums, for example, the Natural History Museum in London, United Kingdom, the Naturalis in Leiden, The Netherlands, and the Lee Kong Chian Natural History Museum in Singapore.

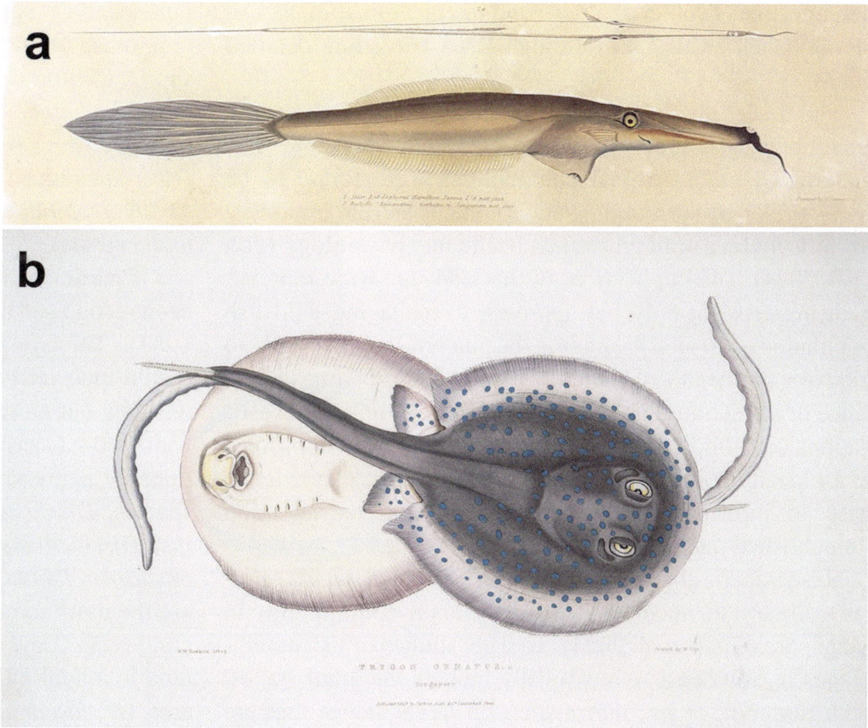

Fig. 8.1 (**a**) Painting of *Balistes barbatus* (now *Anacanthus barbatus*), the first fish species described from Singapore; and (**b**) *Trygon ornatus* (now *Taeniura lymma*), a ray commonly encountered in coral reef habitats in Singapore. (Taken from Hardwicke and Gray 1830–34)

8.3.2 Underwater Fish Surveys

In the mid-1980s, comprehensive coral reef monitoring in Singapore was initiated under the ASEAN-Australia Living Coastal Resources (LCR) Project in response to the noticeable reef degradation across Southeast Asian nations (Chou et al. 1994). The invention of the SCUBA diving apparatus allowed extended underwater time and rapid assessment of reef organisms. Along with coral surveys, reef fish diversity has been periodically monitored by SCUBA divers at various coral reefs until today (Low and Chou 1992; Taira et al. 2018; Loong et al. 2025). Unlike specimen collections, underwater visual survey allows non-invasive quantitative assessment of fishes across locations and time. Belt transects are commonly used but often modified to shorter (e.g. 20 m) and narrower (e.g. 4 m) transects by accounting for small reef areas and poor water visibility. Recently, timed-swim methods have also been used to improve the detectability of specific fish species (Seah et al. 2021; Taira et al. 2024a).

Technological advances in underwater cameras enabled observations of fish assemblages (e.g. Ng et al. 2021) and behaviours (e.g. Guest et al. 2016; Bauman et al. 2017) through the recording of underwater video footage. This method has some advantages over visual surveys carried out by divers. For one, the duration that a diver can remain underwater is limited, and movement and noise from SCUBA equipment can exert some observer bias that influences fish presence and behaviour (Lowry et al. 2012; Emslie et al. 2018). Remote underwater video (RUV) methods, for instance, are often used in studies related to feeding behaviours of diver-shy fishes such as parrotfishes and rabbitfishes (e.g. Guest et al. 2016; Bauman et al. 2017, 2019, 2021; Seah et al. 2021). Baited Remote Underwater Video Systems (BRUVS), in which a camera is trained to focus on a baited bag, attracts larger, more elusive fish species in the vicinity. Thereafter, the footage from these underwater cameras is analysed for the presence of predators such as sharks. However, these methods are typically applied in areas with good water clarity, while limited information exists on their efficacy in poor visibility environments such as Singapore.

8.3.3 Emerging Research Approaches

Ecological studies of coral reef fishes in Singapore are being performed with novel technological tools and allow for better understanding, and therefore more effective resource management strategies. One such endeavour—CODEFISH-SG—aims to barcode mitochondrial genes for marine fishes in Singapore. At present, the sequences of over 300 marine fish species are deposited within this database. Unlike global and open-source DNA databases, barcoded specimens here were always photographed, identified and vouchered. The traceability of all data within the CODEFISH-SG database is an important element in ensuring its strong foundation as a DNA reference library. Since its establishment, the CODEFISH-SG database has aided other research foci, such as studies on the microbiome within fish intestines (Soh et al. 2024) and the identification of fish larvae (Chan et al. 2024) through DNA barcoding detailed below.

The first study that used the CODEFISH-SG DNA reference library, examined 50 fish species from Singapore. The patterns of gut microbial communities were found to be partly driven by surrounding environments and are important towards understanding host fish health and physiology (Soh et al. 2024). The implications of this study are wide-ranging, as it impacts not only our approach to managing wild fish populations, but also leveraging the outcomes of this study to improve the health of cultured fishes. The growing importance of aquaculture in Singapore has also prompted investigations on otoliths to document the age and growth profiles of 14 locally common species that serve as a key knowledge base to inform fisheries management and conservation (unpublished data).

Researchers also used the CODEFISH-SG DNA reference library to understand the fish larval communities in Singapore (Chan et al. 2024). To date, studies of fish assemblages in Southeast Asia typically include the adult stages, with minimal, or no, information on larval stages that are important to marine food webs (Putpipat 2018; Ara et al. 2020). This research gap can be attributed to the drastic ontogenetic morphological transformations in fishes (see Johnson et al. 2009, for example). Identifying fish larvae is a challenge not only due to the differences between larvae and adults, but also their small sizes and similar characteristics between members of the same family. Efforts are underway to understand the spatio-temporal distribution of fish larvae, in particular the connectivity of these fishes between coral reef habitats within Singapore's territorial waters.

8.4 Contemporary Coral Reef Fish Diversity in Singapore

At present, a total of 731 marine fish species have been reported from Singapore (Jaafar et al. 2024), a record culminating from many years of studying marine fishes. Efforts are underway to characterise coral reef fish communities in Singapore, in particular their use of reef habitats with high anthropogenic signatures. In the following, we detail some noteworthy extant reef fish species and functional groups.

8.4.1 Damselfishes—Most Common Fish Family

Damselfishes, from the family Pomacentridae, are among the most common groups of fishes in Singapore's urban reefs (Low and Chou 1992; Taira et al. 2018). Damselfishes consist of diverse functional groups, including omnivores (e.g. *Pomacentrus* spp., *Abudefduf bengalensis*), planktivores (e.g. other *Abudefduf* spp., *Chromis* spp., *Neopomancentrus* spp.), territorial algal farmers (e.g. *Dischistodus* spp., *Hemiglyphidodon plagiometopon*, *Pomacentrus tripunctatus*, *Stegastes obreptus*) and microhabitat (e.g. hard corals, sea anemones) specialists (e.g. *Amphiprion* spp, *Amblyglyphidodon curacao*, *Dascyllus* spp., *Pomacentrus moluccensis*).

Of particular interest, territorial algal farmers have long been recognised as ecosystem engineers (Hixon and Brostoff 1983). These algal farmers modify benthic communities within their territory by cultivating palatable turf algae and weeding out unwanted species of turf/macroalgae (Hata and Kato 2004; Ceccarelli 2007). They also alter species interactions by aggressive behaviours towards intruder fishes. Six species, *Dischistodus chrysopoecilus*, *Dischistodus fasciatus*, *Dischistodus prosopotaenia, Hemiglyphidodon plagiometopon*, *Pomacentrus tripunctatus* and *Stegastes obreptus*, are the most common territorial damselfishes in Singapore's coral reefs (Low et al. 1997; Ng et al. 2021). These species tend to inhabit different reef zones from the lower intertidal area (*D. fasciatus* and *P. tripunctatus*) to the reef flat (*D. chrysopoecilus and S. obreptus*) and crest zones (*D. prosopotaenia* and *H. plagiometopon*) (Low et al. 1997; Ng et al. 2021). Such interspecific spatial partitioning among species could be driven by preferential resource availability as they occupy different trophic niches (Ceccarelli 2007).

Unlike herbivores, they may not contribute to controlling algal abundance in degraded reefs as they selectively cultivate algae. However, there are potential benefits for reef resilience due to their territorial and weeding behaviours. For example, a recent study in the Great Barrier Reef found reduced turf sediment loads within farming damselfish territory compared to those outside the territory (Tebbett et al. 2020), and this could increase algal productivity and potentially facilitate grazing activities by other algal turf-feeding fishes (Ceccarelli et al. 2005). During the 2019 coral bleaching event in Moorea, French Polynesia, the bleached *Pocillopora* corals within *Stegastes nigricans*' territories had higher survivorship and faster recovery, possibly through protection from corallivores and additional food provision (Honeycutt et al. 2023). These benefits are yet to be tested in urban coral reef systems. Given their high abundance, they may contribute to improving reef resistance and resilience during sedimentation disturbances and periodic marine heatwaves.

8.4.2 Habitat and Host Specialists

Fish assemblages in Singapore's coral reefs are dominated by opportunists and habitat generalists. There is also a proportion of reef fishes that depend heavily on other reef organisms for food and/or shelter. The lives of these habitat specialists are intricately intertwined with their host organisms. Therefore, this group of fishes are viewed as highly susceptible to reef degradation and is often used as an indicator of reef health. To date, much of fish research in Singapore has focused on understanding overall patterns of fish assemblages, but relatively few studies have investigated the response of habitat/host specialists to disturbances.

An iconic example of such specificities would be the mutualistic relationship between sea anemones and anemonefishes. There are five anemonefish species (Jaafar et al. 2024) and eight associated host sea anemone species (Yap et al. 2024) in Singapore. Among the most common ones are *Amphiprion frenatus* associated with *Entacmaea quadricolor*, and *Amphiprion ocellaris* residing in *Radianthus magnifica, Stichodactyla gigantica* and *Stichodactyla mertensii* (Fig. 8.2a; Taira et al. 2024a).

There are relatively few obligate coral-dependent fishes. Butterflyfishes from the family Chaetodontidae, especially the genus *Chaetodon*, often have tight associations with coral abundance. Among 11 recorded butterflyfish species, *Chaetodon octofasciatus* is the most common obligate corallivore. This species typically inhabits coral-rich habitats and predominantly feeds on *Acropora* polyps (Srinivasan 2003). Their larvae prefer to settle onto live branching coral colonies, especially *Acropora* and *Pocillopora* corals, and spend most time in their host colonies during the juvenile phase (Fig. 8.2b; Taira et al. 2017a). Despite being a habitat specialist and with the low abundance of branching corals, *C. octofasciatus* thrives in local reefs (Taira et al. 2017a). This could possibly be related to their dietary plasticity, as Singapore's *C. octofasciatus* tends to feed on a wider variety of coral species compared to populations in Malaysia's Tioman Island, where a greater abundance of *Acropora* corals is present (Feary et al. 2018).

In the diverse bottom-dwelling fish group of Gobiidae, some species show host specificities. For example, Broad-barred *Acropora*-goby, *Gobiodon histrio*, is the most commonly encountered coral goby in Singapore and specialises in inhabiting *Acropora* corals (Fig. 8.2c; Munday 2000). Another such host-specific species is the Blackfin coral goby, *Paragobiodon lacunicolus*, which lives within the branches of the host coral *Pocillopora acuta*—only discovered in Singapore in 2023 (Rajendran and Jaafar 2023). Deeper along the reef slopes is the Whip coral goby, *Bryaninops amplus*, solely associated with gorgonian corals. This species is associated with *Junceella juncea* (Fig. 8.2d) and *Dichotella gemmacea* and generally prefers taller host gorgonians for better zooplankton capture (Goh and Chou 1994).

8.4.3 Predators

Large sharks once roamed the local waters, as evidenced by historical records and pictorial documentation. For instance, a 2.7-meter-long (300 kg) shark was recorded off Pulau Brani in 2000 (The New Paper 2000), and a 6-meter-long

Fig. 8.2 Examples of habitat and host specialist: (**a**) *Amphiprion ocellaris* (4 cm TL) on the host anemone, *Stichodactyla mertensii*, (**b**) juvenile *Chaetodon octofasciatus* (3 cm in TL) inhabiting *Acropora* sp., (**c**) *Gobiodon histrio* (2 cm TL) hiding in an *Acropora digitifera* colony and (**d**) *Bryaninops amplus* (2 cm TL) on *Junceella juncea*. (Images by Daisuke Taira)

whale shark was spotted near Pulau Sebarok in 1964 (The Straits Times 1964). Additionally, at least four fatal shark attacks were reported in Singapore between 1861 and 1954 (Coppleson 1958), further suggesting the presence of these giants in the region during that period. Today, however, large predators appear sparse within the coral reefs of Singapore, partly due to low detectability in the poor water visibility conditions.

Advanced techniques, including the use of eDNA, have improved the detection of predatory fishes. For example, only seven species have been sighted within Singapore's marine areas including coral reef habitats in the past two decades. Yet, when eDNA metabarcoding was applied to water samples from nine localities within the territorial waters, it uncovered 14 known shark and ray species, including two species—*Brevitrygon imbricata* and *Pastinachus sephen*—heretofore unreported from Singapore (Ip et al. 2021). This demonstrates the utility of such novel approaches towards effective monitoring of fish communities.

Furthermore, ongoing studies use the BRUVS method to study the predator fish communities in coral reefs within the Singapore Straits. Preliminary results have proven encouraging, capturing footage of predators that are associated with coral reef habitats such as the Blacktip reef shark (*Carcharhinus melanopterus*), and Golden trevally (*Gnathanodon speciosus*) (unpublished data).

8.4.4 Roving Herbivores

Singapore's coral reefs face significant challenges characterised by the dominance of turf and macroalgae, which compete with corals for space (Heery et al. 2018). Thus, top-down control of algal abundance is key to maintaining the resilience of coral reefs. Herbivory research has focused on nominally roving herbivores, including surgeonfishes (Acanthuridae), sea chubs (Kyphosidae), angelfishes (Pomacanthidae), parrotfishes (Labridae) and rabbitfishes (Siganidae), which play significant roles in removing algae. Roving herbivores are divided into different functional groups based on their feeding habit (Choat et al. 2002). Grazers (e.g. some *Pomacanthus* angelfish, surgeonfishes, rabbitfishes) feed primarily on turf algae and fleshy algae. Scrapers (e.g. *Scarus* parrotfishes) similarly graze on turf algae and other organic materials on hard surfaces such as coral skeletons and scrape off the underlying substrate using their specialised fused teeth. Browsers (e.g. rabbitfish, sea chubs), on the other hand, tend to consume canopy-forming macroalgae.

While herbivore abundance and diversity are relatively low in Singapore (Bauman et al. 2017), a one-year herbivory study demonstrated that the current levels of a common canopy-forming alga, *Sargassum ilicifolium*, appear to be maintained, largely by Doublebar rabbitfish (*Siganus virgatus*), and occasionally and to a lesser extent, by Surf parrotfish (*Scarus rivulatus*), the most abundant herbivore in Singapore's coral reefs (Seah et al. 2021). Herbivory rates and behaviours could also be influenced by 'fear' effects such as the presence of predator (Bauman et al. 2019) and territorial damselfish (Taira et al. 2020), as well as shading by overgrown macroalgae (Bauman et al. 2021), but acute fears might have little influence on overall grazing pressure (Tebbett et al. 2024).

8.4.5 Cryptobenthic Fishes

Cryptobenthic reef fishes rarely exceed 50 mm in length when mature or as adults (Brandl et al. 2018) yet are ubiquitous and abundant in coral reef habitats. It is believed that these fishes account for almost half of all reef fish species and are important in coral reef trophodynamics (Ahmadia et al. 2018). Common fishes of this ecoguild in Singapore include the gobies (Gobiidae), blennies (Blennidae), triplefins (Tripterygiidae) and dragonets (Callionymidae). Cryptobenthic reef fishes are typically closely associated with the substrate and are often overlooked in fish visual surveys or studies on fish diversity despite the striking colours in some species such as those in males of Banded triplefin, *Enneaptrygius fasciatus* (Fig. 8.3). Furthermore, due to their small size, identifying them without specimen collection can be a challenge. Increasingly, however, scientists realise their high vulnerability to the changing climate and the importance of their conservation (Brandl et al. 2020). In Singapore, efforts are underway to better understand the cryptobenthic fish communities through their inclusion in current fish surveys and establishing new studies to chart changes in their distribution.

8.5 Anthropogenic Impacts on Fish Diversity

8.5.1 Derelict Fishing Gear

Abandoned, lost or discarded fishing gear (ALDFG) is considered a major component of all marine debris (Löhr et al. 2017). As most ALDFGs are now manufactured from plastic materials, they can cause far-reaching ecological and socio-economic impacts. For example, plastic ALDFGs are especially buoyant and can be moved by prevailing currents far from source points (Vance and McGregor 2019). By virtue of their design with floats at the top and weights at the bottom, abandoned nets drift through the water column like an open curtain, entangling marine organisms (Gall and Thompson 2015). Fishes are prone to being trapped by these derelict

Fig. 8.3 A pair of (**a**) female (28.6 mm SL) and (**b**) male (28.5 mm SL) Banded triplefin, *Enneaptrygius fasciatus*, from Pulau Salu. (Images by Tan Heok Hui)

nets and eventually die of starvation. These trapped fishes also attract predators and scavengers that, in turn, become entangled. In Singapore, the most common ALDFG are discarded monofilament fishing lines and nets (Gajanur and Jaafar 2022). When these ALDFGs remain within the water column, fishes such as the Blacktip reef shark *C. melanopterus* (Chim et al. 2015) are easily entangled. A study conducted in Singapore on abandoned and derelict nets found that fishes experience the highest mortality when entangled, as compared to other marine animals such as crabs and horseshoe crabs (Gajanur and Jaafar 2022).

8.5.2 Coastal Development

Coastal urbanisation is a major threat to coastal habitats (Todd et al. 2019). At least 320,000 km^2 of seafloor have been anthropogenically modified globally, the phenomenon known as 'ocean sprawl' (Bugnot et al. 2021). In particular, extensive transformation of the natural shoreline into engineered infrastructure such as seawalls for coastal defence has occurred in many major cities worldwide, drastically impacting coastal environments (Firth et al. 2016). Compared to coral reefs, seawalls are structurally uniform and simplified, providing fewer resources and habitat features, leading to overall reductions in biodiversity (Aguilera et al. 2014; Henderson et al. 2019; Lai et al. 2018). There is a growing interest in understanding how marine organisms respond to the construction of human-engineered waterfront structures and how the designs of these structures can be improved to enhance their ecological values (Munsch et al. 2017; Burt and Bartholomew 2019). To date, such research primarily focuses on intertidal biodiversity and relatively little effort has been made to advance fish ecology research in urban settings, especially those associated with coral reefs.

Through decades of coastal development, Singapore has armoured over two-thirds of its shoreline, and this has substantially contributed to the historical areal loss of coral reefs (Hilton and Manning 1995; Lai et al. 2015; Tan et al. 2023). Past studies have found that more than 100 reef-associated fish species have colonised marinas (Toh et al. 2016; Ng et al. 2017a) and seawalls (Taira et al. 2018; Ng et al. 2021), but they also report that fish assemblages supported by these human-made structures are distinct from those on coral reefs due to differential three-dimensional framework and benthic communities (Taira et al. 2018; Ng et al. 2021). For example, Taira et al. (2018) found comparable mean fish abundance, and greater mean generic richness and diversity on seawalls, compared to coral reefs (Fig. 8.4). However, the overall species richness was higher on coral reefs, suggesting the proliferation of seawalls may homogenise fish communities across sites, resulting in lower beta diversity.

The presence of such human-made structures adjoining the reefs may influence fish assemblage patterns and alter behaviours. Lower intertidal seawalls harbour a high abundance of algal turfs, which comprise a high density of turf algae, detritus and small invertebrates (Heery et al. 2020). Some turf-feeding reef fishes, including benthic carnivores, scrapers and territorial algal farming damselfishes, seem to migrate to seawalls for foraging with the tides and, in some cases, show a higher feeding rate on seawall turfs than those on reefs (Ng et al. 2021).

Interestingly, the same study found that parrotfish had significantly higher abundance and grazing activities on seawalls next to the reefs. Further investigations discovered high densities of grazing scars made by parrotfishes on gran-

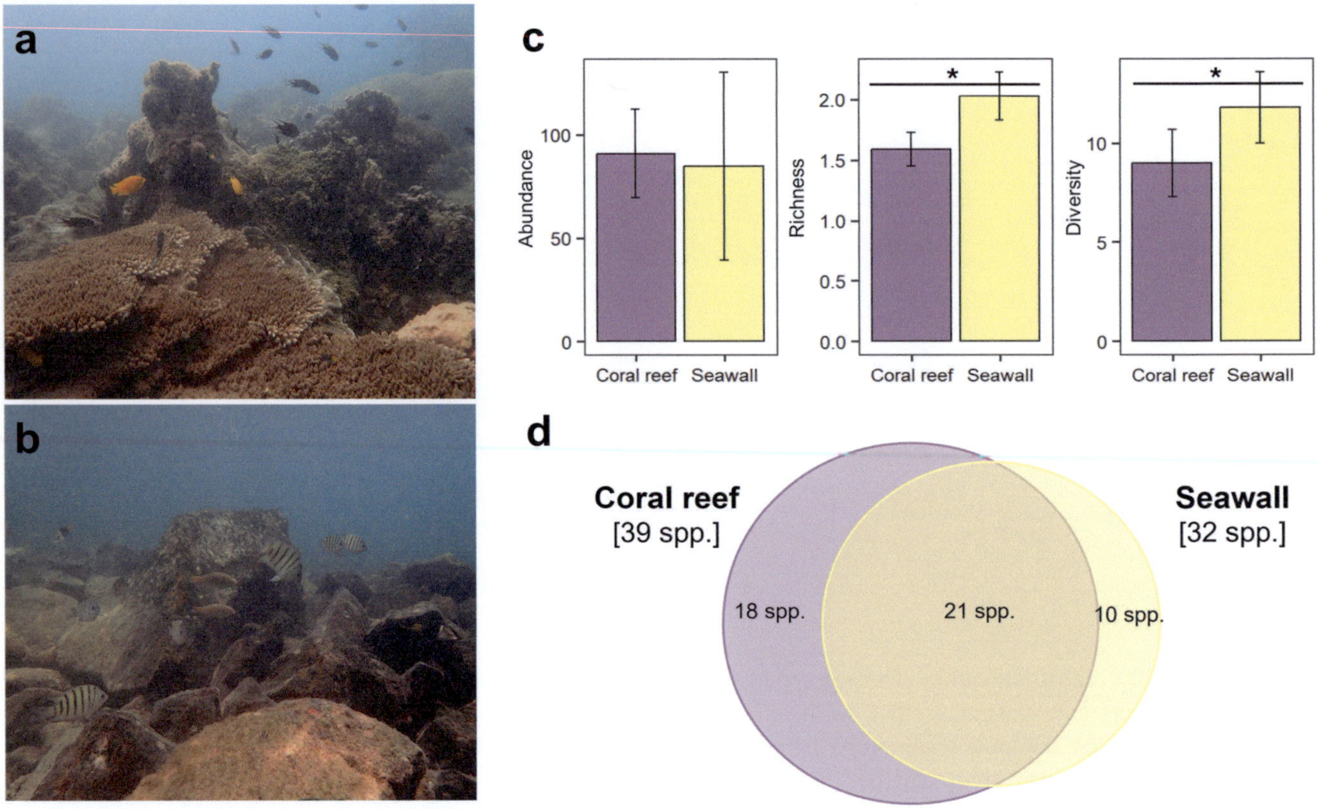

Fig. 8.4 Examples of (**a**) coral reef and (**b**) seawall habitats in Singapore. (**c**) Bar charts showing mean (± standard errors) abundance, generic richness and Shannon's diversity (*H*) per transect (80 m^2) on coral reefs (Kusu Island and Pulau Subar Darat) and seawalls (Lazarus Island and Pulau Seringat). Asterisks (*) indicate significant differences ($p < 0.05$). (**d**) A Venn diagram of 49 fish species recorded on two coral reefs (Kusu Island and Pulau Subar Darat) and two seawalls (Lazarus Island and Pulau Seringat) through underwater visual survey shows that slightly less than half the recorded species can be found on both habitat types, but coral reefs support more unique species than seawalls. (Adapted from Taira et al. (2018). Images photographed by Daisuke Taira)

ite seawall surfaces, likely because parrotfishes prefer seawall algal turfs which are shorter and more cyanobacteria-rich with lower sediment load than those on the reefs (Mark et al. 2024). These observations suggest that shoreline armouring shifted their feeding grounds to the nearby armoured shorelines from their original reef habitats. This altered behaviour could have a significant implication on reef resilience as they play a very important role in maintaining algal abundance (Seah et al. 2021). Seawalls could also be a novel feeding ground for benthic carnivores due to a high abundance of gastropod prey on seawalls (Yeo et al. 2024a). Preying on these gastropods by multiple invertivorous labrids such as *Choerodon anchorago*, *Haliochoeres nigrescens* and *Haliochoeres chloropterus* is evident through RUV (Yeo et al. 2024b) and gut content analysis (unpublished data).

Some coastal infrastructure, such as overwater structures (e.g. docks, piers), casts shadows over a large area and may create permanently dark areas. Dark environments under such structures can impair visual tasks such as feeding and predator vigilance and reduce prey availability during daytime (Ono and Simenstad 2014). Some migratory fishes, such as juvenile salmon, avoid shaded areas created by the structures, which disrupts habitat connectivity by restricting their movements along shorelines (Munsch et al. 2017).

Fishes are known to aggregate or associate with floating structures, such as jetties and pontoons, and many studies on fish responses to floating structures focused on fish aggregation devices (FADs) for fisheries productivity and management (Moreno et al. 2016). While cave ceilings and overhang reef structures offer similar overhead habitat features, floating structures at the scale of marine construction are rare in the natural environment. Interesting observations were made at marinas in Singapore, where some fishes were swimming upside down near the underside of the floating pontoons (Fig. 8.5). A study in Queensland, Australia, (Waltham et al. 2023) reported similar observations where 68% of the fish associated with the floating pontoons were swimming upside down. The same study also suggested that such altered behaviours could be an adaptive response to the novel floating habitat, as orienting according to the substrate increases the chance of benthic prey capture and the ability to detect predators. While these structures may provide additional sur-

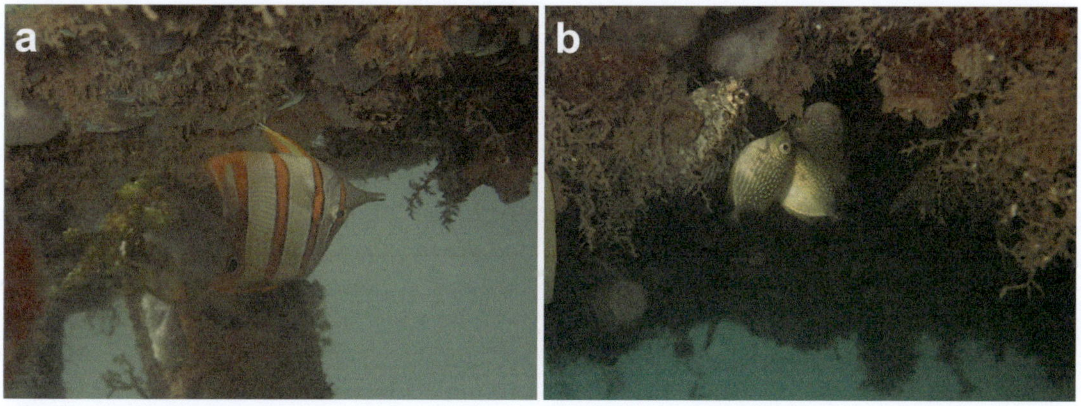

Fig. 8.5 Examples of (**a**) *Chelmon rostratus* (5 cm TL) and (**b**) *Siganus canaliculatus* and *Siganus javus* (3 cm TL) swimming upside down underneath the floating jetty at a marina. (Images photographed by Daisuke Taira)

faces for fish to colonise, further studies are needed to better understand their ecological impacts on fish assemblages and behaviours.

8.5.3 Sedimentation

Human activities, including land reclamation, dredging and shipping traffic, can increase sedimentation and water turbidity in urban coral reefs. As algal turfs are capable of efficiently trapping settling sediments, an excessive amount of sediment can be accumulated within turfs. Consequently, this can increase inorganic materials and hinder algal growth and may negatively influence grazing activities of some turf-feeding fishes such as parrotfish and surgeonfish (Bellwood and Fulton 2008; Gordon et al. 2016; Tebbett et al. 2017). In Singapore, an experiment showed that two territorial damselfish species, *Dischistodus chrysopoecilus* and *Dischistodus fasciatus,* grazed more on algal turfs with reduced sediment loads than those that naturally accumulated over 8 months (Ng et al. 2021). On some occasions, they spit out sediment, an indication of rejected bites of sediment from the mouth (Tebbett et al. 2017), when feeding on the turf treatment with natural sediment loads (Ng et al. 2021). This behaviour suggests that the current level of sediment within turfs growing on reef substrates could already be excessive for some grazing fishes.

The effect of poor water visibility on fish behaviour and diversity remains understudied. Increased suspended sediment levels can hinder visual cues necessary for prey detection and capture (Johansen and Jones 2013) while also altering predator–prey interactions, a key process that regulates fish assemblage structures (Wenger et al. 2013). Although water visibility has drastically decreased from 10 m to 2 m since the 1960s, primarily due to high levels of suspended sediment (Chou 1996), no study has yet explored the negative effects of poor visibility on coral reef fishes in Singapore.

8.5.4 Pollution

Water quality deterioration due to elevated levels of nutrients, various pollutants and debris derived from land-based and marine-based human activities may directly and indirectly influence reef diversity and assemblage structures (Madduppa et al. 2013; Heery et al. 2018). Several incidents of harmful algal blooms, oil and chemical spills have been documented over decades in Singapore and, in some cases, localised mass die-offs of marine organisms including fishes have been reported from fish farms, marinas and beaches (e.g. Gin et al. 2001; Kiong and Saparudin 2007; Ng et al. 2017a; Trottet et al. 2022; Qing 2023). Moreover, there is also growing evidence that both macro- and micro-plastics are accumulating in Singapore's marine environments (Fong et al. 2023; Ledet et al. 2024) and in fish in the surrounding waters (Foo et al. 2022). This warrants further studies to understand the cumulative and interactive effects of environmental pollution and other anthropogenic stressors on fish.

8.5.5 Artificial Lighting at Night (ALAN)

Coastal development and urbanisation largely alter light penetrating into the nearshore waters and may influence reef fishes. ALAN has recently been recognised as a widespread urban stressor that can alter behaviours and species interactions of ALAN-exposed fishes and other communities in coastal waters (Bassi et al. 2022). Excessive artificial light intensity disrupts light cues for regulating daily biological rhythms and alters the vertical migration of zooplankton and consumer fishes at night (Davies et al. 2014). ALAN can attract high densities of larval and juvenile fishes near the water surface but reduce the effectiveness of their anti-predator behaviour, leading to higher mortality from predation (O'Connor et al. 2019). Meanwhile, increased foraging activity under ALAN exerts greater grazing pressure on benthic assemblages (Bolton et al. 2017). Despite the high relevance of the light-related stressors to Singapore's urbanised and heavily modified nearshore environment, no local studies have been conducted to date.

8.5.6 Introduction of Non-Indigenous Species

Due to Singapore being a global shipping hub, Seebens et al. (2013) ranked the island nation among the high-risk countries for bioinvasion. However, in spite of multiple pathways of possible marine bioinvasion, including shipping, commercial fisheries and ornamental trade, few records of non-native or invasive marine fish species have been reported in Singapore (Jaafar et al. 2012; Wells et al. 2019). Nevertheless, there are concerns over increasing sightings of a hybrid grouper (*Epinephelus lanceolatus* × *Epinephelus fuscoguttatus*) underwater (Cabrera et al. 2015; Tan 2015; Taira personal observation) and through recreational fishing (Lise 2020) on and near coral reefs in Singapore. The hybrid grouper is commonly imported from Malaysia and reared as a food fish in Singapore (Tan 2015). The sighted individuals are likely either released or escapees from floating fish farms. Though the long-term ecological effects on the reef ecosystems are not known, the population, when established, may compete with native fishes for food and pose a threat to local food webs.

There are also new sightings of popular ornamental coral reef fish species near publicly accessible shorelines or islands. These were potentially introduced to the local waters via irresponsible pet release or mercy release—the practice of freeing animals into the wild often done in religious or cultural belief as an act of compassion that is believed to bring a person good fortune. Some examples of such dubious species records include Goldtail damselfish, *Chrysiptera parasema* (Taira 2020), Humbug damselfish, *Dascyllus aruanus* (Taira 2021), Mandarinfish, *Synchiropus splendidus* (Tan and Jaafar 2022), Picasso triggerfish, *Rhinecanthus aculeatus* (Adib et al. 2023) and Yellowtail wrasse, *Anampses meleagrides* (Lin 2023). Although Singapore is within the natural geographic ranges of these species, their recent records bear the hallmarks of releases (e.g. occurring singly or in pairs). If the release of non-native species persists, such practices may unintentionally facilitate the establishment of their populations and adversely influence trophic interactions, and thus need to be closely monitored.

8.6 Climate Change Impacts

Climate change-driven marine heatwaves have resulted in several coral bleaching events recorded in Singapore since 1998 (Chou 2001; Guest et al. 2012; Chou et al. 2016; Ng et al. 2020). While bleaching impacts on corals are well-documented in Southeast Asia, there is little information on reef fish response to such large-scale disturbances. The 2016 mass bleaching led to drastic changes in coral cover and species compositions (Ng et al. 2020), as well as flattening of reef framework (Bauman et al. 2022). However, quarterly reef surveys found no clear evidence of fish population changes or declines at impacted reefs (unpublished data), possibly because many local reef fishes are not dependent on live corals and are well-adapted to weedy, low-complexity reef environments.

However, the same severe warming event affected host specialists, anemonefishes. At Kusu Island and Sultan Shoal, host sea anemones, comprising three species (*Entacmaea quadricolor*, *Radianthus magnifica* and *Stichodactyla mertensii*), were affected as 16% of the surveyed sea anemones that host anemonefishes bleached (Taira et al. 2024a). Differential bleaching prevalence among host anemone species (100%, 1% and 0% bleaching of *E. quadricolor*, *R. magnifica* and *S. mertensii*, respectively) disproportionately impacted the tomato anemonefish (*Amphiprion frenatus*) knowingly only associated with *E. quadricolor* (Fig. 8.6a, b). The contrasting recovery rates of bleached *E. quadricolor* among two study sites (88% and 13% of the bleached *E. quadricolor* regained pigmentation at Kusu Island and Sultan Shoal, respectively) also influenced the associated *A. frenatus* populations differently (Fig. 8.6c). At Sultan Shoal, prolonged bleaching of *E. quadricolor* led to substantial size reduction followed by a decrease in resident *A. frenatus* abundance, while early recovery led to host size increase and retained associated anemonefish population at Kusu Island (Fig. 8.6d, e). This result implies that host anemone size is key in regulating anemonefish populations after disturbances. Interestingly, unusual associations of *A. frenatus* with other healthy host anemone species during and after the bleaching events, including *R. magnifica* and *S. mertensii*, suggest the potential flexibility in their host specificity (Taira et al. 2017b, 2024a).

Characterised by the small spatial extent and narrow depth range of the coral reefs, sea-level rise (SLR) is an emerging threat that can cause habitat loss/fragmentation and community shifts (Nguyen et al. 2022; Law and Huang 2023). SLR also increases reproductive isolation for reef fish species and potentially leads to significant population bottlenecks and genetic structuring among various populations (Fauvelot et al. 2003). This underscores the concomitant need to improve our understanding of how species with different ecological specialisations respond to local threats and climate change—crucial for planning effective conservation strategies.

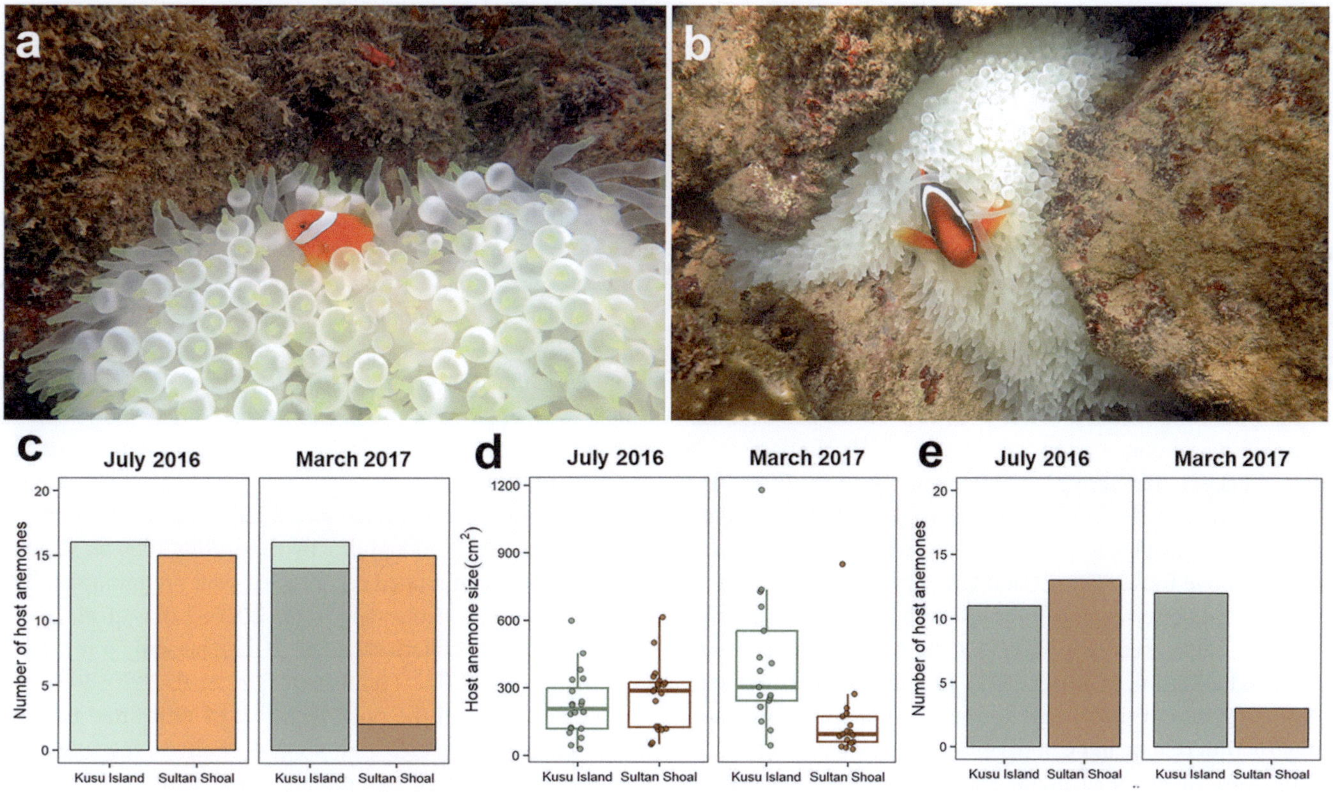

Fig. 8.6 During the 2016 mass bleaching event, all surveyed *Entacmaea quadricolor* at Kusu Island and Sultan Shoal were bleached. Images show (**a**) male (4 cm TL) and (**b**) female *Amphiprion frenatus* (9 cm TL) residing in bleached *E. quadricolor* in July 2016. (**c**) Bar charts show the differential recovery rate of *E. quadricolor* (light = bleached, dark = recovered) between Kusu Island (green) and Sultan Shoal (brown) from July 2016 to March 2017. (**d**) Boxplots with strip plots indicate that host size increased at Kusu Island with a higher recovery rate, while those at Sultan Shoal, with only a few recovered individuals, shrank. (**e**) Bar charts of resident *A. frenatus* presence revealed that host health status and size affect the number of host anemones occupied by the resident anemonefish. (Adapted from Taira et al. (2024a). Images by Daisuke Taira)

8.7 Fish Diversity Conservation and Way Forward

8.7.1 Conservation Status of Coral Reef Fishes

Many fish scientists and naturalists have contributed to the collective knowledge of fishes in Singapore over the past decades. Data from the study of fish diversity, ecology and evolution inform conservation approaches and management strategies. The Singapore Red Data Book (RDB), first published in 1994 (Ng and Wee 1994), details the threatened statuses of plants and animals of Singapore. Marine fishes were first included in the second edition of the Red Data Book (RDB2, Davison et al. 2008). In RDB2, seven species of marine fishes from two genera and two families, all coral-reef associated, were assessed—*Hippocampus kuda* (Spotted seahorse), *Hippocampus comes* (Tiger-tail seahorse), *Amphiprion clarkii* (Yellowtail clownfish), *Amphiprion ocellaris* (Clown anemonefish), *Amphiprion frenatus* (Tomato anemonefish), *Amphiprion peridaraion* (Pink anemonefish) and *Amphiprion polymnus* (Saddleback clownfish). All seven species were considered 'Vulnerable' to extinction primarily driven by collection pressure from indiscriminate capture for the aquaria trade. The third, and most recent edition of the Singapore Red Data Book (RDB3) published in 2024 assessed a total of 731 marine fish species (Jaafar et al. 2024). Of these, 58 species were considered 'Critically Endangered', 49 species 'Endangered' and 149 species 'Vulnerable' to extinction. Still, 227 species were considered 'Data Deficient', and no threat assessments could be made, underpinning the importance to continue with efforts to study and document Singapore's marine fish fauna.

8.7.2 Active Interventions

Coral reef management is complicated by the interplay of global and local stressors, compounded by ongoing coastal urbanisation. In a highly urbanised city-state with limited reef space like Singapore, achieving conservation targets while ensuring a sustainable balance for the coexistence of people and nature is challenging, necessitating innovative and proactive approaches. Various reef habitat mitigation and restoration interventions have been trialled in Singapore and potentially contribute to reef fish conservation.

The use of artificial reefs to restore stony coral and fish populations is a method popularised in the 1960s (Lima et al. 2019). Artificial structures of various materials and shapes are deployed within or adjacent to coral reef areas to encourage habitat rehabilitation. In Singapore, artificial reefs were first deployed in 1989 to study parameters that might enhance fish populations (Chua and Chou 1994). Two types of artificial reefs were constructed, and deployed 10 m apart, at an area adjacent to Terumbu Pempang Tengah. The first structure was made from 50 pieces of 1 m^3 hollowed concrete blocks; 40 pieces serve as a base while 10 pieces were stacked above these. The second structure was made with 42 pieces of rubber tyres in a pyramid configuration, with 50 kg concrete cubes securing them to the bottom. This study attributed differences in diversity and abundance of fishes colonising the artificial reefs to the materials used for their construction (Chua and Chou 1994).

In 2001, a novel artificial reef design was conceptualised—called Reef Enhancement Units (REUs). These dome-shaped structures were made from fibreglass and therefore light, each weighing approximately 10 kg (Loh et al. 2006). They were small for easy handling during deployment, with a base diameter about 70 cm and a height of about 50 cm. Several of these units were then deployed at various coral reef habitats within the Straits of Singapore (Loh et al. 2006). While the REUs experienced limited success as refugia for fishes immediately after deployment, monitoring of these units 10 years post deployment revealed that fishes from the families Labridae and Pomacentridae utilised these structures (Ng et al. 2017b).

In 2018, large artificial reef units were deployed off the western shore of Little Sister's Island, on the seafloor up to 15 m depth (Chou et al. 2023). Eight units were placed 15 m apart in two separate clusters. The first cluster comprised three 7 m-tall units, while the second cluster comprises five 12 m-tall units. Each of the eight units has a concrete base frame as well as metal pipes and platforms. These artificial reef units serve as recipient sites for coral transplantation and, over the years, have become refugia for many reef organisms, including coral reef fishes.

Underwater coral nursery structures temporarily hold a high density of coral nubbins until they are ready for transplantation onto the degraded reefs. These structures can also attract diverse fish species through food and shelter provision (Fig. 8.7; Seraphim et al. 2020; Rinkevich 2021). For example, belt transect surveys recorded comparable fish species richness between the natural reef at Kusu Island (22.0 species 80 m^{-2}) and the nearby sandy areas of Pulau Seringat with coral nursery tables that harboured a variety of coral colonies (18.3 species 80 m^{-2}) (unpublished data). In the same table nursery, coral-dependent juvenile *C. octofasciatus* was found at greater abundance than the neighbouring reef due to higher density of branching *Acropora* and *Pocillopora* colonies (Taira et al. 2017a). This suggests that coral nursery interventions have benefits beyond increasing coral survivorship and growth and, if appropriately designed, can contribute to enhancing diversity of reef-associated fishes (Horoszowski-Fridman et al. 2024).

Given the burgeoning waterfront infrastructure in Singapore, ecological engineering has great potential to increase the ecological values of such human-made structures and provide novel habitats for coral reef fishes. There have been numerous efforts to enhance intertidal seawall biodiversity in Singapore, for example, by retrofitting tiles on seawall surfaces with manipulated complexity, water retaining ability, pH and materials (Loke and Todd 2016; Loke et al. 2019; Hsiung et al. 2020; Hartanto et al. 2022), but only a few have investigated their effects on fishes (Taira et al. 2020, 2024b).

One study found the positive effect of retrofitted tiles on fish abundance, species richness, as well as feeding rate, when comparing fish assemblages and activities between plots with and without tiles on lower intertidal seawalls (Taira et al. 2020). While these tiles are too small for the study fishes to utilise as shelter, their complex and added surface areas enhanced algal cover and increased turf-feeding reef fish diversity and their feeding activity. Though cryptobenthic fishes were not included in this study due to low detectability by the video survey method, the study also noted that some species, such as *Istigobius ornatus,* were observed resting and feeding on these tiles.

A positive relationship between fish diversity and habitat complexity is a well-recognised ecological concept and has been widely applied in conservation and restoration efforts. A study by Taira et al. (2024b) deployed three types of habitat enhancement units called 'fish houses' on seawalls at Pulau Hantu by manipulating the number, size and type of microhabitat features. The more complex fish house treatment supported significantly higher fish abundance, species richness and distinct species compositions of species. This study also assessed fish functional diversity and nocturnal assemblages on ecologically engineered shorelines. The results showed only a modest increase in functional diversity but significantly different compositions of functional traits as the more complex fish house accommodated fish with wider ranges of size and morphology. The study also found that fish use microhabitats differently between day and night, as they prefer larger microhabitats relative to their body size during the daytime, where they can explore food, while most fish used tight-fitting microhabitats at night as they take shelter from predators.

Fig. 8.7 Examples of fishes utilising corals on nursery tables, (**a**) juvenile *Chaetodon octofasciatus* (3 cm TL), (**b**) juvenile *Plectorhinchus chaetodonoides* (3 cm TL), (**c**) juvenile *Chaetodontoplus mesoleucus* (2 cm TL) and (**d**) *Ostorhinchus margaritophorus* (4 cm TL). (Images by Daisuke Taira)

In addition to retrofitting artificial habitat features, attachment of habitat-forming reef organisms, including hard corals, soft corals and sponges on seawalls, have been tested (e.g. Ng et al. 2015; Afiq-Rosli et al. 2017; Toh et al. 2017; Kikuzawa et al. 2021; Sam et al. 2024). It is expected that reef fishes would benefit from additional food resources and microhabitats provided by these organisms in the novel habitats. Such studies primarily focus on improving the survivorship and growth of the transplanted organisms, but the effect on the associated reef organisms, including fishes, has not been assessed. Where coral abundance is high or can be enhanced, seawalls potentially function as a novel habitat for reef-associated fishes. In fact, the seawall at southern Pulau Semakau (31% coral cover and 4 butterflyfish 80 m^{-2}) hosted, on average, higher densities of corals and coral-dependent butterflyfish at the nearby coral reef on the northwestern side of the same island (20% coral cover, 1 butterflyfish 80 m^{-2}) during the survey conducted in 2017 (unpublished data). On the other hand, no significant increase in fish abundance and species richness was found after coral transplantation at Pulau Seringat (unpublished data), where the efforts in April–August 2015 led to a sevenfold increase in coral cover on the seawall (Toh et al. 2017). This result could be partly affected by the coral cover decline during the 2016 mass bleaching event (Ng et al. 2020) and/or perhaps suggests that longer monitoring efforts are needed to detect trends.

All this knowledge contributes to crucial considerations for designing appropriate ecological engineering interventions in the future. To date, most of such interventions have been implemented at small scales, and scaling up is urgently needed to match the scale of local and global disturbances reefs are currently experiencing (Bellwood et al. 2019). Furthermore, when designing coastal defence strategies, the principle of ecological engineering can be applied beyond biodiversity enhancement by incorporating it into broader coastal nature-based solutions to provide a variety of other ecosystem benefits such as coastal protection and tourism (Sloey et al. 2025).

8.7.3 Public Engagement and Citizen Science

The Biophilia hypothesis—the desire and tendency to commune with nature—popularised by the luminary biologist E.O. Wilson, is demonstrated widely in Singapore as its public has increasingly embraced the idea of sustainable development. One clear indication is the sheer number of nature blogs and social media posts focusing on nature-related themes (see blogs such as WildSingapore, Naked Hermit Crabs or Our Wild Neighbours). Every weekend, social media abound with photographs and videos of coral reef organisms by divers and beachgoers. It is not uncommon that academics become aware of new organism records because of these posts, for example, the record of the Whale shark *Rhincodon typus* (Jaafar 2020) and the new record of the Ornate slipper goby *Lubricogobius ornatus* (Tan and Jaafar 2015). Anglers keen on publishing their rarer catches also contribute towards the updates of the fish fauna checklist of Singapore.

More citizen scientists are involved in projects in which their contributions aid in data gathering for scientific studies or for the development of conservation plans; for example Team Seagrass for spatio-temporal patterns of seagrass distribution and Beach Patrol for nesting turtle distribution. Friends of the Sisters' Islands Marine Park (FMP) is an initiative facilitated by the National Parks Board Singapore that provides a platform for stakeholders in the marine environment (e.g. academics, native islanders and recreational users) to meet, discuss and make collective decisions on management aspects of marine areas in Singapore. These sessions, started in 2017 with members meeting every quarter, are well-attended by diverse stakeholders and members of the marine community.

Fostering partnerships between different stakeholders is key to ensuring the success of any conservation measure. The implementation of best practices in activities that directly impact coral reef fishes can only be successfully achieved with the buy-in of users of coral reef habitats, such as recreational diving operators and angling enthusiasts. These stakeholders have been especially amenable to adopting sustainable practices, such as no collection of marine organisms while diving or releasing catches from angling attempts. Indeed, successful stewardship of coral reef habitats and the communities therein, including coral reef fishes, is more likely when stakeholder groups work together with management agencies, and can thus bolster the conservation of coral reef fishes in Singapore.

Acknowledgements We thank all past and present fish scientists and naturalists whose dedicated efforts have contributed significantly towards the understanding of Singapore's coral reef fishes. We are grateful to Tan Heok Hui for the use of his images.

References

Adib A, Sim JH, Lim TK, Heng KS (2023) Biodiversity record: a Picasso triggerfish, *Rhinecanthus aculeatus*, at Marina South. Nat Singap 16:e2023104

Afiq-Rosli L, Taira D, Loke HX, Toh TC, Toh KB, Ng CSL, Cabaitan PC, Chou LM, Song T (2017) In situ nurseries enhance coral transplant growth in sedimented waters. Mar Biol Res 13(8):878–887

Aguilera MA, Broitman BR, Thiel M (2014) Spatial variability in community composition on a granite breakwater versus natural rocky shores: lack of microhabitats suppresses intertidal biodiversity. Mar Pollut Bull 87(1–2):257–268

Ahmadia GN, Tornabene L, Smith DJ, Pezold FL (2018) The relative importance of regional, local, and evolutionary factors structuring cryptobenthic coral-reef assemblages. Coral Reefs 37:279–293

Ara R, Amin SMN, Yusoff FM, Arshad A, Romano N (2020) Larval fish assemblage, diversity and habitat ecology in the Matang Mangrove Ecosystem, Perak, Malaysia. J Environ Biol 41:1316–1325

Bassi A, Love OP, Cooke SJ, Warriner TR, Harris CM, Madliger CL (2022) Effects of artificial light at night on fishes: a synthesis with future research priorities. Fish Fish 23(3):631–647

Bauman AG, Hoey AS, Dunshea G, Feary DA, Low J, Todd PA (2017) Macroalgal browsing on a heavily degraded, urbanized equatorial reef system. Sci Rep 7:8352

Bauman AG, Seah JC, Januchowski-Hartley FA, Hoey AS, Fong J, Todd PA (2019) Fear effects associated with predator presence and habitat structure interact to alter herbivory on coral reefs. Biol Lett 15(10):20190409

Bauman AG, Hoey AS, Dunshea G, Fong J, Chan IZ, Todd PA (2021) Fear effects and group size interact to shape herbivory on coral reefs. Funct Ecol 35(9):1985–1997

Bauman AG, Januchowski-Hartley FA, Teo A, Todd PA (2022) Further flattening of a degraded, turbid reef system following a severe coral bleaching event. Front Mar Sci 9:910085

Bellwood DR, Fulton CJ (2008) Sediment-mediated suppression of herbivory on coral reefs: decreasing resilience to rising sea-levels and climate change? L&O 53:2695–2701

Bellwood DR, Hughes TP, Folke C, Nyström M (2004) Confronting the coral reef crisis. Nature 429(6994):827–833

Bellwood DR, Pratchett MS, Morrison TH, Gurney GG, Hughes TP, Álvarez-Romero JG, Day JC, Grantham R, Grech A, Hoey AS, Jones GP, Pandolfi JM, Tebbett SB, Techera E, Weeks R, Cumming GS (2019) Coral reef conservation in the Anthropocene: confronting spatial mismatches and prioritizing functions. Biol Conserv 236:604–615

Bolton D, Mayer-Pinto M, Clark GF, Dafforn KA, Brassil WA, Becker A, Johnston EL (2017) Coastal urban lighting has ecological consequences for multiple trophic levels under the sea. Sci Total Environ 576:1–9

Boomgaard P, Henley D, Osseweijer M (2005) Muddied waters: historical and contemporary perspectives on management of forests and fisheries in Island Southeast Asia. Leiden KITLV Press

Brander LM, van Beukering P, Cesar HS (2007) The recreational value of coral reefs: a meta-analysis. Ecol Econ 63(1):209–218

Brandl SJ, Goatley CH, Bellwood DR, Tornabene L (2018) The hidden half: ecology and evolution of cryptobenthic fishes on coral reefs. Biol Rev 93:1846–1873

Brandl SJ, Johansen JL, Casey JM, Tornabene L, Morais RA, Burt JA (2020) Extreme environmental conditions reduce coral reef fish biodiversity and productivity. Nat Commun 11:3832

Bugnot AB, Mayer-Pinto M, Airoldi L, Heery EC, Johnston EL, Critchley LP, Strain EMA, Morris RL, Loke LHL, Bishop MJ, Sheehan EV, Coleman RA, Dafforn KA (2021) Current and projected global extent of marine built structures. Nat Sustain 4:33–41

Burt JA, Bartholomew A (2019) Towards more sustainable coastal development in the Arabian Gulf: opportunities for ecological engineering in an urbanized seascape. Mar Pollut Bull 142:93–102

Cabrera B, Ong D, Yaakub SM (2015) Probable breeding pair of grouper. Singapore Biodivers Rec 2015:154–155

Carpenter KE (2007) A short biography of Pieter Bleeker. Raffles Bullet Zool 14:5–6

Ceccarelli DM (2007) Modification of benthic communities by territorial damselfish: a multi-species comparison. Coral Reefs 26(4):853–866

Ceccarelli DM, Jones GP, McCook LJ (2005) Effects of territorial damselfish on an algal-dominated coastal coral reef. Coral Reefs 24:606–620

Chan WWR, Chang JJM, Tan CZ, Ng JX, Ng MHC, Jaafar Z, Huang D (2024) Eyeing DNA barcoding for species identification of fish larvae. J Fish Biol 105(6):1784–1799

Chew SB (1982) Fishermen in flats. Centre of Southeast Asian Studies No. 9, Monash University

Chia LS, Khan H, Chou LM (1988) The coastal environmental profile of Singapore. ICLARM Technical Report No.21. International Centre for Living Aquatic Resources Management (ICLARM), Manila

Chim CK, Lee Y-L, Tong SJW, Chim CK, Lee YL, Tong S, Tay T, Ong R (2015) Blacktip reef sharks caught in trammel nets at Lazarus Island. Singapore Biodivers Rec 2015:158–159

Choat JH, Clements KD, Robbins WD (2002) The trophic status of herbivorous fishes on coral reefs. Mar Biol 140:613–623

Chou LM (1996) Response of Singapore reefs to land reclamation. Galaxea 13(1):85–92

Chou LM (2001) Country Report: Singapore. International Coral Reef Initiative (ICRI)

Chou LM, Low JKY, Loo MGK (1994) The state of coral reefs and coral reef research in Singapore. Chulalongkorn University

Chou LM, Toh TC, Toh KB, Ng CS, Cabaitan P, Tun K, Goh E, Afiq-Rosli L, Taira D, Du RC, Loke HX (2016) Differential response of coral assemblages to thermal stress underscores the complexity in predicting bleaching susceptibility. PLoS One 11(7):e0159755

Chou LM, Srirangam S, Kiong J, Tun K (2023) High-relief artificial reefs at Sister's Islands Marine Park. In: Chan L, Ng D, Lim LJ (eds) Handbook on habitat restoration: general principles and case studies in Singapore. National Parks Board, Singapore

Chua CY, Chou LM (1994) The use of artificial reefs in enhancing fish communities in Singapore. Springer, Netherlands, pp 177–187

Chuang SH (1973) Animal life and nature in Singapore. Singapore University Press

Coppleson VM (1958) Shark attack. Angus and Robertson

Davies TW, Duffy JP, Bennie J, Gaston KJ (2014) The nature, extent, and ecological implications of marine light pollution. Front Ecol Environ 12(6):347–355

Davison G, Ng PKL, Ho HC (2008) The Singapore Red Data book, 2nd edn. The Nature Society, Singapore, p 285

Dawson WR (1946) On the history of Gray and Hardwicke's illustrations of Indian zoology, and some biographical notes on general Hardwicke. J Soc Bibliogr Nat Hist 2(3):55–69

Doropoulos C, Hyndes GA, Abecasis D, Vergés A (2013) Herbivores strongly influence algal recruitment in both coral-and algal-dominated coral reef habitats. Mar Ecol Prog Ser 486:153–164

Eddy TD, Lam VW, Reygondeau G, Cisneros-Montemayor AM, Greer K, Palomares MLD, Bruno JF, Ota Y, Cheung WW (2021) Global decline in capacity of coral reefs to provide ecosystem services. One Earth 4(9):1278–1285

Emslie MJ, Cheal AJ, MacNeil MA, Miller IR, Sweatman HP (2018) Reef fish communities are spooked by scuba surveys and may take hours to recover. PeerJ 6:e4886

Fabricius KE (2005) Effects of terrestrial runoff on the ecology of corals and coral reefs: review and synthesis. Mar Pollut Bull 50(2):125–146

Fauvelot C, Bernardi G, Planes S (2003) Reductions in the mitochondrial DNA diversity of coral reef fish provide evidence of population bottlenecks resulting from Holocene sea-level change. Evolution 57(7):1571–1583

Feary DA, Bauman AG, Guest J, Hoey AS (2018) Trophic plasticity in an obligate corallivorous butterflyfish. Mar Ecol Prog Ser 605:165–171

Firth LB, Knights AM, Bridger D, Evans AJ, Mieszkowska N, Moore P, O'Connor NE, Sheehan E, Thompson RC, Hawkins SJ (2016) Ocean sprawl: challenges and opportunities for biodiversity management in a changing world. Oceanogr Mar Biol 54:189–262

Fong J, Lee SHR, Sun Y, Lim CL, Tan YAJ, Tan YH, Neo ML (2023) Litter traps: a comparison of four marine habitats as sinks for anthropogenic marine macro-litter in Singapore. Mar Pollut Bull 196:115645

Foo YH, Ratnam S, Lim EV, Abdullah M, Molenaar VJ, Tan ASH, Zhang S, Li H, Zanuri NBM (2022) Microplastic ingestion by commercial marine fish from the seawater of Northwest Peninsular Malaysia. PeerJ 10:e13181

Gajanur AR, Jaafar Z (2022) Abandoned, lost, or discarded fishing gear at urban coastlines. Mar Pollut Bull 175:113341

Gall SC, Thompson RC (2015) The impact of debris on marine life. Mar Pollut Bull 92(1–2):170–179

Gin KYH, Huda MK, Lim WK, Tkalich P (2001) An oil spill–food chain interaction model for coastal waters. Mar Pollut Bull 42(7):590–597

Goh NK, Chou LM (1994) The associates of Singapore gorgonians: Crustacea, Mollusca, Echinodermata and Chordata. In: Proceedings of the 3rd ASEAN–Australia symposium on living coastal resources, research papers, vol 2. Chulalongkorn University, Bangkok, Thailand, pp 215–218

Gordon SE, Goatley CHR, Bellwood DR (2016) Low-quality sediments deter grazing by the parrotfish *Scarus rivulatus* on inner-shelf reefs. Coral Reefs 35:285–291

Gray JE (1830) Illustrations of Indian zoology: chiefly selected from the collection of major-general Hardwicke, vol 1. Wurtz, Treuttel, Jun. and Richter, Treuttel

Guest JR, Baird AH, Maynard JA, Muttaqin E, Edwards AJ, Campbell SJ, Yewdall K, Affendi YA, Chou LM (2012) Contrasting patterns of coral bleaching susceptibility in 2010 suggest an adaptive response to thermal stress. PLoS One 7(3):e33353

Guest JR, Vergés A, Bauman AG, Campbell AH, Chou LM, Feary DA, Low JK, Marzinelli EM, Tun K, Steinberg PD (2016) Examining the relationship between fish herbivore biomass, coral and macroalgal cover on Singapore's heavily disturbed reefs. PeerJ Preprints 4:e1907v2

Hartanto RS, Loke LHL, Heery EC, Hsiung AR, Goh MW, Pek YS, Birch WR, Todd PA (2022) Material type weakly affects algal colonization but not macrofaunal community in an artificial intertidal habitat. Ecol Eng 176:106514

Hata H, Kato M (2004) Monoculture and mixed-species algal farms on a coral reef are maintained through intensive and extensive management by damselfishes. J Exp Mar Biol Ecol 313(2):285–296

Heery EC, Hoeksema BW, Browne NK, Reimer JD, Ang PO, Huang D, Friess DA, Chou LM, Loke LHL, Saksena-Taylor P, Alsagoff N, Yeemin T, Sutthacheep M, Vo ST, Bos AR, Gumanao GS, Hussein MAS, Waheed Z, Lane DJW, Johan O, Kunzmann A, Jompa J, Suharsonos TD, Baumana AG, Todd PA (2018) Urban coral reefs: degradation and resilience of hard coral assemblages in coastal cities of East and Southeast Asia. Mar Pollut Bull 135:654–681

Heery EC, Oh RKE, Taira D, Ng D, Kong CC, Hartanto RS, Hsiung AR, Chai TMF, Loke LHL, Yeo HHJ, Todd PA (2020) Human-engineered hydrodynamic regimes as a driver of cryptic microinvertebrate assemblages on urban artificial shorelines. Sci Total Environ 725:138348

Henderson CJ, Gilby BL, Schlacher TA, Connolly RM, Sheaves M, Flint N, Borland HP, Olds AD (2019) Contrasting effects of mangroves and armoured shorelines on fish assemblages in tropical estuarine seascapes. ICES J Mar Sci 76(4):1052–1061

Hilton MJ, Manning SS (1995) Conversion of coastal habitats in Singapore: indications of unsustainable development. Environ Conserv 22:307–322

Hixon MA, Brostoff WN (1983) Damselfish as keystone species in reverse: intermediate disturbance and diversity of reef algae. Science 220(4596):511–513

Honeycutt RN, Holbrook SJ, Brooks AJ, Schmitt RJ (2023) Farmerfish gardens help buffer stony corals against marine heat waves. PLoS One 18(3):e0282572

Horoszowski-Fridman YB, Izhaki I, Katz SM, Barkan R, Rinkevich B (2024) Shifting reef restoration focus from coral survivorship to biodiversity using Reef Carpets. Commun Biol 7(1):141

Hsiung AR, Tan WT, Loke LHL, Firth LB, Heery EC, Ducker J, Clark V, Pek YS, Birch WR, Ang AC, Hartanto RS, Chai TMF, Todd PA (2020) Little evidence that lowering the pH of concrete supports greater biodiversity on tropical and temperate seawalls. Mar Ecol Prog Ser 656:193–205

Ip YCA, Chang JJM, Lim KK, Jaafar Z, Wainwright BJ, Huang D (2021) Seeing through sedimented waters: environmental DNA reduces the phantom diversity of sharks and rays in turbid marine habitats. BMC Ecol Evol 21:166

Jaafar Z (2020) Recent sightings of whale shark in Singapore waters. Singapore Biodivers Rec 2020:129–131

Jaafar Z, Yeo DC, Tan HH, O'Riordan RM (2012) Status of estuarine and marine non-indigenous species in Singapore. Raffles Bullet Zool Suppl 25:79–92

Jaafar Z, Low J, Lim KKP (2024) Checklist of marine fish species with their category of threat status for Singapore. In: Davison GWH, Gan JWM, Huang D, Hwang WS, Lum SKY, Yeo DCJ (eds) The Singapore Red Data book 2024. Red lists of Singapore biodiversity, 3rd edn. National Parks Board, Singapore

Johansen JL, Jones GP (2013) Sediment-induced turbidity impairs foraging performance and prey choice of planktivorous coral reef fishes. Ecol Appl 23(6):1504–1517

Johnson GD, Paxton JR, Sutton TT, Satoh TP, Sado T, Nishida M, Miya M (2009) Deep-sea mystery solved: astonishing larval transformations and extreme sexual dimorphism unite three fish families. Biol Lett 5(2):235–239

Kikuzawa YP, Ng CSL, Sam SQ, Toh TC, Tan KS, Loo PL, Chou LM (2021) Transplanting coral fragments in close contact enhances their survival and growth on seawalls. J Mar Sci Eng 9(12):1377

Kiong J, Saparudin (2007, October 12) Major oil spills in the Straits of Singapore. Singapore Infopedia. https://www.nlb.gov.sg/main/article-detail?cmsuuid=355427db-e45f-4779-90b6-db563455659d. Accessed on 23 Nov 2024

Lai S, Loke LHL, Hilton MJ, Bouma TJ, Todd PA (2015) The effects of urbanisation on coastal habitats and the potential for ecological engineering: a Singapore case study. Ocean Coast Manag 103:78–85

Lai S, Loke LHL, Bouma TJ, Todd PA (2018) Biodiversity surveys and stable isotope analyses reveal key differences in intertidal assemblages between tropical seawalls and rocky shores. Mar Ecol Prog Ser 587:41–53

Law MT, Huang D (2023) Light limitation and coral mortality in urbanised reef communities due to sea-level rise. Clim Change Ecol 5:100073

Ledet J, Tan C, Guan XH, Yong CLX, Ying L, Todd P (2024) Trapping of microplastics and other anthropogenic particles in seagrass beds: ubiquity across a vertical and horizontal sampling gradient. Mar Environ Res 197:106487

Lima JS, Zalmon IR, Love M (2019) Overview and trends of ecological and socioeconomic research on artificial reefs. Mar Environ Res 145:81–96

Lin J (2023) Biodiversity record: first Singapore record of the yellowtail wrasse, *Anampses meleagrides*. Nature Singap 16:e2023002

Lise M (2020) Invasion of the Hybrid Grouper. Marine Stewards Singapore. https://www.marinestewards.org/post/hello-hybrid

Loh TL, Tanzil JTI, Chou LM (2006) Preliminary study of community development and scleractinian recruitment on fibreglass artificial reef units in the sedimented waters of Singapore. Aquat Conserv Mar Freshwat Ecosyst 16:61–76

Löhr A, Savelli H, Beunen R, Kalz M, Ragas A, van Belleghem F (2017) Solutions for global marine litter pollution. Curr Opin Environ Sustain 28:90–99

Loke LHL, Todd PA (2016) Structural complexity and component type increase intertidal biodiversity independently of area. Ecology 97(2):383–393

Loke LHL, Heery EC, Lai S, Bouma TJ, Todd PA (2019) Area-independent effects of water-retaining features on intertidal biodiversity on eco-engineered seawalls in the tropics. Front Mar Sci 6:16

Low JKY, Chou LM (1992) Distribution of coral reef fish in Singapore. Third ASEAN Science & Technology Week Conference Proceedings 6:139–144

Loong S, Taira D, Chan YKS, Chan WWR, Fong J, Kwan V, Low JKY, Ng CSL, Sam SQ, Tan CZ, Tun KPP, Yip ZT, Jaafar Z, Huang D (2025) Covariations of coral, macroalgal, macroinvertebrate and fish communities in an urbanised tropical reef environment. Coral Reefs. https://doi.org/10.1007/s00338-025-02723-9

Low et al (1997) Pomacentrid population dynamics on Singapore coral reefs. Environ Monit Assess 44:53–66

Lowry M, Folpp H, Gregson M, Suthers I (2012) Comparison of baited remote underwater video (BRUV) and underwater visual census (UVC) for assessment of artificial reefs in estuaries. J Exp Mar Biol Ecol 416:243–253

Madduppa HH, Subhan B, Suparyani E, Siregar AM, Arafat D, Tarigant SA, Alimuddin A, Khairudi D, Rahmawati F, Brahmandito A (2013) Dynamics of fish diversity across an environmental gradient in the Seribu Islands reefs off Jakarta. Biodivers J Biol Divers 14(1):17

Mark RY, Taira D, Todd PA (2024) Shoreline armoring potentially diminishes the ecological roles of parrotfishes in urban coral reefs. Ecology 105(3):e4250

Meyer JL, Schultz ET, Helfman GS (1983) Fish schools: an asset to corals. Science 220:1047–1049

Morais RA, Bellwood DR (2019) Pelagic subsidies underpin fish productivity on a degraded coral reef. Curr Biol 29(9):1521–1527

Moreno G, Dagorn L, Capello M, Lopez J, Filmalter J, Forget F, Sancristobal I, Holland K (2016) Fish aggregating devices (FADs) as scientific platforms. Fish Res 178:122–129

Munday PL (2000) Interactions between habitat use and patterns of abundance in coral-dwelling fishes of the genus *Gobiodon*. Environ Biol Fish 58:355–369

Munsch SH, Cordell JR, Toft JD (2017) Effects of shoreline armouring and overwater structures on coastal and estuarine fish: opportunities for habitat improvement. J Appl Ecol 54:1373–1384

Ng PKL, Wee YC (1994) The Singapore Red Data book. Singapore Nature Society, p 343

Ng CSL, Lim SC, Ong JY, Teo LM, Chou LM, Chua KE, Tan KS (2015) Enhancing the biodiversity of coastal defence structures: transplantation of nursery-reared reef biota onto intertidal seawalls. Ecol Eng 30(82):480–486

Ng CSL, Toh KB, Toh TC, Low IZ, Jaafar Z, Leong WKG, Ng NK, Cheo PR, Tun K, Chou LM (2017a) Influence of a tropical marina on nearshore fish communities during a harmful algal bloom event. Raffles Bullet Zool 65:525–538

Ng CSL, Toh TC, Chou LM (2017b) Artificial reefs as a reef restoration strategy in sediment-affected environments: insights from long-term monitoring. Aquat Conserv 27(5):976–985

Ng CSL, Huang D, Toh KB, Sam SQ, Kikuzawa YP, Toh TC, Taira D, Chan YK, Hung LZ, Sim WT, Rashid AR (2020) Responses of urban reef corals during the 2016 mass bleaching event. Mar Pollut Bull 154:111111

Ng D, Taira D, Heery EC, Todd PA (2021) Antagonistic effects of seawalls and urban sedimentation on epilithic algal matrix (EAM)-feeding fishes. Mar Pollut Bull 173:113098

Ng CZ, Jensen OP, Ong JJ (2025) Reconstructing a century of marine landings and fishing effort in Malaysia and Singapore. ICES J Mar Sci 82(2):fsaf008

Nguyen NT, Friess DA, Todd PA, Mazor T, Lovelock CE, Lowe R, Gilmour J, Chou LM, Bhatia N, Jaafar Z, Tun K, Yaakub SM, Huang D (2022) Maximising resilience to sea-level rise in urban coastal ecosystems through systematic conservation planning. Landscape Urban Plan 221:104374

O'Connor JJ, Fobert EK, Besson M, Jacob H, Lecchini D (2019) Live fast, die young: Behavioural and physiological impacts of light pollution on a marine fish during larval recruitment. Mar Pollut Bull 146:08–914

Ono K, Simenstad CA (2014) Reducing the effect of overwater structures on migrating juvenile salmon: an experiment with light. Ecol Eng 71:180–189

Putpipat P (2018) Density and distribution of fish Larvae in Gulf of Thailand. In: SEAFDEC technical seminar 2018. Training Department, Southeast Asian Fisheries Development Center, pp 22–26

Qing A (2023, October 25) Dead fish surface off Raffles Marina after drop in water quality due to Tuas chemical facility fire. The Straits Times https://www.straitstimes.com/singapore/environment/dead-fish-in-raffles-marina-waters-surface-after-drop-in-water-quality-due-to-tuas-chemical-facility-fire. Accessed on 23 Nov 2024

Rajendran N, Jaafar Z (2023) New record of the coralgoby, *Paragobiodon lacuniclous* in Singapore. Nat Singap 16:e2023092

Reeves P, Reeves N (2021) Port-city development and Singapore's inshore and culture fisheries. In: Boomgaard P, Henley DE (eds) Muddied waters: historical and contemporary perspectives on management of forests and fisheries in Island Southeast Asia, vol 200. Brill

Rinkevich B (2021) Ecological engineering approaches in coral reef restoration. ICES J Mar Sci 78(1):410–420

Roberts CM, Bohnsack JA, Gell F, Hawkins JP, Goodridge R (2001) Effects of marine reserves on adjacent fisheries. Science 294(5548):1920–1923

Sam SQ, Ng CSL, Kikuzawa YP, Toh TC, Chou LM (2024) Propagating the stony coral *Echinopora horrida*: evaluating the effects of rearing environment and transplant orientation. Restor Ecol 32(4):e14100

Seah JCL, Bauman AG, Todd PA (2021) Temporal variation in macroalgal removal: insights from an impacted equatorial coral reef system. Mar Biol 168:2

Seebens H, Gastner MT, Blasius B (2013) The risk of marine bioinvasion caused by global shipping. Ecol Lett 16(6):782–790

Seraphim MJ, Sloman KA, Alexander ME, Janetski N, Jompa J, Ambo-Rappe R, Snellgrove D, Mars F, Harborne AR (2020) Interactions between coral restoration and fish assemblages: implications for reef management. J Fish Biol 97(3):633–655

Shimizu H (1997) The Japanese fisheries based in Singapore, 1892–1945. J Southeast Asian Stud 28(2):324–344

Sloey TM, Hildebrandt S, Morris R, Bilskie MV, Bland A, Bushek D, DiPetto G, Elefant D, Encomio V, Familkhalili R, Hladik C, Kreeger D, Paxton AB, Palinkas C, Steele L, Scheld A, Taira D, Toft JD, Ubeda AJ, Whitcraft C, Bilkovic DM (2025) A blueprint to greener shorelines: Advancing the effectiveness, sustainability , and widespread adoption of coastal nature-based solutions through transdisciplinary research. Estuar Coast 48(5): 133

Soh M, Tay YC, Lee CS, Low A, Orban L, Jaafar Z, Seedorf H (2024) The intestinal digesta microbiota of tropical marine fish is largely uncultured and distinct from surrounding water microbiota. npj Biofilms Microbiol 10(1):11

Srinivasan M (2003) Depth distributions of coral reef fishes: the influence of microhabitat structure, settlement, and post-settlement processes. Oecologia 137:76–84

Strona G, Lafferty KD, Fattorini S, Beck PS, Guilhaumon F, Arrigoni R, Montano S, Seveso D, Galli P, Planes S, Parravicini V (2021) Global tropical reef fish richness could decline by around half if corals are lost. P R Soc B-Biol Sci 288(1953):20210274

Sully S, Hodgson G, van Woesik R (2022) Present and future bright and dark spots for coral reefs through climate change. Glob Change Biol 28(15):4509–4522

Taira D (2020) A goldtail damsel, *Chrysiptera parasema*. Singapore Biodivers Rec 2020:170

Taira D (2021) Biodiversity record: sighting of a white-tailed humbug, *Dascyllus aruanus*, off Pulau Hantu. Nat Singap 14:e2021003

Taira D, Toh TC, Sam SQ, Ng CSL, Chou LM (2017a) Coral nurseries as habitats for juvenile corallivorous butterflyfish. Mar Biodivers 47:787–788

Taira D, Toh TC, Afiq-Rosli L, Sim WT, Ng CSL, Chou LM (2017b) First record of the host-specific anemonefish *Amphiprion frenatus* inhabiting *Heteractis magnifica* during a mass bleaching event. Bull Mar Sci 94(1):123–124

Taira D, Poquita-Du RC, Toh TC, Toh KB, Ng CSL, Afiq-Rosli L, Chou LM, Song T (2018) Spatial variability of fish communities in a highly urbanised reef system. Urban Ecosyst 21:85–95

Taira D, Heery EC, Loke LH, Teo A, Bauman AG, Todd PA (2020) Ecological engineering across organismal scales: trophic-mediated positive effects of microhabitat enhancement on fishes. Mar Ecol Prog Ser 656:181–192

Taira D, Ng CSL, Toh TC, Afiq-Rosli L, Sim WT, Chou LM (2024a) Impacts of bleaching on host sea anemones and associated anemonefish in Southeast Asian coral reefs. Reg Environ Chang 24(2):1–12

Taira D, Mark RY, Hsiung AR, Jaafar Z, Todd PA (2024b) Fish responses to manipulated microhabitat complexity in urbanised shorelines. J Appl Ecol 61(5):1095–1108

Tan IK (1966) The Fishing Settlements of Pulau Semakau. Unpublished thesis, Department of Geography, National University of Singapore

Tan HH (2015) Sabah grouper in Singapore. Singapore Biodivers Rec 2015:197–198

Tan G, Jaafar Z (2015) New record of ornate slippery goby *Lubricogobius ornatus* in Singapore. Singapore Biodivers Rec 2015:53

Tan V, Jaafar Z (2022) Biodiversity Record: new Singapore record of the mandarin dragonet, *Synchiropus splendidus*. Nat Singap 15:e2022054

Tan YHJ, Tham JKQ, Paul A, Rana U, Ang HP, Nguyen NTH, Yee ATK, Leong BPI, Drummond S, Tun KPP (2023) Remote sensing mapping of the regeneration of coastal natural habitats in Singapore: implications for marine conservation in tropical cities. Singapore J Trop Geo 44:130–148

Tebbett SB, Goatley CHR, Bellwood DR (2017) The effects of algal turf sediments and organic loads on feeding by coral reef surgeonfishes. PLoS One 12:e0169479

Tebbett SB, Chase TJ, Bellwood DR (2020) Farming damselfishes shape algal turf sediment dynamics on coral reefs. Mar Environ Res 160:104988

Tebbett SB, Faul SI, Bellwood DR (2024) Quantum of fear: herbivore grazing rates not affected by reef shark presence. Mar Environ Res 196:106442

Tham AK (1976) Some aspects of the fisheries biology of Singapore Straits. Nanyang University, Singapore

The New Paper (2000, October 17) The shark man of Pulau Brani. The New Paper

The Straits Times (1964, June 7) Whale shark 'ravaging' fishing stakes shot by marine police. The Straits Times

Todd PA, Heery EC, Loke LHL, Thurstan RH, Kotze DJ, Swan C (2019) Towards an urban marine ecology: characterizing the drivers, patterns and processes of marine ecosystems in coastal cities. Oikos 128:1215–1242

Toh KB, Ng CSL, Leong WKG, Jaafar Z, Chou LM (2016) Assemblages and diversity of fishes in Singapore's marinas. Raffles Bullet Zool 32:85–94

Toh TC, Ng CS, Loke HX, Taira D, Toh KB, Afiq-Rosli L, Du RC, Cabaitan P, Sam SQ, Kikuzawa YP, Chou LM (2017) A cost-effective approach to enhance scleractinian diversity on artificial shorelines. Ecol Eng 99:349–357

Trottet A, George C, Drillet G, Lauro FM (2022) Aquaculture in coastal urbanized areas: a comparative review of the challenges posed by Harmful Algal Blooms. Crit Rev Environ Sci Tecnol 52(16):2888–2929

Vance A, McGregor I (2019) Desert Island dump: the shameful state of Henderson Island. https://interactives.stuff.co.nz/2019/07/henderson-island-rubbish-plastic-ocean-waste/chapter1/. Accessed on 17 Feb 2025

Waltham NJ, Bradley M, Wilson S, Kane K, Langham O, Sheaves M (2023) Remotely operated vehicle reveals fish orientate to the substrate underneath marina floating pontoons. Estuar Coast Shelf S 280:108184

Wells FE, Tan KS, Todd PA, Jaafar Z, Yeo DC (2019) A low number of introduced marine species in the tropics: a case study from Singapore. Manag Biol Invasion 10(1):23

Wenger AS, McCormick MI, McLeod IM, Jones GP (2013) Suspended sediment alters predator-prey interactions between two coral reef fishes. Coral Reefs 32(2):369–374

Yap WLN, Oh RM, Iesa I (2024) Checklist of Cnidaria (non-Scleractinia) species with their category of threat status for Singapore. In: Davison GWH, Gan JWM, Huang D, Hwang WS, Lum SKY, Yeo DCJ (eds) The Singapore Red Data book 2024. Red lists of Singapore biodiversity, 3rd edn. National Parks Board, Singapore

Yeo HH, Loke LH, Todd PA (2024a) Population size and movement ecology of intertidal gastropods on rocky shores and seawalls in Singapore. J Mollus Stud 90:eyae016

Yeo HH, Yeo JY, Todd PA (2024b) Relative predation intensity of an intertidal gastropod on artificial coastal defense structures. Ecol Evol 14(5):e11385

9 Long-Term Trends of Coral Biodiversity in Singapore

Denise A. McIntyre, Aubane L. M. Carré de Malberg, Myra Y. Y. Khoo, James Philip D. V. Cabatbat, Z. K. Medric Tan, J. J. Marc Chang, and Danwei Huang

Abstract

Coral reefs are amongst the world's most biodiverse and economically valuable ecosystems, yet they are increasingly threatened by anthropogenic pressures. In highly urbanised countries such as Singapore, coastal development, land reclamation and recurrent bleaching events have led to a decrease in coral cover and shifts in coral composition. This chapter combines historical museum collections, long-term line intercept transect (LIT) surveys and recent timed-swim surveys to assess temporal changes in scleractinian coral communities at two reef sites, Pulau Hantu and Pulau Satumu, from the 1950s to 2020s. A total of 9740 colonies were observed from across museum collections and field surveys. Including records from previous studies, we document 257 coral species recorded in Singapore (excluding Fungiidae) and show comparable species richness between Pulau Hantu and Pulau Satumu before large-scale reclamation activities began. While phylogenetic and species richness increased over time, functional richness declined until the 2010s, indicating more functionally constrained coral assemblages. Coral communities shifted away from competitive and weedy life-history strategies towards stress-tolerant and generalist species, reflecting increasing environmental disturbances. The 2020s saw an increase in functional richness, suggesting potential recovery from bleaching. These findings highlight the value of integrating museum collections with long-term monitoring in reconstructing ecological baselines and tracking biodiversity changes over multiple decades.

Keywords

Biodiversity baseline · Community change · Functional diversity · Phylogenetic diversity · Scleractinia · Southeast asia · Species richness · Zooxanthellate

9.1 Introduction

Coral reefs are highly biodiverse ecosystems that offer vital resources and perform essential ecosystem services, such as food and other resource provision, tourism and coastal protection, which are critical for supporting human well-being (Hoegh-Guldberg and Smith 1989; Moberg and Folke 1999). Given that an estimated 30% of total fisheries catch consists of reef-associated fish and invertebrates (Geronimo and Cabral 2013; Tacon and Metian 2017), coral reefs play a vital role in global food security. Approximately 30% of reefs also support the tourism industry, generating an estimated US$36 billion in annual revenue (Spalding et al. 2017). Additionally, coral reefs form natural coastal barriers, with shoreline protection valued at approximately US$9 billion per year (Van Zanten et al. 2014). These ecosystem services make them amongst the most economically valuable ecosystems on Earth, worth over US$350,000 per hectare per year (de Groot et al. 2012).

D. A. McIntyre · A. L. M. Carré de Malberg · J. J. M. Chang
Lee Kong Chian Natural History Museum, National University of Singapore, Singapore, Singapore

Department of Biological Sciences, National University of Singapore, Singapore, Singapore

M. Y. Y. Khoo · J. P. D. V. Cabatbat · Z. K. M. Tan
Lee Kong Chian Natural History Museum, National University of Singapore, Singapore, Singapore

D. Huang (✉)
Lee Kong Chian Natural History Museum, National University of Singapore, Singapore, Singapore

Department of Biological Sciences, National University of Singapore, Singapore, Singapore

Tropical Marine Science Institute, National University of Singapore, Singapore, Singapore

Centre for Nature-based Climate Solutions, National University of Singapore, Singapore, Singapore
e-mail: huangdanwei@nus.edu.sg

L. M. Chou, D. Huang (eds.), *Coral Reefs of Singapore's Urbanised Sea*, Coral Reefs of the World 23,
https://doi.org/10.1007/978-981-95-0903-4_9

Despite their ecological and economic values, coral reefs face threats from a wide range of anthropogenic activities such as overfishing, pollution, rising sea surface temperatures and urbanisation (Hughes et al. 2017; Wear 2016; Chou et al. 2019). In highly urbanised countries such as Singapore and Hong Kong, coastal development and the expansion of shipping lanes have led to removal of coral reef habitats and increased sedimentation, which could lead to the smothering of ecosystems (Morton 1994; Chou et al. 2019). Anthropogenic activities have also been the primary drivers of rising global temperatures since the 1800s (IPCC 2021). The ocean has absorbed 93% of the excess heat generated in recent decades, contributing to the increasing frequency, duration and intensity of marine heat waves since the 1970s (Stocker 2013; Hobday et al. 2016; Von Kietzell et al. 2022; Venegas et al. 2023). With increased frequency, the intervals between coral bleaching events are becoming too brief for reefs to fully recover, affecting thermally sensitive species and resulting in reduced coral cover and diversity (Hughes et al. 2018).

In Singapore, due to urbanisation and land reclamation, coastal areas have undergone extensive modifications with 65% of Singapore's shoreline being artificially armoured by seawalls (Hilton and Manning 1995; Lai et al. 2015; Tan et al. 2023). In total, Singapore has increased its land area by 25% through land reclamation (Glaser et al. 1991; Teo et al. 2025), which has both directly and indirectly impacted coral reef areas. Between 1993 and 2011, Singapore experienced a reduction in intertidal coral reef area from 17 km^2 to 9.5 km^2 (Lai et al. 2015). All these anthropogenic impacts have resulted in approximately 76.5% decrease in total coral reef area—from 39.85 km^2 in 1953 to 9.37 km^2 in 2018 (Tan et al. 2023). Extensive coastal development has resulted in habitat loss and increased sedimentation, reducing light availability and leading to reef compression (Chou et al. 2019; Chow et al. 2019; Morgan et al. 2020). Consequently, Singapore's coral reefs are restricted to depths much shallower than 10 m (Chow et al. 2019), with coral cover at depths of 6–7 m declining from 43% to 14% over a 27-year period (Guest et al. 2016b).

Singapore has also experienced mass coral bleaching events in 1998, 2010, 2013, 2016 and 2024 (Chan et al. 2024; Chin 2024). Despite temperatures in 1998 and 2010 being similar, the bleaching experienced in 2010 was less severe than that of 1998 (Guest et al. 2012). The 2010 bleaching affected 60% of corals and saw a 21.5% decrease in coral cover (Guest et al. 2016a). In 2013, a minor coral bleaching event led to bleaching prevalence ranging from 2.3% to 12.4% across different reef sites and an overall incidence of 6.1% (Chou et al. 2016). The 2016 bleaching event impacted between 25.5% and 76.3% of coral colonies across surveyed sites, resulting in a decline in coral cover from approximately 24% to 15% and shifted coral community composition (Jain et al. 2020; Ng et al. 2020).

Life-history strategies can be useful in predicting how coral reefs respond to environmental change. Based on data from 143 species of reef corals, four distinct life-history strategies were identified: competitive, weedy, stress-tolerant and generalist (Darling et al. 2012). Coral communities dominated by competitive species were found to be more susceptible to environmental stress, whereas stress-tolerant and weedy species exhibited greater resilience (Darling et al. 2013). In Singapore, the recurrent bleaching events and shifts in the coral community composition have likely contributed to a reduction in reef resilience (Chan et al. 2024).

Although previous research examined changes in Singapore's coral communities from the 1980s to 2012, focusing on the impacts of coastal development and coral bleaching events in 1998 and 2010 (Guest et al. 2016a, b), no study has explored coral community changes prior to the extensive land reclamation projects that began in the 1960s. A recent donation of thousands of coral specimens collected from Southeast Asia between the 1950s and 1970s by Shou Hwa Chuang (SH Chuang) to the Lee Kong Chian Natural History Museum (LKCNHM) has provided a rare opportunity to investigate Singapore's baseline coral biodiversity and community composition prior to extensive land reclamation. However, while museum collections are invaluable for offering historical perspectives that complement field surveys, such datasets are often limited by gaps in collection records, sampling biases and the inherent restriction to presence-only data (Ponder et al. 2001). For example, collections may overrepresent areas that are more accessible and hence not reflect true species distributions and abundance. Nevertheless, when analysed in combination with field survey data, they can provide valuable historical baselines and insights on biodiversity loss by filling both temporal and taxonomic gaps (Suarez and Tsutsui 2004). Given the limited efforts to resolve the inherent limitations of museum collections, this study utilises data from two sites in Singapore, Pulau (=Island in Bahasa Melayu) Hantu and Pulau Satumu (= Raffles Lighthouse) (Fig. 9.1), to evaluate how specimen collections from the 1980s compare with field surveys in establishing presence-absence data.

Both Pulau Hantu and Pulau Satumu experienced direct impacts from land reclamation activities during the 1970s; however, the scale and ecological impacts were much greater at Pulau Hantu. Approximately 400,000 m^3 of sand was utilised to expand the land area of Pulau Hantu by 9.8 hectares, primarily through the infilling of the reef flat between Pulau Hantu Besar and Pulau Hantu Kechil from March 1974 to April 1975 (Chou 2010). This reclamation buried and thus destroyed much of the two islets' reef habitats. In contrast, reclamation at Pulau Satumu was limited to a 1-hectare development to create a lagoon at the eastern side of the

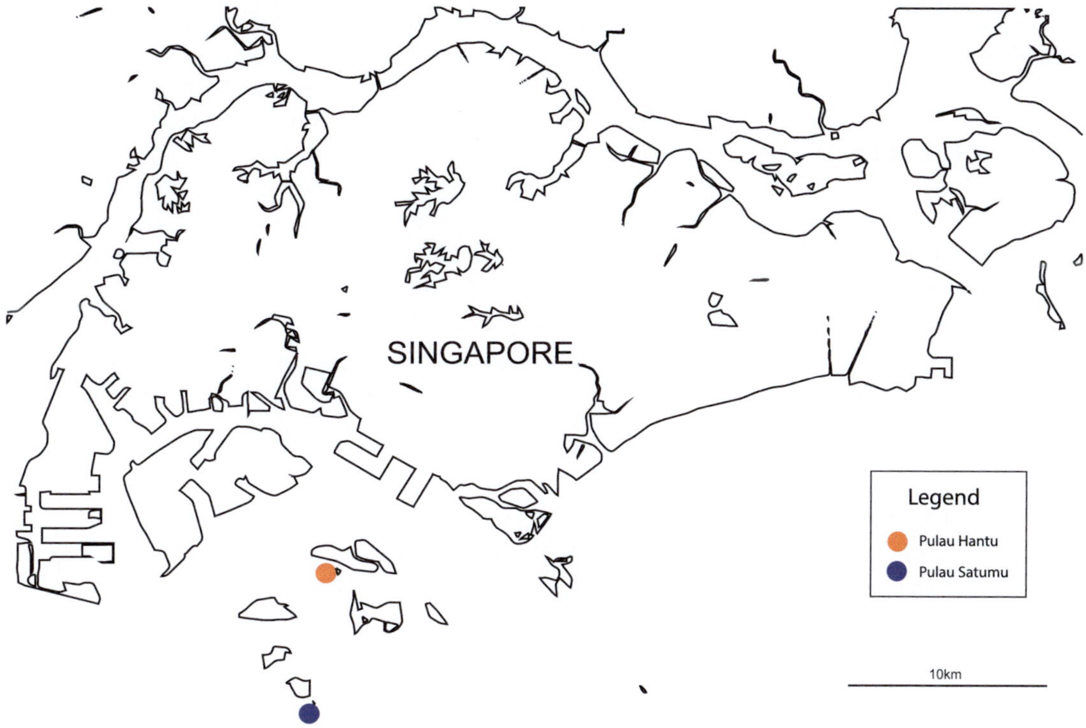

Fig. 9.1 Map illustrating locations of study sites in Singapore

island (Holmberg 1976). Construction of the Raffles Lighthouse on the islet was completed in 1855 (Tan and Low 2022), long before the coral collections and surveys examined here.

Since 2003, Pulau Satumu and the sea area of 300-metre radius around the Raffles Lighthouse, which is located on the island, have become restricted and all vessels are prohibited from entering, anchoring, mooring and transiting (Maritime and Port Authority of Singapore 2003). On the other hand, Pulau Hantu is a site where a wide range of recreational activities including SCUBA diving and fishing are allowed, and hence is exposed to heavier boat traffic (Maritime and Port Authority of Singapore 2017). Pulau Hantu and Pulau Satumu have been subject to consistent study, collections and monitoring since the 1980s (Guest et al. 2016b; Chan et al. 2024; Loong et al. 2025), with coral collections at both sites by SH Chuang from going back to the 1950s, so they are ideal locations for tracking long-term biodiversity change. Specifically, we aim to establish a baseline of scleractinian coral biodiversity at Pulau Hantu and Pulau Satumu prior to large-scale land reclamation in Singapore, based on specimens and species records obtained since the 1950s and local coral reef studies (e.g. Chuang 1961; Loo and Chou 1995), and to examine temporal changes in scleractinian coral community structure, species, phylogenetic and functional richness over the last seven decades.

9.2 Methods

9.2.1 Study Sites

Two coral reef sites at the southern islands of Singapore were chosen for this study—Pulau Hantu and Pulau Satumu (= Raffles Lighthouse) (Fig. 9.1). Both Pulau Hantu and Pulau Satumu have been extensively studied over the last four decades (Guest et al. 2016b; Chan et al. 2024). These sites have accumulated long-term monitoring data from LIT surveys dating back to 1986 and continuing through 2024 (Loong et al. 2025), making them ideal locations for tracking species richness and other diversity metrics over time.

Pulau Hantu, located approximately 5 km from mainland Singapore, is part of Singapore's southern offshore islands. It is relatively sheltered by surrounding islands and experiences weaker currents compared to Pulau Satumu (Chow et al. 2019). Additionally, Pulau Hantu is subject to higher boat traffic due to it being a recreational fishing and SCUBA diving destination (Maritime and Port Authority of Singapore 2017). In contrast, Pulau Satumu, situated approximately 11.5 km from mainland Singapore, faces the open sea and is exposed to stronger currents (Low and Chou 1994). The presence of a 300-metre vessel exclusion zone surrounding Raffles Lighthouse, which is located on Pulau Satumu, results in reduced boat traffic. Due to its greater distance from mainland Singapore, Pulau Satumu has not been affected by anthropogenic stressors such as increased sedimentation from coastal development as much as most other southern islands (Chou et al. 2019; Chow et al. 2019).

9.2.2 Museum Collections

Prior to SH Chuang's donation, the LKCNHM housed approximately 3000 catalogued Scleractinia specimens and even more that were uncatalogued. In 2024, the collections by SH Chuang added nearly 2000 specimens to LKCNHM's holdings. Of these, 734 originated from Singapore and only 128 specimens, excluding Fungiidae species, were collected from Pulau Hantu ($n = 74$) and Pulau Satumu ($n = 54$). These specimens were collected by SH Chuang between the 1950s and 1970s, and were initially identified to genus level before being catalogued into LKCNHM's Zoological Reference Collection (ZRC). The current study further identified all study specimens to the species level. This was achieved using morphological descriptions and images in Wallace (1999) and Veron (2000), as well as Veron and Pichon (1976), Veron et al. (1977), Veron and Pichon (1980), Veron and Pichon (1982) and Veron and Wallace (1984). The use of current taxonomic nomenclature and classification was verified with the World Register of Marine Species (WoRMS Editorial Board 2024).

These historical collections contributed to establishing a baseline understanding of Scleractinia diversity at Pulau Hantu and Pulau Satumu prior to significant coastal development in Singapore, providing a valuable reference for ecological and conservation efforts in the region.

During the 1980s and early 1990s, a total of 1255 Scleractinia specimens, not including specimens from the Fungiidae family, were collected from Pulau Hantu ($n = 783$) and Pulau Satumu ($n = 472$) by the Reef Ecology Study Team (REST) at the National University of Singapore (Chou et al. 1994; Koh and Chou 1989; Goh and Chou 1993; Chou 1988). In the present study, all 1255 specimens were re-identified and their taxonomy updated as above to ensure their taxonomic accuracy and relevance. This dataset offers critical insights into the Scleractinia diversity at Pulau Hantu and Pulau Satumu from the 1980s to early 1990s. By comparing specimen data with field survey data from the same period, this analysis sheds light on the historical coral diversity and composition of these sites, forming an essential reference point for assessing changes in coral reef ecosystems over time.

9.2.3 Field Surveys

Between 1986 and 2024, LIT surveys were carried out at 15 sites across the southern islands of Singapore (Guest et al. 2016b; Chan et al. 2024). These surveys provided critical data on benthic and coral community structures over time. Pulau Hantu and Pulau Satumu were surveyed almost consistently during this period, with the exception of a few years (Table 9.1). As surveys conducted between 1993 and 2012 did not identify coral colonies to species level, data from this time period were omitted from our analyses.

To address the potential omission of scleractinian species in the LIT surveys conducted at specific depths, additional timed-swim surveys were conducted to collect presence-

Table 9.1 Years in which line intercept transect surveys were conducted at each site

Year	Pulau Hantu	Pulau Satumu
1986	×	
1987	×	×
1988	×	×
1989	×	
1990	×	×
1991		×
1992	×	
1993[a]	×	×
1994[a]	×	×
1995[a]	×	×
1996[a]	×	×
1997[a]	×	×
1998[a]	×	×
1999[a]		×
2000[a]		×
2001[a]		×
2002[a]		×
2003[a]	×	×
2004[a]	×	
2005[a]	×	×
2006[a]	×	×
2007[a]	×	
2008[a]	×	×
2009[a]	×	×
2010[a]	×	×
2011[a]	×	
2012[a]	×	×
2013		
2014		
2015		
2016		×
2017	×	×
2018	×	×
2019	×	×
2020	×	×
2021	×	×
2022	×	×
2023	×	×
2024	×	×

× denotes survey conducted while blank denotes no survey conducted that year

[a]Denotes year in which surveys did not identify corals to species level

absence data at Pulau Hantu and Pulau Satumu. This survey followed the methodology described in Huang et al. (2009). Over a 90-min duration, SCUBA divers performed systematic swim searches in a meandering pattern, starting from the deepest point where zooxanthellate corals were observed (i.e. 9.5 m at Pulau Hantu and 10.5 m at Pulau Satumu; Chow et al. 2019) and moving towards the shore, ending at the shallowest sections of the reef flat. Specifically, divers generally swam parallel to the shore. After covering approximately 70–100 m, the divers ascended slightly and reversed direction, maintaining the parallel swim with respect to the shore.

All scleractinian corals encountered during the survey were identified to species level in the field. Corals that could not be identified immediately were photographed for subsequent identification in the laboratory as above.

9.2.4 Data Analyses

Rarefaction curves for museum and LIT survey records from the 1980s were plotted to assess the completeness of each dataset and estimate species richness using iNext Online (Chao et al. 2016). Species diversity was quantified using Simpson's Index (Simpson 1949) and Shannon's Index (Shannon 1948). Simpson's Index estimates the probability that two randomly selected individuals belong to the same species, emphasising species dominance. Shannon's Index incorporates both species richness and evenness, thus measuring community diversity. Additionally, Hill numbers were calculated for $q = 1$, the exponential of Shannon's Index, and $q = 2$, the inverse of Simpson's Index (Hill 1973). The parameter $q = 1$ is sensitive to species abundance, representing diversity with greater emphasis on evenness across species. In contrast, $q = 2$ gives more weight to dominant species, reflecting the influence of the most common taxa in a community. A higher Hill number at $q = 1$ indicates greater evenness and overall diversity, while a lower value suggests either lower richness or strong dominance by few species. For $q = 2$, a higher value implies the presence of several dominant species—indicating evenness amongst abundant species, whereas a lower value reflects the overwhelming dominance of one or very few species. Non-metric multidimensional scaling (NMDS) was employed to visualise dissimilarities between collection and survey datasets (Kruskal 1964) using the vegan package in R (Oksanen et al. 2013).

All species records from Pulau Hantu and Pulau Satumu, including museum collections, LIT survey records and timed-swim survey data, were consolidated into a comprehensive species inventory for these two sites. Additional species records were incorporated from checklists by Loo and Chou (1995), Huang et al. (2009) and Ng et al. (2024). Data were consolidated to characterise long-term changes in coral community structure, species richness, phylogenetic richness and functional richness from the 1950s to the present.

Phylogenetic richness for each year was calculated as the sum of the minimum branch lengths on a phylogenetic tree that includes all species recorded per year. This approach followed the method used by Chan et al. (2024), which was based on Faith's (1992) phylogenetic diversity (PD) metric. The calculations drew on 1000 time-calibrated Bayesian posterior trees of reef coral species reconstructed by Huang and Roy (2015).

Functional richness was calculated as the volume of a convex hull encompassing coral species within a multidimensional trait space, following the methods in Wong et al. (2018) and Chan et al. (2024). This trait space was built using 12 coral traits—including life history strategies, growth forms, and growth rates—sourced from the Coral Trait Database (https://www.coraltraits.org) and additional sources collated by Chan et al. (2024). The trait space was generated using principal coordinate analysis (PCoA) based on Gower's dissimilarity matrix, with negative eigenvalues corrected using the Cailliez method (Gower and Legendre 1986).

9.3 Results

9.3.1 Species Richness

A total of 1383 coral specimens representing 17 scleractinian families collected from Pulau Hantu and Pulau Satumu and deposited at the Lee Kong Chian Natural History Museum (LKCNHM; Zoological Reference Collection, ZRC) were examined (Fig. 9.2). These specimens were identified as 169 unique species (Fig. 9.3), represented by 128 specimens dating back to 1954–1975 from the collections of SH Chuang and 1255 specimens from the 1980s to early 1990s in the existing ZRC holdings (Fig. 9.2).

LIT surveys conducted between 1986 and 2024 recorded 8046 coral colonies, representing 186 unique species. Additionally, timed-swim surveys conducted in 2024 documented 311 colonies, representing 89 unique species. To supplement the dataset, 42 additional species were incorporated from the checklists of Chuang (1961), Loo and Chou (1995), Huang et al. (2009) and Ng et al. (2024), bringing the total number of unique species assessed in this study to 257.

9.3.2 Records from 1950s to 1970s

Specimens from this period comprised primarily SH Chuang's collections totalling 128 specimens from Pulau Hantu and Pulau Satumu. The rarefaction curve shows that it is far from reaching a plateau (Fig. 9.4), indicating that sam-

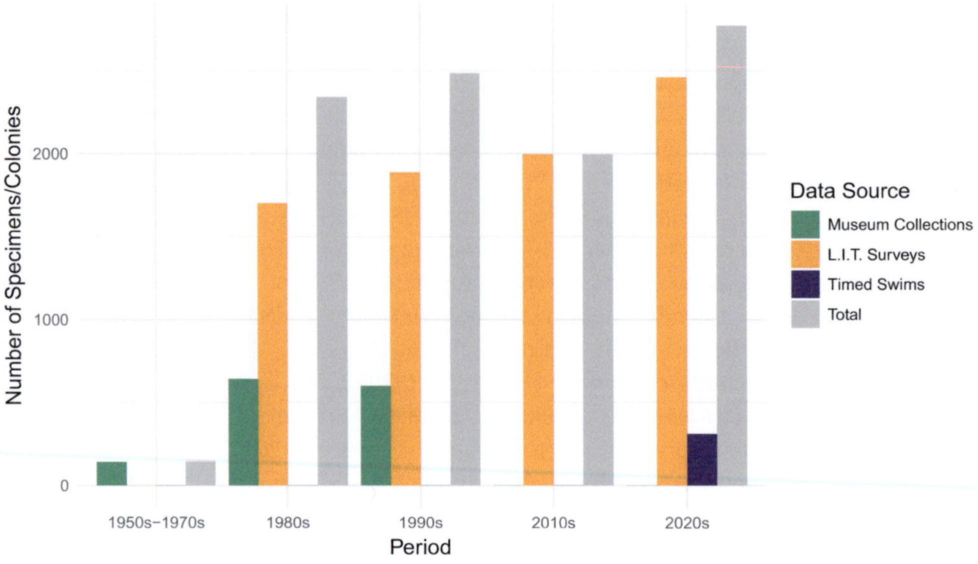

Fig. 9.2 Number of specimens or colonies recorded during each time period according to data source

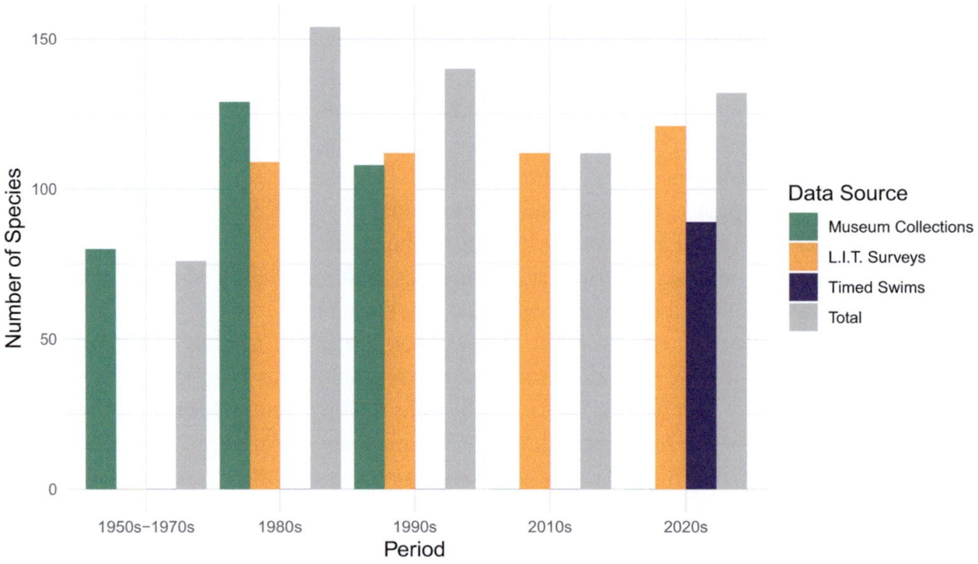

Fig. 9.3 Number of species recorded during each time period according to data source

pling was inadequate. As a result, the 1950s–1970s collections alone cannot be used to establish a baseline for Singapore's coral biodiversity at that time.

A total of 74 specimens were collected from Pulau Hantu and 54 specimens from Pulau Satumu. While neither rarefaction curve has reached a plateau, the curve for Pulau Satumu exhibits a shallower slope compared to that of Pulau Hantu (Fig. 9.5). The extrapolation curves suggest that species diversity at Pulau Hantu may be higher than at Pulau Satumu during this period.

9.3.3 Records from 1980s to 1990s

Rarefaction curves from 1980s and 1990s were close to plateauing, indicating that species richness estimates for those time periods were representative. The 1980s recorded a total of 2343 specimens and colonies, while the 1990s recorded 2486. Despite there being more records from the 1990s and 2020s (Fig. 9.2), the rarefaction curves suggest that species richness was higher in the 1980s (Fig. 9.4). Earlier decades did not reach saturation in the respective rarefaction curves, likely due to limited sampling effort, meaning the estimated species richness may not be precise for those periods.

A total of 642 specimens from the 1980s museum collections (ZRC) were recorded, while the LIT surveys recorded 1702 colonies. Despite this, the ZRC hosted 129 unique species, compared to 109 from the LIT surveys (Fig. 9.3). Similarly, a total of 599 specimens from the 1990s museum collections (ZRC) were recorded, while the LIT surveys recorded 1887 colonies, more than three times that of the museum collections. Despite this, the ZRC hosted 108

Fig. 9.4 Rarefaction curves of species recorded according to time period

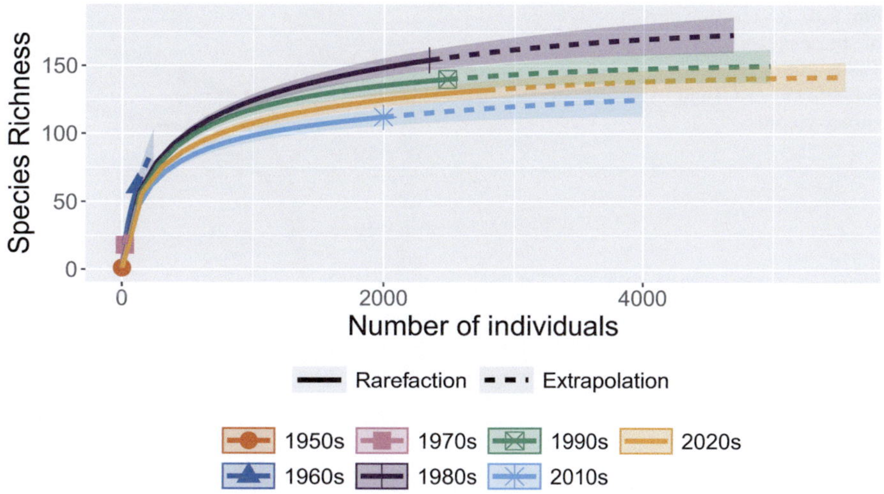

Fig. 9.5 Rarefaction curves of species recorded in the 1950–1970s museum collections (ZRC) according to site (HA = Pulau Hantu, RL = Raffles Lighthouse, Pulau Satumu)

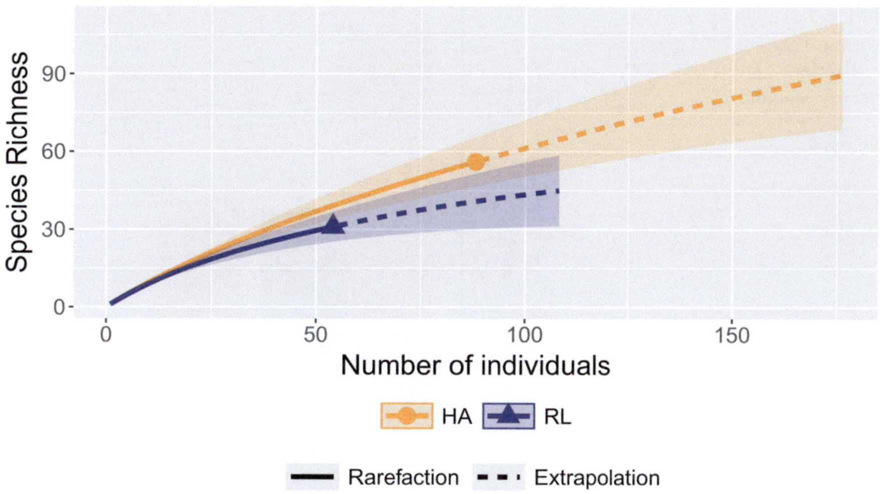

unique species, just 3 less than the 112 from LIT surveys (Fig. 9.3). This suggests that museum collections have a higher potential to capture species richness more comprehensively. Rarefaction curves show that the LIT surveys reached saturation at lower richness compared to museum collections, and hence, fewer new species were detected as sampling proceeded (Fig. 9.6). In contrast, the museum collections' extrapolated trend suggests high potential that additional sampling could reveal even more species.

The computed diversity metrics reinforce this pattern (Fig. 9.7). Museum collections from 1980s to 1990s exhibited higher Shannon's Index and Hill numbers, supporting the hypothesis that museum records contained greater species diversity than LIT records. However, Simpson's Index was similar for both data sources, indicating that dominant species were evenly distributed across datasets.

The non-metric multidimensional scaling further highlights that museum collections capture a broader range of species, while survey data consistently documented similar species, reflecting lower variability in species composition (Fig. 9.8).

In the 1980s, the rarefaction curve for Pulau Hantu plateaued with more specimens or colonies than that of Pulau Satumu. Although their extrapolated sections overlapped, the extrapolated curve for Pulau Hantu extended further (Fig. 9.9). This suggests that species richness was comparable between the two sites, with Pulau Hantu exhibiting slightly higher species richness.

In the 1990s, there was a shift in species accumulation patterns (Fig. 9.10). The rarefaction curve for Pulau Satumu plateaued with further sampling, and while the extrapolated sections overlapped, this occurred at a later stage in the

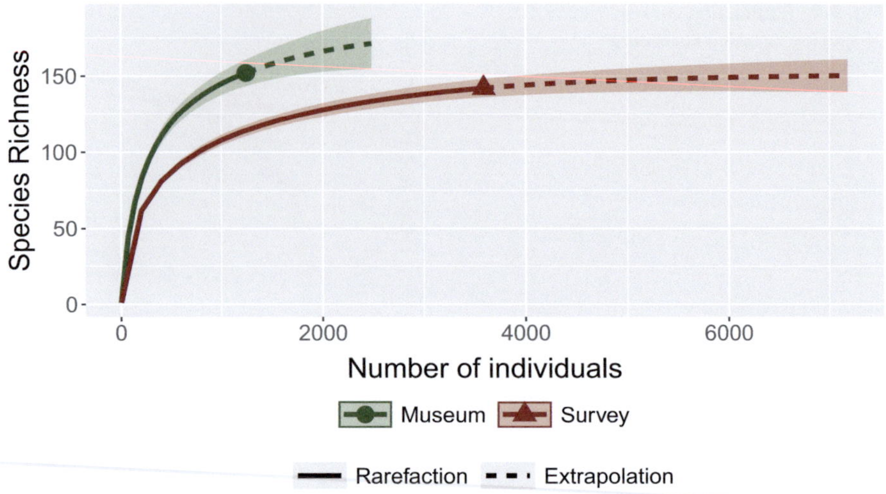

Fig. 9.6 Rarefaction curves of species recorded in the 1980s–1990s museum collections (ZRC) and LIT survey records

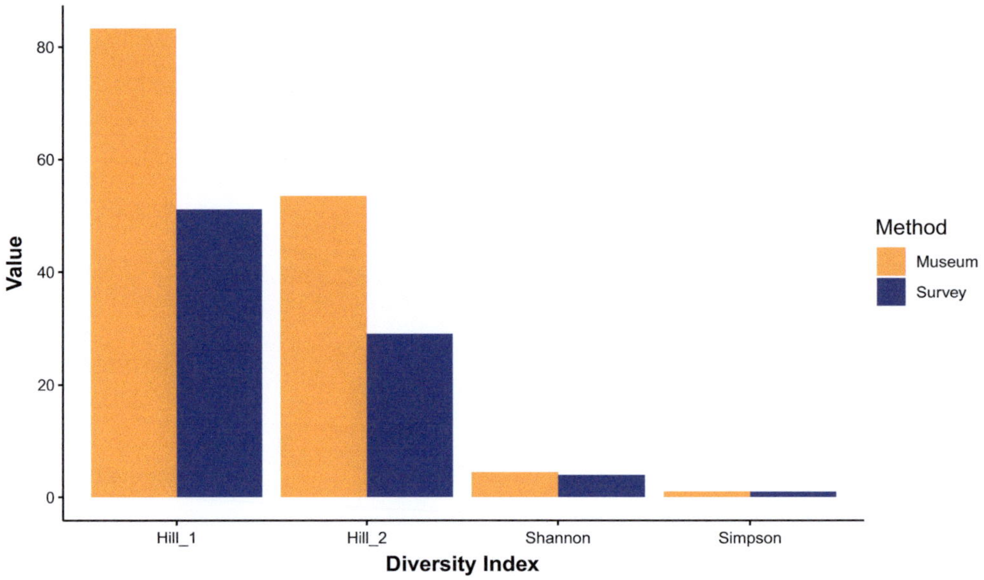

Fig. 9.7 Diversity measures for the 1980s–1990s museum collections (ZRC) and LIT survey records

curve, indicating that Pulau Satumu had a higher estimated species richness during this period.

9.3.4 Records from 2010s to 2020s

The LIT surveys conducted in the 2010s recorded a total of 1996 coral colonies, while the combined LIT surveys and timed swims in the 2020s documented 2952 colonies (Fig. 9.2). For both time periods, the rarefaction curves plateaued with further sampling (Fig. 9.4), indicating that the data produced reasonable species richness estimates. However, the rarefaction curves for both the 2010s and 2020s peaked earlier than the 1980s curves, and neither curve—even when extrapolated—overlapped with that of the 1980s, suggesting a decline in species richness over time.

The rarefaction curve for timed-swim surveys in the 2020s did not plateau, and its extrapolation exceeded that of LIT surveys, indicating the potential for additional species records through timed-swim surveys (Fig. 9.11). The LIT surveys only recorded 124 species despite the 2461 colonies observed while the timed-swim surveys added another 10 species (Fig. 9.2). The total of 134 species recorded in the 2020s is fewer than the 156 species recorded in the 1980s (Fig. 9.3). This further suggests that species richness was higher in the 1980s than it is in the present time period.

The rarefaction curves for the 2010s to 2020s show that the curve for Pulau Satumu plateaued later and extended further in its extrapolation compared to Pulau Hantu (Fig. 9.12). This suggests that species richness at Pulau Satumu remained higher than that of Pulau Hantu during this period.

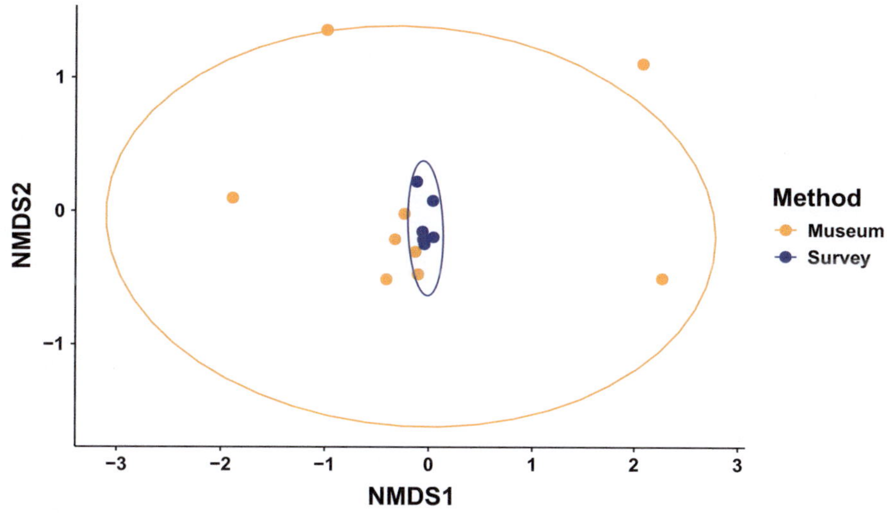

Fig. 9.8 Non-metric multidimensional scaling showing the dissimilarities between the 1980s and 1990s museum collections (ZRC) and LIT survey records

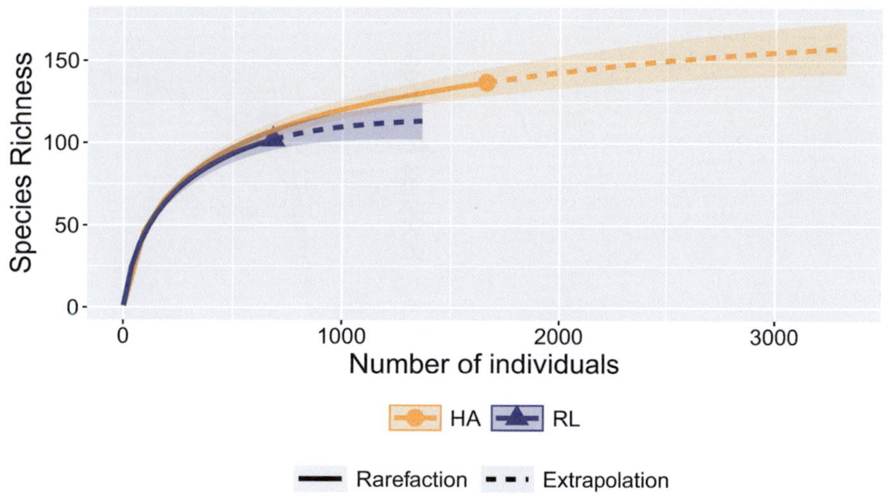

Fig. 9.9 Rarefaction curves of species recorded in the 1980s integrating records from museum collections (ZRC) and LIT survey records according to site (HA = Pulau Hantu, RL = Raffles Lighthouse, Pulau Satumu)

9.3.5 Community Structure and Diversity

From the 1950s to the 1970s, species richness was low, with Pulau Hantu exhibiting higher species richness than Pulau Satumu (Fig. 9.13). This trend remained relatively stable until the 1990s, when species richness at Pulau Satumu surpassed that of Pulau Hantu. Species richness was the highest at both sites during the 1990s. In the 2010s, species richness at Pulau Hantu declined, falling well below that of Pulau Satumu, which also experienced a slight decrease in species richness. By the 2020s, species richness at Pulau Hantu showed a modest increase but remained lower than Pulau Satumu, which experienced a slight decline. Overall, species richness aggregated across sites declined since the 1990s.

Phylogenetic richness was initially low from the 1950s to the 1970s, with Pulau Hantu exhibiting higher phylogenetic richness than Pulau Satumu (Fig. 9.14). This trend continued into the 1980s. In the 1990s, phylogenetic richness at Pulau Hantu declined slightly, whereas Pulau Satumu continued to show an increase. From the 1990s to 2020s, Pulau Satumu maintained a higher phylogenetic richness than Pulau Hantu. Both sites experienced a rise in phylogenetic richness in the 2010s. By the 2020s, phylogenetic richness at Pulau Hantu increased, though it remained lower than Pulau Satumu, which experienced a slight decline. Overall, both sites exhibited a general increase in phylogenetic richness from the 1950s to the present time period.

Functional richness was initially high at Pulau Hantu from the 1950s to 1970s but steadily declined until the 2010s, before showing an increase in the 2020s (Fig. 9.15). Throughout this period, functional richness remained consistently higher at Pulau Hantu than at Pulau Satumu. In contrast, Pulau Satumu exhibited low functional richness in the 1950s, followed by a gradual increase until the 1990s. However, this was followed by a decline in the 2010s, and while there was a slight increase in the 2020s, functional richness at Pulau Satumu remained below the levels observed in the 1990s.

On coral colony growth forms and life history strategies, from the 1950s to the 2020s, the relative proportions of sub-

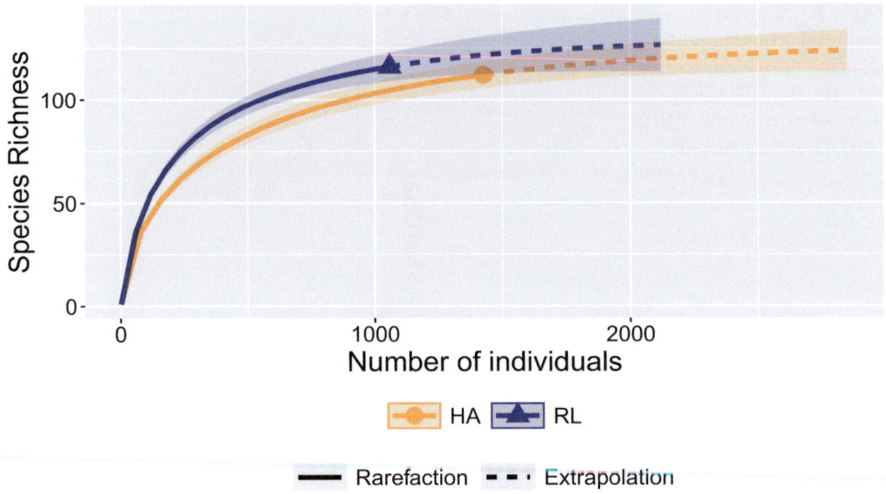

Fig. 9.10 Rarefaction curves of species recorded in the 1990s integrating records from museum collections (ZRC) and LIT survey records according to site (HA = Pulau Hantu, RL = Raffles Lighthouse, Pulau Satumu)

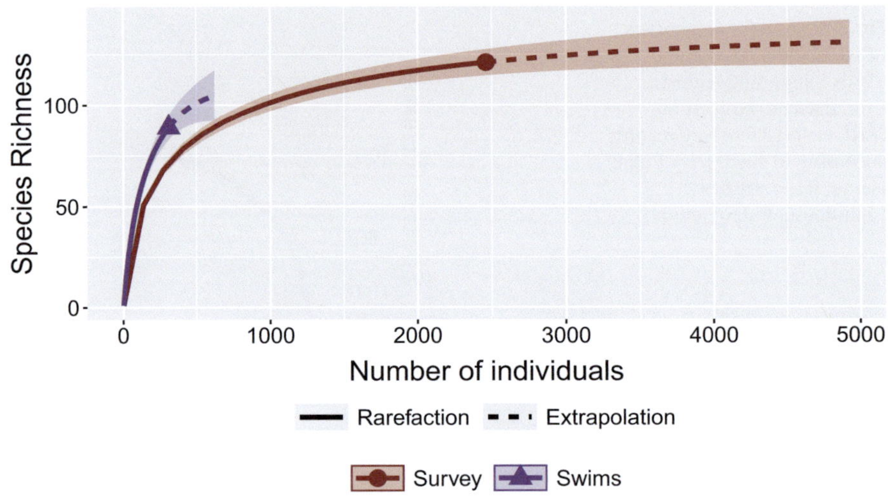

Fig. 9.11 Rarefaction curves of species recorded in the 2020s by survey method

massive and laminar growth forms have remained stable at approximately 10% and 20%, respectively (Fig. 9.16). In contrast, the branching growth form experienced the sharpest decline, decreasing by approximately 67% over six decades from around 30% in the 1950s–1970s to approximately 10% in the 2020s. The proportion of corals with massive growth form exhibited fluctuations over time but consistently ranged between 30% and 50%. Notably, the encrusting growth form was the only category to increase over the six-decade period, rising from just under 20% in the 1950s–1970s to slightly over 20% in the 2020s.

Results indicate a substantial decline in corals with competitive and weedy life histories, decreasing from 50% to nearly 0% and from 32% to 0%, respectively (Fig. 9.17). In contrast, both stress-tolerant and generalist corals exhibited an increasing trend, each rising from 25% to 50%. As a result, these two life history strategies now dominate coral communities at Pulau Hantu and Pulau Satumu. This represents a shift from the more even distribution of all four life history strategies observed in the 1950s–1970s.

9.4 Discussion

9.4.1 Coral Species Records in Singapore

This study reports that, prior to Singapore's development, and excluding species from the Fungiidae family, Singapore had a recorded total of 257 species of scleractinian corals. The museum collections dating to 1980s–1990s at the LKCNHM ZRC exhibited greater species richness compared to LIT survey data from the same period, with the species rarefaction curve plateauing with more specimens or collections (Fig. 9.6). The more variable and distinct species assemblages held in museum collections (Fig. 9.8) further support the notion that museum records can serve as a reliable source for estimating species diversity and richness. This finding aligns with previous studies that demonstrated the effectiveness of museum collections in estimating species diversity (Shaffer et al. 1998; Suarez and Tsutsui 2004; Ward 2012).

Fig. 9.12 Rarefaction curves of species recorded in the 2010s–2020s integrating records from museum collections (ZRC) and LIT survey records according to site (HA = Pulau Hantu; RL = Raffles Lighthouse, Pulau Satumu)

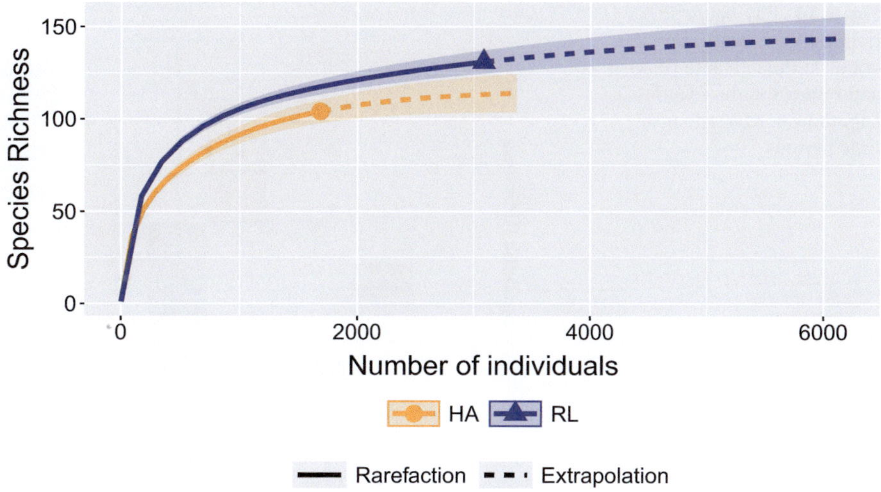

Fig. 9.13 Species richness of scleractinian corals at Pulau Hantu (HA) and Pulau Satumu (Raffles Lighthouse, RL) amongst time periods

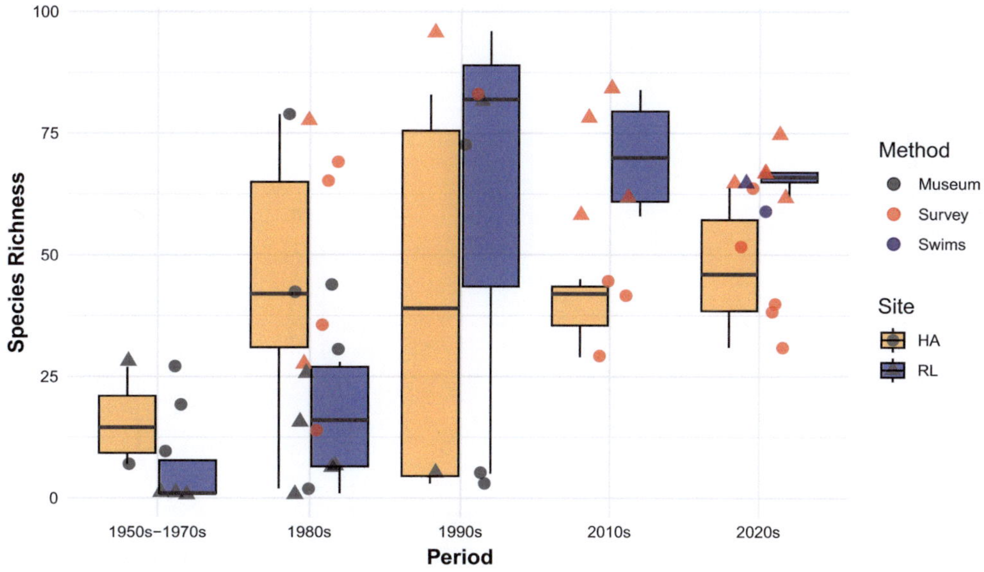

Comparatively, the 1950s–1970s collection was much smaller, and the rarefaction curve for this time period did not plateau, indicating that this collection may not render an accurate representation of the species richness and diversity during that time. Despite this limitation, the collections from that period provide valuable information on the species present at Pulau Hantu and Pulau Satumu in the past. When combined with the more extensive 1980s museum collections, they contribute to a more comprehensive picture of historical coral diversity in Singapore.

The 3rd edition of the Singapore Red Data Book (Davison et al. 2024) serves as a comprehensive checklist of Singapore's biodiversity, aiming to include as many species as possible, whether threatened or not. Excluding Fungiidae and azooxanthellate corals, the Red Data Book listed 240 species of scleractinian corals, with only 1 species listed as locally extinct, and 121 species listed as Data Deficient (Ng et al. 2024). This chapter has integrated species records since the 1950s and identified 17 additional species that were not listed in the publication.

While Pulau Satumu is currently recognised as having the highest coral cover and diversity in Singapore (Wong et al. 2018; Ng et al. 2020; Loong et al. 2025), our findings suggest that Pulau Hantu previously exhibited comparable species richness. This raises the possibility that other reef sites, such as Pulau Jong, Kusu Island and Pulau Semakau, may have also supported similar levels of biodiversity in the past. Indeed, in the 2000s, both Pulau Jong and Kusu Island recorded higher species richness than Pulau Hantu, while Pulau Semakau recorded an equivalent number of species (Huang et al. 2009).

Pulau Hantu and Pulau Satumu are amongst the most extensively studied reef sites in Singapore. However, additional data from other locations may account for certain species not captured in this study but present at others (e.g. *Acropora florida*, *Acropora glauca*, *Gardineroseris planu-*

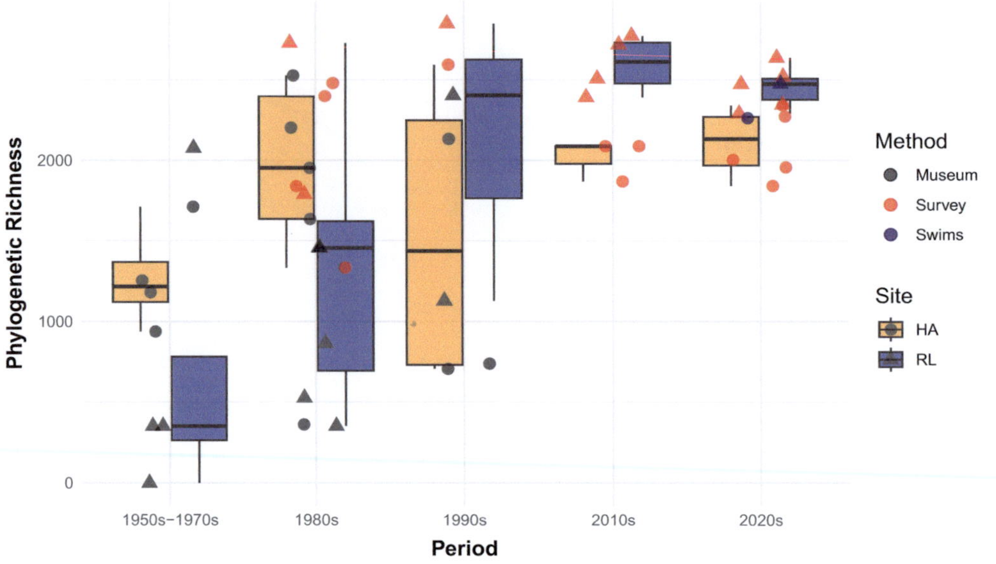

Fig. 9.14 Phylogenetic richness of scleractinian corals at Pulau Hantu (HA) and Pulau Satumu (Raffles Lighthouse, RL) amongst time periods

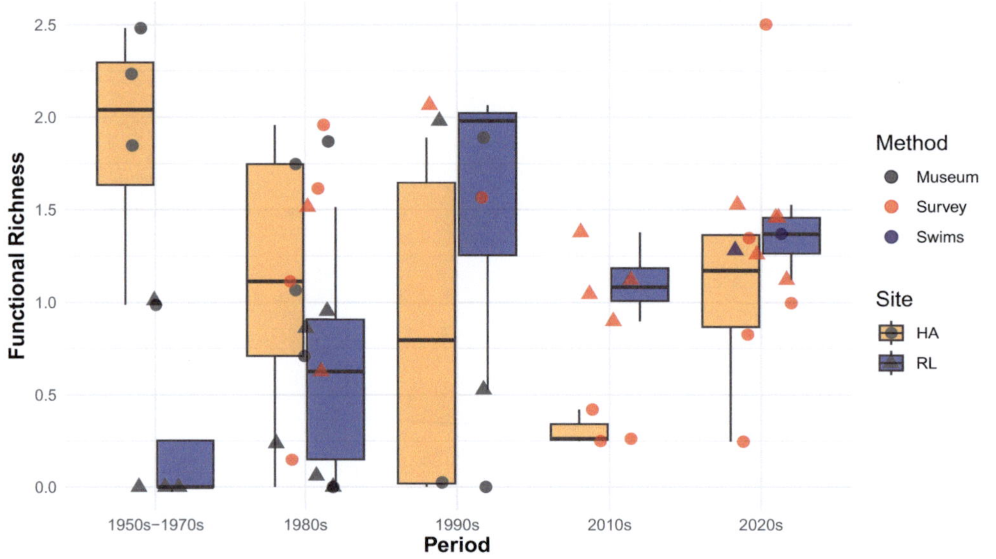

Fig. 9.15 Functional richness of scleractinian corals at Pulau Hantu (HA) and Pulau Satumu (Raffles Lighthouse, RL) amongst time periods

lata and *Lobophyllia valenciennesii*). Additionally, the LKCNHM ZRC and SH Chuang collections were identified based on skeletal morphology. However, species-level identification of scleractinian corals from skeletons has always been challenging (Kitahara et al. 2016; Ramírez-Portilla et al. 2022), and is also subjected to human error. To address this, this study captured detailed images of all 128 specimens from the SH Chuang collections, along with over 80% of the 1255 specimens from the 1980s–1990s ZRC collection. These high-resolution images would facilitate future verification of species identification, particularly in light of potential taxonomic revisions (Quek et al. 2023).

The absence of physical specimens or photographs for the more recent LIT survey records made it impossible to verify the accuracy of species identifications. It is thus challenging to confirm whether species recorded in the surveys, but absent from the 3rd edition of the Singapore Red Data Book or other checklists, were accurately identified. To mitigate this, photographs of coral colonies recorded during LIT surveys are now captured to facilitate accurate species identification through comparisons with morphological descriptions (including original species diagnoses). More generally, images and physical specimens collected during surveys contribute to building a comprehensive reference database that facilitates future coral species identifications, especially when species identifications are challenging. Furthermore, increasing the number of surveyors or conducting additional timed-swim surveys could potentially improve species detection rates, as supported by 10 unique species recorded during the timed-swims but were not captured by the LIT surveys conducted in the 2020s (Fig. 9.3). Rarefaction curves for both methods in the 2020s indicate that the timed-swim sur-

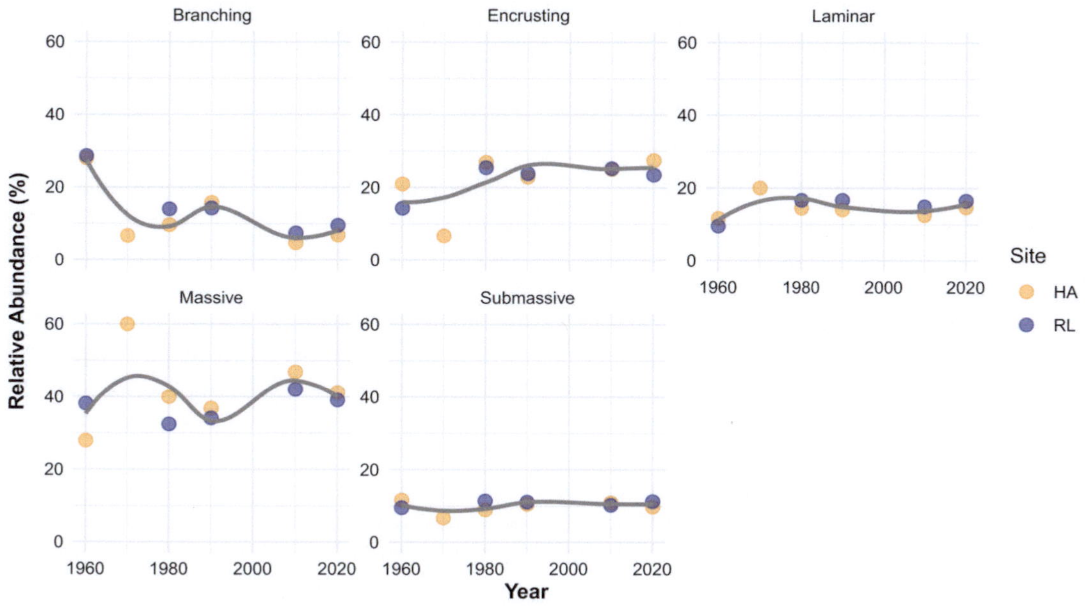

Fig. 9.16 Temporal trends in the growth forms of coral assemblages at Pulau Hantu (HA) and Pulau Satumu (Raffles Lighthouse, RL)

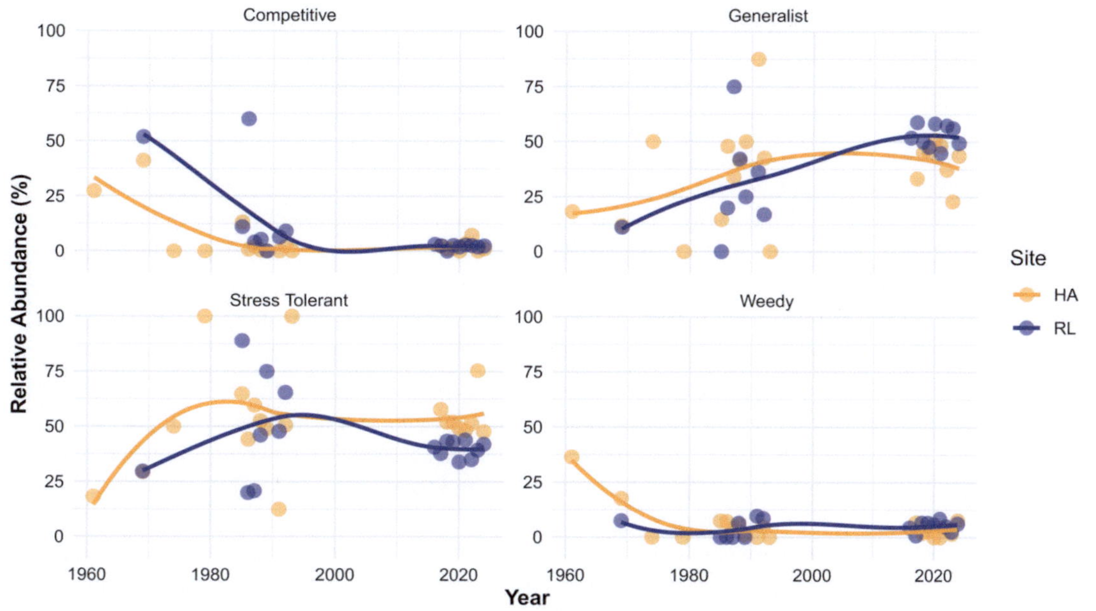

Fig. 9.17 Temporal trends in the life histories of coral assemblages at Pulau Hantu (HA) and Pulau Satumu (Raffles Lighthouse, RL)

vey rarefaction curve did not plateau, suggesting that additional surveys could yield more unique species (Fig. 9.11).

As the coral taxonomic classification evolves and species delimitation increases in precision due to genetic and genomic advances in coral phylogenetic studies (Quek et al. 2023), the use of morphological descriptions and images for species identification can prove insufficient, especially for species complexes (Kitahara et al. 2016; Bongaerts et al. 2021). For example, the conventional reliance on morphological methods led to misidentifications of *Acropora tenuis* in various regions, leading to overrepresentation of its species distribution (Bridge et al. 2023). Subsequent molecular genetics and population studies revealed multiple genetically distinct lineages within what was considered *Acropora tenuis*. Therefore, integrating the collection of coral tissue subsamples in field surveys for subsequent genetic analyses and DNA sequencing can enhance taxonomic accuracy by reducing reliance on subjective morphological identifications and minimising human error (Hebert et al. 2003). Furthermore, this approach has the potential to reveal additional cases of cryptic diversity (Ramírez-Portilla et al. 2022; Bridge et al. 2023; Burgess et al. 2024), Rassmussen et al. 2025).

9.4.2 Variations in Community Structure and Diversity

From the 1950s to the early 1990s, Pulau Hantu exhibited higher species and richness than Pulau Satumu (Fig. 9.13). As the records from this time period are a mixture of museum specimens and LIT survey records, this could possibly be due to lower sampling effort at Pulau Satumu due to nearly twice as many specimens recorded at Pulau Hantu compared to Pulau Satumu (Fig. 9.2). However, the species rarefaction curves (Fig. 9.9) for both sites exhibit similar peak values, and their extrapolated trends overlap, suggesting comparable historical species richness at both sites.

In the 1990s, species richness at Pulau Satumu appeared to rise, while it remained relatively stable at Pulau Hantu. During this period, data primarily originated from LIT surveys, which employed standardised methodology and consistent survey effort across sites, ensuring that the observed differences in species richness reflect ecological patterns rather than variations in sampling effort. This trend continued into the 2010s–2020s, with more species records made at Pulau Satumu than at Pulau Hantu. The observed increase in species richness at Pulau Satumu in the 1990s, followed by a slight decline in the 2010s, aligns with findings reported by Chan et al. (2024). In contrast, species richness at Pulau Hantu has remained relatively stable since the 1980s.

The rarefaction curves from the 1950s to 1990s suggest that Pulau Hantu and Pulau Satumu likely had similar species diversity during those time periods (Figs. 9.5, 9.9, and 9.10). However, between the 2010s and 2020s, the rarefaction curves suggest that Pulau Satumu has higher species diversity than Pulau Hantu (Fig. 9.12). The higher species richness at Pulau Satumu in more recent times may be attributed to Pulau Hantu's closer proximity to mainland Singapore, rendering it more susceptible to coastal urbanisation impacts, such as increased sedimentation. Pulau Hantu is also sheltered from strong currents, leading to elevated sedimentation rates (Sin et al. 2016; Chow et al. 2019). Its proximity to Pulau Semakau—where extensive land reclamation occurred in 1995, followed by the installation of renewable energy systems in 2006 (Tan et al. 2009; Chua 2016)—likely contributed further to increased sedimentation in the surrounding waters. In contrast, Pulau Satumu, being further offshore, is subjected to stronger currents, which facilitates the dispersal of suspended sediment (Chow et al. 2019). While sea surface light levels at Pulau Hantu and Pulau Satumu are comparable, light attenuation occurred at a faster rate at Pulau Hantu, likely due to the higher sedimentation rates (Chow et al. 2019). Increased sedimentation affects species richness by smothering coral colonies, reducing available suitable substrate for coral recruitment and limiting light availability (Erftemeijer et al. 2012; Browne et al. 2014; Poquita-Du et al. 2019a).

This study also observed higher functional richness at Pulau Satumu than at Pulau Hantu from the 1990s to 2020s (Fig. 9.15). This is supported by previous research that also registered higher functional richness at Pulau Satumu than Pulau Hantu (Chan et al. 2024). Moreover, it has been suggested that corals near mainland Singapore and coastal development tend to be tolerant against highly sedimented conditions, resulting in species with similar stress-tolerant traits, and thus lower functional richness (Wong et al. 2018). This may explain why Pulau Hantu, being closer to the mainland, now exhibits lower functional richness than Pulau Satumu. This reduction in functional richness can also be accounted for by the life history composition of coral assemblages shifting from competitive and weedy species to a dominance of stress-tolerant and generalist species (Fig. 9.12). Less disturbed reefs are typically dominated by competitive species, while a transition towards stress-tolerant and generalist species is indicative of increasing environmental disturbances, such as those associated with coastal urbanisation (Darling et al. 2012). This pattern aligns with the observed decline in functional richness, further supporting the conclusion that reef ecosystems are experiencing considerable ecological shifts. Indeed, the sharp functional decline observed in the 2010s at both sites may have been exacerbated by mass bleaching events that occurred in 2010, 2013 and 2016 (Chou et al. 2016; Guest et al. 2016a; Ng et al. 2020; Chan et al. 2024). Meanwhile, the slight increase of functional richness in the early 2020s (prior to the 2024 mass bleaching event) could suggest a potential recovery from these bleaching events; however, further research is required to confirm this trend.

We note that life history data were only available for 69 out of the 257 recorded species, potentially limiting the representativeness of the functional richness data. Further, research can help expand the available trait dataset, including life history strategies and other functional characteristics, which will provide a more comprehensive understanding of coral functional richness in the region.

This study further found that at both Pulau Hantu and Pulau Satumu, phylogenetic richness exhibited a general increase, while functional richness showed an overall decline (Figs. 9.10 and 9.11). These trends are consistent with the findings of Chan et al. (2024), which reported an increase in species and phylogenetic richness but a concurrent decline in functional richness. The increase in phylogenetic richness is likely attributed to greater sampling effort over time, resulting in the detection of a higher number of unique coral species, as evidenced by the rising number of colonies recorded during LIT surveys from 1980s to 2020s (Fig. 9.2). Conversely, the declining trend in functional richness is due to environment disturbances selecting for stress-tolerant and generalist species (Darling et al. 2012).

Despite the increase in species and phylogenetic richness, the assemblages are occupying a narrower range of functional roles, suggesting that Singapore's reefs may be becoming increasingly vulnerable to anthropogenic stressors (Guest et al. 2016b; Ng et al. 2020; Chan et al. 2024). With continuing urbanisation and land reclamation, it is crucial to implement effective sediment management strategies to mitigate further declines in functional richness and support reef resilience.

9.5 Conclusions

This study recorded 257 species of scleractinian corals based on historical museum collections, long-term LIT surveys and recent timed-swim surveys. The museum collections have played a crucial role in documenting local extirpations, such as the loss of *Seriatopora hystrix* and *Stylophora pistillata* (Poquita-Du et al. 2019b). Without these historical collections, tracking species decline and extirpation over time would not have been possible.

Although the 1950s–1970s collection was considerably smaller, it contributed valuable presence data and provided insights into potential species losses over time. Furthermore, museum records have enhanced our understanding of species, phylogenetic and functional richness over a seven-decade timescale, offering critical historical context on the impacts of urbanisation on coral reefs—information that would have been inaccessible through field surveys alone. While museum collections may sometimes lack detailed metadata such as dates, depths and collection methods, they remain valuable for species records. The 1980s collection was particularly extensive and recorded higher species richness compared to surveys. Additionally, the correlation between collection and survey datasets suggests that when collections are sufficiently large, they can also serve as indicators of species abundance.

The study observed that despite an increase in species and phylogenetic richness, the long-term decline in functional richness suggests that coral assemblages are becoming more functionally constrained, potentially making reefs more vulnerable to environmental stressors. The shift from competitive and weedy species to stress-tolerant and generalist species aligns with patterns seen in disturbed reef systems (Aronson et al. 2004; Alvarez-Filip et al. 2011; Chan et al. 2024), reinforcing the influence of urbanisation on coral communities (Darling et al. 2012; Heery et al. 2018). To mitigate further declines in functional richness, effective coastal management is essential, particularly in light of continuing coastal urbanisation and land reclamation. The negative consequences on coral diversity over decades due to these impacts in the vicinity of coral reef habitats highlight the urgent need for environmental impact assessments and mitigation, as well as reef restoration to enhance coral resilience and support the recovery of degraded reef systems.

Acknowledgements We thank all past and present members of the Reef Ecology Study Team for their tireless efforts in studying Singapore's coral biodiversity, and Tashfia Raquib for assisting with access to and information on the Zoological Reference Collection at the Lee Kong Chian Natural History Museum (LKCNHM). This study would not have been possible without the kind donation of coral collections by Professor Chuang Shou Hwa to the LKCNHM. We are also grateful to the following sponsors for their generous contributions in support of the Comprehensive Marine Biodiversity Survey II: Dalio Philanthropies, GSK-EDB Trust Fund, HSBC, ExxonMobil Asia Pacific, and Shell Eastern Trading (Pte) Ltd.

References

Alvarez-Filip L, Dulvy NK, Côté IM, Watkinson AR, Gill JA (2011) Coral identity underpins architectural complexity on Caribbean reefs. Ecol Appl 21(6):2223–2231. https://doi.org/10.1890/10-1563.1

Aronson RB, Macintyre IG, Wapnick CM, O'Neill MW (2004) Phase shifts, alternative states, and the unprecedented convergence of two reef systems. Ecology 85(7):1876–1891. https://doi.org/10.1890/03-0108

Bongaerts P, Cooke IR, Ying H, Wels D, Haan SD, Hernandez-Agreda A, Brunner CA, Dove S, Englebert N, Eyal G, Forêt S, Grinblat M, Hay KB, Harii S, Hayward DC, Lin Y, Mihaljević M, Moya A, Muir P, Sinniger F, Smallhorn-West P, Torda G, Ragan MA, van Oppen MJH, Hoegh-Guldberg O (2021) Morphological stasis masks ecologically divergent coral species on tropical reefs. Curr Biol 31(11):2286–2298.e8. https://doi.org/10.1016/j.cub.2021.03.028

Bridge TCL, Cowman PF, Quattrini AM, Bonito VE, Sinniger F, Harii S, Head CEI, Hung JY, Halafihi T, Rongo T, Baird AH (2023) A tenuis relationship: traditional taxonomy obscures systematics and biogeography of the '*Acropora tenuis*' (Scleractinia: Acroporidae) species complex. Zool J Linnean Soc 202(1):zlad062. https://doi.org/10.1093/zoolinnean/zlad062

Browne NK, Precht E, Last KS, Todd PA (2014) Photo-physiological costs associated with acute sediment stress events in three near-shore turbid water corals. Mar Ecol Prog Ser 502:129–143. https://doi.org/10.3354/meps10714

Burgess SC, Turner AM, Johnston EC (2024) Niche breadth and divergence in sympatric cryptic coral species (*Pocillopora* spp.) across habitats within reefs and among algal symbionts. Evol Appl 17:e13762. https://doi.org/10.1111/eva.13762

Chan YKS, Ng CSL, Tun KPP, Chou LM, Huang D (2024) Decadal decline of functional diversity despite increasing taxonomic and phylogenetic diversity of coral reefs under chronic urbanisation stress. Ecol Indic 164:112143. https://doi.org/10.1016/j.ecolind.2024.112143

Chao A, Ma KH, Hsieh TC (2016) iNEXT (iNterpolation and EXTrapolation) online: software for interpolation and extrapolation of species diversity. Program and User's Guide. http://chao.stat.nthu.edu.tw/wordpress/software_download/inext-online/

Chin HS (2024) Singapore corals showing signs of recovery amid global coral bleaching event. The Straits Times. Available at: https://www.straitstimes.com/singapore/singapore-corals-showing-signs-of-recovery-amid-global-coral-bleaching-event. Last accessed: 26 Mar 2025

Chou LM (1988) The impact of human influence on the fringing reef of Pulau Hantu, Singapore. In: Proceedings of the Sixth International Coral Reef Symposium, vol 2, pp 201–205

Chou LM, Low JKY, Loo MGK (1994) The state of coral reefs and coral reef research in Singapore. In: Wilkinson C, Sudara S, Chou LM (eds) Third ASEAN-Australia Symposium on Living Coastal Resources, Chulalongkorn University, Bangkok, Thailand, 16-20 May 1994, pp 77–88

Chou LM (2010) Anticipated impacts of climate change on marine biodiversity: using field situations that simulate climate change in Singapore. In: Sajise PE, Ticsay MV, Saguiguit GC Jr (eds) Moving forward: southeast Asian perspectives on climate change and biodiversity. ISEAS/SEARCA, Singapore, pp 131–140

Chou LM, Toh TC, Toh KB, Ng CSL, Cabaitan P, Tun K, Goh E, Afiq-Rosli L, Taira D, Poquita-Du RC, Loke HX, Khalis A, Li J, Song T (2016) Differential response of coral assemblages to thermal stress underscores the complexity in predicting bleaching susceptibility. PLoS One 11:e0159755. https://doi.org/10.1371/journal.pone.0159755

Chou LM, Huang D, Tan KS, Toh TC, Goh BPL, Tun K (2019) Singapore. In: Sheppard CRC (ed) World seas: an environmental evaluation. Volume II: the Indian Ocean to the Pacific. Academic Press, pp 539–558. https://doi.org/10.1016/b978-0-08-100853-9.00031-2

Chow GSE, Chan YKS, Jain SS, Huang D (2019) Light limitation selects for depth generalists in urbanised reef coral communities. Mar Environ Res 147:101–112. https://doi.org/10.1016/j.marenvres.2019.04.010

Chua G (2016) Case study: Singapore—Pulau Semakau. Ministry of National Development, Singapore. Centre for Liveable Cities. Available at: https://isomer-user-content.by.gov.sg/50/a7668b86-9a47-4cef-9b33-fc71217552fb/case-study-singapore-pulau-semakau.pdf. Last accessed: 1 Apr 2025

Chuang SH (1961) On Malayan Shores. Muwu Shosa, Singapore

Darling ES, Alvarez-Filip L, Oliver TA, McClanahan TR, Côté IM (2012) Evaluating life-history strategies of reef corals from species traits. Ecol Lett 15(12):1378–1386. https://doi.org/10.1111/j.1461-0248.2012.01861.x

Darling ES, McClanahan TR, Côté IM (2013) Life histories predict coral community disassembly under multiple stressors. Glob Chang Biol 19:1930–1940. https://doi.org/10.1111/gcb.12191

Davison GWH, Gan JWM, Huang D, Hwang WS, Lum SKY, Yeo DCJ (eds) (2024) The Singapore Red Data Book: red lists of Singapore biodiversity, 3rd edn. National Parks Board, Singapore

De Groot R, Brander L, Van Der Ploeg S, Costanza R, Bernard F, Braat L, Christie M, Crossman N, Ghermandi A, Hein L, Hussain S, Kumar P, McVittie A, Portela R, Rodriguez LC, Brink PT, Van Beukering P (2012) Global estimates of the value of ecosystems and their services in monetary units. Ecosyst Serv 1(1):50–61. https://doi.org/10.1016/j.ecoser.2012.07.005

Erftemeijer PLA, Riegl B, Hoeksema BW, Todd PA (2012) Environmental impacts of dredging and other sediment disturbances on corals: A review. Mar Pollut Bull 64:1737–1765. https://doi.org/10.1016/j.marpolbul.2012.05.008

Faith DP (1992) Conservation evaluation and phylogenetic diversity. Biol Conserv 61:1–10. https://doi.org/10.1016/0006-3207(92)91201-3

Geronimo RC, Cabral RB (2013) Fish production in the coral triangle: status, trends, and challenges. In: Economics of fisheries and aquaculture in the Coral Triangle. Consultant's Report submitted to the Asian Development Bank, Manila, pp 5–29

Glaser R, Haberzettl P, Walsh RPD (1991) Land reclamation in Singapore, Hong Kong and Macau. GeoJournal 24(4):365–373. https://doi.org/10.1007/bf00578258

Goh NKC, Chou LM (1993) The coral reef community of Pulau Satumu (Raffles Lighthouse), Singapore, with emphasis on the hard corals. J Singapore Natl Acad Sci 20–21:51–57

Gower JC, Legendre P (1986) Metric and Euclidean properties of dissimilarity coefficients. J Classif 3:5–48. https://doi.org/10.1007/bf01896809

Guest JR, Baird AH, Maynard JA, Muttaqin E, Edwards AJ, Campbell SJ, Yewdall K, Affendi YA, Chou LM (2012) Contrasting patterns of coral bleaching susceptibility in 2010 suggest an adaptive response to thermal stress. PLoS One 7:e33353. https://doi.org/10.1371/journal.pone.0033353

Guest JR, Low J, Tun K, Wilson B, Ng C, Raingeard D, Ulstrup KE, Tanzil JTI, Todd PA, Toh TC, McDougald D, Chou LM, Steinberg PD (2016a) Coral community response to bleaching on a highly disturbed reef. Sci Rep 6:20717. https://doi.org/10.1038/srep20717

Guest JR, Tun K, Low J, Vergés A, Marzinelli EM, Campbell AH, Bauman AG, Feary DA, Chou LM, Steinberg PD (2016b) 27 years of benthic and coral community dynamics on turbid, highly urbanised reefs off Singapore. Sci Rep 6:36260. https://doi.org/10.1038/srep36260

Hebert PDN, Cywinska A, Ball SL, deWaard JR (2003) Biological identifications through DNA barcodes. Proc R Soc B Biol Sci 270(1512):313–321. https://doi.org/10.1098/rspb.2002.2218

Heery EC, Hoeksema BW, Browne NK, Reimer JD, Ang PO, Huang D, Friess DA, Chou LM, Loke LHL, Saksena-Taylor P, Alsagoff N, Yeemin T, Sutthacheep M, Vo ST, Bos AR, Gumanao GS, Hussein MAS, Waheed Z, Lane DJW, Johan O, Kunzmann A, Jompa J, Suharsono, Taira D, Bauman AG, Todd PA (2018) Urban coral reefs: degradation and resilience of hard coral assemblages in coastal cities of East and Southeast Asia. Mar Pollut Bull 135:654–681. https://doi.org/10.1016/j.marpolbul.2018.07.041

Hill MO (1973) Diversity and evenness: A unifying notation and its consequences. Ecology 54(2):427–432. https://doi.org/10.2307/1934352

Hilton MJ, Manning SS (1995) Conversion of coastal habitats in Singapore: indications of unsustainable development. Environ Conserv 22:307–322. https://doi.org/10.1017/S0376892900034883

Hobday AJ, Alexander LV, Perkins SE, Smale DA, Straub SC, Oliver ECJ, Benthuysen JA, Burrows MT, Donat MG, Feng M, Holbrook NJ, Moore PJ, Scannell HA, Gupta AS, Wernberg T (2016) A hierarchical approach to defining marine heatwaves. Prog Oceanogr 141:227–238. https://doi.org/10.1016/j.pocean.2015.12.014

Hoegh-Guldberg O, Smith GJ (1989) The effect of sudden changes in temperature, light and salinity on the population density and export of zooxanthellae from the reef corals *Stylophora pistillata* Esper and *Seriatopora hystrix* Dana. J Exp Mar Biol Ecol 129(3):279–303. https://doi.org/10.1016/0022-0981(89)90109-3

Holmberg J (1976) Islands in the sun: $60 million plan to turn Singapore's Southern Islands into recreational centres: Sentosa was the test case. New Nation, 3 December 1976, pp 10–11. https://eresources.nlb.gov.sg/newspapers/Digitised/Article/newnation19761203-1.2.69. Accessed 25 Apr 2025

Huang D, Roy K (2015) The future of evolutionary diversity in reef corals. Philos Trans R Soc B Biol Sci 370:20140010. https://doi.org/10.1098/rstb.2014.0010

Huang D, Tun KP, Chou LM, Todd PA (2009) An inventory of zooxanthellate scleractinian corals in Singapore, including 33 new records. Raffles Bull Zool Suppl 22:69–80

Hughes TP, Kerry JT, Álvarez-Noriega M, Álvarez-Romero JG, Anderson KD, Baird AH, Babcock RC, Beger M, Bellwood DR, Berkelmans R, Bridge TC, Butler IR, Byrne M, Cantin NE, Comeau S, Connolly SR, Cumming GS, Dalton SJ, Diaz-Pulido G, Eakin CM, Figueira WF, Gilmour JP, Harrison HB, Heron SF, Hoey AS, Hobbs J-PA, Hoogenboom MO, Kennedy EV, Kuo C-Y, Lough JM, Lowe RJ, Liu G, McCulloch MT, Malcolm HA, McWilliam MJ, Pandolfi JM, Pears RJ, Pratchett MS, Schoepf V, Simpson T,

Skirving WJ, Sommer B, Torda G, Wachenfeld DR, Willis BL, Wilson SK (2017) Global warming and recurrent mass bleaching of corals. Nature 543:373–377. https://doi.org/10.1038/nature2170
Hughes TP, Kerry JT, Baird A, Connolly SR, Dietzel A, Eakin CM, Heron SF, Hoey AS, Hoogenboom MO, Liu G, McWilliam M, Pears R, Pratchett MS, Skirving WJ, Stella J, Torda G (2018) Global warming transforms coral reef assemblages. Nature 556(7702):492–496. https://doi.org/10.1038/s41586-018-0041-2
IPCC (2021) In: Masson-Delmotte V et al (eds) Climate change 2021: the physical science basis. Contribution of working group I to the sixth assessment report of the intergovernmental panel on climate change. Cambridge University Press. https://doi.org/10.1017/9781009157896
Jain SS, Afiq-Rosli L, Feldman B, Levy O, Phua JW, Wainwright BJ, Huang D (2020) Homogenization of endosymbiont communities hosted by equatorial corals during the 2016 mass bleaching event. Microorganisms 8(9):1370. https://doi.org/10.3390/microorganisms8091370
Kitahara MV, Fukami H, Benzoni F, Huang D (2016) The new systematics of Scleractinia: integrating molecular and morphological evidence. In: Goffredo S, Dubinsky Z (eds) The Cnidaria, past, present and future. Springer International Publishing, Cham, pp 41–59. https://doi.org/10.1007/978-3-319-31305-4_4
Koh EGL, Chou LM (1989) The mushroom corals of Singapore. National University of Singapore, Singapore
Kruskal JB (1964) Multidimensional scaling by optimizing goodness of fit to a nonmetric hypothesis. Psychometrika 29(1):1–27. https://doi.org/10.1007/bf02289565
Lai S, Loke LHL, Hilton MJ, Bouma TJ, Todd PA (2015) The effects of urbanisation on coastal habitats and the potential for ecological engineering: A Singapore case study. Ocean Coast Manag 103:78–85. https://doi.org/10.1016/j.ocecoaman.2014.11.006
Loo MGK, Chou LM (1995) Environmental status of Singapore's coral reefs. In: Second workshop of the EAS-35 project: Integrated Management of Watersheds in Relation to the Coastal and Marine Environment, Bangkok, Thailand, 3–7 April 1995, pp 1–20
Loong S, Taira D, Chan YKS, Chan WWR, Fong J, Kwan V, Low JKY, Ng CSL, Sam SQ, Tan CZ, Tun KPP, Yip ZT, Jaafar Z, Huang D (2025) Covariations of coral, macroalgal, macroinvertebrate and fish communities in an urbanised tropical reef environment. Coral Reefs. https://doi.org/10.1007/s00338-025-02723-9
Low JKY, Chou LM (1994) Sedimentation rates in Singapore waters. In: Proceedings of the 3rd ASEAN-Australia Symposium on Living Coral Resources, vol 2, pp 697–701
Maritime and Port Authority of Singapore (2003) Prohibited area at Pulau Satumu (Raffles lighthouse) and waters surrounding a radius of 300 metres around the lighthouse. Port Marine Circular No. 22 of 2003. Available at: https://www.mpa.gov.sg/media-centre/details/prohibited-area-at-pulau-satumu-(raffles-lighthouse)-and-waters-surrounding-a-radius-of-300-metres-around-the-lighthouse. Last accessed: 20 Dec 2024
Maritime and Port Authority of Singapore (2017) Underwater diving operations within the port. Port Marine Circular No. 23 of 2017. Available at: https://www.mpa.gov.sg/PDFViewer/web/viewer.html?file=/docs/mpalibraries/circulars-and-notices/port-marine-circulars/-pc17-23. Last accessed: 20 Dec 2024
Moberg F, Folke C (1999) Ecological goods and services of coral reef ecosystems. Ecol Econ 29:215–233. https://doi.org/10.1016/s0921-8009(99)00009-9
Morgan KM, Moynihan MA, Sanwlani N, Switzer AD (2020) Light limitation and depth-variable sedimentation drives vertical reef compression on turbid coral reefs. Front Mar Sci 7:571256. https://doi.org/10.3389/fmars.2020.571256
Morton B (1994) Hong Kong's coral communities: status, threats and management plans. Mar Pollut Bull 29(1–3):74–83. https://doi.org/10.1016/0025-326x(94)90429-4
Ng CSL, Huang D, Toh KB, Sam SQ, Kikuzawa YP, Toh TC, Taira D, Chan YKS, Hung LZT, Sim WT, Rashid AR, Afiq-Rosli L, Ng NK, Chou LM (2020) Responses of urban reef corals during the 2016 mass bleaching event. Mar Pollut Bull 154:111111. https://doi.org/10.1016/j.marpolbul.2020.111111
Ng LCS, Chan YKS, Poquita-Du RC, Huang D (2024) Checklist of Cnidaria (Scleractinia) species with their category of threat status for Singapore. In: Davison GWH, Gan JWM, Huang D, Hwang WS, Lum SKY, Yeo DCJ (eds) The Singapore Red Data Book: red lists of Singapore biodiversity, 3rd edn. National Parks Board, Singapore, pp 527–534
Oksanen J, Blanchet FG, Kindt R, Legendre P, Minchin PR, O'Hara RB, Simpson GL, Solymos P, Stevens MHH, Wagner H (2013) Vegan: community ecology package. R Package Version 2.0-8. https://CRAN.R-project.org/package=vegan
Ponder WF, Carter GA, Flemons P, Chapman RR (2001) Evaluation of museum collection data for use in biodiversity assessment. Conserv Biol 15(3):648–657. https://doi.org/10.1046/j.1523-1739.2001.015003648.x
Poquita-Du RC, Huang D, Chou LM, Mrinalini T, P. A. (2019a) Short term exposure to heat and sediment triggers changes in coral gene expression and photo-physiological performance. Front Mar Sci 6:121. https://doi.org/10.3389/fmars.2019.00121
Poquita-Du RC, Quek ZBR, Jain SS, Schmidt-Roach S, Tun K, Heery EC, Chou LM, Todd PA, Huang D (2019b) Last species standing: loss of Pocilloporidae corals associated with coastal urbanization in a tropical city state. Mar Biodivers 49(4):1727–1741. https://doi.org/10.1007/s12526-019-00939-x
Ramírez-Portilla C, Baird AH, Cowman PF, Quattrini AM, Harii S, Sinniger F, Flot J-F (2022) Solving the coral species delimitation conundrum. Syst Biol 71:461–475.https://doi.org/10.1093/sysbio/syab077
Quek ZBR, Jain SS, Richards ZT, Arrigoni R, Benzoni F, Hoeksema BW, Carvajal JI, Wilson NG, Baird AH, Kitahara MV, Seiblitz IGL, Vaga CF, Huang D (2023) A hybrid-capture approach to reconstruct the phylogeny of Scleractinia (Cnidaria: Hexacorallia). Mol Phylogenet Evol 186:107867. https://doi.org/10.1016/j.ympev.2023.107867
Rassmussen SH, Cowman PF, Baird AH, Crosbie AJ, Quattrini AM, Bonito V, Sinniger F, Harii S, Cabaitan PC, Fadli N, Tan C-H, Hung JY-H, Rongo T, Huang D, Halafihi T, Bridge TCL (2025) The tables have turned: taxonomy, systematics and biogeography of the *Acropora hyacinthus* (Scleractinia: Acroporidae) complex. Invertebr Syst 39:IS24049. https://doi.org/10.1071/is24049
Shaffer HB, Fisher RN, Davidson C (1998) The role of natural history collections in documenting species declines. Trends Ecol Evol 13(1):27–30. https://doi.org/10.1016/s0169-5347(97)01177-4
Shannon CE (1948) A mathematical theory of communication. Bell Syst Tech J 27(3):379–423. https://doi.org/10.1002/j.1538-7305.1948.tb01338.x
Simpson EH (1949) Measurement of diversity. Nature 163(4148):688. https://doi.org/10.1038/163688a0
Sin TM, Ang HP, Buurman J, Lee AC, Leong YL, Ooi SK, Steinberg P, Teo SL-M (2016) The urban marine environment of Singapore. Reg Stud Mar Sci 8:331–339. https://doi.org/10.1016/j.rsma.2016.01.011
Spalding M, Burke L, Wood SA, Ashpole JE, Hutchison JE, Ermgassen PSEZ (2017) Mapping the global value and distribution of coral reef tourism. Mar Policy 82:104–113. https://doi.org/10.1016/j.marpol.2017.05.014
Stocker TF (2013) The ocean as a component of the climate system. Int Geophys Ser:3–30. https://doi.org/10.1016/b978-0-12-391851-2.00001-5
Suarez AV, Tsutsui ND (2004) The value of museum collections for research and society. BioScience 54(1):66–74. https://doi.org/10.1641/0006-3568(2004)054

Tacon AGJ, Metian M (2017) Food matters: fish, income, and food supply—a comparative analysis. Rev Fish Sci Aquacul 26(1):15–28. https://doi.org/10.1080/23308249.2017.1328659

Tan SK, Low MEY (2022) An inventory of the molluscs of Pulau Satumu (Raffles Lighthouse), with notes on the etymology and history of the islet. Lee Kong Chian Natural History Museum, Singapore

Tan YS, Lee TJ, Tan K (2009) Clean, green and blue: Singapore's journey towards environmental and water sustainability. Institute of Southeast Asian Studies, Singapore

Tan YHJ, Tham JKQ, Paul A, Rana U, Ang HP, Nguyen NTH, Yee ATK, Leong BPI, Drummond S, Tun KPP (2023) Remote sensing mapping of the regeneration of coastal natural habitats in Singapore: implications for marine conservation in tropical cities. Singap J Trop Geogr 44(1):130–148. https://doi.org/10.1111/sjtg.12454

Teo J, Hussin M, Poh K, Lim SC, Low MEY, Huang D (2025) Archipelago of islands: natural and other histories of Singapore. Lee Kong Chian Natural History Museum, Singapore

Van Zanten BT, Van Beukering P, Wagtendonk AJ (2014) Coastal protection by coral reefs: A framework for spatial assessment and economic valuation. Ocean Coastal Manag 96:94–103. https://doi.org/10.1016/j.ocecoaman.2014.05.001

Venegas RM, Acevedo J, Treml EA (2023) Three decades of ocean warming impacts on marine ecosystems: a review and perspective. Deep-Sea Res II Top Stud Oceanogr 212:105318. https://doi.org/10.1016/j.dsr2.2023.105318

Veron JEN, Pichon M (1976) Scleractinia of Eastern Australia. Part I. Families Thamnasteriidae, Astrocoeniidae, Pocilloporidae. Australian Institute of Marine Science, Townsville. http://archive.org/details/ScleractiniaEas00JENVB

Veron JEN, Pichon M (1980) Scleractinia of Eastern Australia. Part III. Families Agariciidae, Siderastreidae, Fungiidae, Oculinidae, Merulinidae, Mussidae, Pectiniidae, Caryophylliidae, Dendrophylliidae. Australian Institute of Marine Science, Townsville. http://archive.org/details/ScleractiniaEas00JENVD

Veron JEN, Pichon M (1982) Scleractinia of Eastern Australia. Part IV. Family Poritidae. Australian Institute of Marine Science, Townsville. http://archive.org/details/ScleractiniaEas00JENVA

Veron JEN, Wallace CC (1984) Scleractinia of Eastern Australia. Part V. Family Acroporidae. Australian Institute of Marine Science, Townsville. http://archive.org/details/ScleractiniaEas00JENVC

Veron JEN, Pichon M, Wijsman-Best M (1977) Scleractinia of Eastern Australia. Part II. Families Faviidae, Trachyphylliidae. Australian Institute of Marine Science, Townsville. http://archive.org/details/ScleractiniaEas00JENV

Veron JEN (2000) Corals of the world. Australian Institute of Marine Science, Townsville

von Kietzell A, Schurer A, Hegerl GC (2022) Marine heatwaves in global sea surface temperature records since 1850. Environ Res Lett 17(8):084027. https://doi.org/10.1088/1748-9326/ac81db

Wallace CC (1999) Staghorn corals of the world. CSIRO Publishing, Collingwood

Ward DF (2012) More than just records: Analysing natural history collections for biodiversity planning. PLoS One 7(11):e50346. https://doi.org/10.1371/journal.pone.0050346

Wear SL (2016) Missing the boat: critical threats to coral reefs are neglected at global scale. Mar Policy 74:153–157. https://doi.org/10.1016/j.marpol.2016.09.009

Wong JSY, Chan YKS, Ng CSL, Tun KPP, Darling ES, Huang D (2018) Comparing patterns of taxonomic, functional and phylogenetic diversity in reef coral communities. Coral Reefs 37(3):737–750. https://doi.org/10.1007/s00338-018-1698-6

WoRMS Editorial Board (2024) World Register of Marine Species. https://www.marinespecies.org. Accessed 20 Dec 2024

Conservation Status of Corals in Singapore

10

Chin Soon Lionel Ng, Yong Kit Samuel Chan, Rosa Celia Poquita-Du, Karenne Tun, and Danwei Huang

Abstract

Throughout much of Singapore's modern history, impacts from intensive coastal urbanisation and climate change have resulted in significant changes to the structure and function of its coral reefs. Here, we assembled information from contemporary reef survey data, museum records and the literature reviews to determine the conservation statuses of all 276 recorded stony coral species in Singapore. Threat statuses for 142 species (1 presumed Nationally Extinct, 1 Critically Endangered, 7 Endangered, 20 Vulnerable, 26 Near Threatened and 87 Least Concern) are now provided, representing a substantial increase from just two species assessed from the last report in 2008. One of them, *Seriatopora hystrix*, is now presumably extirpated owing to its last specimen collected before the 1960s, while the other, *Stylophora pistillata*, remains Critically Endangered. These updated conservation statuses serve to inform strategies for enhancing the recovery of Singapore's corals and to improve habitat management for sustained reef ecosystem functioning. However, nearly half of all species lack sufficient data for a formal assessment and are considered Data Deficient, underscoring the urgent need for more detailed surveys to increase detection and quantify species trends. Mitigating impacts from development and climate change remains critical to reduce the risks of further coral species extirpations.

Keywords

Climate change · Coral cover · Coral reef management · Population trends · Scleractinia · National Red List · Threat assessment · Urban reefs

10.1 Introduction

Singapore is situated in the southwestern reaches of the South China Sea, a region that boasts extraordinary scleractinian coral diversity (Huang et al. 2015). Since the nineteenth century, numerous marine science researchers have examined and documented the country's rich diversity of corals (e.g. Brüggemann 1878; Studer 1881; Chuang 1961; Chou and Teo 1983; Loo and Chou 1995), which are mainly found in the shallow seas of the Singapore Strait. Notably, Huang et al. (2009)'s extensive review—based on prior literature, reef surveys and museum records—updated Singapore's scleractinian inventory to a total of 255 species, nearly a third of the global diversity. This inventory represented a substantial increase from the 197 species reported in the 1990s, and has helped shape coral research and coastal resource management in the new millennium. Since then, researchers and citizen scientists have also broadened the spatial extent of coastal habitats surveyed, contributing

C. S. L. Ng (✉)
Tropical Marine Science Institute, National University of Singapore, Singapore, Singapore

Y. K. S. Chan
Australian Institute of Marine Science, Townsville, QLD, Australia

AIMS@JCU, Division of Research and Innovation, James Cook University, Townsville, QLD, Australia

R. C. Poquita-Du
Senckenberg Biodiversity and Climate Research Center, Frankfurt am Main, Germany

K. Tun
National Parks Board, Singapore, Singapore

D. Huang
Tropical Marine Science Institute, National University of Singapore, Singapore, Singapore

Lee Kong Chian Natural History Museum, National University of Singapore, Singapore, Singapore

Department of Biological Sciences, National University of Singapore, Singapore, Singapore

Centre for Nature-based Climate Solutions, National University of Singapore, Singapore, Singapore

L. M. Chou, D. Huang (eds.), *Coral Reefs of Singapore's Urbanised Sea*, Coral Reefs of the World 23,
https://doi.org/10.1007/978-981-95-0903-4_10

towards a better understanding of the distribution of corals across Singapore (Wong et al. 2018; Chow et al. 2019; Lim et al. 2020; Loong et al. 2025).

The past half-century has seen Singapore's land area increasing by more than 20% due to coastal reclamation works to cater for economic and population growth (Chou et al. 2019). The intense coastal development and other anthropogenic activities have led to significant reduction in natural reef area and the degradation of remaining reef habitats, such that Singapore's reefs have been considered by some researchers to be marginal (i.e. low coral cover) and extreme (i.e. suboptimal environmental conditions) (Heery et al. 2018; Schoepf et al. 2023). Many nearshore reefs have been buried and replaced by man-made infrastructure such as marinas, lagoons and jetties (Chou et al. 2023); in particular, coastal defence structures now comprise more than 60% of the coastline (Lai et al. 2015). Maintenance dredging of shipping channels also takes place regularly to ensure that the port waters are deep enough for vessel traffic, particularly for the passage of ultra large container vessels (ULCV) and very large crude carriers (VLCC). While the reefs here may have been subjected to turbid conditions throughout much of the Holocene (Chan et al. 2025), the present anthropogenic sediment input would have reduced light penetration in the water column and the photic zone for hermatypic organisms to the top 5–8 m of the water column, thereby affecting their photosynthetic abilities (Browne et al. 2014; Browne et al. 2015). High sedimentation levels also smother sessile benthic organisms, and increase their energy expenditure to clear excess sediment (Erftemeijer et al. 2012; Junjie et al. 2014). Even though coastal defence structures such as seawalls can provide stable habitats for coral recruitment and establishment (Ng et al. 2012; Kikuzawa et al. 2020), it is clear that the cumulative impacts from habitat loss and coastal urbanisation have resulted in local species extirpations (Poquita-Du et al. 2019). Additionally, just like other parts of the world, reefs in Singapore have not been spared by climate change. Over the past two decades, numerous major and minor mass bleaching episodes have led to sharp declines in coral cover, although recovery has subsequently been observed (Toh et al. 2022). At the point of writing, the fourth global bleaching event is underway (Reimer et al. 2024), and has also affected large swathes of Singapore's reefs (Begum 2024a, b).

While the recorded species diversity of corals in Singapore has grown over the years—in part due to increased reef survey efforts and expertise in coral taxonomy and field identification—anthropogenic and climate impacts have collectively resulted in an overall decline of reef functional diversity, potentially threatening overall reef resilience (Chan et al. 2024). However, these stressors have also enabled a deeper understanding of how corals in Singapore respond to urbanisation (e.g. Poquita-Du et al. 2019; Prasetia et al. 2020; Ng et al. 2021) and climate change (e.g. Chou et al. 2016; Ng et al. 2020; Law and Huang 2023), and highlighted how species-specific responses can influence community change and ecosystem functioning (Guest et al. 2016; Chan et al. 2024; Loong et al. 2025). These changes critically underscore the need for assessments of sufficient resolution to determine conservation priorities and guide resource allocation.

At the global level, the International Union for Conservation of Nature (IUCN) considers approximately a third of the world's coral species to be at heightened risk of extinction (Carpenter et al. 2008; see also Gutierrez et al. 2024). As part of Singapore's international obligations to the Convention of Biological Diversity, assessments have similarly been carried out to guide development and enhance the conservation of its flora and fauna (Davison et al. 2008). The Singapore Red Data Book, which presents the Red Lists of Singapore's biodiversity, is now in its third edition (Davison 2024). Prior to the current run, only two corals (*Seriatopora hystrix* and *Stylophora pistillata*) had been listed in the second edition of the national Red List, and both had been deemed Critically Endangered (Chou and Tan 2008). The lack of information on all other coral species was attributed to the difficulty of surveying marine habitats to understand population size and rates of decline, as well as the nascency of conservation techniques and supporting policy tools at the point of assessment (Chou and Tan 2008).

At just over 10 km^2, Singapore's present reef area is less than 40% of that recorded in the 1950s (Tun 2012; Chou et al. 2019; Tan et al. 2023). Yet, the reefs remain valued for supporting high marine biodiversity and providing ecosystem goods and services such as recreational fishing and SCUBA diving (Jaafar et al. 2018; Chou et al. 2019; Ip et al. 2019; Ip et al. 2021). With the ongoing synergistic risks of local and global stressors compounded by projected habitat loss from coastal development plans over the coming decades (e.g. 'Long Island'; Ng and Begum 2024), it is even more pertinent to thoroughly evaluate how Singapore's corals may fare in the face of such impacts. High-resolution assessments, especially at the species level, will be particularly important to provide strong scientific basis for enforcing current legislation to protect Singapore's corals and further strengthen the environmental impact assessment (EIA) process to guide coastal development and mitigation (Lye 2024). This will also enable proactive measures (e.g. mitigation measures, species recovery programmes, habitat creation and restoration and maritime regulation) to be implemented to protect all remaining coral species and ensure that ecosystem functioning is sustained. Here, as part of the assessment for the third edition of the Singapore Red Data Book, we aim to update Singapore's reef scleractinian species inventory, assess the national conservation status of each species and propose strategies required for their preservation.

10.2 Materials and Methods

Records of stony corals (order Scleractinia) in Singapore from 1878 to 2021 were consolidated from a range of sources that included contemporary reef survey data, museum records and reports from published and grey literature. These data contained information on the first and last known digital and/or physical records of all coral species recorded in Singapore to date. Species records were derived from Huang et al. (2009), the Lee Kong Chian Natural History Museum (LKCNHM) collections, as well as reef surveys conducted between 1986 and 2021 by the Reef Ecology Study Team (REST) at the National University of Singapore that mainly focused on the reef crest (~3–4 m below relative sea level), where coral cover and diversity are typically highest (Fig. 10.1). Verification of coral taxonomic classification, including the synonymisation of outdated nomenclature, was also carried out based on the World Register of Marine Species (WoRMS 2024).

Each coral species was evaluated using criteria from Singapore's National Red List system (Davison 2024) and the IUCN Red List (IUCN 2012). We considered all relevant stressors in Singapore such as habitat loss, habitat modification, declining water quality, sedimentation, ocean warming and acidification as ongoing threats and therefore were causes of decline that are still operating and impacting most of its corals. Criteria used to assess other taxonomic groups, such as whether they were migrants, natural hybrids, cryptic species complexes or endemic species, were not applicable to corals and therefore not considered.

To evaluate if a species is threatened, the IUCN Red List uses five criteria: population size reduction; limited geographic range in terms of occurrence or occupancy; small population size and decline of mature individuals; very small or restricted population of mature individuals; and quantitative analysis into the probability of extinction in the wild (IUCN 2012). As population sizes and numbers of mature individuals were difficult to ascertain for colonial organisms, we based our assessment for Singapore's corals primarily on coral cover as a proxy for change in population size over time. The first step of our process ('Part I') comprised collating survey data across the history of reef monitoring in Singapore (i.e. from 1986 to 2019; Chan et al. 2024), when the same method (i.e. the line intercept transect; English et al. 1997) had been used to identify corals to species. For species with available coral cover data spanning 30 years (i.e. three generations of 10 years), we focused on three components: absolute change in coral cover (estimated by the difference in mean species cover between two major survey periods 1986–1992 and 2015–2019), modelled rate of change (using linear mixed effects models, with site as random effect) and model significance (as an indication of the certainty of change in coral cover). For other species with coral cover information spanning less than 30 years, we used the same metrics with a caveat on the amount of data analysed when interpreting the results. While species declines were not necessarily linear, the modelled rate of change represented a means to estimate the change in the cover of each species over time with the available quantitative data to give a broad indication of species-specific reef trends. Combining

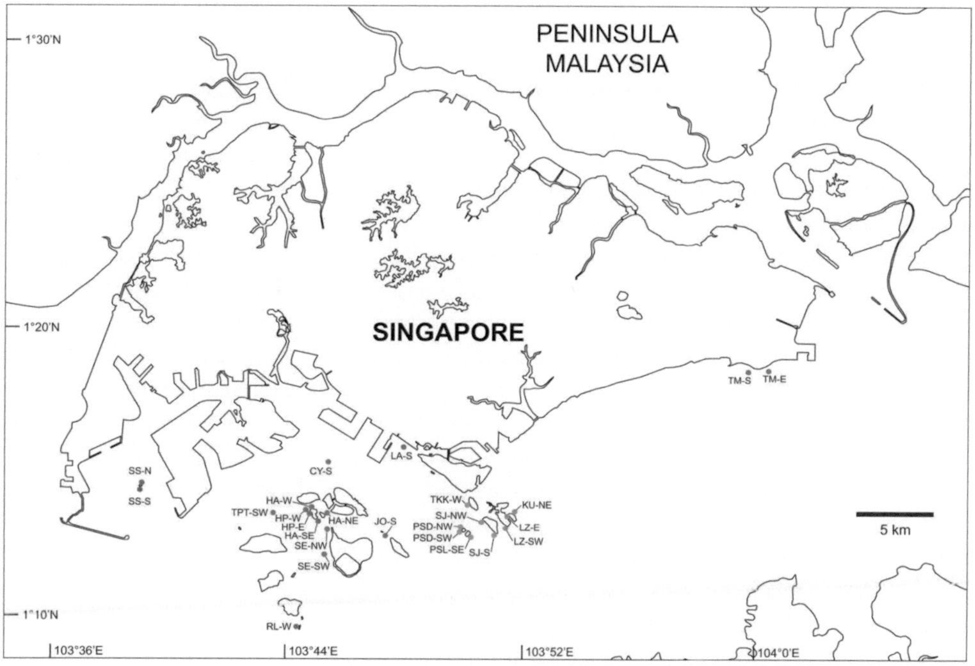

Fig. 10.1 Map of Singapore and survey sites with available scleractinian data for the current Red List assessment. Colours indicate the legal protection accorded at the respective sites: red (under maritime regulation and off-limits to the public), blue (Sisters' Islands Marine Park), orange (proposed second marine park) and purple (not under maritime or public park regulations). Information on the sites is available as Supplementary Data (https://doi.org/10.5281/zenodo.13284027)

the extent of species-level population change, site occurrence data and our collective expertise (especially for 'special cases'), a tentative national status for each species was derived (Fig. 10.2).

The next step ('Part II') entailed a revision of our initial assessments by considering species ranges and the possibility of Singapore's reefs receiving propagules from other reefs in the region (Tay et al. 2012, 2015; Huang et al. 2015). For most species in Singapore, propagule immigration was expected from neighbouring reefs, and thus, the initial assessment was downgraded to arrive at the final threat status of the species (Fig. 10.2).

10.3 Results and Discussion

10.3.1 Species Records

A total of 276 coral species from 68 genera and 19 families were documented in Singapore's waters. This represented an increase of 21 species from Huang et al. (2009), with additional species such as *Goniopora tenuidens*, *Hydnophora grandis* and *Micromussa amakusensis* only recorded in recent years (Wong et al. 2018; Ng et al. 2019), along with others included from collections at the Lee Kong Chian Natural History Museum (e.g. *Acropora pulchra*, *Leptoseris hawaiiensis*). In the past decade (2014–2024), 189 species have been documented in the scientific literature, having been encountered in reef surveys or used in experimental studies. Fifty species, however, have not been sighted since 2005. The full list of species and details of observations are provided in the Supplementary Data (https://doi.org/10.5281/zenodo.13284027).

10.3.2 Threat Assessments and Conservation Statuses

Compared to the previous Red List for Singapore's corals, which relied on expert opinion due to the relative lack of quantitative data for a comprehensive assessment (Chou and Tan 2008), we here leverage substantially on reef survey data to refine the assessment methodology and enhance its utility for future threat assessments. Notably, from only two species that were considered highly threatened in the previous assessment, 28 species are now assigned threatened statuses (i.e. Critically Endangered, Endangered and Vulnerable) and one species is presumed to be Nationally Extinct. On the other hand, 113 species are currently not considered threatened (i.e. Near Threatened and Least Concern), and 134 others are Data Deficient (Figs. 10.3 and 10.4). While the increase in threatened species is a direct result of the availability of new data from burgeoning coral research efforts over the last three decades, the assessments also highlight that these corals are at highest risk of loss unless targeted conservation measures are undertaken.

Crucially, *Seriatopora hystrix* was designated as Critically Endangered in the previous assessment but is now presumed nationally extinct. Having not been observed on Singapore's reefs since 1961, at the latest (Chuang 1961; Poquita-Du et al. 2019), it has exceeded the local-scale extinction threshold of 50 years (Davison 2024). Despite its presence on other reefs in the region (Huang et al. 2015), even in neighbouring Peninsula Malaysia (Shahbudin et al. 2017), and the possibility of propagules migrating to Singapore from these areas, the species has not re-established, strongly suggesting that local environmental pressures have led to its extirpation (Poquita-Du et al. 2019).

Stylophora pistillata retained its status as Critically Endangered. Documented in the 1970s (Chuang 1961), only one colony had been seen at Pulau Satumu in the 2000s (Tun K, pers obs). Similarly, despite being also found on reefs in the region (Huang et al. 2015), colonies have not been observed on Singapore reefs since the last one perished from the 2010 mass coral bleaching event. Its status was therefore not downgraded after the initial assessment in Part I. Importantly, the national statuses of *Seriatopora hystrix* and *Stylophora pistillata* are in stark contrast to their global statuses of Least Concern and Near Threatened, respectively, highlighting the importance of local assessments to ensure relevance for species conservation.

Seven species, consisting mainly of *Acropora* spp. and *Trachyphyllia geoffroyi*, are listed as Endangered. These species were documented only at a few sites in Singapore and were designated Critically Endangered in our initial assessment—either because of significant population decline or a lack of information to determine recent population change—but subsequently downgraded to Endangered because of their presence on regional reefs that potentially support propagule immigration. This is followed by the Vulnerable species, comprising 20 species from nine families. Originally assessed as Endangered for reasons ranging from their limited presence at few sites, negative (albeit non-significant) modelled rate of change or significant population decline, their threat statuses were also downgraded based on the potential of propagule immigration from regional reefs.

Twenty-six Near Threatened coral species were initially designated Vulnerable because of significant declines, non-significant decline but negative absolute change in cover or were relatively rare; these were however of lower severity or significance than the Endangered corals. Because of their regional presence, their threat statuses were downgraded. The Least Concern category includes a) species originally designated Near Threatened because of their uncommonness or limited declines in coral cover, but were subsequently downgraded; and b) species already defined as Least Concern

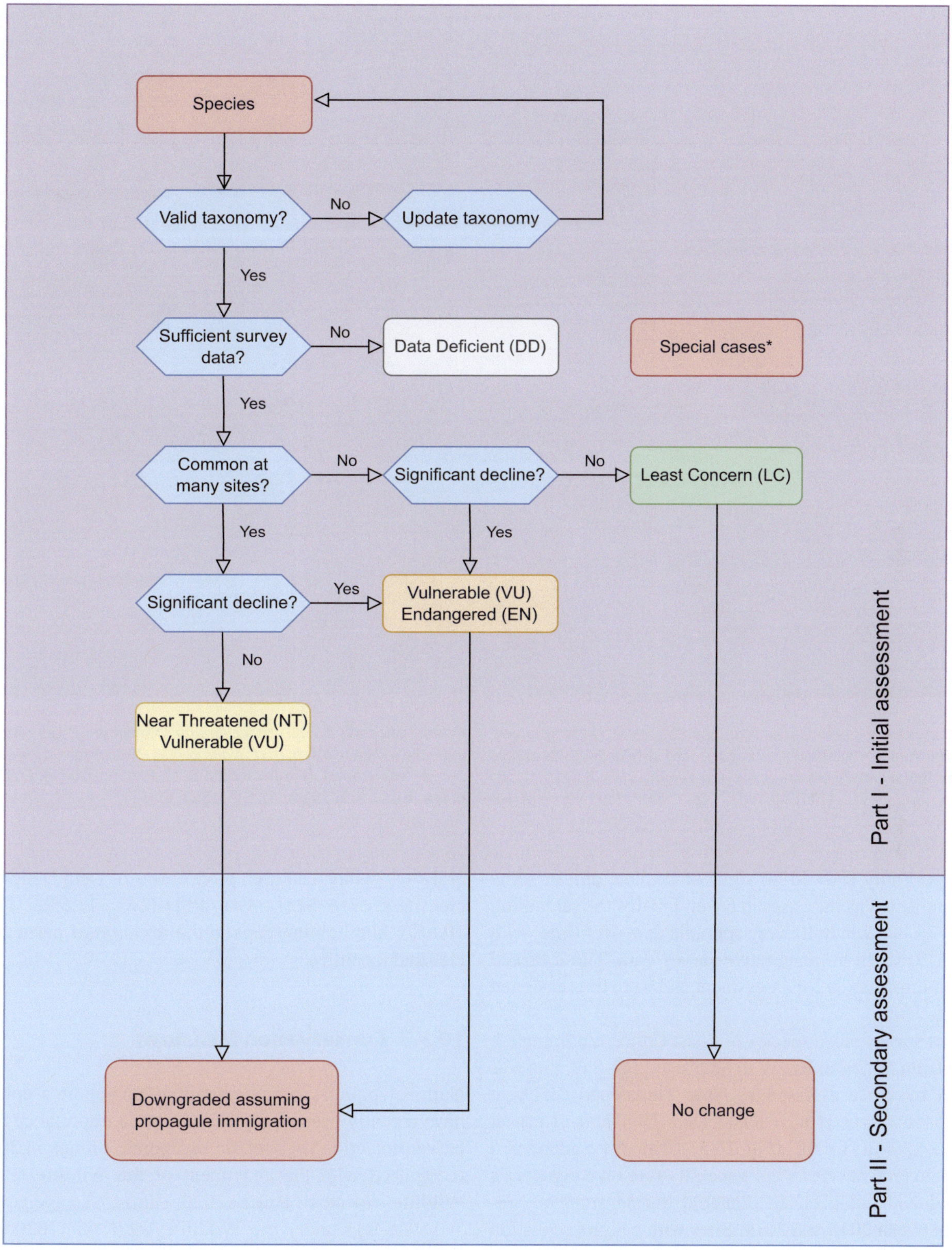

Fig. 10.2 Flowchart for assessing threat statuses of Singapore's corals. Species were first assessed using criteria in Part I (Initial Assessment), followed by Part II (Secondary Assessment). Some species that were not assessable with these guidelines (e.g. population declines observed in the subtidal zone but known to occupy intertidal niches such as reef flats) were treated separately ('special cases') and assigned a threat status based on the co-authors' collective expertise

Fig. 10.3 A selection of threatened corals in Singapore: (**a**) *Seriatopora hystrix* (presumed Nationally Extinct); (**b**) *Stylophora pistillata* (Critically Endangered); (**c**) *Trachyphyllia geoffroyi* (Endangered); (**d**) *Acropora millepora* (Endangered); (**e**) *Fimbriaphyllia divisa* (Vulnerable); (**f**) *Leptoseris explanata* (Vulnerable); (**g**) *Scapophyllia cylindrica* (Vulnerable); (**h**) *Pavona cactus* (Vulnerable); (**i**) *Lobophyllia hataii* (Vulnerable). (Photo credits: R. C. Poquita-Du (**a**), D. Huang (**b**, **e**, **f**, **g**, **h**), C. S. L. Ng (**c**, **i**), S. Q. Sam (**d**))

as they generally showed no signs of decline, and in some cases manifested as increases in cover. Finally, the remaining 134 species had generally very sporadic data over time, with less than 10 years of quantitative survey data. These lacked sufficient information for assessment and were thus assigned Data Deficient.

Across survey sites, species of Least Concern dominated, ranging from 66.7% at RL-W to 86.5% at PSD-SW; this was followed by those assessed as Near Threatened (2.7% at TM-E to 16.1% at HA-NE) and Data Deficient (1.8% at HP-E to 14.3% at HA-SE) (Fig. 10.5). Critically Endangered and Endangered species were generally rare on Singapore's reefs and had not been encountered during surveys conducted between 2015 and 2019. Sites with a high proportion of threatened species (i.e. Vulnerable) such as KU-NE (11.6%) and RL-W (8.5%) tended to also register considerable proportions of Near Threatened species (KU-NE: 12.8%; RL-W: 14.5%). Sites with lower richness but high proportions of threatened corals (e.g. CY-S [10.3%; n = 3 Vulnerable species] and SJ-NW [8.8%; n = 5 Vulnerable species]) may stand to lose even more ecosystem functions in the event of severe environmental impact. Of note were several sites, which had high proportions of Data Deficient species (e.g. HA-SE:14.3%; PSD-NW: 11.5%; TPT-SW: 10.5%), highlighting gaps in the research of particular species and localities.

10.3.3 Conservation Measures

National policies on safeguarding Singapore's coral reefs have recently been strengthened, as the importance of legislative protection for specific taxonomic groups in 2020 was recognised with the enactment of the Wildlife (Protected Wildlife Species) Rules 2020 (https://sso.agc.gov.sg/SL/WA1965-R1) under the Wildlife Act 1965 (2020 Revised Edition) (https://sso.agc.gov.sg/Act/WA1965). Under this Act, all species listed in the Schedule of Protected Wildlife are protected from intentional killing, trapping, taking or keeping in any place without the Director-General's written approval. This includes marine invertebrates under the higher taxon of Scleractinia, as well as Tridacnidae, Antipatharia, Helioporidae, Tubiporidae, Milleporidae and Stylasteridae, among others.

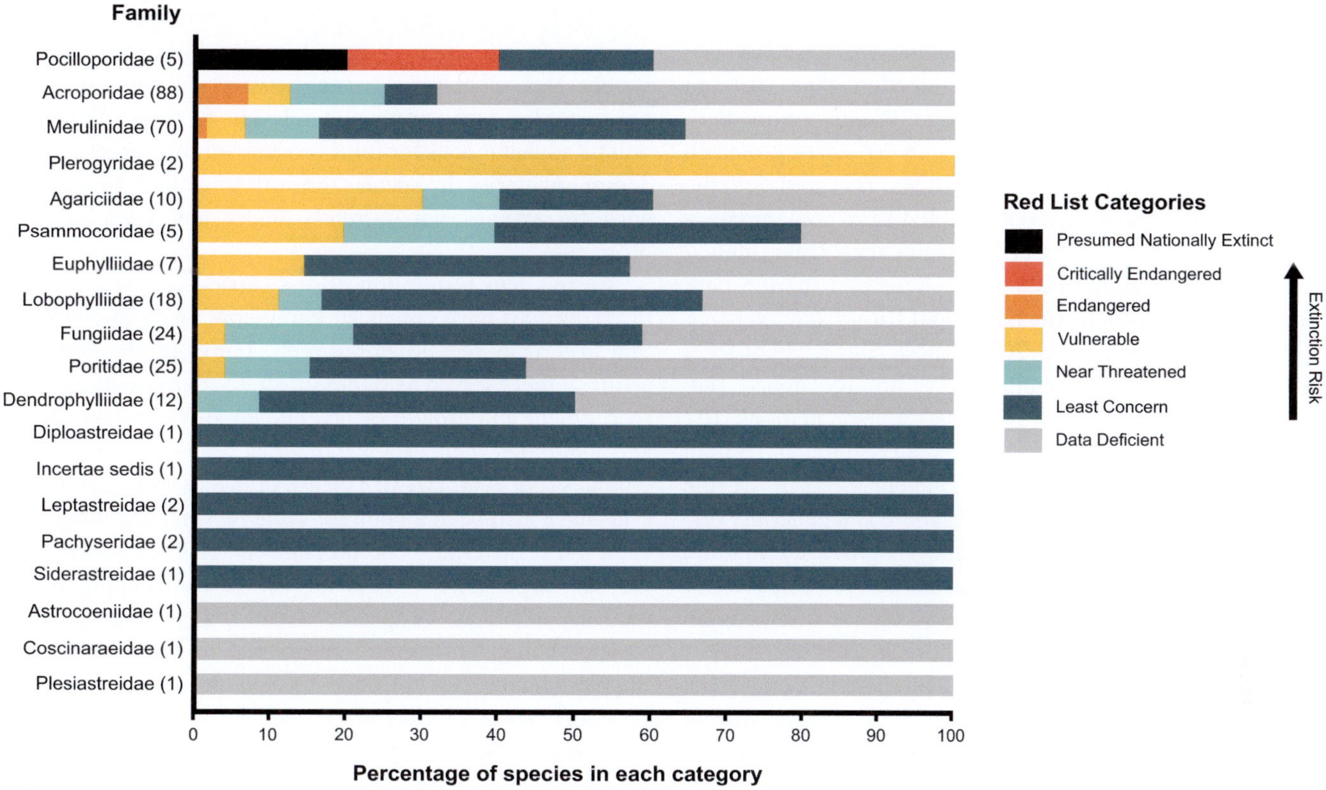

Fig. 10.4 Red List Categories of Singapore's coral species, grouped according to scleractinian families

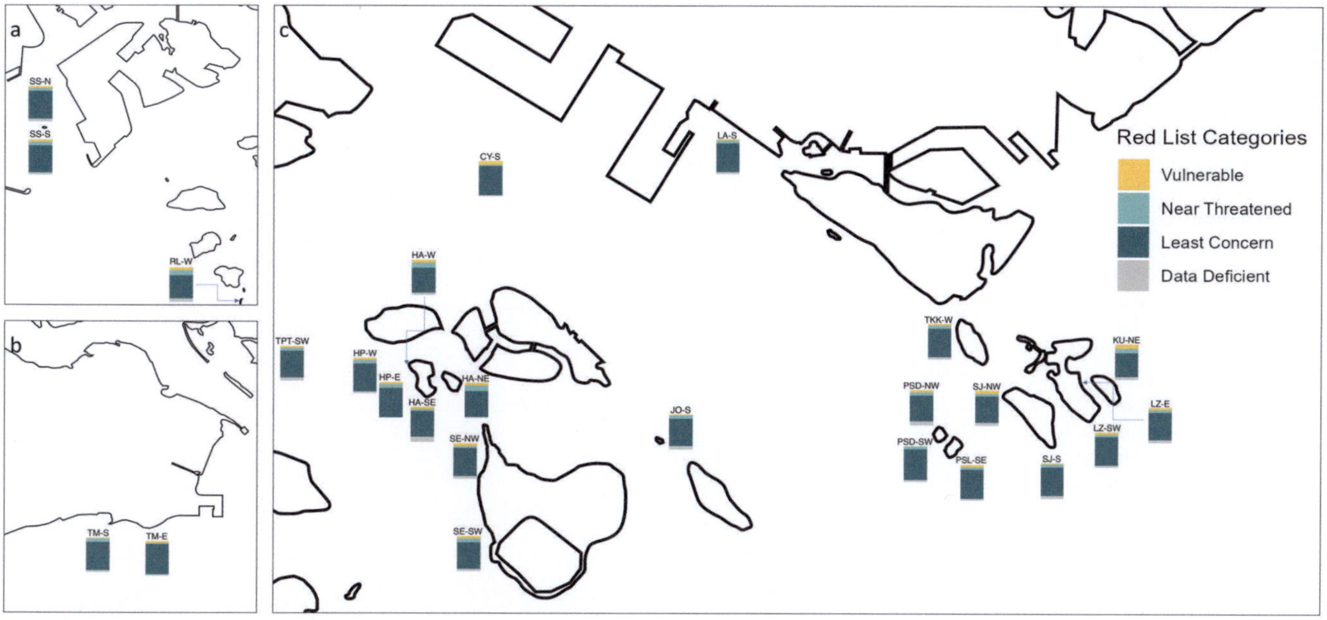

Fig. 10.5 Distribution of threat categories for Singapore's stony corals at (**a**) maritime regulated sites on the southwestern coast, (**b**) sites along the eastern coast and (**c**) all other sites at the southern islands. Information on the sites is available as Supplementary Data (https://doi.org/10.5281/zenodo.13284027)

More specifically, by virtue of their occurrence on certain reefs where access is regulated, several species have also been accorded some level of legal protection (Figs. 10.1 and 10.6). The flora and fauna at the Sisters' Islands Marine Park (SIMP), established as a public park in 2014, are protected under the Parks and Trees Act (Lye 2024). This applies to the 113 species that have been documented from contemporary reef surveys at the SIMP (see Supplementary Data; https://doi.org/10.5281/zenodo.13284027). A second marine park comprising parts of Lazarus Island and Kusu Island is being planned (Begum and Leow 2024); these sites account for a total of 98 species, wherein 11.2% are threatened (i.e. Vulnerable). Similarly, because of maritime regulations, 127 species documented on the reefs at Pulau Satumu and Sultan Shoal—both under the ambit of agencies such as the Maritime and Port Authority of Singapore and generally off-limits to the public—have been indirectly protected. In total, these area-based regulations contribute to the protection of 148 species (53.6% of Singapore's known scleractinian diversity). Overlaps exist between these areas, with 76 scleractinians having been documented in all of them, comprising six Vulnerable (e.g. *Montipora grisea*, *Platygyra lamellina*), six Near Threatened (e.g. *Hydnophora grandis*, *Pectinia lactuca*), 60 Least Concern (e.g. *Porites lutea*, *Pachyseris speciosa*) and four Data Deficient (e.g. *Lobophyllia flabelliformis*, *Plesiastrea peroni*) species (Fig. 10.6). Some other species were also only recorded from individual areas—among those assessed as Vulnerable, for instance, four species each were documented from SIMP (e.g. *Lobophyllia hataii*) and the proposed second Marine Park (e.g. *Scapophyllia cylindrica*), as well as one from the maritime restricted areas (*Pavona cactus*). Seven other species, including *Porites cylindrica* (Vulnerable) and *Heliofungia actiniformis* (Near Threatened), were however only found at sites not accorded legal protection relevant to scleractinians. It will be advantageous for future conservation efforts to ensure complementarity in species prioritisation and address the gaps for species not currently under any form of areal protection.

Various reef restoration efforts carried out over the past three decades have enhanced know-how on coral conservation in Singapore's urbanised reefs. For example, fibreglass artificial reefs served as stable substrates that encouraged the recruitment and growth of juvenile corals (Loh et al. 2006) and facilitated their establishment as sexually mature coral colonies (Ng et al. 2017). The use of in situ and ex situ nurseries have also been a boon for coral rearing, enabling sufficient coral material to be propagated for transplantation at various target sites (Ng et al. 2016), contributing to the recovery of at least 48 species to date and promoting a better understanding of species responses to restoration (see Supplementary Data; https://doi.org/10.5281/zenodo.13284027). More recently, 10 coral species including *Trachyphyllia geoffroyi* (now assessed as Endangered) were also identified for targeted species recovery programmes at the SIMP, with the aim of furthering research and conservation efforts to enhance reef resilience and ecosystem functions such as habitat for other marine organisms (Tan 2021; Davison 2024). The species recovery efforts will be further supported under the recently announced 100k Corals Initiative (Begum 2024c), a whole-of-nation effort led by the National Parks Board (NParks). This programme

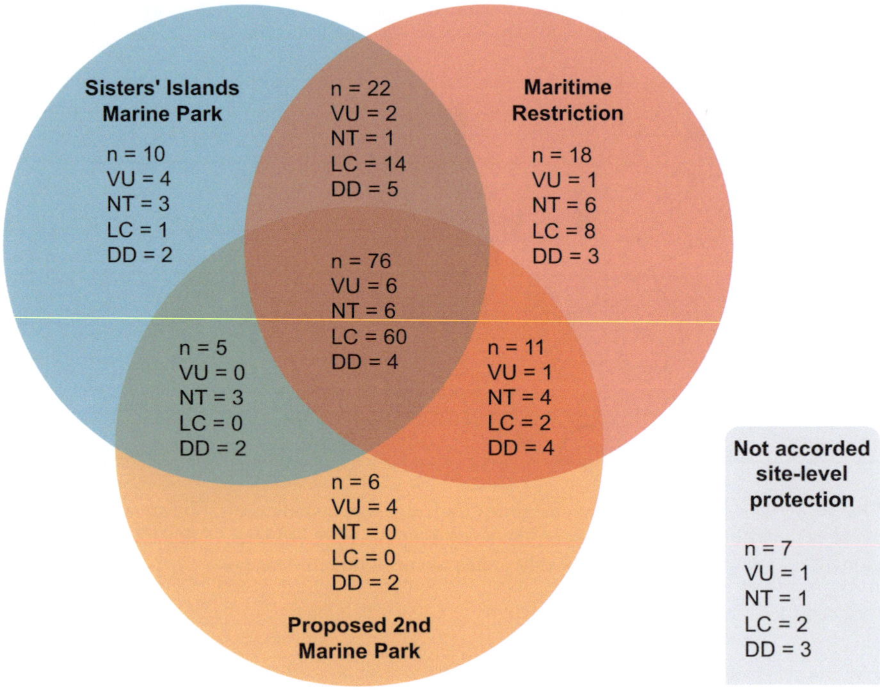

Fig. 10.6 Distribution of species (as documented from surveys conducted in 2015–2019) across different threat categories at sites accorded legal protection—the Sisters' Islands Marine Park, proposed second Marine Park and maritime restricted areas (Pulau Satumu and Sultan Shoal)—and sites without regulations relevant to scleractinian conservation. Species assessed as Endangered and Critically Endangered are not included here, given their limited occurrence on Singapore's reefs (n = total number of species in each subcategory, VU = Vulnerable, NT = Near Threatened, LC = Least Concern, DD = Data Deficient)

aims to transplant 100,000 corals in Singapore's marine environment over the next decade and beyond and represents the country's largest coral reef restoration effort to date. In addition, the upcoming plans for the second Marine Park will contribute towards refining existing restoration strategies and innovating new ones to achieve better conservation outcomes for Singapore's corals.

Cognisant that the current Red List assessment can complement existing coral conservation efforts, we therefore propose the following measures based on the final threat statuses of Singapore's corals. For species assigned as Near Threatened or Vulnerable, we propose developing species-specific conservation plans that are practicable, taking into consideration Singapore's sea space context, and undertaken in consultation with all stakeholders and relevant regulatory agencies, to reduce the risks of an escalation in their threat statuses. Species that are designated Vulnerable and which can be found on artificial structures such as seawalls (e.g. *Pavona cactus*, *Lobophyllia hataii* and *Plerogyra sinuosa*) would benefit from restoration or creation of additional habitat to improve their establishment. Species recovery programmes, including innovative propagation methods that cater to different life history stages, are urgently needed for the eight corals classified as Endangered and Critically Endangered.

However, large research gaps are immediately apparent with nearly half of all recorded species (spanning 13 families) listed as Data Deficient. These have not been studied specifically or recorded adequately via commonly used survey techniques (e.g. line intercept transects) over time to infer population change. Particularly, we note that the majority of the family Acroporidae (60 of 88 species, or 68.1%) are categorised as Data Deficient. This stems from a lack of information for many species since their initial records in the 1980s and 1990s. In addition, taxonomic challenges abound both in identification and species delimitations, especially for *Acropora* spp. as outlined in the recent literature (Cowman et al. 2020; Bridge et al. 2023). As *Acropora* spp. are generally rare on Singapore's reefs (Wong et al. 2018; Chan et al. 2024; Loong et al. 2025), assessing their statuses remains difficult. Yet, with the lack of species-specific information impeding future threat assessments, we strongly highlight the need for increased research focus and a better understanding of these Data Deficient species to inform future conservation planning. This could include modifying reef survey methods to enhance detection of species distribution patterns (e.g. the point count transect method; Roberts et al. 2016) or incorporating information from different reef zones (e.g. species may exhibit varying abundances due to light attenuation with depth; Chow et al. 2019). These can be supplemented with more lab- or field-based studies to understand species-specific responses to stressors (e.g. bleaching susceptibility) or characterise species traits (e.g. growth or carbonate production rates), which are useful proxies of key ecological processes (Ng et al. 2020, 2021). Similarly, citizen scientists can also be equipped with more skills to assist with recording species occurrences (Lim et al. 2020), such as by ensuring that corals observed during intertidal walks or dives are photographed with scale bars and documented at resolutions that are sufficient for identification to at least morpho-species level.

These threat assessments also represent a common source for ensuring that environmental impact assessments (EIAs) by consultants are comparable across sites (Davison 2024). Future EIAs should require species-level assessments to form part of the tender specifications for an accurate assessment of impacts from development. This will be especially important in strengthening the conservation and management of Singapore's reefs in light of the scale of potential human impacts in the coming decades, such as the 'Long Island' project in the East Coast area and the installation of renewable energy farms at Pulau Satumu, among others (Begum 2023; Yap and Tang 2024). It is also essential that the responses of species—especially Data Deficient ones—to coral relocation, an oft-used strategy to mitigate reef biodiversity loss from large scale coastal development, are documented by environmental consultants. Longer term species-level responses to these impacts also need to be identified to refine management strategies that promote reef resilience. Consolidation of such data will enable NParks to develop impactful species recovery programmes and ensure optimal usage of conservation resources.

10.4 Conclusions

While urbanisation may be necessary for the growth of Singapore's economy and population, sustained coastal development inevitably threatens reefs and puts more species under further risk of loss and potential extirpation. Although efforts to enhance the resilience of Singapore's corals to climate change are underway (Chan 2023), this will take time to bear fruit. Importantly, additional safeguards should be put in place to mitigate further environmental damage from impending development; otherwise, the re-emergence of the 87 species unobserved since 2014 may become more unlikely. It is therefore imperative for the current Red List assessment to go beyond that of an academic exercise and be taken seriously to ensure the protection of Singapore's biodiversity. We urge the relevant stakeholders to consider these implications in the light of impacts from intensifying climate change and impending development, or the extirpation of the pocilloporids could very well be the proverbial canary in the coal mine.

Acknowledgements We thank all past and present members of the Reef Ecology Study Team for their tireless efforts in documenting Singapore's coral reefs, and Iffah Iesa for contributing information on the LKCNHM scleractinian records.

References

Begum S (2023) Ensure long island project doesn't harm marine life at east coast, Southern Islands: experts. In: The Straits Times. https://www.straitstimes.com/singapore/ensure-long-island-project-doesn-t-harm-marine-life-at-east-coast-southern-islands-experts Accessed 16 Oct 2024

Begum S (2024a) Coral bleaching in S'pore's shallow areas rises to 40%; scientists on alert for impacts from oil spill. In: The Straits Times. https://www.straitstimes.com/singapore/coral-bleaching-in-shallow-areas-rises-to-40-scientists-on-alert-for-additional-impacts-from-oil-spill Accessed 16 Oct 2024

Begum S (2024b) S'pore waters see slight coral bleaching amid global event. In: The Straits Times. https://www.straitstimes.com/singapore/s-pore-waters-see-slight-coral-bleaching-amid-global-event. Accessed 16 Oct 2024

Begum S (2024c) Singapore's drive to restore 100,000 corals begins with growing coral fragments in HDB-style tanks. In: The Straits Times. https://www.straitstimes.com/singapore/singapores-drive-to-restore-100000-corals-begins-with-baby-corals-grown-in-hdb-style-tanks Accessed 11 Dec 2024

Begum S, Leow WX (2024) S'pore's proposed second marine park in Lazarus South, Kusu is home to rare rocky shore, coral habitats. In: The Straits Times. https://www.straitstimes.com/singapore/s-pore-s-proposed-second-marine-park-in-lazarus-south-kusu-is-home-to-rare-rocky-shore-coral-habitats. Accessed 16 Oct 2024

Bridge TCL, Cowman PF, Quattrini AM, Bonito VE, Sinniger F, Harii S, Head CEI, Hung JY, Halafihi T, Rongo T, Baird AH (2023) A *tenuis* relationship: traditional taxonomy obscures systematics and biogeography of the '*Acropora tenuis*' (Scleractinia: Acroporidae) species complex. Zool J Linn Soc 202(1):zlad062

Browne NK, Precht E, Last KS, Todd PA (2014) Photo-physiological costs associated with acute sediment stress events in three near-shore turbid water corals. Mar Ecol Prog Ser 502:129–143

Browne NK, Tay JK, Low J, Larson O, Todd PA (2015) Fluctuations in coral health of four common inshore reef corals in response to seasonal and anthropogenic changes in water quality. Mar Environ Res 105:39–52

Brüggemann F (1878) Ueber einige Steinkorallen von Singapore. Jahresberichte Naturwissenschaftliger Verein Bremen 3:539–549

Carpenter KE, Abrar M, Aeby G, Aronson RB, Banks S, Bruckner A, Chiriboga A, Cortés J, Delbeek C, DeVantier L, Edgar GJ, Edwards AJ, Fenner D, Guzmán HM, Hoeksema BW, Hodgson G, Johan O, Licuanan WY, Livingstone SR, Lovell ER, Moore JA, Obura DO, Ochavillo D, Polidoro BA, Precht WF, Quibilan MC, Reboton C, Richards ZT, Rogers AD, Sanciangco J, Sheppard A, Sheppard C, Smith S, Stuart S, Turak E, Veron JEN, Wallace C, Weil E, Wood E (2008) One-third of reef-building corals face elevated extinction risk from climate change and local impacts. Science 321(5888):560–563

Chan AM (2023) Two new marine climate projects awarded grants under $25m research programme. In: The Straits Times. https://www.straitstimes.com/singapore/two-new-marine-climate-projects-awarded-grants-under-25m-research-programme. Accessed 18 Oct 2024

Chan YKS, Ng CSL, Tun KPP, Chou LM, Huang D (2024) Decadal decline of functional diversity despite increasing taxonomic and phylogenetic diversity of coral reefs under chronic urbanisation stress. Ecol Indic 164:112143

Chan YKS, Ang ACF, Choo M, Oh RM, Morgan KM, Todd PA, O'Leary MJ, Huang D (2025) Holocene coral assemblages reveal similarities to living communities in Singapore's urban reef environment. Palaeogeogr Palaeoclim Palaeoecol 669:112930

Chou LM, Tan KS (2008) Corals, worms and molluscs (Cnidaria, Platyhelminthes, Nemertea, Annelida, Brachiopoda, Mollusca). In: Davison GWH, Ng PK, Ho HC (eds) The Singapore Red Data book, 2nd edn. Nature Society Singapore, Singapore, pp 39–61

Chou LM, Teo YH (1983) The hard coral community of a *Sargassum*-type reef in Singapore. In: 15th Pacific Science Congress, Dunedin, New Zealand, 1–11 February 1983, pp 1–29

Chou LM, Toh TC, Toh KB, Ng CSL, Cabaitan P, Tun K, Goh E, Afiq-Rosli L, Taira D, Poquita-Du RC, Loke HX, Khalis A, Li J, Song T (2016) Differential response of coral assemblages to thermal stress underscores the complexity in predicting bleaching susceptibility. PLoS One 11(7):e0159755

Chou LM, Huang D, Tan KS, Toh TC, Goh BPL, Tun K (2019) Singapore. In: Sheppard CRC (ed) World seas: an environmental evaluation. Volume II: The Indian Ocean to the Pacific. Academic, London, pp 539–558

Chou LM, Ng CSL, Toh KB, Tun K, Cheo PR, Ng JY (2023) Marinas can co-function as biodiversity sanctuaries. In: Ting DSK, Stagner JA (eds) Nourishing tomorrow: clean engineering and nature-friendly living. World Scientific, Singapore, pp 179–212

Chow GSE, Chan YKS, Jain SS, Huang D (2019) Light limitation selects for depth generalists in urbanised reef coral communities. Mar Environ Res 147:101–112

Chuang SH (1961) On Malayan Shores. Muwu Shosa, Singapore

Cowman PF, Quattrini AM, Bridge TCL, Watkins-Colwell GJ, Fadli N, Grinblat M, Roberts TE, McFadden CS, Miller DJ, Baird AH (2020) An enhanced target-enrichment bait set for Hexacorallia provides phylogenomic resolution of the staghorn corals (Acroporidae) and close relatives. Mol Phylogenet Evol 153:106944

Davison GWH (2024) Applying the Red List categories. In: Davison GWH, Gan JWM, Huang D, Hwang WS, Lum SKY, Yeo DCJ (eds) The Singapore Red Data book: Red Lists of Singapore biodiversity, 3rd edn. National Parks Board, Singapore, pp 1–9

Davison GWH, Ng PK, Ho HC (eds) (2008) The Singapore Red Data book, 2nd edn. Nature Society Singapore, Singapore

English SA, Wilkinson C, Baker VJ (1997) Survey manual for tropical marine resources, 2nd edn. Australian Institute of Marine Science, Townsville

Erftemeijer PL, Riegl B, Hoeksema BW, Todd PA (2012) Environmental impacts of dredging and other sediment disturbances on corals: a review. Mar Pollut Bull 64(9):1737–1765

Guest JR, Tun K, Low J, Vergés A, Marzinelli EM, Campbell AH, Bauman AG, Feary DA, Chou LM, Steinberg PD (2016) 27 years of benthic and coral community dynamics on turbid, highly urbanised reefs off Singapore. Sci Rep 6:36260

Gutierrez L, Polidoro B, Obura D, Cabada-Blanco F, Pettersson E, Pearce-Kelly P, Kemppinen K, Linardich C, Alvarado JJ, Alvarez-Filip L, Banaszak AT, Casado de Amezua P, Crabbe MJC, Croquer A, Feingold J, Goergen L, Goffredo S, Hoeksema BW, Huang D, Kennedy E, Kitahara MV, Kružić P, Miller M, Nunes F, Quimbayo JP, Rivera-Sosa A, Rodríguez-Martínez R, Santodomingo N, Sweet MJ, Vermeij M, Villamizar E, Aeby G, Alliji K, Couce E, Cowburn B, Porter S, Shlesinger T, Bayley D, Wilson B, Nuñez Lendo CI, Samimi-Namin K (2024) Half of Atlantic reef-building corals at elevated risk of extinction due to climate change and other threats. PLoS One 19(11):e0309354

Heery EC, Hoeksema BW, Browne NK, Reimer JD, Ang PO, Huang D, Friess DA, Chou LM, Loke LHL, Saksena-Taylor P, Alsagoff N, Yeemin T, Sutthacheep M, Vo ST, Bos AR, Gumanao GS, Hussein MAS, Waheed Z, Lane DJW, Johan O, Kunzmann A, Jompa J, Suharsono TD, Bauman AG, Todd PA (2018) Urban coral reefs:

degradation and resilience of hard coral assemblages in coastal cities of East and Southeast Asia. Mar Pollut Bull 135:654–681

Huang D, Tun KPP, Chou LM, Todd PA (2009) An inventory of zooxanthellate scleractinian corals in Singapore, including 33 new records. Raffles B Zool S22:69–80

Huang D, Licuanan WY, Hoeksema BW, Chen CA, Ang PO, Huang H, Lane DJW, Vo ST, Waheed Z, Affendi YA, Yeemin T, Chou LM (2015) Extraordinary diversity of reef corals in the South China Sea. Mar Biodivers 45:157–168

Ip YCA, Tay YC, Gan SX, Ang HP, Tun K, Chou LM, Huang D, Meier R (2019) From marine park to future genomic observatory? Enhancing marine biodiversity assessments using a biocode approach. Biodivers Data J 7:e46833

Ip YCA, Tay YC, Chang JJM, Ang HP, Tun KPP, Chou LM, Huang D, Meier R (2021) Seeking life in sedimented waters: environmental DNA from diverse habitat types reveals ecologically significant species in a tropical marine environment. Environ DNA 3:654–668

IUCN (2012) In: IUCN Red List Categories and Criteria: version 3.1, 2nd edn. IUCN, Gland/and Cambridge. iv + 32pp. https://portals.iucn.org/library/sites/library/files/documents/RL-2001-001-2nd.pdf. Accessed 17 Oct 2024

Jaafar Z, Huang D, Tanzil JTI, Ow YX, Yap NWL (2018) The Singapore Blue Plan 2018. Singapore Institute of Biology, Singapore, p 219. Available from: http://singaporeblueplan2018.blogspot.com. Accessed 17 Oct 2024

Junjie RK, Browne NK, Erftemeijer PL, Todd PA (2014) Impacts of sediments on coral energetics: partitioning the effects of turbidity and settling particles. PLoS One 9(9):e107195

Kikuzawa YP, Ng CSL, Toh TC, Sam SQ, Lee YL, Loo PL, Chua YZ, Tan KS, Chou LM (2020) Diversity of subtidal benthic and hard coral communities on sloping and vertical seawalls in Singapore. Mar Biodivers 50:95

Lai S, Loke LHL, Hilton MJ, Bouma TJ, Todd PA (2015) The effects of urbanisation on coastal habitats and the potential for ecological engineering: a Singapore case study. Ocean Coast Manage 103:78–85

Law MT, Huang D (2023) Light limitation and coral mortality in urbanised reef communities due to sea-level rise. Clim Change Ecol 5:100073

Lim LJW, Loh JBY, Lim AJS, Tan BYX, Ip YCA, Neo ML, Tan R, Huang D (2020) Diversity and distribution of intertidal marine species in Singapore. Raffles Bull Zool 68:396–403

Loh T-L, Tanzil JTI, Chou LM (2006) Preliminary study of community development and scleractinian recruitment on fibreglass artificial reef units in the sedimented waters of Singapore. Aquat Conserv Mar Freshw Ecosyst 16:61–76

Loo MGK, Chou LM (1995) Environmental status of Singapore's coral reefs. In: Second Workshop of the EAS-35 Project: Integrated Management of Watersheds in Relation to the Coastal and Marine Environment, Bangkok, Thailand, 3–7 April 1995, pp 1–20

Loong S, Taira D, Chan YKS, Chan WWR, Fong J, Kwan V, Low JKY, Ng CSL, Sam SQ, Tan CZ, Tun KPP, Yip ZT, Jaafar Z, Huang D (2025) Covariations of coral, macroalgal, macroinvertebrate and fish communities in an urbanised tropical reef environment. Coral Reefs. https://doi.org/10.1007/s00338-025-02723-9

Lye LH (2024) Restoring Singapore's biodiversity: from a Garden City to a City in Nature—Laws, policies and environmental governance. In: Davison GWH, Gan JWM, Huang D, Hwang WS, Lum SKY, Yeo DCJ (eds) The Singapore Red Data book: Red Lists of Singapore biodiversity, 3rd edn. National Parks Board, Singapore, pp 10–18

Ng KG, Begum S (2024) Long Island to be reclaimed off East Coast could add 800ha of land, create Singapore's 18th reservoir. In: The Straits Times. https://www.straitstimes.com/singapore/long-island-to-be-reclaimed-off-east-coast-could-add-800ha-of-land-and-singapore-s-18th-reservoir. Accessed 16 Oct 2024

Ng CSL, Chen D, Chou LM (2012) Hard coral assemblages on seawalls in Singapore. In: Tan KS (ed) Contributions to marine science. Tropical Marine Science Institute, National University of Singapore, Singapore, pp 75–79

Ng CSL, Toh TC, Chou LM (2016) Coral restoration in Singapore's sediment-challenged sea. Reg Stud Mar Sci 8:422–429

Ng CSL, Toh TC, Chou LM (2017) Artificial reefs as a reef restoration strategy in sediment-affected environments: insights from long-term monitoring. Aquat Conserv Mar Freshw Ecosyst 27:976–985

Ng CSL, Jain SS, Nguyen NTH, Sam SQ, Kikuzawa YP, Chou LM, Huang D (2019) New genus and species record of reef coral *Micromussa amakusensis* in the southern South China Sea. Mar Biodivers Rec 12:17

Ng CSL, Huang D, Toh KB, Sam SQ, Kikuzawa YP, Toh TC, Taira D, Chan YKS, Hung LZT, Sim WT, Rashid AR, Afiq-Rosli L, Ng NK, Chou LM (2020) Responses of urban reef corals during the 2016 mass bleaching event. Mar Pollut Bull 154:111111

Ng CSL, Chan YKS, Nguyen NTH, Kikuzawa YP, Sam SQ, Toh TC, Mock AYJ, Chou LM, Huang D (2021) Coral community composition and carbonate production in an urbanized seascape. Mar Environ Res 168:105322

Poquita-Du RC, Quek ZBR, Jain SS, Schmidt-Roach S, Tun K, Heery EC, Chou LM, Todd PA, Huang D (2019) Last species standing: loss of Pocilloporidae corals associated with coastal urbanization in a tropical city state. Mar Biodivers 49:1727–1741

Prasetia R, Lim ZW, Teo A, Shlesinger T, Loya Y, Todd PA (2020) Population dynamics and growth rates of free-living mushroom corals (Scleractinia: Fungiidae) in the sediment-stressed reefs of Singapore. Adv Mar Biol 87(1):115–140

Reimer JD, Peixoto RS, Davies SW, Traylor-Knowles N, Short ML, Cabral-Tena RA, Burt JA, Pessoa I, Banaszak AT, Winters RS, Moore T, Schoepf V, Kaullysing D, Calderon-Aguilera LE, Wörheide G, Harding S, Munbodhe V, Mayfield A, Ainsworth T, Vardi T, Eakin CM, Pratchett MS, Voolstra CR (2024) The fourth global coral bleaching event: where do we go from here? Coral Reefs 43:1121–1125

Roberts TE, Bridge TC, Caley MJ, Baird AH (2016) The point count transect method for estimates of biodiversity on coral reefs: improving the sampling of rare species. PLoS One 11(3):e0152335

Schoepf V, Baumann JH, Barshis DJ, Browne NK, Camp EF, Comeau S, Cornwall CE, Guzman HM, Riegl B, Rodolfo-Metalpa R, Sommer B (2023) Corals at the edge of environmental limits: a new conceptual framework to re-define marginal and extreme coral communities. Sci Total Environ 884:163688

Shahbudin S, Akmal KMF, Faris S, Normawaty M-N, Mukai Y (2017) Current status of coral reefs in Tioman Island, Peninsular Malaysia. Turk J Zool 41:294–305

Studer T (1881) Beitrag zur Fauna der Steinkorallen von Singapore. Mitteilungen der Naturforschenden Gesellschaft in Bern [aus dem Jahre] 1880:15–53

Tan A (2021) NParks adds 10 hard coral species on front-lines of global warming to recovery programme. In: The Straits Times https://www.straitstimes.com/singapore/environment/nparks-adds-10-hard-coral-species-on-frontlines-of-global-warming-to-recovery. Accessed 17 Oct 2024

Tan YJ, Tham JK, Paul A, Rana U, Ang HP, Nguyen NT, Yee ATK, Leong BPI, Drummond S, Tun KPP (2023) Remote sensing mapping of the regeneration of coastal natural habitats in Singapore: implications for marine conservation in tropical cities. Singap J Trop Geogr 44:130–148

Tay YC, Todd PA, Rosshaug PS, Chou LM (2012) Simulating the transport of broadcast coral larvae among the Southern Islands of Singapore. Aquat Biol 15:283–297

Tay YC, Noreen AME, Suharsono CLM, Todd PA (2015) Genetic connectivity of the broadcast spawning reef coral *Platygyra sinensis* on impacted reefs, and the description of new microsatellite markers. Coral Reefs 34:301–311

Toh TC, Huang D, Chan YKS, Ng CSL, Tun K, Chou LM (2022) Singapore. In: Kimura T, Chou LM, Huang D, Tun K, Goh E (eds) Status and trends of East Asian coral reefs: 1983–2019. Global Coral Reef Monitoring Network, East Asia Region, pp 174–179

Tun KPP (2012) Optimisation of reef survey methods and application of reef metrics and biocriteria for the monitoring of sediment impacted reefs. Dissertation, National University of Singapore

Wong JSY, Chan YKS, Ng CSL, Tun KPP, Darling ES, Huang D (2018) Comparing patterns of taxonomic, functional and phylogenetic diversity in reef coral communities. Coral Reefs 37:737–750

WoRMS (2024) Scleractinia. https://www.marinespecies.org/aphia.php?p=taxdetails&id=1363. Accessed 27 Aug 2024

Yap J, Tang L (2024) Singapore looks to build first-of-its-kind renewable energy farm in waters around Raffles Lighthouse. In: Channel News Asia https://www.channelnewsasia.com/singapore/raffles-lighthouse-waters-build-renewable-energy-solar-tidal-farm-4537406. Accessed 17 Oct 2024

Biological Connectivity and Conservation of Singapore's Urbanised Seascape

11

Jing Hui Ong, Benjamin J. Wainwright, and Danwei Huang

Abstract

The integration of data on biological connectivity between different sites and habitats in spatial conservation planning (SCP) has become increasingly important as conservation moves beyond focusing on species representation. This chapter describes connectivity patterns across Singapore's seascape inferred from multiple marine metazoans, ranging from corals to fishes, with comparisons to studies of similar species in the region. Hydrodynamic patterns have been found to drive genetic connectivity patterns in Singapore to a degree, with the strength of this influence associated with life history traits such as having a planktonic life stage. Migration patterns are species-specific, and where they cannot be entirely explained by current hydrodynamics, they may be a result of other traits like larval behaviour or adult mobility. Future research with integrative analyses looking into the combined effects of hydrodynamics, life history, behavioural traits and other such biological factors across different life stages will provide a more holistic insight into the drivers of realised connectivity. Incorporating genetic connectivity in SCP reveals the importance of accounting for connectivity in conservation decisions, where an important source site with no at-risk habitats would have been overlooked when preserving lower proportions of habitat area. With gene flow occurring over evolutionary time, the integration of genetic connectivity into conservation planning and management, such as in SCP or in strategies such as resilience enhancement projects, can enhance conservation effectiveness for the long term.

Keywords

Single-nucleotide polymorphisms · Population genomics · Landscape ecology · Tropical ecology · Spatial conservation planning · Urban ecosystems

J. H. Ong (✉)
Department of Biological Sciences, National University of Singapore, Singapore, Singapore

Tropical Marine Science Institute, National University of Singapore, Singapore, Singapore
e-mail: ong.jinghui@u.nus.edu

B. J. Wainwright
Department of Biological Sciences, National University of Singapore, Singapore, Singapore

Yale-NUS College, National University of Singapore, Singapore, Singapore

D. Huang
Department of Biological Sciences, National University of Singapore, Singapore, Singapore

Tropical Marine Science Institute, National University of Singapore, Singapore, Singapore

Lee Kong Chian Natural History Museum, National University of Singapore, Singapore, Singapore

Centre for Nature-based Climate Solutions, National University of Singapore, Singapore, Singapore

11.1 Introduction

Tropical marine ecosystems are spatially heterogenous and comprise different habitats such as coral reefs, mangroves, seagrass beds and rocky shores across the vast seascape. Within and between these different habitats, organic and inorganic materials are constantly being transported through physical, biological and chemical processes. The dynamic nature of such fluxes creates complexity in understanding and predicting ecological connectivity across the seascape, even more so when accounting for different spatial and temporal scales.

One of the most direct ways in which different localities are connected is through the physical movement of organisms, but there are numerous, varied reasons why and how organisms move about. For example, ontogenetic niche shifts (Werner and Gilliam 1984) describe how species can have different patterns of resource use (e.g. diet, habitat) at

L. M. Chou, D. Huang (eds.), *Coral Reefs of Singapore's Urbanised Sea*, Coral Reefs of the World 23,
https://doi.org/10.1007/978-981-95-0903-4_11

different life stages, such as how coral reef fishes may reside in nearshore habitats (e.g. seagrass beds and mangroves) as juveniles to minimise predation risk and mortality (Grol et al. 2011) but move to coral reefs in deeper waters as adults for resources such as larger prey in greater abundance (Kimirei et al. 2013). In other cases, factors such as temperature, seasonality and resource availability drive the migration of marine megafauna over long distances to areas for specific behaviours (e.g. mating, breeding and feeding), as exemplified by the migratory patterns of whales (Acevedo et al. 2017; Meynecke et al. 2021) and sharks (Bansemer and Bennett 2011). Aside from the active movement of marine organisms, they can also be passively transported by wave action and ocean currents. This is especially so for the dispersal of marine plants (e.g. seagrass and mangrove trees (Ruiz-Montoya et al. 2012; Lai et al. 2020)), as well as animals that have at least one planktonic life stage (Cowen and Sponaugle 2009; Koehl and Hadfield 2010). Hydrodynamic patterns are driven by the multifaceted interactions between bathymetry and topography of the marine environment, wind, tidal motions and oceanic currents (Lambrechts et al. 2008) across different spatial and temporal scales.

Ontogenetic migration between habitats is poorly understood in Singapore's waters. For example, comprehensive surveys of the coastal environments for the sand-sifting sea star *Archaster typicus* showed that individuals were larger in Singapore than the rest of the Central Indo-Pacific region (Chan et al. 2018). Interestingly, no individuals smaller than 28 mm were uncovered, and no juveniles were recorded in mangrove and seagrass habitats, which are known to be nursery grounds for the species elsewhere.

With the physical movement of organisms, seeds/propagules or gonads (as in the case of spawning organisms that release egg or sperm bundles directly into the water) between different locations, genetic material is also transported. Genetic connectivity, defined as 'the degree to which gene flow affects evolutionary processes within subpopulations' (Lowe and Allendorf 2010), affects ecological processes such as gene flow (Toonen et al. 2011; Afiq-Rosli et al. 2021) and is believed to contribute to population persistence as well as resilience and recovery against disturbances (Zubillaga et al. 2008; Bonin et al. 2016; Padrón et al. 2018; McManus et al. 2021). Understanding how the transport of genetic material is facilitated or hindered across the seascape will help form the groundwork for assessments of vulnerability to disturbances and the evolutionary rescue potential of populations, leading to more effective conservation strategies.

Another dimension of ecological connectivity is microbial connectivity, which has not been explored and examined extensively. Microbial connectivity is important because marine microbes are strongly associated with the health and functioning of the planet, with major biogeochemical fluxes being mediated by marine microbes (Falkowski et al. 2008). For instance, marine phytoplankton, in particular, is responsible for half of the planetary primary production and mediating the biological carbon pump, among other functions (Field et al. 1998; Basu and Mackey 2018). Research on sinking microbial particles has uncovered the functional vertical connectivity between surface waters and the deep sea (Mestre et al. 2018; Chen et al. 2024), while community and network analysis found recurrent seasonal dynamics in temperate zones (Comte et al. 2018; Deutschmann et al. 2023). In the Atlantic Ocean, patterns of mutual exclusion in communities of free-living bacterioplankton and micro-eukaryotes were also found to play an important role in shaping marine bacterial community compositions (Milici et al. 2016). Microbial connectivity can also provide insight into how easily or quickly bacterial pathogens can spread, and research on such patterns is often done through biophysical modelling (Kough et al. 2015; Cantrell et al. 2020).

11.2 Connectivity in Singapore

11.2.1 Patterns of Circulation

Singapore is situated close to the equator between the Pacific Ocean to the east and the Indian Ocean to the west at the southern tip of the Malay Peninsula. The Johor Strait lies between Singapore and Malaysia and is further divided into separate water bodies (east and west) by the Johor-Singapore Causeway, with no mixing between the two. To the south of Singapore, the Singapore Strait separates Singapore from the Riau Islands of Indonesia. The Johor River of Malaysia drains into the eastern edge of both the Johor and Singapore Straits.

Tidal action and major currents driven by the trade wind system, combined with the seasonal monsoon influences, are the main drivers of hydrodynamic patterns in Singapore's waters (Chen et al. 2005; Sin et al. 2016). Singapore is connected to the Pacific Ocean via the South China Sea and to the Andaman Sea of the Indian Ocean via the Malacca Strait, while water from the Malacca and Singapore Straits drains into the Java Sea to the south (Sin et al. 2016). Currents in the Singapore Strait generally flow from the east to the west and become stronger during the Northeast Monsoon (December to March), while the Southwest Monsoon (May to September) reduces the westbound throughflow (Sin et al. 2016); intense storms in the South China Sea during these months can even result in a short-term net eastward current flow (Kurniawan et al. 2015).

11.2.2 Coral Reefs and Beyond

Coral reefs are richly biodiverse, and Singapore's reefs are no exception. As a system, the state of individual species present around reefs can affect the overall health and functioning of the reefs, thus necessitating the understanding of the population structure of a variety of organisms, as well as the barriers or corridors to genetic connectivity in Singapore's seascape, for a more nuanced approach to conservation planning. In the following, we summarise findings from the past decade on a broad range of marine metazoan species in Singapore, with comparisons to the same or similar species in the region. Unless otherwise stated, analyses were performed on genomic data—specifically, single-nucleotide polymorphisms obtained by double-digest RADseq (Peterson et al. 2012).

11.2.3 Corals

Corals in Singapore's waters generally form panmictic populations, with some exhibiting some degree of substructuring. Microsatellite analysis of the brain coral (*Platygyra sinensis*) revealed three potential genetic lineages present in the Southern Islands of Singapore, but low genetic structuring was detected (Tay et al. 2015). This was attributed to the coral's high reproductive capacity and its reproductive strategy as a broadcast-spawning coral that disperses its gametes into the water, allowing it to disperse effectively.

With the recent application of high-throughput sequencing, nextRAD genotyping-by-sequencing of samples morphologically identified as the pore coral (*Porites* sp.), another broadcast spawner, showed three genetic lineages with subtle fine-scale population differentiation (Afiq-Rosli et al. 2021). Colonies of *Porites* sp. from the inshore reefs of Tanah Merah and Sultan Shoal (see Fig. 11.1) were primarily genetically assigned to the Blue cluster, while central-offshore reefs (Pulau Semakau and Pulau Jong) were assigned to the Orange cluster; colonies from the peripheral-offshore sites (Kusu Island, Pulau Satumu, Sisters' Islands and Pulau Hantu) had genetic assignments from both Blue and Orange clusters. This distribution pattern was attributed to contemporary surface current regimes, where the oscillation of residual surface currents in the Singapore Strait during the spawning and post-spawning months of April to June (Sin et al. 2016) leads to higher retention of propagules within the central-offshore sites and a weak barrier between the inshore and central-offshore sites. The Blue cluster could be further separated into two subclusters (Dark Blue and Light Blue), though there was no evident structuring when comparing both Blue subclusters. Each of the three clusters was found to be a panmictic population. Self-recruitment was found to be high at all sites for all three clusters (~60%–80% of nonmigrants), and evidence of contemporary migration among the inshore sites for the Blue cluster was found. Kusu Island was also revealed to be an important source site. The contrasting result of the *Porites* sp. with the previous study on *Platygyra sinensis* (Tay et al. 2015), which had revealed a highly connected, panmictic population, was thought to be because of the limited resolution of microsatellite markers used in the latter, or because they had different reproductive traits (*Porites* sp. is gonochoric, but *Platygyra sinensis* is a hermaphrodite).

In the same analysis for the cauliflower coral (*Pocillopora acuta*) (Afiq-Rosli et al. 2021), a brooding coral, where a limited dispersal range might have been expected, genetic differentiation was instead found only between Pulau Satumu and the other islands. When excluding Satumu, all other sites formed a panmictic population with high degrees of self-recruitment as well as migrant exchange. As with the *Porites* corals, Kusu Island was also an important source site for *Pocillopora acuta*. These results were attributed to *Pocillopora acuta* having a longer pelagic larval duration than *Porites* sp., with *Pocillopora acuta* larvae being able to survive twice as long in its pelagic state compared with those of *Porites* sp. (Richmond 1987; Polato et al. 2010).

In the analysis of the coral species complex, the castle coral (*Pachyseris speciosa*), another broadcast spawning gonochore, nextRAD genotyping-by-sequencing showed the samples forming two genetic lineages (Blue and Green) that were not site-specific (Feldman et al. 2022). The Green lineage could also further form two subclusters (Dark Green and Light Green). Colonies from the Blue lineage formed a single panmictic population, with no evidence for geographic structuring, while for the Green lineage, St. John's Island comprised exclusively of colonies from the Dark Green genetic cluster, while Pulau Semakau and Terumbu Pempang Tengah had colonies mainly from the Light Green genetic cluster. For both Blue and Green lineages, a dispersal corridor at the southern periphery of the islands along the Singapore Strait was identified, but connectivity was found mainly between the eastern sites and Pulau Satumu and Pulau Hantu, with most other islands being isolated. While it is not clear why such population structure patterns exist for *Pachyseris speciosa* in Singapore, it is hypothesised that variations in the timing of gamete release, combined with the current regimes at the time of spawning, may be important drivers.

11.2.4 Sea Anemones and Symbionts

Sea anemones commonly form symbiotic relationships with other organisms. In Singapore, *Entacmaea quadricolor* (bubble-tipped anemone) is typically found in subtidal coral reefs and forms a symbiotic relationship with *Amphiprion*

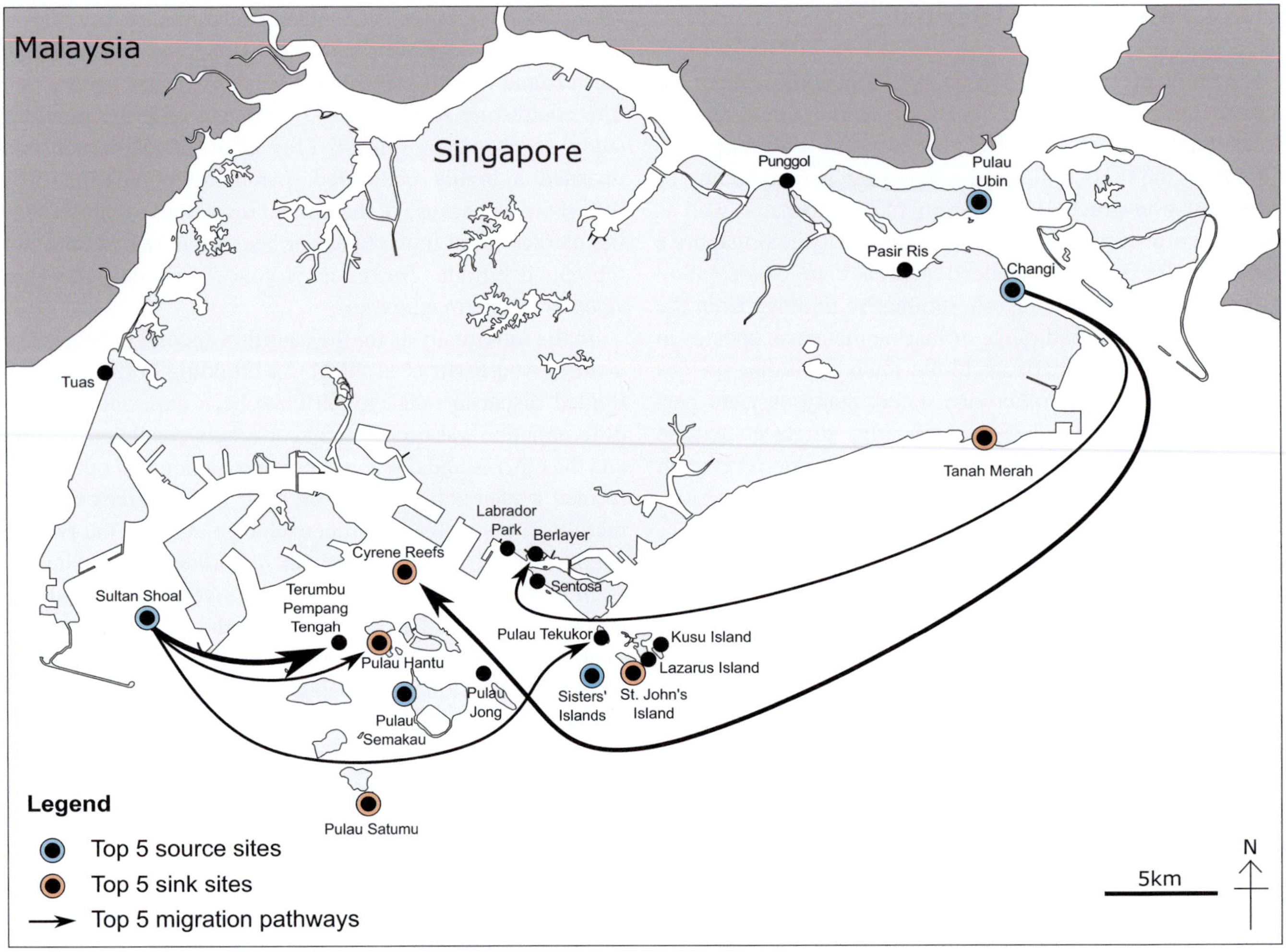

Fig. 11.1 Map of Singapore showing sites with data on genetic connectivity and the high migration pathways. Sites with a blue outline are key source sites, while sites with an orange outline are key sink sites in Singapore's seascape. The top five migration pathways based on aggregated contemporary migration rates are depicted by the arrows, with the width of each arrow being proportional to the migration rate and the direction of the arrow showing the direction of migration

frenatus (tomato clownfish), while *Stichodactyla haddoni* (Haddon's carpet anemone) mainly inhabits seagrass beds and sandy substrate in the intertidal zone and forms a symbiont pair with the anemone shrimp *Periclimenes* cf. *obscurus* (Richardson et al. 1997b; Chan et al. 2020; Pwa and Park 2020). The sea anemones, both of which are broadcast spawners, exhibited high genetic diversity (Chan et al. 2020) and all four species formed panmictic populations in Singapore (Chan et al. 2020).

No obvious genetic structuring was identified for *Periclimenes* cf. *obscurus* (ovigerous (Pwa and Park 2020)), while *Amphiprion frenatus* (a benthic spawner (Roux et al. 2019)) formed two genetic clusters, though not because of geographical structuring (Pwa 2016). Analysis of contemporary migration patterns revealed that the main source site for *Entacmaea quadricolor* was Sisters' Islands (south of mainland Singapore) while the source site of its symbiont *Amphiprion frenatus* was Sultan Shoal (southwest of mainland Singapore). For *Stichodactyla haddoni* and *Periclimenes* cf. *obscurus*, their main source sites were both located in northeast Singapore at Changi Beach and Pulau Ubin, respectively.

Contrasting with the apparent non-geographic structuring of the *Amphiprion frenatus* population in Singapore, studies of the genetic population structure of the false clown anemonefish (*Amphiprion ocellaris*) in Indonesia using six microsatellite loci revealed low but significant population structuring in regions where sites were 10–1000 km apart, with structuring becoming more evident when examining the entire Indo-Malay Archipelago (Timm et al. 2012). Extending this analysis, Timm et al. (2017) examined the mitochondrial control region and six microsatellite loci of *Amphiprion ocellaris* in the Spermonde Archipelago located off the southwest coast of Sulawesi (distance between sampling sites: 10–200 km apart), and hierarchical AMOVA analysis found extensive geographical subdivisions within the Archipelago, with differentiation among proximate sites increasing when moving southward and shoreward. These results concurred with their findings from their analysis on the sea squirt *Polycarpa aurata* in the same study, suggesting local bathymetry and current features were driving connectivity patterns; the deeper sea basin in the north was subjected to strong currents from the Indonesian Throughflow, while the shallower shelf in the south and close to shore experienced more localised mixing. Timm et al. (2017) also found that larval behaviour can influence realised connectivity patterns; the higher degree of connectivity in the sea squirt than the false clown anemonefish was attributed to the more passive planktonic drifting of sea squirt larvae in contrast to the ability of anemonefish larvae to actively swim towards suitable habitat, particularly their natal reefs (Jones et al. 2005).

At a greater spatial scale, similar to the analysis of *Amphiprion ocellaris* in the Spermonde Archipelago, the low but significant population structure around Sulawesi (distance between sites: 40–1100 km) was attributed to major current patterns around the island, with the strong Indonesian Throughflow along the western coastline facilitating connectivity (Timm et al. 2012).

11.2.5 Echinoderms

Protoreaster nodosus (knobbly sea star) is a broadcast spawner and can be found not only in coral reefs but also seagrass and shallow sandy intertidal habitats in Singapore (Bos et al. 2008; Tay et al. 2016). It is considered locally Endangered in Singapore because of land reclamation as well as the potential threat of illegal collection for aquaria or as curio, with its local persistence deemed unlikely without habitat protection (Davison et al. 2024). Genomic data for *Protoreaster nodosus* revealed high genetic diversity and low genetic structure (Tay et al. 2016), but with high levels of self-recruitment in the three sites it was sampled from (Ubin, Semakau, Cyrene Reefs) accompanied by limited contemporary migration (J. H. Ong, unpublished data).

Similarly, in the northwest Pacific, *Protoreaster nodosus* was found to have high genetic diversity in the Ryukyu Archipelago upon microsatellite analysis. Across a spatial scale spanning over 2500 km, *Protoreaster nodosus* also has low genetic structure when comparing between the Ryukyu Archipelago (Japan) and the Philippines, but there was enough genetic differentiation between sites to observe an isolation-by-distance effect (Nakajima et al. 2024). Currents facilitating larval dispersal were thought to contribute to the few genetic barriers between the Philippines and the Ryukyu Archipelago.

Like *Protoreaster nodosus, Meridiastra calcar* (carpet sea star, endemic to Australia) is another broadcast spawning sea star, while, contrastingly, *Parvulastra exigua* (dwarf cushion star) lays benthic eggs (Byrne 2006). Using a combination of mitochondrial sequences as well as microsatellites, a comparative analysis of these sea stars that exist in sympatry in Australia showed that *Meridiastra calcar* had lower levels of genetic structure than *Parvulastra exigua* (Puritz et al. 2017), demonstrating how reproductive life histories can contribute to different population structuring patterns.

11.2.6 Molluscs

Peronia verruculata (onch slug) inhabits the intertidal zone and their eggs hatch into free-swimming larvae (Deshpande et al. 1980). In Singapore, this species has been found to comprise cryptic species complexes (Chang et al. 2018). Upon analysis of the most abundant species, Pulau Ubin was found to be a vital source site for *Peronia verruculata*; hydrodynamics is possibly an important driver for this connectivity pattern because the onch slug is likely to spawn during the Northeast Monsoon period, when the Singapore Strait has a strong westward flow (J. H. Ong, unpublished data).

Littoraria pallescens (the periwinkle snail) is a mangrove species that lays pelagic eggs which then hatch into planktotrophic veliger larvae (Reid 1985; Torres et al. 2008; Reid et al. 2010). Like *Peronia verruculata*, the breeding seasonality of *Littoraria pallescens* in Singapore is not fully understood, but its spawning periods are thought to coincide with periods of both the Southwest and Northeast Monsoons. Since these monsoon periods have contrasting effects on currents in the Singapore Strait, this could explain why Pasir Ris (northeast of Singapore) and Pulau Semakau (south off the coast of mainland Singapore) are both important source sites for *Littoraria pallescens*. As expected for a species with planktonic larvae, *Littoraria pallescens* was found to be a panmictic population in Singapore (J. H. Ong, unpublished data). In the South China Sea, analysis of 12 microsatellite markers found that the congeneric species *Littoraria sinensis* along the coast of China experiences high gene flow, though with signatures of low genetic differentiation between the northern population from the rest, consistent with a phylogeographic split because of historical glaciation (Li et al. 2022). Outside of the Central Indo-Pacific, *Littoraria pallescens* from the western coast of the Indian Ocean was similarly shown to be a panmictic population, but *Littoraria subvittata* was found to be more sensitive to isolation-by-distance effects (Nehemia et al. 2019). The difference in population differentiation in the Western Indian Ocean was attributed to differences in both larval dispersal potential and habitat preference of adult snails, with *Littoraria subvittata* being found closer inland, while *Littoraria pallescens* inhabits the zone closer to the sea (Torres et al. 2008; Nehemia et al. 2019).

Tridacna squamosa (the fluted giant clam) is one of three remaining giant clam species in Singapore and is locally Critically Endangered according to the Singapore Red Data Book (Davison et al. 2024). Analysis using the cytochrome c oxidase subunit I gene revealed a single haplotype that was present in half of the analysed samples, suggesting possible panmixia and connectivity around the Southern Islands where they were surveyed (Neo and Todd 2012). Larval transport simulations suggested that residual hydrodynamic flows and spawning time drive connectivity, with low residual flow during spawning resulting in the greatest chances of successful settlement (Neo et al. 2013), in particular because the larvae of *Tridacna squamosa* can take more than a week before settling (Neo et al. 2015). However, there is likely poor larval connectivity from regional reefs to Singapore because of strong surface currents during monsoons, transporting giant clam larvae out of the Singapore Strait before they might settle (Neo et al. 2013).

11.2.7 Arthropods

Ligia exotica and *Ligia* sp. are sea slaters that inhabit the rocky shore and brood their eggs within females, which hatch into non-planktonic juveniles (Tsai and Chen 1997). All populations of both *Ligia* species exhibited high levels of self-recruitment with low migration between populations (J. H. Ong, unpublished data). Since these species spend a significant portion of their time on intertidal benthic substrates, any marine barrier would have a stronger isolating effect for them.

The sea slater (*Halobates hayanus*) is hemimetabolous and mainly resides on submerged rocks or vegetation (Andersen and Cheng 2004). As with *Peronia verruculata*, *H. hayanus* was also found to comprise cryptic species complexes (J. J. M. Chang, unpublished data). When the most abundant species was analysed, it was shown to have a westward migration signature with Sisters' Islands as a source site, which was to be expected with the prevailing current direction along the Singapore Strait. However, there was also a strong migration signature from Changi Beach to Berlayer Creek despite the relatively large distance. The latter may result from stepping stones along the coastline and neuston, since adults locomote by skimming the sea surface, as their name suggests. While *H. hayanus* has been reported from potential stepping stones, such as Tanah Merah, they have not yet been observed during sampling for genetic studies.

Carcinoscorpius rotundicauda (the mangrove horseshoe crab) and *Tachypleus gigas* (the coastal horseshoe crab) are two of four extant horseshoe crab species and are both categorised as Vulnerable according to the Singapore Red Data Book (Davison et al. 2024). As their common names suggest, *C. rotundicauda* inhabits mangrove fringes and nearby mudflats, while *Tachypleus gigas* inhabits sandy beaches (Cartwright-Taylor et al. 2011; Behera et al. 2015). Tang et al. (2021) found that *C. rotundicauda* had a finer population structure along a northwest-southeast cline, while *Tachypleus gigas* exhibited panmictic population structure. The spatial differentiation observed for *C. rotundicauda* was determined to be likely because of isolation-by-distance effects as a result of depth-limited dispersal. Additionally, Tang et al. (2021) hypothesised that active dispersal of adults rather than passive dispersal of planktonic larvae drove the differences in dispersal distances between the two horseshoe crab species, with the smaller *C. rotundicauda* having a more limited dispersal ability and the larger *Tachypleus gigas*, a strong recolonisation potential. Size-influenced dispersal ability was therefore another reason for the finer population structure of *C. rotundicauda*.

11.3 Genetic Connectivity in Singapore's Seascape

To understand genetic connectivity across Singapore's seascape, it is not sufficient to only examine migration patterns of each species in isolation. A multi-species approach is necessary to comprehensively capture the dynamics across the seascape. Based on genomic data of 11 marine metazoan species and three putative lineages of *Porites* species, it was found that species that dispersed their eggs freely into the water column or had free-swimming larvae generally had panmictic populations, while species with non-planktonic juveniles had more genetically distinct populations that were geographically structured.

From aggregated contemporary migration rates across all species and putative genetic lineages, where the mean migration rate for each pair of sites (normalised based on number of sampling sites) was calculated, the key source sites across Singapore's seascape were, in descending order: Sultan Shoal, Changi Beach, Pulau Ubin, Sisters' Islands and Pulau Semakau (Fig. 11.1). These sites had some of the highest migration rates across all sampling sites. These key source sites were also generally important source sites for many of the individual species.

The key sink sites were, in descending order: Cyrene Reef, Tanah Merah Beach, Pulau Satumu, St. John's Island and Pulau Hantu (Fig. 11.1). Of note is the lack of overlap between the key source and sink sites. Instead of having the highest migration rates as with the key source sites, the key sink sites receive migrants from many sites.

Three of the five migration pathways with the highest aggregated migration rates originated from Sultan Shoal, the predominant key source site in Singapore's seascape, while two originated from Changi Beach, the key source site with the second highest total emigration rate (Fig. 11.1). Two key sink sites—Cyrene Reef and Pulau Hantu—were also part of these high-migration pathways. While the direction of connectivity from Changi followed the predominant westbound throughflow along the Singapore Strait, connectivity from Sultan Shoal was directed oppositely.

Genetic connectivity patterns in Singapore are driven by hydrodynamics to a degree, especially for organisms with a reproductive life history with stages where dispersal is influenced by current direction and strength (e.g. planktonic eggs or larvae). Some migration patterns observed in Singapore have been attributed to current patterns, with migration direction being explained by spawning periods that coincide with monsoon periods that influence current flow, such as in the case of the sea anemones *Entacmaea quadricolor* and *Stichodactyla haddoni* and the molluscs *Peronia verruculata* and *Littoraria pallescens*. Of the top five migration pathways in Singapore's seascape (Fig. 11.1), the migration rates of *Stichodactyla haddoni* contributed to both migration pathways originating from Changi Beach, while *H. hayanus* additionally contributed to the Changi-Berlayer migration pathway. Where hydrodynamics drive connectivity, it is to be noted that timing is critical: factors such as when eggs are laid or juveniles hatch, how long it takes before larvae settle (i.e. pelagic larval duration) and the length of the settlement competency period, combined with the hydrodynamic patterns during these periods, can determine where the next generation is located. This is exemplified by larval dispersal simulation models (as reviewed in Swearer et al. (2019)), for which such characteristics are important to consider. For instance, Carson et al. (2010) observed how differences in spawning period resulted in vastly opposing source–sink dynamics for two congeneric mussels—*Mytilus galloprovincialis*, primarily bay-dwelling, and *Mytilus californianus*, primarily inhabiting open coasts—along the coastline of San Diego, California, because of seasonal reversals of surface currents.

However, there are other genetic connectivity patterns in Singapore that cannot be explained by current hydrodynamics and reproductive phenology. *Protoreaster nodosus* exhibited high levels of self-recruitment and limited migration despite being a broadcast spawner. *Amphiprion frenatus*, having its most important source site in the southwest in Sultan Shoal, had a strong eastward migration pattern signature that did not correspond with that of its host (*Entacmaea quadricolor*), which had an important source site at Sisters' Islands and a migration pattern following the general westward throughflow along the Singapore Strait. Even though there could be instances of net eastward current flow during the Southwest Monsoon (Kurniawan et al. 2015), tropical *Amphiprion* species typically spawn year-round (Ross 1978; Richardson et al. 1997a), and so its migration pattern would not be the result of the interaction between hydrodynamics and reproductive season. The migration rates of *Amphiprion frenatus* contributed to all three of the top migration pathways that originated from Sultan Shoal (Fig. 11.1), while the Blue *Porites* lineages contributed to the Sultan Shoal-Pulau Hantu migration pathway and *Entacmaea quadricolor* contributed to the Sultan Shoal-Cyrene Reefs migration pathway.

Where hydrodynamics is unable to fully explain migration patterns, other traits such as egg/larval size, larval behaviour (e.g. vertical migration or active swimming affecting horizontal transport), as well as pre- and post-settlement mortality have been observed to influence dispersal patterns and population connectivity (e.g. Gerlach et al. 2007; Gary et al. 2020; see also review by Swearer et al. 2019). In addition, most studies focus on larval or juvenile dispersal as a determining factor of genetic connectivity, but there are other upstream variables affecting gene flow that may also warrant attention, depending on the species in question. Indeed, Frisk et al. (2014) argued that adult-mediated population connectivity is under-appreciated in the current literature, which largely focuses on the interactions between physical hydrodynamics and biological traits during early life history. Adult-mediated population connectivity would therefore be particularly important for species where such a dispersal paradigm does not apply. In line with this argument, adult mobility and the appearance of transient individuals in a population have been found to be able to promote gene flow (Kiester et al. 1982). Besides that, fecundity and reproductive success can also be affected by anthropogenic stressors (Sierra-Flores et al. 2015; Foo and Byrne 2017; Pirotta et al. 2019), which in turn affect the quantity and quality of the next generation. There is thus a need for more integrative work examining the combined effects of hydrodynamics, life history, behaviour and other relevant traits across different life stages to describe and explain the complex patterns of realised connectivity, as well as to improve the accuracy of predicting connectivity in simulations.

11.4 Genetic Connectivity and Conservation

Genetic connectivity is essential to population persistence, and it therefore follows that preserving well-connected sites can help in conservation efforts to safeguard marine species. Singapore's waters are home to rich biodiversity, with 255 reef-building coral species (Huang et al. 2009), 35 true mangrove species (of which eight are Endangered and another eight are Critically Endangered) (Yang et al. 2011) and 618 fish species (Wells et al. 2019). However, the coastal habitats of Singapore face pressures from ongoing urbanisation and coastal development over the decades, as well as the high maritime traffic (Chou 2006; Chou et al. 2019), which can induce sediment remobilisation and increase sediment smothering of corals (Weber et al. 2006; Bessell-Browne et al. 2017). Beyond sediment resuspension, ship-induced waves can also have other effects, such as partially or completely uprooting macrophytes (Asplund and Cook 1997), displacing epifauna (Bishop 2008; Gabel et al. 2012) and affecting habitat use patterns of fish, including their larvae (Stoll and Beeck 2012; reviewed in Gabel et al. 2017).

Singapore's total reef area has decreased by 70% since 1953 (Jaafar et al. 2018), while approximately 95% of the original mangrove cover has been lost (Chou 2008). Projections made in 2015 following Singapore's 2008 Master Plan and 2011 Concept Plan estimate that a further 10% of coral reefs and 33% of mangroves may be lost, increasing up to an alarming 47% for sand and mudflats (Lai et al. 2015). While not all of this loss has been realised at this time, the updated Singapore 2019 Master Plan (planning for the next 10–15 years) (Urban Redevelopment Authority 2023) and 2021 Long-Term Plan (planning for the next 50 years) (Urban Redevelopment Authority 2021) have either continued to leave areas with such habitats available for development or earmarked additional such areas for possible future reclamation.

Given the threats that Singapore's coastal habitats face, it is paramount to identify ecologically important sites—which include sites that are important for genetic connectivity—that require greater conservation attention. The incorporation of genetic connectivity data in conservation is not yet commonly done, with only 11% of 739 marine protected areas from countries/regions with advanced systematic priorities conservation planning having included demographic or genetic connectivity as one of their criteria (Balbar and Metaxas 2019). Within Southeast Asia, the conservation priorities in focus during the design of marine protected areas such as the Tun Mustapha Marine Park in Malaysia and the Raja Ampat Marine Park in Indonesia have been habitat representation, species representation and access to fishing grounds (Grantham et al. 2013; Jumin et al. 2018). In the first study of its kind in Southeast Asia, the genetic connectivity patterns of 11 marine metazoans species and three putative lineages of *Porites* species were incorporated into spatial conservation planning (SCP) software Marxan (Ball et al. 2009) through the use of Marxan Connect (Daigle et al. 2020), with the target features to be conserved being each of six habitat types (rocky shore, mud/sand/rubble, mangrove, seagrass meadow, lagoon and coral reef) while the associated costs were based on land/maritime use of space (J. H. Ong, unpublished data). Results revealed that without the incorporation of genetic connectivity data in SCP, ecologically important areas may not be selected by Marxan, especially at proportions of lower habitat area conserved. This was exemplified by how Sultan Shoal, the predominant key source site in Singapore's waters with the highest contemporary emigration rate out of the site, was only selected frequently in the connectivity-informed SCP scenario. It was likely not a high priority in the SCP scenario without genetic connectivity incorporated because the site is a small islet comprising rocky bunds and sandy shore, neither of which is apparently an at-risk habitat. This demonstrates how conservation design needs to go beyond conserving habitat and species representation, especially along urban coasts where urbanisation and other anthropogenic effects may lead to further environmental degradation.

Similar to protected area planning, resilience enhancement projects should also enhance intervention strategies—e.g. selective breeding (Humanes et al. 2024) and microbiome manipulation (Peixoto et al. 2017), as in the case for coral reefs (reviewed in National Academies of Sciences, Engineering, and Medicine 2019)—by incorporating considerations for genetic connectivity when selecting target sites. Under Singapore's Marine Climate Change Science Programme, there is an ongoing project (*Enhancing urban coral reef resilience to climate change across biological scales*) where the site-selection for microbially enhanced corals involves active consideration of genetic connectivity patterns in Singapore's seascape. Other potential ways of integrating genetic connectivity into conservation planning include prioritising source areas with high functional genetic diversity or where populations possess adaptive alleles that confer traits beneficial for future conditions through evolutionary rescue (Riginos and Beger 2022), though these will require additional assessments to identify the relevant and appropriate loci that confer the desired phenotypes. Conservation strategies can also focus on creating or maintaining corridors of connectivity, especially for more isolated populations or populations with low genetic diversity. For example, in the study by Tang et al. (2021) on horseshoe crabs, it was suggested that it may be more cost-effective to conserve existing mangroves while habitat restoration efforts should be targeted and aim to connect isolated populations.

Aside from conservation planning, conservation management can also benefit from knowledge of genetic connectivity. Understanding the interconnectedness of different sites and populations across the seascape will assist researchers and protected area managers in quickly determining if specific sites may require greater attention in the event of disturbances, especially when a particular site is more vulnerable or in cases of localised disturbances at a specific site. This will facilitate faster intervention, which will in turn speed up recovery.

It is imperative to keep in mind that gene flow occurs over evolutionary time, and the benefits of genetic connectivity in spreading adaptive genetic variation and conferring population resilience require time to be realised. Incorporating genetic connectivity in conservation will be critical for the preservation of nature in the long term and should be complemented by other short-term to mid-term strategies for effective conservation action. Given the time scale in which gene flow acts, further work should also be done to project how barriers and corridors may change and assess how vulnerable different corridors of connectivity may be to climate change and other anthropogenic impacts so as to facilitate future-ready conservation decisions.

Acknowledgements This research is supported by the National Research Foundation and National Parks Board, Singapore, under the Marine Climate Change Science programme (Award No.: NRF-MCCS21-1-5-0001).

References

Acevedo J, Aguayo-Lobo A, Allen J et al (2017) Migratory preferences of humpback whales between feeding and breeding grounds in the eastern South Pacific. Mar Mamm Sci 33:1035–1052. https://doi.org/10.1111/mms.12423

Afiq-Rosli L, Wainwright BJ, Gajanur AR et al (2021) Barriers and corridors of gene flow in an urbanized tropical reef system. Evol Appl 14:2502–2515. https://doi.org/10.1111/eva.13276

Andersen N, Cheng L (2004) The marine insect *Halobates* (Heteroptera: Gerridae): biology, adaptations, distribution, and phylogeny. Oceanogr Mar Biol Annu Rev 42:119–180. https://doi.org/10.1201/9780203507810.ch5

Asplund TR, Cook CM (1997) Effects of motor boats on submerged aquatic Macrophytes. Lake Reserv Manag 13:1–12. https://doi.org/10.1080/07438149709354290

Balbar AC, Metaxas A (2019) The current application of ecological connectivity in the design of marine protected areas. Glob Ecol Conserv 17:e00569. https://doi.org/10.1016/j.gecco.2019.e00569

Ball IR, Possingham HP, Watts ME (2009) Marxan and relatives: software for spatial conservation prioritization. In: Moilanen A, Wilson KA, Possingham HP (eds) Spatial conservation prioritization: quantitative methods and computational tools. Oxford University Press, pp 185–195

Bansemer CS, Bennett MB (2011) Sex- and maturity-based differences in movement and migration patterns of grey nurse shark, *Carcharias taurus*, along the eastern coast of Australia. Mar Freshw Res 62:596–606. https://doi.org/10.1071/MF10152

Basu S, Mackey KRM (2018) Phytoplankton as key mediators of the biological carbon pump: their responses to a changing climate. Sustainability 10:869. https://doi.org/10.3390/su10030869

Behera S, Tripathy B, Sivakumar K et al (2015) Distribution and abundance of two sympatric species of horseshoe crabs along the Odisha Coast, India. In: Carmichael RH, Botton ML, Shin PKS, Cheung SG (eds) Changing global perspectives on horseshoe crab biology, conservation and management. Springer International Publishing, Cham, pp 181–191

Bessell-Browne P, Negri AP, Fisher R et al (2017) Impacts of turbidity on corals: the relative importance of light limitation and suspended sediments. Mar Pollut Bull 117:161–170. https://doi.org/10.1016/j.marpolbul.2017.01.050

Bishop MJ (2008) Displacement of epifauna from seagrass blades by boat wake. J Exp Mar Biol Ecol 354:111–118. https://doi.org/10.1016/j.jembe.2007.10.013

Bonin MC, Harrison HB, Williamson DH et al (2016) The role of marine reserves in the replenishment of a locally impacted population of anemonefish on the Great Barrier Reef. Mol Ecol 25:487–499. https://doi.org/10.1111/mec.13484

Bos AR, Gumanao GS, Alipoyo JCE, Cardona LT (2008) Population dynamics, reproduction and growth of the Indo-Pacific horned sea star, *Protoreaster nodosus* (Echinodermata; Asteroidea). Mar Biol 156:55–63. https://doi.org/10.1007/s00227-008-1064-2

Byrne M (2006) Life history diversity and evolution in the Asterinidae. Integr Comp Biol 46:243–254. https://doi.org/10.1093/icb/icj033

Cantrell DL, Groner ML, Ben-Horin T et al (2020) Modeling pathogen dispersal in marine fish and shellfish. Trends Parasitol 36:239–249. https://doi.org/10.1016/j.pt.2019.12.013

Carson HS, López-Duarte PC, Rasmussen L et al (2010) Reproductive timing alters population connectivity in marine metapopulations. Curr Biol 20:1926–1931. https://doi.org/10.1016/j.cub.2010.09.057

Cartwright-Taylor L, Yap V, Hsu C, Lou S (2011) Distribution and abundance of horseshoe crabs *Tachypleus gigas* and *Carcinoscorpius rotundicauda* around the main Island of Singapore. Aquat Biol 13:127–136. https://doi.org/10.3354/ab00346

Chan YKS, Toh TC, Huang D (2018) Distinct size and distribution patterns of the sand-sifting sea star, *Archaster typicus*, in an urbanised marine environment. Zool Stud 57. https://doi.org/10.6620/ZS.2018.57-28

Chan WWR, Tay YC, Ang HP et al (2020) Reproduction in urbanised coastal waters: shallow-water sea anemones (*Entacmaea quadricolor* and *Stichodactyla haddoni*) maintain high genetic diversity and panmixia. Diversity 12:467. https://doi.org/10.3390/d12120467

Chang JJM, Tay YC, Ang HP, Tun KPP, Chou LM, Meier R, Huang D (2018) Molecular and anatomical analyses reveal that Peronia verruculata (Gastropoda: Onchidiidae) is a cryptic species complex. Contributions to Zoology 87:149–165

Chen M, Murali K, Khoo B-C et al (2005) Circulation modelling in the Strait of Singapore. J Coast Res 215:960–972. https://doi.org/10.2112/04-0412.1

Chen S, Xie Z-X, Yan K-Q et al (2024) Functional vertical connectivity of microbial communities in the ocean. Sci Adv 10:eadj8184. https://doi.org/10.1126/sciadv.adj8184

Chou LM (2006) Marine habitats in one of the world's busiest harbours. In: Wolanski E (ed) The environment in Asia Pacific harbours. Springer, Dordrecht, pp 377–391

Chou LM (2008) Nature and sustainability of the marine environment. In: Wong T-C, Yuen B, Goldblum C (eds) Spatial planning for a sustainable Singapore. Springer, Dordrecht, pp 169–182

Chou LM, Huang D, Tan KS et al (2019) Singapore. In: Hamel J-F (ed) World seas: an environmental evaluation. Volume II: The Indian Ocean to the Pacific. Academic, London, pp 539–558

Comte J, Culley AI, Lovejoy C, Vincent WF (2018) Microbial connectivity and sorting in a High Arctic watershed. ISME J 12:2988–3000. https://doi.org/10.1038/s41396-018-0236-4

Cowen RK, Sponaugle S (2009) Larval dispersal and marine population connectivity. Annu Rev Mar Sci 1:443–466. https://doi.org/10.1146/annurev.marine.010908.163757

Daigle RM, Metaxas A, Balbar AC et al (2020) Operationalizing ecological connectivity in spatial conservation planning with Marxan Connect. Methods Ecol Evol 11:570–579. https://doi.org/10.1111/2041-210X.13349

Davison GWH, Gan JWM, Huang D et al (2024) The Singapore Red Data book: Red Lists of Singapore biodiversity, 3rd edn. National Parks Board, Singapore

Deshpande UD, Nagabhushanam R, Hanumante MM (1980) Reproductive ecology of the marine pulmonate, *Onchidium verruculatum*. Hydrobiologia 71:83–85. https://doi.org/10.1007/BF00005830

Deutschmann IM, Krabberød AK, Latorre F et al (2023) Disentangling temporal associations in marine microbial networks. Microbiome 11:83. https://doi.org/10.1186/s40168-023-01523-z

Falkowski PG, Fenchel T, Delong EF (2008) The microbial engines that drive earth's biogeochemical cycles. Science 320:1034–1039. https://doi.org/10.1126/science.1153213

Feldman B, Afiq-Rosli L, Simon-Blecher N et al (2022) Distinct lineages and population genomic structure of the coral *Pachyseris speciosa* in the small equatorial reef system of Singapore. Coral Reefs 41:575–585. https://doi.org/10.1007/s00338-021-02160-4

Field CB, Behrenfeld MJ, Randerson JT, Falkowski P (1998) Primary production of the biosphere: integrating terrestrial and oceanic components. Science 281:237–240. https://doi.org/10.1126/science.281.5374.237

Foo SA, Byrne M (2017) Marine gametes in a changing ocean: impacts of climate change stressors on fecundity and the egg. Mar Environ Res 128:12–24. https://doi.org/10.1016/j.marenvres.2017.02.004

Frisk MG, Jordaan A, Miller TJ (2014) Moving beyond the current paradigm in marine population connectivity: are adults the missing link? Fish Fish 15:242–254. https://doi.org/10.1111/faf.12014

Gabel F, Garcia XF, Schnauder I, Pusch MT (2012) Effects of ship-induced waves on littoral benthic invertebrates. Freshw Biol 57:2425–2435. https://doi.org/10.1111/fwb.12011

Gabel F, Lorenz S, Stoll S (2017) Effects of ship-induced waves on aquatic ecosystems. Sci Total Environ 601–602:926–939. https://doi.org/10.1016/j.scitotenv.2017.05.206

Gary SF, Fox AD, Biastoch A et al (2020) Larval behaviour, dispersal and population connectivity in the deep sea. Sci Rep 10:10675. https://doi.org/10.1038/s41598-020-67503-7

Gerlach G, Atema J, Kingsford MJ et al (2007) Smelling home can prevent dispersal of reef fish larvae. Proc Natl Acad Sci 104:858–863. https://doi.org/10.1073/pnas.0606777104

Grantham HS, Agostini VN, Wilson J et al (2013) A comparison of zoning analyses to inform the planning of a marine protected area network in Raja Ampat, Indonesia. Mar Policy 38:184–194. https://doi.org/10.1016/j.marpol.2012.05.035

Grol MGG, Nagelkerken I, Rypel AL, Layman CA (2011) Simple ecological trade-offs give rise to emergent cross-ecosystem distributions of a coral reef fish. Oecologia 165:79–88. https://doi.org/10.1007/s00442-010-1833-8

Huang D, Tun KPP, Chou LM, Todd PA (2009) An inventory of zooxanthellate scleractinian corals in Singapore, including 33 new records. Raffles Bull Zool Suppl 22:69–80

Humanes A, Lachs L, Beauchamp E et al (2024) Selective breeding enhances coral heat tolerance to marine heatwaves. Nat Commun 15:8703. https://doi.org/10.1038/s41467-024-52895-1

Jaafar Z, Huang D, Tanzil JTI et al (2018) The Singapore Blue Plan 2018. Singapore Institute of Biology, Singapore

Jones GP, Planes S, Thorrold SR (2005) Coral reef fish larvae settle close to home. Curr Biol 15:1314–1318. https://doi.org/10.1016/j.cub.2005.06.061

Jumin R, Binson A, McGowan J et al (2018) From Marxan to management: ocean zoning with stakeholders for Tun Mustapha Park in Sabah, Malaysia. Oryx 52:775–786. https://doi.org/10.1017/S0030605316001514

Kiester AR, Schwartz CW, Schwartz ER (1982) Promotion of gene flow by transient individuals in an otherwise sedentary population of box turtles (*Terrapene carolina triunguis*). Evolution 36:617–619. https://doi.org/10.2307/2408108

Kimirei IA, Nagelkerken I, Trommelen M et al (2013) What drives ontogenetic niche shifts of fishes in coral reef ecosystems? Ecosystems 16:783–796. https://doi.org/10.1007/s10021-013-9645-4

Koehl MAR, Hadfield MG (2010) Hydrodynamics of larval settlement from a larva's point of view. Integr Comp Biol 50:539–551. https://doi.org/10.1093/icb/icq101

Kough AS, Paris CB, Behringer DC, Butler MJ IV (2015) Modelling the spread and connectivity of waterborne marine pathogens: the case of PaV1 in the Caribbean. ICES J Mar Sci 72:i139–i146. https://doi.org/10.1093/icesjms/fsu209

Kurniawan A, Tay SHX, Ooi SK et al (2015) Analyzing the physics of non-tidal barotropic sea level anomaly events using multi-scale numerical modelling in Singapore regional waters. J Hydro-Environ Res 9:404–419. https://doi.org/10.1016/j.jher.2014.10.005

Lai S, Loke LHL, Hilton MJ et al (2015) The effects of urbanisation on coastal habitats and the potential for ecological engineering: a Singapore case study. Ocean Coast Manag 103:78–85. https://doi.org/10.1016/j.ocecoaman.2014.11.006

Lai S, Cheok ZY, Bouma TJ et al (2020) The relative importance of light and hydrodynamics in the decay and transport of vegetative seagrass fragments. Estuar Coast Shelf Sci 232:106430. https://doi.org/10.1016/j.ecss.2019.106430

Lambrechts J, Hanert E, Deleersnijder E et al (2008) A multi-scale model of the hydrodynamics of the whole Great Barrier Reef. Estuar Coast Shelf Sci 79:143–151. https://doi.org/10.1016/j.ecss.2008.03.016

Li M, Li Y, Xing T et al (2022) Microsatellite marker development and population genetic analysis revealed high connectivity between populations of a periwinkle *Littoraria sinensis* (Philippi, 1847). J Oceanol Limnol 40:1097–1109. https://doi.org/10.1007/s00343-021-1079-9

Lowe WH, Allendorf FW (2010) What can genetics tell us about population connectivity? Mol Ecol 19:3038–3051. https://doi.org/10.1111/j.1365-294X.2010.04688.x

McManus LC, Forrest DL, Tekwa EW et al (2021) Evolution and connectivity influence the persistence and recovery of coral reefs under climate change in the Caribbean, Southwest Pacific, and Coral Triangle. Glob Change Biol 27:4307–4321. https://doi.org/10.1111/gcb.15725

Mestre M, Ruiz-González C, Logares R et al (2018) Sinking particles promote vertical connectivity in the ocean microbiome. Proc Natl Acad Sci 115:E6799–E6807. https://doi.org/10.1073/pnas.1802470115

Meynecke J-O, de Bie J, Barraqueta J-LM et al (2021) The role of environmental drivers in humpback whale distribution, movement and behavior: a review. Front Mar Sci 8. https://doi.org/10.3389/fmars.2021.720774

Milici M, Deng Z-L, Tomasch J et al (2016) Co-occurrence analysis of microbial taxa in the Atlantic Ocean reveals high connectivity in the free-living bacterioplankton. Front Microbiol 7. https://doi.org/10.3389/fmicb.2016.00649

Nakajima Y, Yasuda N, Matsuki Y et al (2024) Genetic connectivity and diversity between tropical and subtropical populations of the tropical horned sea star *Protoreaster nodosus* in the Northwest Pacific. Mar Biol 171:142. https://doi.org/10.1007/s00227-024-04461-y

National Academies of Sciences, Engineering, and Medicine (2019) A research review of interventions to increase the persistence and resilience of coral reefs. National Academies Press, Washington, DC

Nehemia A, Ngendu Y, Kochzius M (2019) Genetic population structure of the mangrove snails *Littoraria subvittata* and *L. pallescens* in the Western Indian Ocean. J Exp Mar Biol Ecol 514–515:27–33. https://doi.org/10.1016/j.jembe.2019.03.005

Neo M, Todd P (2012) Population density and genetic structure of the giant clams *Tridacna crocea* and *T. squamosa* on Singapore's reefs. Aquat Biol 14:265–275. https://doi.org/10.3354/ab00400

Neo ML, Erftemeijer PLA, van Beek JKL et al (2013) Recruitment constraints in Singapore's fluted giant clam (*Tridacna squamosa*) population–a dispersal model approach. PLoS One 8:e58819. https://doi.org/10.1371/journal.pone.0058819

Neo ML, Vicentuan K, Teo SL-M et al (2015) Larval ecology of the fluted giant clam, *Tridacna squamosa*, and its potential effects on dispersal models. J Exp Mar Biol Ecol 469:76–82. https://doi.org/10.1016/j.jembe.2015.04.012

Padrón M, Costantini F, Bramanti L et al (2018) Genetic connectivity supports recovery of gorgonian populations affected by climate change. Aquat Conserv Mar Freshw Ecosyst 28:776–787. https://doi.org/10.1002/aqc.2912

Peixoto RS, Rosado PM, de Leite DCA et al (2017) Beneficial microorganisms for corals (BMC): proposed mechanisms for coral health and resilience. Front Microbiol 8:341. https://doi.org/10.3389/fmicb.2017.00341

Peterson BK, Weber JN, Kay EH et al (2012) Double digest RADseq: an inexpensive method for de novo snp discovery and genotyping in model and non-model species. PLoS One 7:e37135. https://doi.org/10.1371/journal.pone.0037135

Pirotta E, Mangel M, Costa DP et al (2019) Anthropogenic disturbance in a changing environment: modelling lifetime reproductive success to predict the consequences of multiple stressors on a migratory population. Oikos 128:1340–1357. https://doi.org/10.1111/oik.06146

Polato NR, Concepcion GT, Toonen RJ, Baums IB (2010) Isolation by distance across the Hawaiian Archipelago in the reef-building coral *Porites lobata*. Mol Ecol 19:4661–4677. https://doi.org/10.1111/j.1365-294X.2010.04836.x

Puritz JB, Keever CC, Addison JA et al (2017) Life-history predicts past and present population connectivity in two sympatric sea stars. Ecol Evol 7:3916–3930. https://doi.org/10.1002/ece3.2938

Pwa KH (2016) Fine-scale phylogeography and diversity of *Periclimenes* anemone shrimps (Crustacea: Caridea: Palaemonidae) in Singapore. BSc thesis, National University of Singapore, Singapore

Pwa KHV, Park J-H (2020) New record of shrimp, *Periclimenes* cf. *obscurus*, on carpet anemone. Singap Biodivers Rec:204–205

Reid DG (1985) Habitat and zonation patterns of *Littoraria species* (Gastropoda: Littorinidae) in Indo-Pacific mangrove forests. Biol J Linn Soc 26:39–68. https://doi.org/10.1111/j.1095-8312.1985.tb01551.x

Reid DG, Dyal P, Williams ST (2010) Global diversification of mangrove fauna: a molecular phylogeny of *Littoraria* (Gastropoda: Littorinidae). Mol Phylogenet Evol 55:185–201. https://doi.org/10.1016/j.ympev.2009.09.036

Richardson D, Harrison P, Harriott V (1997a) Timing of spawning and fecundity of a tropical and subtropical anemonefish (Pomacentridae: *Amphiprion*) on a high latitude reef on the east coast of Australia. Mar Ecol Prog Ser 156:175–181. https://doi.org/10.3354/meps156175

Richardson DL, Harriott VJ, Harrison PL (1997b) Distribution and abundance of giant sea anemones (Actiniaria) in subtropical eastern Australian waters. Mar Freshw Res 48:59. https://doi.org/10.1071/MF96020

Richmond RH (1987) Energetics, competency, and long-distance dispersal of planula larvae of the coral *Pocillopora damicornis*. Mar Biol 93:527–533. https://doi.org/10.1007/BF00392790

Riginos C, Beger M (2022) Incorporating genetic measures of connectivity and adaptation in marine spatial planning for corals. In: Van Oppen MJH, Aranda Lastra M (eds) Coral reef conservation and restoration in the omics age. Springer International Publishing, Cham, pp 7–33

Ross RM (1978) Reproductive behavior of the anemonefish *Amphiprion melanopus* on Guam. Copeia 1978:103–107. https://doi.org/10.2307/1443829

Roux N, Salis P, Lambert A et al (2019) Staging and normal table of postembryonic development of the clownfish (*Amphiprion ocellaris*). Dev Dyn 248:545–568. https://doi.org/10.1002/dvdy.46

Ruiz-Montoya L, Lowe RJ, Van Niel KP, Kendrick GA (2012) The role of hydrodynamics on seed dispersal in seagrasses. Limnol Oceanogr 57:1257–1265. https://doi.org/10.4319/lo.2012.57.5.1257

Sierra-Flores R, Atack T, Migaud H, Davie A (2015) Stress response to anthropogenic noise in Atlantic cod *Gadus morhua* L. Aquac Eng 67:67–76. https://doi.org/10.1016/j.aquaeng.2015.06.003

Sin TM, Ang HP, Buurman J et al (2016) The urban marine environment of Singapore. Reg Stud Mar Sci 8:331–339. https://doi.org/10.1016/j.rsma.2016.01.011

Stoll S, Beeck P (2012) Larval fish in troubled waters—is the behavioural response of larval fish to hydrodynamic impacts active or passive? Can J Fish Aquat Sci 69:1576–1584. https://doi.org/10.1139/f2012-086

Swearer SE, Treml EA, Shima JS (2019) A review of biophysical models of marine larval dispersal. In: Hawkins SJ, Allcock AL, Bates AE et al (eds) Oceanography and marine biology: an annual review. CRC Press/Taylor and Francis Group, pp 325–356

Tang Q, Shingate P, Wardiatno Y et al (2021) The different fates of two Asian horseshoe crab species with different dispersal abilities. Evol Appl 14:2124–2133. https://doi.org/10.1111/eva.13271

Tay YC, Noreen AME, Suharsono et al (2015) Genetic connectivity of the broadcast spawning reef coral *Platygyra sinensis* on impacted reefs, and the description of new microsatellite markers. Coral Reefs 34:301–311. https://doi.org/10.1007/s00338-014-1206-6

Tay YC, Chng MWP, Sew WWG et al (2016) Beyond the coral triangle: high genetic diversity and near panmixia in Singapore's populations of the broadcast spawning sea star *Protoreaster nodosus*. R Soc Open Sci 3:160253. https://doi.org/10.1098/rsos.160253

Timm J, Planes S, Kochzius M (2012) High similarity of genetic population structure in the false clown anemonefish (*Amphiprion ocellaris*) found in microsatellite and mitochondrial control region analysis. Conserv Genet 13:693–706. https://doi.org/10.1007/s10592-012-0318-1

Timm J, Kochzius M, Madduppa HH et al (2017) Small scale genetic population structure of coral reef organisms in Spermonde Archipelago, Indonesia. Front Mar Sci 4:294. https://doi.org/10.3389/fmars.2017.00294

Toonen RJ, Andrews KR, Baums IB et al (2011) Defining boundaries for ecosystem-based management: a multispecies case study of marine connectivity across the Hawaiian Archipelago. J Mar Biol 2011:1–13. https://doi.org/10.1155/2011/460173

Torres P, Alfiado A, Glassom D et al (2008) Species composition, comparative size and abundance of the genus *Littoraria* (Gastropoda: Littorinidae) from different mangrove strata along the East African coast. Hydrobiologia 614:339–351. https://doi.org/10.1007/s10750-008-9518-6

Tsai M-L, Chen H-C (1997) Variations of reproductive traits of two semi-terrestrial isopods, *Ligia exotica* and *L. taiwanensis* (Crustacea: Ligiidae), in Southern Taiwan. Zool Stud 36:33

Urban Redevelopment Authority (2021) Long-term plan review 2021. In: Long-Term Plan Rev. https://www.ura.gov.sg/Corporate/Planning/Long-Term-Plan-Review. Accessed 30 Oct 2024

Urban Redevelopment Authority (2023) Master plan 2019. In: Master Plan 2019. https://www.ura.gov.sg/maps/?service=MP. Accessed 23 Aug 2023

Weber M, Lott C, Fabricius KE (2006) Sedimentation stress in a scleractinian coral exposed to terrestrial and marine sediments with contrasting physical, organic and geochemical properties. J Exp Mar Biol Ecol 336:18–32. https://doi.org/10.1016/j.jembe.2006.04.007

Wells FE, Tan KS, Todd PA et al (2019) A low number of introduced marine species in the tropics: a case study from Singapore. Manag Biol Invasions 10:23–45. https://doi.org/10.3391/mbi.2019.10.1.03

Werner EE, Gilliam JF (1984) The ontogenetic niche and species interactions in size-structured populations. Annu Rev Ecol Evol Syst 15:393–425. https://doi.org/10.1146/annurev.es.15.110184.002141

Yang S, Lim RLF, Sheue C-R, Rong JWH (2011) The current status of mangrove forests in Singapore. In: Proceedings of NSS conference: nature conservation for a sustainable Singapore. Proceedings of Nature Society, Singapore, Singapore

Zubillaga AL, Márquez LM, Cróquer A, Bastidas C (2008) Ecological and genetic data indicate recovery of the endangered coral *Acropora palmata* in Los Roques, southern Caribbean. Coral Reefs 27:63–72. https://doi.org/10.1007/s00338-007-0291-1

Community-Based Conservation of Coral Reefs in Singapore

12

Tai Chong Toh, Hui Hui Tham, Rui Xuan Veralyn Toh, Shu Qin Sam, Rachel Yuen Yee Mark, and Chin Soon Lionel Ng

Abstract

Extensive land reclamation and coastal development have inevitably impacted Singapore's coral reefs. Despite the chronic stressors, the reefs have shown signs of recovery and continue to support high marine biodiversity. This chapter traces the evolution of how the local community has been directly involved in coral reef conservation and distils the lessons arising from these efforts. Our review of the literature underlined how capacity building in the early years has supported reef conservation efforts in Singapore and the importance of partnership among government, academia and the wider community. In recent years, the emergence of interconnected multistakeholder networks has proved vital in addressing contemporary challenges. Responding to a 2024 collision between two ships, which released 400 tonnes of oil in the Singapore Strait, the networks aggregated resources and information to support the clean-up and helped to develop action plans with the government to protect and restore the impacted coastal habitats. As the community's interests and aspirations diversify, we strongly advocate using participatory design in conservation planning and establishing 'network of networks' to promote inclusive and effective community-based reef conservation programmes in Singapore.

Keywords

Community involvement · Community leadership · Community networks · Civil society · Civic participation · Marine conservation

T. C. Toh (✉)
Tropical Marine Science Institute, National University of Singapore, Singapore, Singapore

College of Alice & Peter Tan, National University of Singapore, Singapore, Singapore

Chua Thian Poh Community Leadership Centre, National University of Singapore, Singapore, Singapore
e-mail: Taichong.toh@nus.edu.sg

H. H. Tham · C. S. L. Ng
Tropical Marine Science Institute, National University of Singapore, Singapore, Singapore

R. X. V. Toh
Methodist Girls' School, Ministry of Education, Singapore, Singapore

S. Q. Sam
Tropical Marine Science Institute, National University of Singapore, Singapore, Singapore

NUS College, National University of Singapore, Singapore, Singapore

R. Y. Y. Mark
Tropical Marine Science Institute, National University of Singapore, Singapore, Singapore

National Biodiversity Centre, National Parks Board, Singapore, Singapore

12.1 Introduction

After its independence in 1965, Singapore has developed rapidly into a modern metropolis. The island nation underwent extensive land reclamation, increasing the total land area from 581 km^2 in the 1960s to 735 km^2 in 2024 (Tun 2012; Singapore Department of Statistics 2024). To mitigate the impacts of erosion and sea level rise, almost two-thirds of the country's shoreline has been modified for coastal protection (Lai et al. 2015). The state's territorial waters are mostly managed for port activities to support the maritime industry, and the remaining are used for a range of sectors that include the military, petrochemical industry, aquaculture, recreation, conservation and education (Chou et al. 2019).

The extensive coastal development and utilisation of its coastal waters have inevitably affected Singapore's marine environment and biodiversity. The total coral reef area has declined from 39 km^2 in the 1950s to 10 km^2 in 2018 (Tun 2012; Tan et al. 2023). High sedimentation rates have increased coral mortality and decreased underwater visibility

L. M. Chou, D. Huang (eds.), *Coral Reefs of Singapore's Urbanised Sea*, Coral Reefs of the World 23,
https://doi.org/10.1007/978-981-95-0903-4_12

(Low and Chou 1994; Chou et al. 2019). The reef substrate is unstable, and the prevalence of loose rubble pieces has compromised the post-settlement survival of coral larvae (Bauman et al. 2015; Ng and Chou 2018). Despite these challenges, Singapore still hosts multiple marine ecosystems and supports high biodiversity within its limited sea space (Chou et al. 2019). Coral cover has gradually increased after its initial decline from the 1980s to the 2000s (Chan et al. 2024). Conservation efforts are widely employed to promote species recovery and habitat enhancement (Low and Lim 2012).

While coral reef conservation in Singapore has, by and large, been led by government agencies and academics (Tan 2023), local communities have had a growing involvement in such efforts, especially over the last three decades (Chou and Goh 1998; Low and Lim 2012). Examples of community members include Non-Governmental Organisations (NGOs), registered societies, charities or associations, informal community groups and individual volunteers. In this chapter, we define community-based conservation as the 'natural resources or biodiversity protection by, for, and with the local community' (Western and Wright 1994) that encompasses a range of ground-up participation (Berkes 2007). Taking stock of community-based efforts provides critical insights into the management and protection of coral reefs (White and Courtney 2004). We thus aim to chronicle the development of community-based reef conservation in Singapore and reflect on how this evolving space will unfold in the future.

12.2 Capacity Building Among the Local Community (1980s–1990s)

Singapore's marine science landscape was in its nascent stage in the 1980s. Locally, coral reefs were increasingly affected by developmental work in the Singapore Strait (Chou et al. 2019) and the unregulated harvesting of corals (The Straits Times 1985). The Reef Ecology Study Team was established at the National University of Singapore (NUS) in 1986 to focus on coral reef research. A significant project undertaken was the 'ASEAN-Australia Marine Science Project: Living Coastal Resources'. From 1986 to 1994, this project provided the resources and expertise to train the community in coral survey techniques (Chou et al. 1994).

The capacity-building efforts played a central role in Singapore's first community-based reef conservation project. Local marine scientists initiated the Reef Survey and Conservation Project (RSCP), a collaborative community project anchored by three organisations: the Republic of Singapore Yacht Club (RSYC), the Singapore Institute of Biology (SIBiol) and the Singapore Underwater Federation (SUF) (Singapore Reef Survey and Conservation Project 1991), from 1987 to 1991. With a shared concern that Singapore's coral cover could continue to decline, this project provided the first country-wide assessment of the status of coral reefs in Singapore (Chou and Goh 1998) and raised awareness of these habitats among the public and the government. Funded by RSYC, sponsors and commercial organisations, at an estimated project value of S$20,000 (US$12,000), the marine biologists from the SIBiol provided scientific training and SUF organised field training for over 150 volunteer recreational divers to survey over 65 sites on 41 coral reefs in Singapore (Chou and Goh 1998).

Through the surveys, the RSCP identified four coral reef areas in Singapore that were of conservation importance. These areas included the island group around St. John's Island (Pulau Sekijang Bendara), Pulau Hantu, patch reefs around Pulau Semakau and the island group around Pulau Sudong and Raffles Lighthouse (Pulau Satumu). The community submitted a conservation proposal that detailed the findings and recommendations to the relevant government authorities in 1991 (Chou and Goh 1998). After extensive consultation with local agencies and the public, these four coral reef areas were included in a conservation action programme in the Singapore Green Plan (SGP) 1992 (Ministry of Environment 1993). The programme strengthened policies on commercial harvesting of corals, and recommended monitoring of coastal development projects around coral reefs to minimise pollution and siltation. In 1995, the proposed St. John's Island marine conservation area was officially designated as a Marine Nature Area (Chou and Goh 1998) and the National Parks Board outlined its commitment to 'keep the corals and marine life safe from reclamation work or other land-based operations' (The Straits Times 1996a). This episode was among the most consequential examples of how local NGOs, companies and community volunteers have influenced marine conservation policies in Singapore.

12.3 The Emergence of Community-Led Conservation Efforts (1990s–2000s)

The capacity-building programmes in the 1980s and the heightened awareness of environmental issues (Chew 2015) prompted a series of community-led conservation efforts. In 1991, the potential impact of the planned reclamation between Sentosa and Buran Darat to increase the land area for mixed-use development led the Nature Society (Singapore)'s Marine Conservation Group (NSS-MCG) to galvanise the community in Singapore's pioneering coral relocation effort (The Straits Times 1991a). The NSS-MCG initiated the Reef Rescue Project to relocate corals and selected reef organisms from Buran Darat, an islet east of

Sentosa (Pulau Blakang Mati), to two sites on the western end of Sentosa (Chou and Goh 1998; Ng and Chou 2018). From August to November 1991, over 140 volunteer divers from 20 SCUBA diving clubs and the Republic of Singapore Navy relocated corals and reef organisms covering 200 m^2 of the reef in the impacted site. Using chisels, corals were removed from the seabed, placed in baskets and pulled in semi-submerged sledges attached to the side of a boat, before being wedged between the rocks at the recipient site located 1.5 km away (The Straits Times 1991b).

The NSS-MCG reported high coral survival rates of 80% 1 year after relocation (The Straits Times 1992b), and volunteers shared how the project was a means to train 'environmental divers' who can 'help preserve some of the corals' (The Straits Times 1991a; The Straits Times 1991b). The Chairman of the NSS-MCG declared the S$10,000 (US$7100) project a success because of the 'survival of the corals and the increased conservation awareness' (The Straits Times 1992a; The Straits Times 1992b).

While well-intended, not all community-led conservation efforts were equally successful. Following its first initiative, the NSS-MCG embarked on a second relocation exercise that was said to be the largest scale in the world at that time (The Straits Times 1993). In 1993, the Society announced plans to relocate a proportion of the corals and reef organisms from the 120,000 m^2 Pulau Ayer Chawan reef to Sentosa, in response to the impending reclamation of the island for the petrochemical industry (The Straits Times 1993; The Straits Times 1994a). This project was funded with S$51,000 (US$35,400) from Hong Kong Bank's Care for Nature Trust Fund and S$12,000 (US$85,000) from Exxon Chemicals (Singapore) (The Straits Times 1996a, b, c; Chou and Goh 1998). The increased private sector funding and greater community involvement signalled a rapidly growing interest in reef conservation.

Between 1993 and 1995, the NSS-MCG led 450 volunteers from six local diving clubs and spent 10,000 h of underwater time for this exercise (The Straits Times 1996b). The NGO used methods similar to their first initiative, except that 'delicate' corals with a lower chance of survival were omitted. The corals were also transported by boat due to the longer distance between the source and recipient sites (The Straits Times 1993). Collectively, they relocated corals and reef organisms from 500 m^2 of the reef off Pulau Ayer Chawan. However, only 35% of the corals survived 1 year after the relocation exercise (Chou and Tun 1997). The NSS-MCG highlighted the difficulties in ensuring competencies among the volunteer divers and financial constraints 'to obtain underwater cement to anchor the corals securely' (The Straits Times 1994a). Marine biologists who conducted the monitoring efforts further concluded that differences in donor and recipient site conditions, strong waves and currents and inter- and intra-specific competition contributed to the reduced coral survivorship (Chou and Tun 1997). For such technical exercises, a science-based approach should be adopted to optimise the methods and enhance the survival of the relocated corals (Chou and Tun 1997; Chou and Goh 1998).

Collaboration between community groups and scientists has a range of benefits, including public education and capacity building. In 1992, the SUF collaborated with the marine biologists from SIBiol to launch the 'Coral Reef Naturalist' programme to equip recreational divers with a foundational understanding of coral reef ecology and conservation, and skills to identify reef organisms in Singapore (Chou and Goh 1998). This, in turn, generated a pool of volunteers who could support research and conservation efforts. From 1995 to 1996, the SUF also initiated regular programmes for volunteer divers and snorkellers to remove marine debris from the seabed, particularly from coral reefs off Pulau Hantu and St. John's Island. Held in partnership with corporate companies, these events garnered interest among the public, with volunteer numbers ranging from 150 to 220 attendees for each event (The Straits Times 1995; The Straits Times 1996c). By timing the clean-ups with significant environmental events such as 'Clean and Green Week' or 'Earth Day' celebrations, the press coverage bolstered the awareness of marine-related issues in Singapore among the public (Chou and Goh 1998).

During this period, community-led conservation efforts also received greater support from the public and private sectors. The National Council of the Environment (NCE; now the Singapore Environment Council, or SEC) in 1994 initiated the installation of 10 mooring buoys off Pulau Hantu, a popular island frequented by boaters, snorkellers and SCUBA divers (The Straits Times 1994b). The S$35,000 (US$25,000) initiative aimed to discourage the use of boat anchors and reduce the impact on the coral reefs (Chou and Goh 1998). Rallying support from the community, volunteer divers deployed and maintained the buoys, academics were involved in monitoring the impact on the coral reefs and funding was provided by Shell Eastern Petroleum (The Straits Times 1994b). While the outcomes of this intervention were not established, it successfully garnered support from the government. Sidek Saniff, who was the Minister of State for Education, reaffirmed the importance of marine conservation in Singapore at the launch of the initiative (The Straits Times 1994b) that preceded a larger scale effort.

In 1997, a project was established to progressively install up to 200 mooring buoys to reduce anchor damage on the coral reefs around the islands earmarked for conservation in the Singapore Green Plan 1993. Led by the SUF, and supported by the SIBiol, RSYC and NSS, the community groups collaborated with two government agencies—the Maritime and Port Authority of Singapore and the National Parks Board. An initial funding of S$50,000 (US$36,000), provided by Shell Eastern Petroleum, was used to install 15 buoys around Pulau Hantu. A SUF spokesman mentioned that the buoys had ‘proven to be very effective in reducing coral damage due to anchors and anchor chains’, and highlighted that shops selling dive equipment and diving tours were involved in maintaining the buoys (The Straits Times 1997). This episode signalled the diversification of the stakeholders and a greater involvement from the people, public and private sectors.

12.4 A Fresh Wave of Community Leadership (2000s–2010s)

The community in the 2000s displayed a growing interest in reef conservation, in part because of the government's leadership and commitment through the SGP 1992 and SGP 2012. Before that, most NGOs involved in reef conservation were established with broad, non-reef conservation goals. For example, the SUF was established in 1973 as a national sports association with a focus on promoting underwater sports (Singapore Underwater Federation 2024), the SiBiol was founded in 1974 to promote the advancement of biology and education (Singapore Institute of Biology 2024), and the SEC was established in 1995 to coordinate environmental and sustainability efforts in Singapore (Singapore Environment Council 2024). While the NSS was established in 1921, the NGO's diverse interests included a range of natural habitats and species and possessed notable strengths in avian biology and ecology (Wee and Hale 2008). Its marine conservation arm (NSS-MCG) was established as an interest group in 1985 (Nature Society Singapore 2024), and it was possibly the only NGO that had a clear focus on reef conservation during that time.

The 2000s witnessed the emergence of reef conservation groups with specific interests. Many of these were ground-up initiatives that were not formally registered as legal entities (Table 12.1) and have developed programmes to actively collaborate with marine scientists. The Blue Water Volunteers (BWV) was established by marine scientists in 2002 to complement research activities through the training of volunteer divers (Blue Water Volunteers 2024). The BWV provided training in coral reef surveys and public education programmes that contributed to the pool of citizen scientists in Singapore. Collaborating with government agencies such as the National Parks Board and research institutions like NUS, the volunteers conducted regular surveys of Singapore's coral reefs and contributed to scientific studies (Low and Lim 2012). Working in close collaboration with BWV to develop the ‘Reef Friends Xplore!’ diver training programme, The Hantu Bloggers' (THB) goal was to provide guided tours around Pulau Hantu and environmental education for divers (The Hantu Bloggers 2024). Originally established as an online platform in 2003 to document and showcase the marine biodiversity associated with the coral reef around Pulau Hantu, THB has since built up a volunteer base and evolved to take on various roles. Over the years, THB has collaborated with marine biologists to conduct coral surveys, partnered with government agencies to conduct underwater clean-ups, and teamed up with other NGOs to monitor coral reef health following environmental impacts such as oil spills and mass bleaching events (The Hantu Bloggers 2024).

Table 12.1 List of local community groups that are actively engaged in marine and coral reef conservation in Singapore

Community groups in Singapore	Founding year	Status
Groups working on coral reef conservation		
Nature Society Singapore (NSS) Marine Conservation Group	1985	Registered non-profit organisation
Blue Water Volunteers	2002	Non-registered community group
Hantu Bloggers	2003	Non-registered community group
WildSingapore	2003	Non-registered community group
Our Singapore Reefs	2017	Non-registered community group
Marine Stewards	2019	Registered non-profit organisation
Other marine conservation groups		
International Coastal Cleanup Singapore	1992	Non-registered community group
Sea Angel	2002	Non-registered community group
Team Seagrass	2007	Non-registered community group
The Naked Hermit Crab	2007	Non-registered community group
Otterwatch	2010	Non-registered community group
Project: FIN	2010	Non-registered community group
Project Driftnet	2011	Non-registered community group
Mangrove Action Squad	2012	Non-registered community group
Little Green Men	2016	Non-registered community group
Small Change	2019	Non-registered community group
Coastal Natives	2019	Registered non-profit organisation

In 2003, WildSingapore (http://www.wildsingapore.com) was established as an online repository of environment-related news, nature events, marine conservation issues and resources on marine biodiversity. One of the key driving factors for its founding dates back to 2001, when the government announced plans to reclaim Chek Jawa, a wetland with six major ecosystems—sandy beach, rocky shore, seagrass lagoon, coral rubble, mangrove and coastal forest. The initial plan was deferred when the community stepped forward to catalogue the marine diversity and subsequently generated public awareness of the rich biodiversity of the wetlands through guided walks and online resources (Wee and Hale 2008). The impetus for online resources and discussion led to the establishment of WildSingapore, and its role has continued to evolve since its founding. The volunteers behind WildSingapore conducted regular surveys that provided valuable insights into the long-term changes of intertidal reefs and expert knowledge on the location of key species (Lim et al. 2020), which in turn contributed to scientific advancement (e.g. Fautin et al. 2015). It continued to train local youths through its intertidal surveys and internship programmes, and collaborated with scientists to initiate new programmes, such as Project Driftnet, to document the impacts of lost and abandoned fishing gear (Project Driftnet 2024). Working alongside government agencies such as the National Parks Board, WildSingapore provided valuable inputs for local publications and development projects (The Straits Times 2012).

In 2017, a group of marine scientists and diving enthusiasts established Our Singapore Reefs (OSR), a community initiative that galvanises volunteer divers in Singapore to remove marine debris from local waters. In 5 years, they conducted over 40 underwater dive clean-ups and removed over 1200 kg of trash from the reefs around Singapore (The Straits Times 2022a). The data collected were submitted and collated online and provided the much-needed information for research and policy formulation. In partnership with government agencies such as the Maritime and Port Authority of Singapore and the National Parks Board, OSR created platforms for environmental education through their clean-ups, exhibitions, film screenings, public talks and guided tours (Toh and Tham 2018). The close partnership between OSR and the government agencies proved to be critical in responding to environmental emergencies, such as the removal of a 100 m abandoned fishing net of Pulau Hantu that could endanger wildlife and recreational divers (The Hantu Bloggers 2021).

In 2019, a group of sea sports enthusiasts established Marine Stewards (MS) to promote sustainable fishing in Singapore. Their signature programme, FishX, was developed through the partnership among stakeholders from the fishing industry, marine biologists and government agencies, and it aimed to promote sustainable fishing. More recently, the ReefX programme was launched to train divers to conduct reef surveys to support local research and conservation efforts (Marine Stewards 2024). The volunteers in MS also aided the uncovering of the illegal sale of endangered marine animals in a local fish market (Marine Stewards 2024).

Interestingly, a sizeable number of reef conservation groups between the 2000s and 2010s were started by individuals who were trained in marine science or were active community volunteers in marine conservation. This can be attributed to the capacity-building and public education efforts in the 1980s and 1990s, which laid the foundation for community leadership in reef conservation. The profound growth in the number of reef conservation groups during this period also paralleled the emergence of groups in Singapore that focused on broader marine issues (Fig. 12.1). It was estimated that the number of marine conservation groups grew from 2 to at least 17 over the past four decades (Toh and Tham 2018). The diversity of marine conservation groups in Singapore provided a wider range of platforms for the local community to create partnerships with each other, as well as with government agencies, private companies and academics.

12.5 Strengthening Partnerships to Support Reef Conservation (2000s–2010s)

As Singapore's marine science capacity strengthened, members of the community were involved in various capacities to support reef conservation efforts that were more technical, such as coral restoration.

In 2001, the Singapore Tourism Board (STB) and NUS initiated a project involving community volunteers to install 76 artificial reef modules off two islands, Raffles Lighthouse and St. John's Island (Loh et al. 2006). Known as Reef Enhancement Units (REUs), these structures provided stable substrates to facilitate coral colonisation. Given the scale of the project, volunteer divers were recruited and trained to support the installation of the REUs so that the deployment would minimise damage to the natural reef. These artificial reefs were assessed to be suitable substrates (Loh et al. 2006) and prompted a second phase, supported by Sentosa Development Corporation (SDC) and funding from public donations (The Straits Times 2005). Community volunteers continued their support as the researchers installed more REUs off Lazarus, Kusu and Sisters' Islands (Ng et al. 2017).

Amidst growing interest in reef restoration, another restoration project led by NUS and the National Parks Board was initiated in 2007 to culture 'corals of opportunity', a term used to describe coral fragments lying loose on the reef and recruits settled on coral rubble (Ng et al. 2016). With funding of S$250,000 (US$165,500) provided by Keppel Corporation, this project established an in situ coral nursery off Pulau Semakau (NParks 2009). More than 660 'corals of opportunity' were collected and cultivated on 20 nursery tables, for

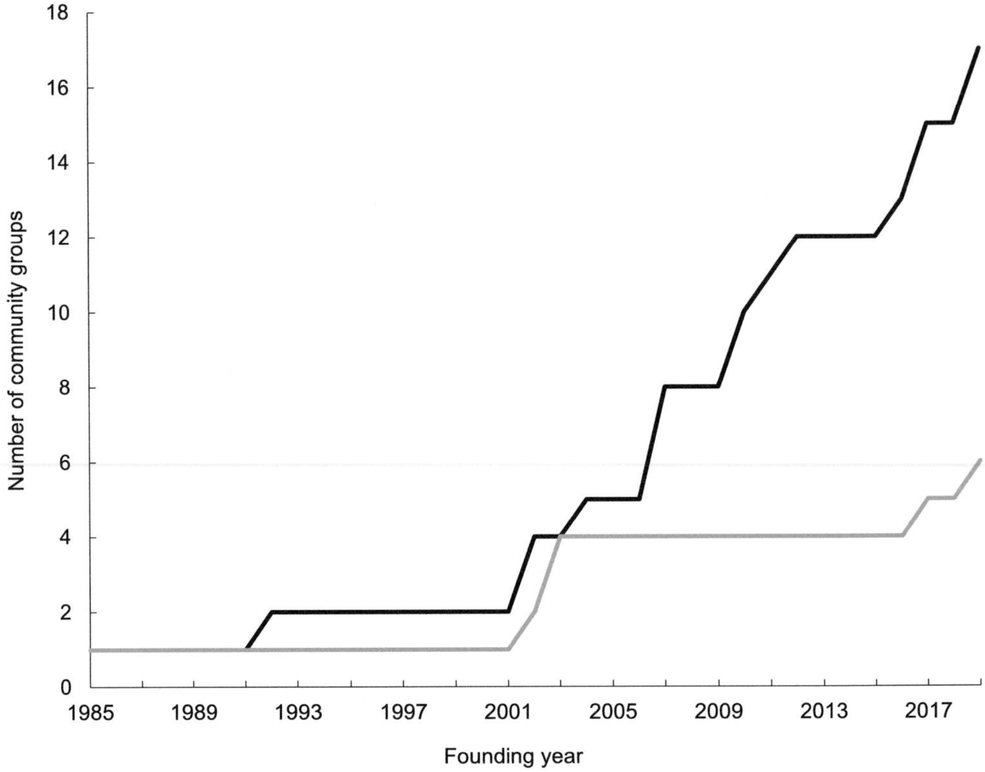

Fig. 12.1 Graph illustrating the number of local community groups working on broader marine conservation issues (black) and a subset of these groups working on subtidal and intertidal coral reefs (grey) in Singapore between 1985 and 2019

subsequent transplantation to the reefs around the island. Researchers trained and partnered with volunteer divers in maintaining the nursery and monitoring coral health, and the project reported high coral survival rates of 70–90% (Ng and Chou 2014). It was evident from these examples that having donations and a trained group of community volunteers can provide much-needed support for larger scale conservation efforts.

In 2013, the Maritime and Port Authority of Singapore funded a S$6 million (US$4.3 million) coral relocation project to mitigate the impact of port development on the coral reefs off Sultan Shoal. Over 5 years, the government agency engaged a commercial company and marine biologists to relocate more than 2300 coral colonies and cultivated more than 1100 coral fragments for transplantation (Maritime and Port Authority of Singapore 2015). To support this large-scale relocation effort, 50 volunteer divers from local conservation groups were trained by the marine biologists to maintain, transplant and monitor the coral fragments (Maritime and Port Authority of Singapore 2015). Involving trained volunteers greatly reduced the labour and cost of coral restoration efforts, and high survival rates of more than 80% could be achieved (Toh et al. 2017). Furthermore, it improved the awareness and interest of volunteers in supporting future marine conservation efforts, which were essential in encouraging public ownership of Singapore's natural heritage.

The prevalence of such partnerships proved to be valuable in shaping reef conservation in Singapore. In 2008, more than 20 members from community groups and academia came together to put up the Singapore Blue Plan (SBP) 2009 (The Singapore Blue Plan 2009). Unlike the first SBP that was developed in 2001, the SBP 2009 had greater involvement from the community and scientists and it was intended to complement the SGP 1992. The plan outlined the challenges of Singapore's marine environment and provided specific recommendations to be considered (The Singapore Blue Plan 2009). The SBP 2009 was submitted to the Minister for National Development and Minister for the Environment and Water Resources, and the suggestions were carefully studied by the relevant government agencies. The collaborative effort proved to be an effective approach for encouraging larger scale marine conservation efforts in Singapore.

For instance, the National Parks Board adopted one of the recommendations and commissioned the Comprehensive Marine Biodiversity Survey to study the marine organisms around coastal areas in Singapore from 2010 to 2015. Co-led by marine biologists, the survey engaged more than 350 volunteers and international scientists, with funding support from corporate donations. This enabled the discovery of more than 40 marine species new to science and more than 120 species previously unknown to Singapore (Tan et al. 2016). Another significant milestone was the establishment of Singapore's first marine park at the Sisters' Islands in

2014, which were areas of high marine biodiversity identified in the SBP 2009 (The Singapore Blue Plan 2009). A decade later, the third SBP was put together by more than 100 contributors from the academics, marine community and key stakeholders (The Singapore Blue Plan 2018), and it was presented to the Minister for National Development in 2018 (Ministry of National Development 2018). In his speech, the minister underscored the importance of 'strong and active partnership among government agencies, nature community, the academic and research fraternity and the wider community', and the government agencies will study the recommendations in detail and continue to collaborate with the community to realise common goals (Ministry of National Development 2018). Following the launch of SBP 2018, the National Parks Board expanded the list of marine species in the Species Recovery Programme in 2021 and announced plans for a second marine park in 2024 (The Straits Times 2024a). The SBP and its outcomes were a testament to the close partnership between the community, academia and government agencies. The collaboration between stakeholders, coupled with constructive engagement with the government, has facilitated sustained and effective conservation management measures.

Evidently, cultivating partnerships between research institutions, government agencies and private companies was an effective mode of coral reef conservation in Singapore (Chou et al. 2009). Excluding any of these parties in community-led conservation efforts can impact the outcomes and two such projects provided critical learning points. In 2006, the SUF proposed Project NOAH (Nurturing our Aquatic Heritage), which aimed to 'build a coral reef in the lagoon enclosed by Pulau Hantu' (Channel NewsAsia 2006). With support from SEC and NSS-MCG, S$100,000 (US$73,000) was raised to support the project across three phases (The Straits Times 2006). The first phase focused on recruiting and training 20 volunteer divers to survey the reefs around Pulau Hantu to identify organisms of conservation interest. The second phase involved mitigation measures to reduce sedimentation in the lagoon and relocation of corals to Pulau Hantu, including deepening of the lagoon and installation of a sediment filtration system. The last phase included long-term public education programmes to raise awareness of the biodiversity and conservation value of Pulau Hantu (The Straits Times 2006; Wee and Hale 2008). However, the project drew strong reactions from the public and NGOs (Ng 2006a) because deepening the lagoon would instead damage the habitat (Wee and Hale 2008). Eventually, the project was cancelled (Ng 2006b) and NSS acknowledged the drawback of not working closely with marine biologists (Wee and Hale 2008).

A subsequent attempt by the community in 2007 reinforced the importance of working with stakeholders with appropriate expertise. The SEACIL project was a large-scale deployment of 40 artificial reefs at the rocky shore along the Labrador Nature Reserve. Led by a maritime lecturer from a local polytechnic and developed by a maritime consultant, artificial reefs called SEACILs were installed to host corals relocated from Sentosa and Labrador Park (The Straits Times 2007a). The S$145,000 (US$111,000) project involved 68 staff, students and community volunteers, who supported the fabrication and installation of the artificial reefs made from a composite of concrete and PVC pipes. One of the project leads had also planned to install lights on the SEACILs 'to draw more creatures to them so the garden will establish itself faster' (The Straits Times 2007a). Later that year, members of the public and nature community raised questions about the disintegrating structures that littered the rocky shore, which had caused visible collateral damage to the habitat. In response, the National Parks Board ordered the project team to remove the structures from the reserve (The Straits Times 2007b; Wild Singapore 2011). Both Project NOAH and the SEACIL project demonstrated that partnership between government agencies, private companies and research institutions with the appropriate expertise was critical for reef conservation projects in Singapore. It has also shone the spotlight on a community that was more aware and educated about marine conservation and could proactively debate, challenge ideas and influence decisions.

12.6 Community Partnership Through 'Networks of Networks' (2010s to Present)

By the 2010s, academic institutions, government agencies and companies in Singapore had grown to be more adept and confident in working with volunteers and community groups (Low and Lim 2012). Academic institutions continue to engage the community through research projects such as the Marine Debris Monitoring Programme, 2017, which trained volunteer citizen scientists in data collection and analyses (Ministry of Sustainability and Environment 2022). Augmented with inputs from different community groups, they supported the state's development of the National Action Strategy on Marine Litter (Ministry of Sustainability and the Environment 2022). Government agencies, on the other hand, have also developed increasingly diverse platforms for the community to support reef conservation. These included the 'Plant-A-Coral, Seed-A-Reef' (PACSAR) programme and the 100,000 Corals Initiative that provided opportunities for contributing towards reef conservation projects through donations and citizen science (National Parks Board 2016; National Parks Board 2024a). Since 2012, the Biodiversity Roundtable, a community-based initiative that includes more than 20 NGOs involved with biodiversity issues, has also partnered with the National Parks Board to organise the annual 'Festival of Biodiversity', which brings together local community groups, academic institutions and government agencies to enhance public education (National Parks Board 2024b). Commercial companies have also started to increase their support for reef conservation through financial or infrastructure support for community groups. One°15 Marina, for example, established coral nurseries within the marina in 2018 in partnership with academics and community groups to facilitate research and education on reef conservation (The International SeaKeepers Society 2018).

A notable practice that emerged in the 2010s was the establishment of 'networks' that comprise diverse stakeholder groups. Following the establishment of the Sisters' Islands Marine Park in 2014, the National Parks Board established the Friends of the Marine Park (FMP) in 2017. Its members comprise representatives from academic institutions, nature groups, businesses and other government agencies (National Parks Board 2024c). The community meets to discuss emerging marine conservation issues, create guidelines and manage a community space called Bendara Bay, located on St. John's Island. Whenever there are major development projects, such as the upcoming reclamation work along the eastern coast of Singapore, the members of the FMP would be engaged to provide feedback as part of the development's design considerations (The Straits Times 2022b). For community groups that are involved in clean-ups, the National Environmental Agency's Public Hygiene Council initiated the 'Reach-Inspire-Synergise-Empower' (R.I.S.E) Champions network in 2019 (R.I.S.E. 2024). Its membership is equally diverse and although they may not focus on reef conservation, nature groups that are involved in underwater clean-ups around the reefs in Singapore are also included in this network. However, such networks are not solely led by government agencies. Depending on the goals, community groups form their ground-up networks. One such example is the 'SG Clean-up Organisers', an informal network of groups involved in clean-ups that actively share resources, experiences and information. Leveraging social messaging applications, the network comprises over 20 representatives from different groups to coordinate coastal and underwater clean-up efforts around Singapore (Fig. 12.2). In a 2024 oil spill incident, 400 tonnes of oil were released into the Singapore Straits that took 3 months to clean up. The government agencies tapped into these formal and informal networks to collate ground observations of the impacted sites and piloted oil spill clean-up protocols intended for public volunteers (Public Hygiene Council 2024). The community was also involved in developing action plans to protect and restore the impacted coastal habitats (The Straits Times 2024b). In turn, the coordinating agency, the Maritime and Port Authority of Singapore, would provide regular updates on the clean-up efforts for these networks to update their members (Maritime and Port Authority of Singapore 2024).

As the networks grow and diversify, community groups in Singapore have started sending representatives to join other networks or to participate in events of interest. The reef conservation community in Singapore now resembles a constellation of networks, with at least one member of a network actively connected to other networks. Globally, such 'networks of networks' aim to connect different knowledge systems, integrate complementary capabilities (Biodiversity and Ecosystem Services Network 2024) and enhance the effectiveness of the solutions (de Oliveira et al. 2018). In the case of Singapore, we observed that community members readily tap into the resources, ideas and information afforded by the members in the respective networks, which in turn enable groups to remain adaptive and collaborative. It also promotes a shared consciousness and systemic understanding (McChrystal et al. 2015), such that networks can better appreciate what others are doing and avoid duplication of efforts. The advantages are exemplified during the government's public consultation exercises. Invited community members can propose representatives from other networks to attend the consultation if their expertise and interests are relevant to the topic of discussion. This diversification of representation adds rigour to the consultation by increasing accessibility to a wider range of views to address various environmental issues (Public Utilities Board 2024).

It is worth noting that working in close partnerships does not imply that ideas and perspectives are homogenous. In 2022, when the Singapore Food Agency announced plans to set aside three areas in the southern islands for potential development of aquaculture (The Straits Times 2022b), some community groups raised concerns that two of these areas were close to coral reefs and requested more detailed studies to be conducted before proceeding with the development plan. Others highlighted how the farms may attract predators like sharks and pose dangers to recreational users, while the business communities lauded the plan as it would increase the amount of sea space for farm operations (The Straits Times 2022b). In response, the agency tapped into different networks and assembled a multi-stakeholder group comprising representatives from businesses, nature groups and government agencies to deliberate on the issue. Eventually, only one area located off the Pulau Bukom oil refinery was selected for future development (The Straits Times 2023). Through months of discussion, this 'network of networks' discussed their differing aspirations and concerns with aquaculture development and co-created the inaugural Singapore Aquaculture Plan in 2024 that sought to balance the commercial and conservation needs for this sector (Singapore Food Agency 2024).

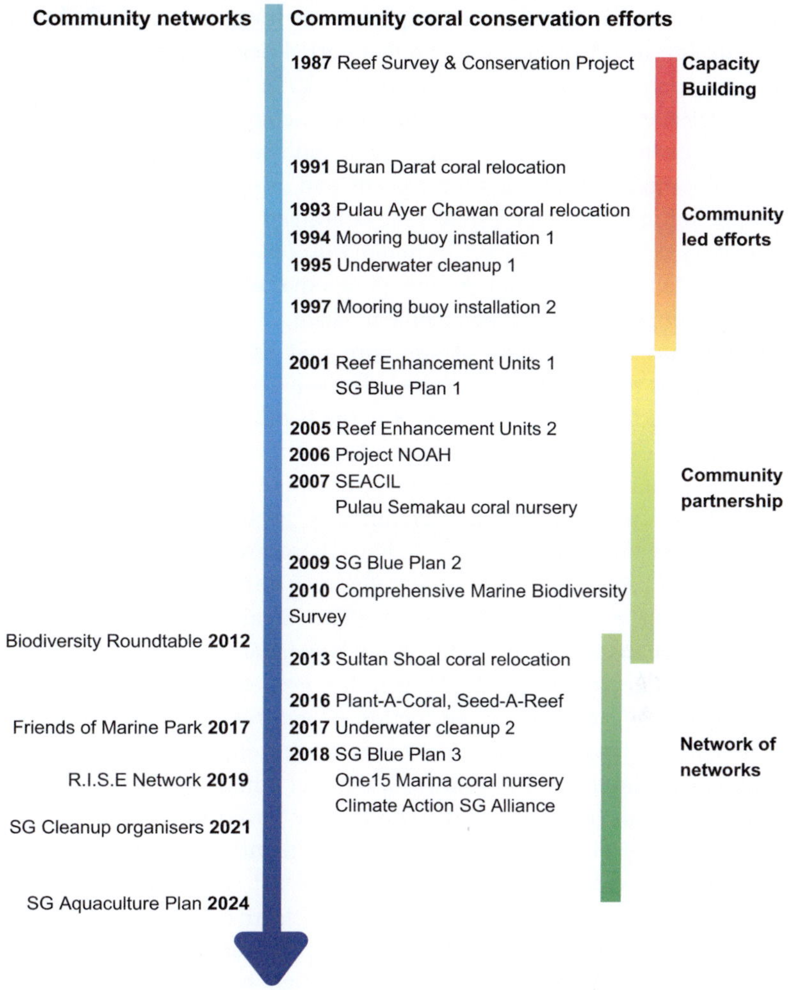

Fig. 12.2 Timeline indicating the establishment of community networks and the commencement of community-based reef conservation projects in Singapore

12.7 The Unfolding Future

The future of reef conservation in Singapore faces two major challenges in the next few decades. As climate change continues unabated, the mean tidal heights in Singapore have risen by 3.32 mm per year, and are projected to rise by 0.75 m by 2050 (Chou et al. 2019; Law and Huang 2023). Singapore's coastal areas and offshore islands are at risk of being eroded and submerged, and the corals in the southern islands will be affected. In addition, coral bleaching events caused by rising sea temperatures have increased in frequency over the last two decades (Taira et al. 2017; Toh et al. 2018), and the intervals between mass coral bleaching are shortening between each episode (Ng et al. 2020). In response to the threats of climate change and sea level rise, Singapore has developed a series of mitigation measures that will inevitably require further development and modifications to Singapore's existing ecosystems. The natural coastlines around Singapore will be augmented with artificial shoreline protection to reduce erosion (Public Utilities Board 2024), and land reclamation will continue in the southern islands where most of Singapore's coral reefs are found to support development in land-scarce Singapore (Urban Redevelopment Agency 2024).

Beyond coral reefs, these impacts will also affect the local populace. Coastal flooding and erosion will increase the likelihood of property damage, endanger public safety and impact quality of life (Chinowsky and Helman 2021). Hence, the community's involvement and support are needed to ensure that the mitigation measures remain inclusive and effective. As such, government agencies in Singapore have actively involved community groups in their national strategy to mitigate the impacts of climate change and development. In the recent Marine Climate Change Science (MCCS) programme 2021, there was a greater emphasis on nature-based solutions and community-driven climate resilience planning (Centre for Urban Greenery and Ecology 2024). Community groups have also increased their engagement in public consultation exercises for coastal protection projects, such as the 'Long Island' project (Urban Redevelopment Agency 2024). These engagements signalled a shift towards sustainable approaches that balance the needs of natural systems and economic aspirations. More importantly, the involvement of the community highlighted the preference for participatory design in urban planning, so that varied perspectives are considered in future development works in Singapore (Urban Redevelopment Agency 2024). The complexity of climate change and development in a land-scarce city-state, coupled with the divergent needs and aspirations of people in Singapore, necessitates the continued use of participatory design principles. Community groups will likely play a greater role in shaping Singapore's marine environment through increased engagement, such as in public consultation, co-designing of solutions and bridging the state–public relationship and knowledge gaps.

A second trend that will likely emerge is the continued emergence of conservation groups. As the awareness of environmental impacts increases, Singapore has witnessed the formation of new community groups, especially in the domain of climate action. These groups tend to focus on broader shifts in policy that may enhance reef conservation efforts in Singapore (The Straits Times 2022c). For instance, six youth group organisations submitted a statement in 2022 to call for greater climate action, which included the protection of natural habitats (The Straits Times 2022c). However, the emergence of groups with interests different from the reef conservation groups may further segment the community into different factions. They may choose to work independently, which reduces the efficiency of community-based work in Singapore and may even result in conflicts that would endanger the strong partnerships built over the years (Koh and Soon 2015). As these groups continue to emerge organically, the establishment of 'network of networks' becomes increasingly important as it creates more touchpoints for stakeholders and community groups with different interests to connect and engage with one another. The Climate Action SG Alliance (CASA; Climate Action SG Alliance 2024), for instance, brings together representatives from businesses, government agencies, grassroots, climate action and conservation groups. Leveraging the diversity of its networks, CASA embarked on interdisciplinary, cross-sectoral collaborations (Climate Action SG Alliance 2024) to enhance climate resilience in Singapore. These 'network of networks' can be achieved through formal settings, such as the recent workgroup to promote sustainable aquaculture or the multi-stakeholder FMP network that onboards various business, academic and conservation networks so that they can collaborate on topics of shared interests. These touch points can also be informal ground-up initiatives for networks to share information and resources. Recognising and growing the 'network of networks' will be critical in fostering a collaborative ecosystem to mitigate future challenges confronting reef conservation in Singapore.

References

Bauman AG, Guest JR, Dunshea G, Low J, Todd PA, Steinberg PD (2015) Coral settlement on a highly disturbed equatorial reef system. PLoS One 10(5):e0127874

Berkes F (2007) Community-based conservation in a globalized world. Proc Natl Acad Sci USA 104(39):15188–15193

Biodiversity and Ecosystem Services Network (2024) Network. https://www.besnet.world/network/. Accessed 31 Dec 2024

Centre for Urban Greenery and Ecology (2024) Marine Climate Change Science (MCCS) Programme. National Parks Board. https://www.nparks.gov.sg/Cuge/Programmes-Schemes/Research%20Programmes/MCCS%20Programme. Accessed 1 Oct 2024

Chan YS, Ng CSL, Tun KPP, Chou LM, Huang D (2024) Decadal decline of functional diversity despite increasing taxonomic and phylogenetic diversity of coral reefs under chronic urbanisation stress. Ecol Indic 164:112143

Channel NewsAsia (2006) Singapore Underwater Federation to build coral reef at Pulau Hantu. Channel NewsAsia, 30 March 2006, Singapore

Chew GK (2015) Community engagement to promote environmental ownership and secure our future. In: Tan YS (ed) 50 years of environment. World Scientific, Singapore, pp 193–220

Chinowsky P, Helman J (2021) Protecting infrastructure and public buildings against sea level rise and storm surge. Sustainability 13(19):10538

Chou LM, Goh BPL (1998) Singapore coral reefs – balancing development and conservation. In: Morton B (ed) The marine biology of the South China Sea. Proceedings of the Third International Conference on the Marine Biology of the South China Sea, Hong Kong, 28 October–1 November 1996, pp 355–368

Chou LM, Tun KPP (1997) Coral transplantation as a reef conservation tool – the Singapore experience. In: Pacific Congress on Marine Science and Technology, PACON International, Hong Kong, 6–8 August 1997, pp 528–537

Chou LM, Wilkinson CR, Licuanan WY, Aliño P, Cheshire AC, Loo MGK, Tangjaitrong S, Sudara S, Ridzwan AR, Soekarno (1994) Status of coral reefs in the ASEAN region. In: Wilkinson C, Sudara S, Chou LM (eds) Third ASEAN-Australia Symposium on Living Coastal Resources, Chulalongkorn University, Bangkok, Thailand, 16–20 May 1994, pp 1–10

Chou LM, Yeemin T, Abdul Rahim BGY, Vo ST, Aliño PM, Suharsono (2009) Coral reef restoration in the South China Sea. Galaxea 11:67–74

Chou LM, Huang D, Tan KS, Toh TC, Goh BPL, Tun K (2019) Singapore. In: Sheppard CRC (ed) World seas: an environmental evaluation. Volume II: The Indian Ocean to the Pacific. London, Academic Press, pp 539–558

Climate Action SG Alliance (2024) About. https://www.climateactionsg.org/about. Accessed 1 Oct 2024

de Oliveira Roque F, Uehara-Prado M, Valente-Neto F, Quintero JM, Ribeiro KT, Martins MB, de Lima MG, Souza FL, Fischer E, da Silva Jr UL, Ishida FY (2018) A network of monitoring networks for evaluating biodiversity conservation effectiveness in Brazilian protected areas. Perspect Ecol Conserv 16(4):177–185

Fautin DG, Tan R, Liang W, Yap N, Hee TS, Crowther A, Goodwill R, Sanpanich K, Chieh TY (2015) Sea anemones (Cnidaria: Actiniaria) of Singapore: shallow-water species known also from the Indian subcontinent. Raffles Bull Zool Supplement 31:44–59

Koh G, Soon D (2015) Developing civil society in Singapore. In: Chan D (ed) 50 years of social issues in Singapore. Singapore, pp 205–228

Lai S, Loke LH, Hilton MJ, Bouma TJ, Todd PA (2015) The effects of urbanisation on coastal habitats and the potential for ecological engineering: a Singapore case study. Ocean Coast Manag 103:78–85

Law MT, Huang D (2023) Light limitation and coral mortality in urbanised reef communities due to sea-level rise. Clim Change Ecol 5:100073

Lim LJW, Loh JBY, Lim AJS, Tan BYX, Ip YCA, Neo ML, Tan R, Huang D (2020) Diversity and distribution of intertidal marine species in Singapore. Raffles Bull Zool 68:396–403

Loh T-L, Tanzil JTI, Chou LM (2006) Preliminary study of community development and scleractinian recruitment on fibreglass artificial reef units in the sedimented waters of Singapore. Aquat Conserv Mar Freshw Ecosyst 16:61–76

Low JKY, Chou LM (1994) Sedimentation rates in Singapore waters. In: Sudara S, Wilkinson CR, Chou LM (eds) Proceedings. Third ASEAN-Australia Symposium on Living Coastal Resources, vol 2. Australian Institute of Marine Science, Townsville, pp 697–701

Low J, Lim LJ (2012) Conservation of our natural heritage: the Singapore experience. In: Proceedings of the 12th International Coral Reef Symposium, Cairns, Australia, pp 10–13

Marine Stewards (2024) About us. https://www.marinestewards.org/about. Accessed 1 Oct 2024

Maritime and Port Authority of Singapore (2015) Relocated corals thriving at new sites. Maritime and Port Authority of Singapore. https://www.mpa.gov.sg/media-centre/details/relocated-corals-thriving-at-new-sites#:~:text=The%20Maritime%20and%20Port%20Authority,the%20impact%20of%20the%20Tuas. Accessed 1 Oct 2024

Maritime and Port Authority of Singapore (2024) Focused cleaning in full swing. Call for volunteers for next phase. Maritime and Port Authority of Singapore. https://www.mpa.gov.sg/media-centre/details/focused-cleaning-in-full-swing-call-for-volunteers-for-next-phase. Accessed 1 Oct 2024

McChrystal GS, Collins T, Silverman D, Fussell C (2015) Team of teams: new rules of engagement for a complex world. Penguin, p 304

Ministry of Environment (1993) The Singapore green plan – action programmes. Ministry of the Environment, Singapore, p 80

Ministry of National Development (2018) Speech by 2M Desmond Lee at the Launch of the Singapore Blue Plan 2018. Ministry of National Development, Singapore. https://www.mnd.gov.sg/newsroom/speeches/view/speech-by-2m-desmond-lee-at-the-launch-of-the-singapore-blue-plan-2018. Accessed 1 Oct 2024

Ministry of Sustainability and the Environment (2022) Singapore launches National Action Strategy on Marine Litter. Ministry of Sustainability the Environment. https://www.mse.gov.sg/latest-news/singapore-launches-national-action-strategy-on-marine-litter. Accessed 1 Oct 2024

National Parks Board (2009) A helping hand for local reefs. NParks Buzz 3:3

National Parks Board (2016) Plant-A-Coral, Seed-A-Reef Programme. National Parks Board. https://www.nas.gov.sg/archivesonline/data/pdfdoc/20160521003/4.%20Factsheet%20C%20-%20Plant-A-Coral%20Seed-A-Reef%20programme.pdf. Accessed 1 Oct 2024

National Parks Board (2024a) NParks launches Singapore's most extensive coral restoration effort, tapping ontechnology and research to plant 100,000 corals in Singapore's waters. National Parks Board.https://www.nparks.gov.sg/news/2024/12/nparks-launchessingapore%E2%80%99s-most-extensive-coral-restoration-effort,--tapping-on-technology-and-research-to-plant-100,000-corals--in-singapore%E2%80%99s-waters. Accessed 1 Oct 2024

National Parks Board (2024b) Festival of Biodiversity. National Parks Board. https://fob.nparks.gov.sg/about-thefestival/#:~:text=Initiated%20in%202012%2C%20the%20Festival,biggest%20biodiversity%20festival%20in%20Singapore. Accessed 31 Dec 2024

National Parks Board (2024c) Friends of Marine Park. https://fotp.nparks.gov.sg/fomp/. Accessed 1 Oct 2024

Nature Society Singapore (2024) History of Nature Society Singapore. https://www.nss.org.sg/about.aspx?id=2. Accessed 1 Oct 2024

Ng D (2006a) NSS conservation chat discusses project NOAH. Pulau Hantu A Celebration of Marine Life. http://habitatnews.nus.edu.sg/news/pulauhantu/2006/04/nss-conservation-chat-discusses.html. Accessed 1 Oct 2024

Ng D (2006b) Update: project NOAH sunk. Pulau Hantu A Celebration of Marine Life. http://habitatnews.nus.edu.sg/news/pulau-hantu/2006/05/update-project-noah-sunk.html. Accessed 1 Oct 2024

Ng CSL, Chou LM (2014) Rearing juvenile "corals of opportunity" in *in situ* nurseries – a reef rehabilitation approach for sediment-impacted environments. Mar Biol Res 10:833–838

Ng CSL, Chou LM (2018) Coral reef restoration in Singapore—past, present and future. In: Sustainability matters: environmental management in the Anthropocene, Singapore, pp 3–23

Ng CSL, Toh TC, Chou LM (2016) Coral restoration in Singapore's sediment-challenged sea. Reg Stud Mar Sci 8:422–429

Ng CSL, Toh TC, Chou LM (2017) Artificial reefs as a reef restoration strategy in sediment-affected environments: insights from long-term monitoring. Aquat Conserv: Mar Freshw Ecosyst 27:976–985

Ng CSL, Huang D, Toh KB, Sam SQ, Kikuzawa YP, Toh TC, Taira D, Chan YKS, Hung LZT, Sim WT, Rashid ARR, Afiq-Rosli L, Ng NK, Chou LM (2020) Responses of urban reef corals during the 2016 mass bleaching event. Mar Pollut Bull 154:111111

Project Driftnet (2024) About. https://projectdriftnet.blogspot.com/p/about.html Accessed 1 Oct 2024

Public Hygiene Council (2024) The Public Hygiene Council mobilises volunteers for oil spill clean-up efforts at East Coast Park. Public Hygiene Council. https://www.publichygienecouncil.sg/files/Media_Release_for_ECP_Oil_Spill_Clean_Up.pdf. Accessed 1 Oct 2024

Public Utilities Board (2024) Our coastal conversation. https://www.pub.gov.sg/Public/KeyInitiatives/Flood-Resilience/Get-Involved-for-Coastal-Protection/Our-Coastal-Conversation. Accessed 1 Oct 2024

R.I.S.E (2024) RISE Champions Network.https://www.publichygienecouncil.sg/initiatives/rise/. Accessed 1 Oct 2024

Singapore Department of Statistics (2024) https://www.singstat.gov.sg/find-data/search-by-theme/society/environment/latest-data. Accessed 1 Oct 2024

Singapore Environment Council (2024) About SEC. https://www.sec.org.sg/knowing-sec.html. Accessed 1 Oct 2024

Singapore Food Agency (2024) Singapore Aquaculture Plan to uplift local aquaculture sector. Singapore Food Agency. https://www.ourfoodfuture.gov.sg/uplifting-aquaculture-industry/sg-aquaculture-plan/#:~:text=To%20uplift%20the%20local%20aquaculture,future%20of%20aquaculture%20in%20Singapore. Accessed 1 Oct 2024

Singapore Institute of Biology (2024) About us. https://sibiol.org.sg/. Accessed 1 Oct 2024

Singapore Reef Survey and Conservation Project (1991) A proposal for the conservation of Coral Reefs in Singapore, 42 pp

Singapore Underwater Federation (2024) About SUF. https://suf.org.sg/about/. Accessed 1 Oct 2024

Taira D, Toh TC, Ng CSL, Loke HX, Afiq-Rosli L, Cabaitan PC, Toh KB, Poquita-Du RCP, Chou LM, Song T (2017) Relocating bleached Platygyra sinensis facilitates recovery from thermal stress during a minor bleaching event. Mar Freshw Behav Physiol 50(5–6):375–385

Tan A (2023) Science journals: waves of change for Singapore's coral reef research. The Straits Times, 11 December 2023, Singapore

Tan KS, Koh KS, Ng JY, Goh JYL (2016) The Comprehensive Marine Biodiversity Survey Singapore Strait International Workshop 2013. Raffles Bull Zool Suppl 34:1–7

Tan YJ, Tham JK, Paul A, Rana U, Ang HP, Nguyen NT, Yee ATK, Leong BPI, Drummond S, Tun KP (2023) Remote sensing mapping of the regeneration of coastal natural habitats in Singapore: implications for marine conservation in tropical cities. Singap J Trop Geogr 44(1):130–148
The Hantu Bloggers (2021) Working together to save our seas. https://www.pulauhantu.sg/working-together-to-save-our-seas/#:~:text=The%20excellent%20team%20patiently%20and,the%20reefs%20at%20Pulau%20Hantu. Accessed 1 Oct 2024
The Hantu Bloggers (2024) About Us. https://www.pulauhantu.sg/about/. Accessed 1 Oct 2024
The International Seakeepers Society (2018) Coral garden and ECO exhibition. https://www.seakeepers.org/sc-global-one15-marina-and-seakeepers-asia-2/. Accessed 1 Oct 2024
The Singapore Blue Plan (2009) The Singapore Blue Plan 2009. http://nss.preview.orion5.iwi.com.sg/old/pdf/blueplan-final.pdf. Accessed 1 Oct 2024
The Singapore Blue Plan (2018) The Singapore Blue Plan 2018. https://sibiol.org.sg/wp-content/uploads/2020/10/The_Singapore_Blue_Plan_2018.pdf. Accessed 1 Oct 2024
The Straits Times (1985) Corals, dredging killing off corals. The Straits Times, 19 May 1985, Singapore
The Straits Times (1991a) Nature Society seeks 200 divers to save corals near Sentosa. The Straits Times, 11 August 1991, Singapore
The Straits Times (1991b) Volunteer Divers help save corals off Sentosa by relocating them. The Straits Times, 8 September 1991, Singapore
The Straits Times (1992a) Transplanted coral surviving well off the Sentosa Coast. The Straits Times, 28 January 1992, Singapore
The Straits Times (1992b) Save-the-corals plan a success. The Straits Times, 13 November 1992, Singapore
The Straits Times (1993) Massive project to save Pulau Ayer Chawan corals may be world's biggest. The Straits Times, 3 May 1993, Singapore
The Straits Times (1994a) Time running out for rescue of coral at Pulau Ayer Chawan. The Straits Times, 23 June 1994, Singapore
The Straits Times (1994b) Buoy, this is a good idea. The Straits Times, 6 November 1994, Singapore
The Straits Times (1995) 150 Volunteer divers help with underwater clean-up. The Straits Times, 10 Oct 1995, Singapore
The Straits Times (1996a) Saving Singapore's slowly-deteriorating coral reefs. The Straits Times, 14 June 1996, Singapore
The Straits Times (1996b) Coral rescue: Only a third survived move to Sentosa site. The Straits Times, 2 Oct 1996, Singapore
The Straits Times (1996c) Earth Day clean-up at St John's island. The Straits Times, 22 April 1996, Singapore
The Straits Times (1997) Keeping anchors away from corals. The Straits Times, 12 July 1997, Singapore
The Straits Times (2005) Adopt a reef, and save the coral. The Straits Times, 28 June 2005, Singapore
The Straits Times (2006) First step to create marine sanctuary on Pulau Hantu. The Straits Times, 31 March 2006, Singapore
The Straits Times (2007a) Undersea garden takes root. The Straits Times, 29 Oct 2007, Singapore
The Straits Times (2007b) Labrador Beach: Nature lovers fear coral project will cause damage. The Straits Times, 26 Nov 2007, Singapore
The Straits Times (2012) Singapore's 'surprising green face' in new book. The Straits Times, 7 Nov 2012, Singapore
The Straits Times (2022a) These S'poreans removed over 1,800kg of trash from our waters. The Straits Times, 1 Dec 2022, Singapore
The Straits Times (2022b) Public consultation for potential fish farms near coral reefs south of S'pore extended by 2 weeks. The Straits Times, 3 June 2022, Singapore
The Straits Times (2022c) Singapore youth give 18 recommendations for tackling environmental crisis. The Straits Times, 6 April 2022, Singapore
The Straits Times (2023) SFA delays plans to set up fish farms at 2 sites over coral reef concerns. The Straits Times, 5 July 2023, Singapore
The Straits Times (2024a) Singapore's proposed second marine park in Lazarus South, Kusu is home to rare rocky shore, coral habitats. The Straits Times, 9 May 2024, Singapore
The Straits Times (2024b) Action plan to protect and restore coastal habitats after oil spill in the works: Desmond Lee. The Straits Times, 25 June 2024, Singapore
Toh TC, Tham HH (2018) Marine habitats and their civil advocates. The Straits Times, 15 March 2018, Singapore
Toh TC, Ng CSL, Loke HX, Taira D, Toh KB, Afiq-Rosli L, Poquita-Du RC, Cabaitan P, Sam SQ, Kikuzawa YP, Chou LM, Song T (2017) A cost-effective approach to enhance scleractinian diversity on artificial shorelines. Ecol Eng 99:349–357
Toh TC, Huang D, Tun K, Chou LM (2018) Summary of coral bleaching from 2014 to 2017 in Singapore. In: Status of coral reefs in East Asian Seas region, pp 21–23
Tun KPP (2012) Optimisation of reef survey methods and application of reef metrics and biocriteria for the monitoring of sediment-impacted reefs. PhD thesis, National University of Singapore, Singapore, 208 pp
Urban Redevelopment Agency (2024) Master Plan. Urban Redevelopment Agency. https://www.ura.gov.sg/Corporate/Planning/Master-Plan. Accessed 1 Oct 2024
Wee YC, Hale R (2008) The nature society (Singapore) and the struggle to conserve Singapore's nature areas. Nat Singapore 1:41–49
Western D, Wright RM (1994) The background to community-based conservation. In: Natural connections: perspectives in community-based conservation. Island Press, Washington, p 7
White AT, Courtney CA (2004) Policy instruments for coral reef management and their effectiveness. In: Economic valuation and policy priorities for sustainable management of coral reefs. World Fish Center, Penang, p 141
Wild Singapore (2011) Learning about the SEACIL project. https://wildshores.blogspot.com/2011/07/learning-about-seacil-project.html. Accessed 1 Oct 2024

13 Integrated Urban Coastal Management in Singapore: Balancing Sustainable Use with Conservation

Karenne Tun and Hor Yee Collin Tong

Abstract

In small, highly urbanised island nations like Singapore, nature conservation and development walk a tight rope in balancing utilisation and conservation to ensure that current needs are addressed without jeopardising the needs of future generations. Rapid coastal transformation since Singapore's independence in 1965 has dramatically altered coastlines and impacted the coastal and marine environment. Despite this, coastal biodiversity has shown stubborn resilience and continues to persist to this day, albeit with altered habitat baselines, in part due to various coastal management measures that were implemented over the decades. An in-depth review of the evolution of Singapore's marine conservation and management since 1965 revealed four defining phases that collectively shaped the contemporary landscape. The history traces back to several pivotal projects in the mid-1980s that laid the groundwork for adopting a science-based approach to coastal resource conservation and management, ultimately leading to the development and adoption of the Integrated Urban Coastal Management (IUCM) strategy, which strives to balance sustainable development with marine conservation. IUCM has allowed Singapore to remain agile and adapt to the changing geopolitical and socioeconomic realities of the day. Along the way, key iconic programmes were initiated to meet marine conservation goals and promote active stewardship of nature.

Keywords

Climate change · Coastal resource management · Ecosystem resilience · Habitat restoration · Marine conservation · Marine park · Southeast Asia · Urbanisation

K. Tun (✉) · H. Y. C. Tong
National Parks Board, Singapore, Singapore
e-mail: Karenne_TUN@nparks.gov.sg; tong_hor_yee@nparks.gov.sg

13.1 Introduction

Singapore is a densely populated and highly urbanised small island nation that faces evolving challenges in balancing sustainable coastal development and coastal protection with natural resource management and conservation. With a land area of just over 735 km^2 and population density of around 8200 per km^2 (Singstat 2024), the realities for providing a safe and sustainable economy while maintaining high living standards have necessitated the modification and transformation of coastal areas through outward expansion of land into the sea to accommodate a diverse range of industries, housing, recreation and other public infrastructure (Tay et al. 2018; Wong et al. 2008). Although 90% of Singapore's primeval forests had been cleared by the early 1900s to accommodate the expansion of settlements (Yee et al. 2011), the seaward expansion that buried swaths of shallow coastal habitats only intensified post-1965 independence, initially focusing on reclamation of shallow coastal mudflats and reefs before expanding to deeper waters as reclamation technologies advanced (Hilton and Manning 1995; Lai et al. 2015). Currently, around 70% of Singapore's natural coastline has been altered through land reclamation and coastal modifications into hard, manmade infrastructure like concrete sea walls and stone embankments that protect the coast from erosion and hinterlands from seawater intrusion. Development and other anthropogenic pressures particularly over the last six decades have impacted Singapore's coastal and marine environment to varying degrees and these impacts are summarised in Chia (1992), Chou and Goh (1998), Lee and Hamid (2009), Tan et al. (2016a), Jaafar et al. (2018), Tay et al. (2018) and Chou (2023), amongst others.

Decades of development and expansion of marine-related industries invariably led to the loss of natural coastal and

L. M. Chou, D. Huang (eds.), *Coral Reefs of Singapore's Urbanised Sea*, Coral Reefs of the World 23,
https://doi.org/10.1007/978-981-95-0903-4_13

marine habitats; by 2018, an estimated 9.79 km^2 of intertidal (8.35 km^2) and subtidal (1.44 km^2) coral reefs remained from a historical estimate of 41.55 km^2 (32.20 km^2 of intertidal and 9.35 km^2 of subtidal coral reefs), representing an estimated 76.43% decline in total coral reef area (Tan et al. 2023). Of these, almost a third of the remaining subtidal coral reefs are found growing on artificial riprap seawalls, some of which extend to deeper than 8 m below chart datum (see Tan et al. 2025). Similarly, mangroves and intertidal mudflats and seagrass areas declined by 87.69% and 79.49% from historical estimates of 75 km^2 and 36.28 km^2 to 9.56 km^2 and 7.44 km^2, respectively.

Despite the significant reduction in area, coral reefs in Singapore are documented to support close to 30% of global scleractinian species (Huang et al. 2009), together with the complement of other tropical coral reef associated biodiversity including marine fish, echinoderms, sponges, molluscs and crustaceans (Tan et al. 2016a; Lim et al. 2020; Ip et al. 2023). Although the coral reefs of Singapore do not provide similar primary goods and services like fisheries as those in many other coral reef nations (Shen et al. 2021), their intrinsic value and importance, together with those of other associated mangroves and tidal seagrass habitats, are recognised as core natural assets that are valued and need to be protected (Lee and Hamid 2009; Friess 2017; Jaafar et al. 2018; Freiss et al. 2020).

By virtue of its small size and the interconnectedness of land and sea spaces, it is not surprising that Singapore does not have an equivalent of an integrated ocean policy that many other countries with more expansive territorial waters have. Instead, marine conservation and management are intrinsically tied to the broader sustainable development and nature conservation strategies, with regulatory and management regimes drawing on different primary and subsidiary legislation and regulatory policies administered by several government agencies and implemented through the IUCM framework (TCCME 2018).

More recently, under the City in Nature pillar of the Singapore Green Plan 2030, the Singapore government has committed to steward natural green and blue spaces, guided by science, to strengthen the conservation efforts while navigating the delicate balance of nature conservation and development (Singapore Green Plan 2021). In a written answer to a parliamentary query on measures in place to protect marine ecosystems from rising temperatures and effects of El Niño in May 2024, the then-Minister of National Development, Demond Lee, responded that '*NParks adopts a science-based approach towards protecting our marine ecosystems, which encompasses proactive monitoring and mitigation, and research*' (MND 2024).

These remarks give credence to the decades-long marine conservation and management efforts which started in the mid-1980s. An in-depth review of Singapore marine conservation and management landscape post-1965 independence revealed four defining phases that collectively shaped the contemporary landscape. These developments are summarised in Fig. 13.1 and elaborated in the next section.

13.2 Evolution of Singapore's Science-Based Approach to Marine Conservation and Management

13.2.1 1985–1994: Laying a Science-Based Foundation

The first defining phase between 1985 and 1994 saw the implementation of three initiatives—the ASEAN-Australia Living Coastal Resources Project (Chou 1991), the ASEAN/United States Coastal Resources Management Project (Chia and Chou 1991; Chia 1992) and the Reef Survey and Conservation Project by the Republic of Singapore Yacht Club, Singapore Institute of Biology and Singapore Underwater Federation (RSYC et al. 1991). These projects collectively laid the foundation for a science-based approach to marine conservation and management, built local expertise and established a strong and collaborative partnership between marine stakeholders and the government.

13.2.2 1990s: Advancing Marine Science in Singapore

The second phase in the 1990s saw the scaling up of efforts by the government and marine stakeholders in advancing marine science in Singapore. Central to that was the establishment of the National Parks Board (NParks) in 1990 and the subsequent merger with the Parks and Recreation Department in 1996, which created a single body to oversee biodiversity conservation and management matters, and this helped facilitate dialogue and establish long-term partnerships with marine stakeholders.

During this decade, two pivotal visioning plans, one by civil society and the other by the government, were released. The 'Master Plan for the Conservation of Nature in Singapore' (Malayan Nature Society (Singapore Branch) 1990) was the first comprehensive document to systematically identify and propose the protection of both established and lesser known natural areas across the country at a time when conservation was not yet a mainstream policy concern. The first 'Singapore Green Plan' by the then Ministry of the Environment (now the Ministry of Sustainability and the Environment or MSE) in 1992 (Yeo 1994) outlined Singapore's first environmental blueprint to balance environmental and developmental needs. Both documents helped

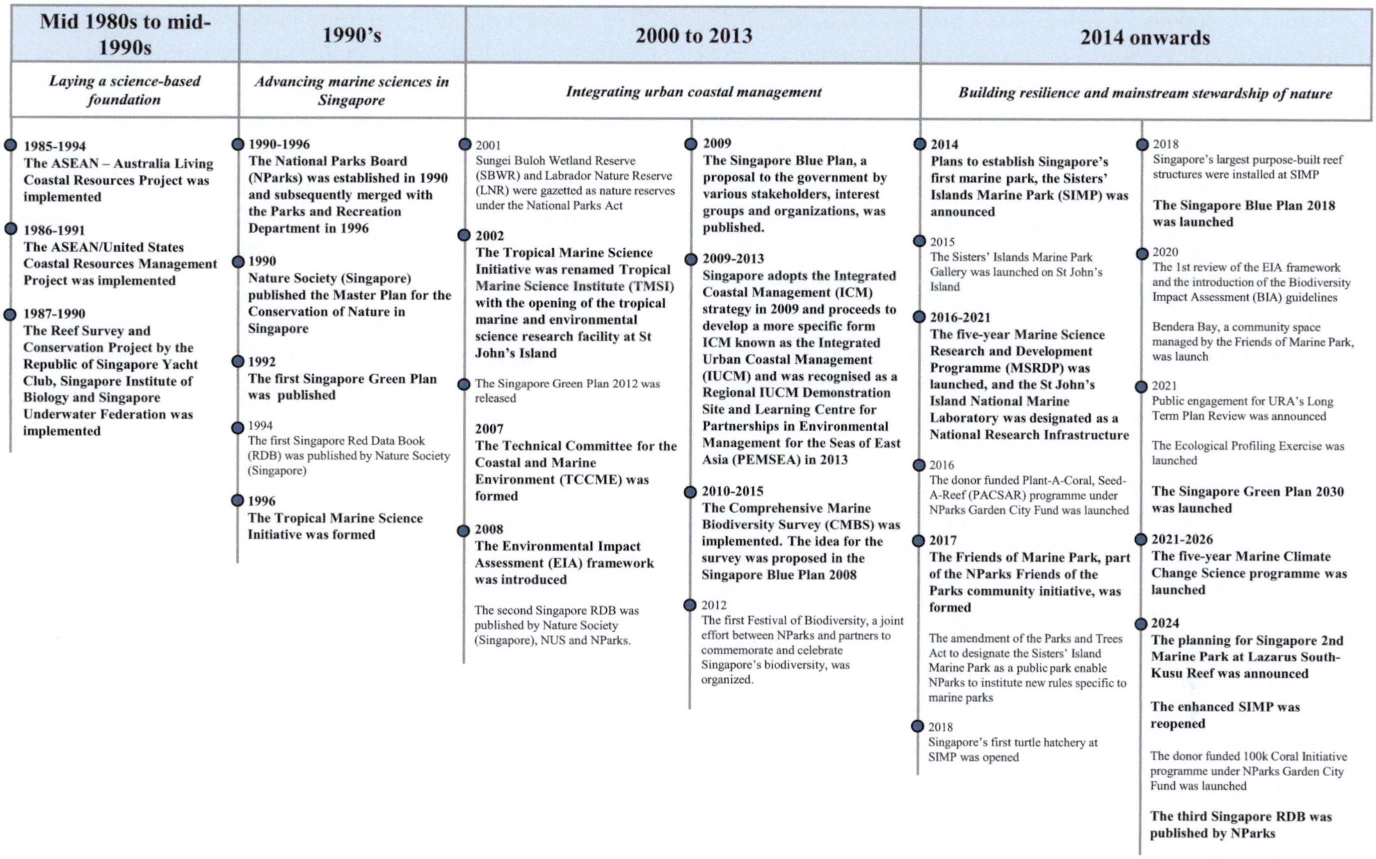

Fig. 13.1 Timeline of marine conservation and management in Singapore from the mid-1980s to 2025, grouped into four defining phases with key milestones, initiatives and events that were pivotal for each phase highlighted in bold

raise awareness on the importance of and the lack of baseline information on Singapore's marine environment.

Riding on the back of these two plans, academic leaders sought and secured support from the government to invest in tropical marine science, leading to the establishment of the Tropical Marine Science Initiative in 1996 and later as an Institute in 2002, with the opening of Singapore's only field marine research facility at St. John's Island (Tan et al. 2012). This set the nation on a path to becoming a leading focal point for marine science research and education in Southeast Asia that is bearing fruit today with a pipeline of young domain experts coming through the ranks to spearhead multi- and transdisciplinary research as Singapore confronts complex and multidimensional environmental challenges of tomorrow.

13.2.3 2000–2013: Integrating Urban Coastal Management

The third phase between 2000 and 2013 produced two game-changing initiatives in the marine conservation and management space. Firstly, the establishment of a multi-institutional Technical Committee on the Coastal and Marine Environment (TCCME) in 2007, co-chaired by NParks and National Environment Agency (NEA) and tasked to provide technical expertise to support Singapore's interest in local and international coastal and marine environment issues (Chou 2018; TCCME 2018), led to the adoption of the integrated coastal management (ICM) strategy in 2009 (Chua et al. 2006), the development of a more specific form of ICM known as IUCM in 2013, and the recognition of Singapore as a Regional IUCM Demonstration Site and Learning Centre (NParks 2018). Singapore's IUCM is dynamic and reiterative, based on establishing continuous baselines and feedback loops supported by broad stakeholder consensus to ensure agility to adapt to ongoing changes in the coastal and marine environment.

Secondly, the introduction of the EIA framework in 2008 allowed Singapore to adopt a systematic framework to assess and mitigate potential impacts on the environment by institutionalising requirements for development projects to undergo a thorough evaluation process that addresses the development's potential impact on traffic, public health, heritage and the environment (Chou 2008; URA 2025a). Although not codified into law, the EIA framework provides robust coverage by leveraging existing administrative processes, sector-specific laws and inter-agency coordination to achieve similar objectives as a legislated process (TCCME 2018). The EIA framework was reviewed in 2020 to improve rigour, strengthen enforcement of environmental measures imposed and improve transparency of environmental study findings (Tan 2020). Nature stakeholders and legal scholars however continue to call for the EIA process to be strengthened and codified into law (Chun 2004), although there is broad acknowledgment of the improvements that have been made in this space (Ananthanarayanan and Ang 2024; Jaafar et al. 2018).

On the civil society front, the Singapore Blue Plan, a proposal developed by marine biologists and marine stakeholders, was published in 2009 to share a collective vision to guide government and the public on conserving and rehabilitating Singapore's coastal and marine heritage (Lee and Hamid 2009). The government saw merit in the objective, science-based proposal and secured funding to kickstart a 5-year Comprehensive Marine Biodiversity Survey (CMBS), anchored on a public–private–people partnership that galvanised over 400 scientists and volunteers to assess and document the diversity of marine life in Singapore's waters. When it concluded in 2015, CMBS collected approximately 72,000 specimens, described over 37 species new to science and discovered more than 300 species new to Singapore (Tan et al. 2016b). Data from CMBS have since contributed to establishing marine biodiversity baselines, screening of development projects under the EIA framework and guiding long-term sea space planning.

Importantly, CMBS established a new and stronger working relationship between marine stakeholders and the government, shifting from a sectoral driven to a more integrated, collaborative and multi-stakeholder-driven model, which is a central tenet of the IUCM approach.

13.2.4 2014 Onwards: Building Resilience and Mainstream Stewardship of Nature

The fourth phase saw an explosion of marine-related activities, kicking off with one of the most significant moments for marine conservation in Singapore—the establishment of Singapore's first marine park, the Sisters' Islands Marine Park (SIMP) in 2014 (Cornelius 2016), hailed as a flagship moment for marine conservation, research and public outreach, and a big victory for the marine community (Jaafar et al. 2018). Leveraging current research findings, including the rich scientific information in the Singapore Blue Plan 2018 report prepared by marine biologists with inputs from marine stakeholders and members of the public (Jaafar et al. 2018), NParks undertook an ecological profiling exercise (EPE) for marine areas and identified the northern reefs of Kusu Island and southern Lazarus as important ecological connectors (Afiq-Rosli et al. 2021). The findings supported the case for designating the areas as Singapore's second marine park in 2024, a decade after the first was established (Begum and Leow 2024).

This phase also saw widespread acknowledgement that existing anthropogenic impacts on the environment will be exacerbated by the growing threats of climate change, and the need to undertake multi- and trans-disciplinary research to understand and mitigate against these threats and find practicable solutions. Two five-year S$25-million marine research programmes were initiated—the Marine Science and Research Development Programme (MSRDP) in 2016 (NUS 2016; Todd et al. 2021) and the ongoing Marine Climate Change Science Programme (MCCS) in 2021 (NParks 2021a; Tan 2021). These programmes helped fill critical knowledge gaps while training a pipeline of young marine scientists to support Singapore's future research landscape, strengthen the country's knowledge-based maritime economy and address pressing environmental and societal challenges relevant to tropical coastal cities worldwide (Todd et al. 2021).

Community stewardship in marine conservation was amplified in 2017 with the formation of the Friends of Marine Park (FMP) community, a collaborative, ground-led initiative that brings together diverse stakeholders from the research, education, recreation, fishing and island heritage communities to promote stewardship, conservation and responsible use of SIMP and all other marine spaces. FMP is viewed as a trusted partner in the marine space and has leaned in to support numerous conservation and management initiatives like providing feedback on development projects and inputs to the Marine Park management plan. The publication of the third edition of the Singapore Red Data Book in 2024 bears witness to the strengthening partnership between the government and civil society with 120 experts, including specialists from government agencies, academia and the public sector, contributing to the detailed conservation assessments (Davison et al. 2024).

Guiding Singapore into the next phase is the Singapore Green Plan 2030, a comprehensive national strategy that can fold the marine conservation agenda into its broader sustainability agenda through targeted initiatives across the key pillars of City in Nature, Resilient Future and Sustainable Living (Singapore Green Plan 2021).

13.3 IUCM: The Turning Point

The adoption of the Integrated Urban Coastal Management (IUCM) framework in 2013 was an important turning point that influenced marine conservation and management in Singapore by harmonising previous sectoral approaches under a single strategy. The coastal and marine landscape is continually changing, influenced by geopolitical, socio-economic and environmental changes. The inherent agility of the IUCM approach that promotes regular review and the adoption of adaptive management strategies will allow Singapore to meet the challenges of the day head-on. The streamlined and centralised structure of the Singapore government with clear lines of authority and few layers between decision-makers and implementers naturally reduces the challenges associated with implementing IUCM's adaptive management approach, compared to countries with multi-layered governance structures.

Singapore's IUCM framework leverages five key instruments—administrative processes, legal framework, coordinating mechanisms, supporting structures and public–private–people partnerships to manage Singapore's coastal and marine environment (TCCME 2018). Table 13.1 summarises these instruments and the tools and processes that support them. The IUCM approach has been implemented for a diverse range of projects and situations, from high-level master planning for nationally important initiatives like coastal protection, e.g. the Public Utilities Board's coastal protection strategy; to situational responses, e.g. oil spills; and planning for safeguarding marine spaces for conservation, e.g. establishing SIMP and planning for the second marine park at Lazarus South-Kusu Reef.

Although the 2013 IUCM publication did not include an indicator system to evaluate the progress of implementing such an approach in the context of Singapore's coastal environment and management, Chng et al. (2022) developed a set of 40 urban coastal zone environmental performance indicators and undertook a semi-quantitative analysis to determine whether they are applicable and beneficial for managing the coastal resources of Singapore. This study determined that Singapore met performance expectations in many core areas such as biodiversity protection, habitat management plan, water quality management and environmental disaster management, but could improve in key indicators such as impacts of land use on coastal ecosystems and solid waste accumulation on beaches. By undertaking such regular reviews, Singapore can continue to learn and adapt her management approaches to extend conservation efforts, increase resilience and expand whole-of-nation stewardship of the blue spaces.

13.4 Current and Future Directions

The next defining phase of Singapore's marine conservation and management journey will likely begin before the end of this decade and coincide with the start of the coastal protection journey. With acute space constraints, Singapore will need to make compromises and find opportunities to ensure that the marine environment can continue to thrive and be sustained. To do so, Singapore must be open to accepting evolving ecological baselines that may not have the same 'look and feel' as those of today but retain similar ecological function and provision of goods and services.

If Singapore continues to leverage IUCM to overcome unseen future challenges, there is every chance that the Singapore Green Plan 2030 'City in Nature' vision to transform the nation into a green, liveable and climate-resilient city can be achieved by expanding green and blue spaces, enhancing biodiversity, and integrating nature into urban environments through initiatives summarised in Table 13.2. The future of Singapore's marine conservation landscape will be defined by dynamic and adaptive management, scaling up of habitat restoration and applied research, expanding partnerships, and regional leadership by sharing best practices in IUCM and contributing to international marine conservation efforts.

13.5 Conclusion

Even though we live in a geopolitically and socio-economically uncertain world, there is certainty that current challenges, which the marine environment faces, will be exacerbated by the holy trinity of climate change impacts—sea-level rise, elevated sea-surface temperature and ocean acidification. As daunting as these challenges may be, there are strategies and solutions that can and are being imple-

Table 13.1 The five IUCM instruments and their supporting tools and processes

IUCM Instrument	Tools/Processes	Description
Administrative processes	Sea-space/marine spatial planning	Sea-space planning is coordinated under the broader planning umbrella of the Urban Redevelopment Authority (URA) to optimise the use of the sea space in a manner sensitive to the marine environment by co-locating uses, matching appropriate uses to suitable sites and harnessing technology (URA 2025b)
	EIA framework	The EIA framework is administered by URA, where proposed development projects near sensitive areas, such as Nature Reserves, Nature Areas, marine and coastal areas, other areas of significant biodiversity or with potential trans-boundary impact, are subject to greater scrutiny, with planning approvals only granted to proposals who meet requirements imposed by regulatory agencies (URA 2025a)
	Area management plans	Parks and reserves, including the Sisters' Islands Marine Park, managed by the National Parks Board (NParks), are guided by comprehensive internal frameworks and technical guidelines that serve the function of management plans. These frameworks ensure that each park is planned, developed, and managed according to its unique ecological, recreational, and community needs
	Risk management and emergency response plans	The Maritime and Port Authority of Singapore (MPA) is the lead agency for all maritime emergencies and has established the Marine Emergency Action Procedure (MEAP) to deal with marine emergencies, including oil spills, and adopts Integrated Urban Coastal Management (IUCM) to coordinate response across various agencies during such emergencies (Sea Alarm Foundation 2012). For NParks managed coastal and marine areas, NParks internal emergency response procedures are concurrently implemented
Legal frameworks	Environmental laws and policies related to IUCM	The legal framework that supports the implementation of IUCM for the conservation and management of the coastal and marine environment and covering the domains of biodiversity and protection of the coastal and marine environment, marine resource management, land- and sea-based pollution control, waste and hazard management, marine activities like merchant shipping, tourism, recreation and heritage conservation, draw on various legislation, regulations and rules under the purview of the following government agencies: NParks, MPA, National Environment Agency (NEA), URA, the Public Utilities Board (PUB), the Singapore Food Agency (SFA), the Singapore Land Authority (SLA), National Heritage Board (NHB), the Building and Construction Authority (BCA), the Singapore Civil Defence Force (SCDF) and the Sentosa Development Corporation (SDC). A comprehensive summary of key legislation is summarised in the 2018 IUCM publication (TCCME 2018) as well as the Singapore Blue Plan 2018 (Jaafar et al. 2018). The key amendments not captured in both documents relate to recent changes in various legislations following the post-2018 statutory board reorganisation where the non-food related functions of the previous Agri-Veterinary Authority (AVA) was carved out and moved to NParks, under a new cluster Animal and Veterinary Service (AVS), including all Acts and corresponding regulations therein. The food-related functions of AVA were reorganised under SFA and moved under the purview of Ministry of Sustainability and the Environment (MSE). Other key changes include the transfer of coastal protection functions that were previously under BCA to the water agency PUB, and the amendments to the Parks and Trees Regulations relating to the administration of the Sisters' Islands Marine Park, which came into effect in 2014
Coordinating mechanisms	Consultative planning	The URA is responsible for the long-term planning of Singapore's land and sea spaces, done in two regularly reviewed segments—the Long-Term Plan and the Master Plan—to cater for the evolving needs of the nation, allowing Singapore's overall development pace to be moderated based on prevailing circumstances. A key success factor in the planning process is public consultation and feedback, which gives citizens and stakeholders a say and a stake in Singapore's long-term development vision and goals. Since its inception and up to the Concept Plan 2011, the planning process focused on the terrestrial environment with limited focus on sea space planning. Fast forward 10 years and the evolving planning process is evident in the 2021 Long-Term Plan (LTP) 2021 review where blue spaces are included and captured under the theme 'Steward the Green and Blue', which recognises the important role that green and blue spaces play in enabling a liveable and resilient city, especially against the impact of urbanisation and climate change. Central to this theme are the three sub-themes of stewardship of our natural capital, multifunctional green and blue spaces, and climate resilience through nature (URA 2022)
	Coordinated policy planning	The Coastal and Marine Environment Policy Committee (CMEPC) is an inter-ministerial committee led by the Ministry of Sustainability and the Environment (MSE) which provides strategic policy direction for coastal and marine-related issues (TCCME 2018). The CMEPC was instrumental in supporting the development of the IUCM framework and Singapore's National Action Strategy for Marine Litter (NASML). The Inter-Ministerial Committee on Climate Change (IMCCC) (NCCS 2025) is a high-level committee that coordinates climate change policies to ensure that Singapore is prepared for the impacts of climate change, through four working groups. The Resilience Working Group (RWG) is tasked to study Singapore's vulnerability to the effects of climate change and develop long-term plans that ensure the nation's resilience to future environmental changes, including the marine environment

	Whole of Government (WoG) approach	Singapore's whole-of-government approach to IUCM combines inter-agency coordination, adaptive policies, and science-based strategies to balance development with environmental sustainability. Examples are the pilot centralisation and management of EIA consultancy services under NParks aimed at standardising processes, improving timelines, reducing cost, while safeguarding biodiversity conservation (NParks 2024) and the oil spill contingency plan under MPA's MEAP, which convenes an Emergency Operations Committee (EOC) that brings together government agencies and other stakeholders to coordinate all stages of the response (ITOPF 2020). The study on the response and management of the June 2024 incident between the Netherlands-flagged dredger Vox Maxima and the Singapore-flagged bunker tanker Marine Honour at Pasir Panjang Terminal provides a good overview of the WoG approach under the IUCM framework (PSD Challenge 2025)
Supporting structures	Inter-agency working groups	The multi-institutional Technical Committee on the Coastal and Marine Environment (TCCME), co-chaired by NParks and NEA, was formed in 2007 and tasked to provide technical expertise to support Singapore's interest in local and international coastal and marine environment issues (TCCME 2018). As coastal and marine issues started to gain traction, other topic-specific ad hoc or ephemeral inter-agency working groups were formed depending on the situation
	Data management	NParks Biodiversity and Environment Database System (BIOME) is Singapore's central repository for biodiversity and environmental data (BIOME 2025) and provides data to other data platforms like MPA's GeoSpace-Sea, a national marine spatial data platform that integrates Singapore's marine and coastal geospatial data and information from various sources into a single integrated marine spatial data platform (MPA 2019a)
	Survey and monitoring programmes	National survey and monitoring programmes for coral reefs and intertidal areas (including seagrass beds) are undertaken by a range of government agencies, academic institutions and interest groups, either independently or in collaboration with other entities (Jaafar et al. 2018) and centralised via NParks research permit system (NParks 2021b). Key marine programmes include long-term coral reef monitoring by NUS and NParks (Chan et al. 2023; Chan et al. 2024), and the volunteer programmes Seagrass Watch, Intertidal Watch, Biodiversity Beach Patrol (NParks 2025) and privately led WildSingapore intertidal surveys (Lim et al. 2020)
	Supporting plans and strategies	Government-led plans and strategies: 1. The National Biodiversity Strategy and Action Plan (NBSAP) (NParks 2018) provides a framework to guide biodiversity conservation efforts in Singapore and includes Singapore's 18 national targets for biodiversity conservation and related issues 2. The Nature Conservation Master Plan (NCMP) (NParks 2015) operationalises the biodiversity conservation efforts outlined in Singapore's NBSAP 3. The Marine Conservation Action Plan (MCAP), a blueprint that steers the direction of Singapore's conservation efforts (Tan 2015) to implement the various thrusts under the NCMP 4. The Singapore Green Plan 2030, a whole-of-nation movement to advance Singapore's national agenda on sustainable development and includes plans to conserve and extend Singapore's natural capital under the City in Nature pillar by expanding the Nature Park Network, intensifying nature in gardens and parks, restoring nature into the urban landscape and strengthening connectivity between green and blue spaces (Singapore Green Plan 2021) 5. Singapore's National Action Strategy for Marine Litter (NASML) (MSE 2022) outlines a comprehensive approach to tackle marine litter across the six priority areas of reducing land-based and sea-based sources of litter, promoting a circular economy, supporting research and development, strengthening outreach and engagement and fostering international collaboration 6. The Singapore Geospatial Masterplan 2024–2033 goes the next mile with the IUCM approach by integrating geospatial strategies across the land and sea continuum to protect Singapore's coastlines and environment and grow the maritime industry in a safe, sustainable and efficient way (Singapore Geospatial Masterplan 2024) 7. The Singapore Aquaculture Plan was developed to boost fish production while minimising impacts to the marine environment from marine-based aquaculture activities by adopting sustainable technologies, enhancing farm practices and strengthening regulations (SFA 2024) Civil society-led plans and strategies: The Singapore Blue Plan 2009 (Lee and Hamid 2009) and 2018 (Jaafar et al. 2018) are the only marine-focused action plans prepared by civil society outlining key recommendations for the conservation and management of the coastal and marine environment, and presented to the government

(continued)

Table 13.1 (continued)

IUCM Instrument	Tools/Processes	Description
	Research programmes	Prior to 2016, investments in marine science research, although extensive, were not coordinated, were sectoral and institutionally separated. However, with the push for multi- and trans-disciplinary applied research to tackle more complex environmental problems, larger multi-institutional research programmes were developed and funded under various National Research Foundation's (NRF) grant programmes. Key programmes since 2016 are: 1. The $25-million Marine Science Research and Development Programme (MSRDP) from 2016 to 2021 under the National University of Singapore (Todd et al. 2021) 2. The Natural Capital Singapore project from 2018 to 2021 under ETH Zurich (ETH Zurich 2018) 3. The ongoing $25-million Marine Climate Change Science (MCCS) Programme from 2021 to 2026 under NParks (NParks 2021a) 4. The ongoing $125-million Coastal Protection and Flood Management Research Programme (CFRP) under PUB (PUB 2025)
Partnerships	Public sector	As owners and custodians of Singapore's land and sea spaces, the government plays a central role in conserving and managing Singapore's marine environment. Although conservation comes under the purview of NParks, management of the coastal and marine environment draws on various legislation, regulations and rules under the purview of various government agencies (see row 5 under Legal frameworks). Since 2011, NParks has established the Communities in Nature initiative, a nationwide movement designed to unite people from all walks of life to conserve Singapore's natural heritage through citizen science programmes, habitat enhancement and nature education programmes and activities (NParks 2025). Other agencies like MPA work with marine stakeholders to undertake underwater clean-ups to reduce anthropogenic impacts and promote the importance of Singapore's marine biodiversity (MPA 2019b)
	Private sector	Singapore's private sector has played an increasingly significant role in marine conservation, primarily through donations and sponsorships to registered charities like the Garden City fund (GCF), institutes of higher learning (IHLs) like NUS, and non-governmental organisations (NGOs) like WWF to support habitat restoration, public education, research and sustainable business practices
	People sector	The people or community sector involved in marine conservation, environmental sustainability, outreach and education has increased exponentially in Singapore over the last two decades and has contributed to a wide range of activities from data collection and habitat restoration to public outreach and advocacy, working alongside government agencies, NGOs and research institutions (Jaafar et al. 2018). More recently, several peer-reviewed scientific papers that used data collected by citizen scientists or under citizen science programmes have been published, highlighting growing confidence in the quality of such data and their contribution to science and management (Oh et al. 2019; Lim et al. 2020; Loy et al. 2024)

Table 13.2 Initiatives and actions to achieve a thriving marine environment under the Singapore's Green Plan 2030 vision

Initiative/Action	Description	Singapore Green Plan Pillar
Restoration of marine/coastal habitats	Restore marine/coastal habitats, including mangroves and seagrasses	City in Nature
Nature-based solutions for coastal protection	Develop hybrid green-grey solutions for shoreline defence and biodiversity enhancement	Resilient future
Emissions reduction in maritime sector	Incentives for green shipping	Sustainable Living, Green Economy, Energy Reset
Research and monitoring	MCCS and other programmes to study, monitor and manage marine ecosystem health	City in Nature, Resilient Future
Community engagement	Education, public awareness and partnership programmes for marine stewardship	City in Nature, Resilient Future

mented today to hedge against the uncertainties of tomorrow. Singapore cannot take the foot off the research pedal and will continue adopting science- and evidence-based approaches in conservation and management decisions and policies while actively engaging and involving the community.

Acknowledgement The authors would like to acknowledge the many National Parks Board colleagues, past and present, and all marine stakeholders from government agencies, academia and nature communities for demonstrating tenacity in championing the marine agenda and commitment to building long-lasting collaborative relationships that helped craft today's vibrant and robust marine conservation and management landscape.

References

Afiq-Rosli L, Wainwright BJ, Gajanur AR, Lee AC, Ooi SK, Chou LM, Huang D (2021) Barriers and corridors of gene flow in an urbanized tropical reef system. Evol Appl 14:2502–2515. https://doi.org/10.1111/eva.13276

Ananthanarayanan S, Ang A (eds) (2024) Singapore terrestrial conservation plan 2024. ISBN: 978-981-18-8794-9

Begum S, Leow WX (2024) Singapore's proposed second marine park in Lazarus South, Kusu is home to rare rocky shore, coral habitats. In: Straits Times. https://www.straitstimes.com/singapore/s-pore-s-proposed-second-marine-park-in-lazarus-south-kusu-is-home-to-rare-rocky-shore-coral-habitats. Accessed 31 Mar 2025

BIOME (2025) National Parks Board Singapore. https://biome.nparks.gov.sg/. Accessed 31 Mar 2025

Chan YKS, Affendi YA, Ang PO, Baria-Rodriguez MV, Chen CA, Chui APY, Giyanto GM, Huang H, Kuo C-Y, Kim SW, Lam VYY, Lane DJW, Lian JS, Lin SMNN, Lunn Z, Nañola CL, Nguyen VL, Park HS, Suharsono, Sutthacheep M, Vo ST, Vibol O, Waheed Z, Yamano H, Yeemin T, Yong E, Kimura T, Tun K, Chou LM, Huang D (2023) Decadal stability in coral cover could mask hidden changes on reefs in the East Asian Seas. Commun Biol 6:630. https://doi.org/10.1038/s42003-023-05000-z

Chan YKS, Ng CSL, Tun KPP, Chou LM, Huang D (2024) Decadal decline of functional diversity despite increasing taxonomic and phylogenetic diversity of coral reefs under chronic urbanisation stress. Ecol Indic 164:112143. https://doi.org/10.1016/j.ecolind.2024.112143

Chia LS (1992) Singapore's urban coastal area: strategies for management. ICLARM Tech Rep 31

Chia LS, Chou LM (1991) Urban coastal area management: the experience of Singapore. In: Proceedings of the Singapore National Workshop on Urban Coastal Area Management, Republic of Singapore, 9–10 November 1989. ICLARM Conference Proceedings (25), p 128

Chng LC, Chou LM, Huang D (2022) Environmental performance indicators for the urban coastal environment of Singapore. Reg Stud Mar Sci 49:102101. https://doi.org/10.1016/j.rsma.2021.102101

Chou LM (1991) Contributions of the ASEAN-Australia coastal living resources project to coastal living resource assessments in Singapore. In: Alcala AC, Chou LM, Miclat R, Kastoro W, Fortes M, Gong WK, Sasekumar A, Bina R, Tridech S (eds) Proceedings of the Regional Symposium on Living Coastal Areas, 30 Jan–1 Feb 1989. UP Marine Science Institute, Manila, pp 569–572

Chou LM (2008) Nature and sustainability of the marine environment. In: Wong TC, Yuen B, Goldblum C (eds) Spatial planning for a sustainable Singapore. Springer, Dordrecht, pp 169–182

Chou LM (2018) Integrated urban coastal management: the Singapore model. In: Chua TE, Chou LM, Jacinto G, Ross SA, Bonga D (eds) Local contributions to global sustainable agenda: case studies in Integrated Coastal Management in the East Asian Seas Region. Partnerships in Environmental Management for the Seas of East Asia (PEMSEA) and Coastal Management Center (CMC), Quezon City, pp 447–452

Chou LM (2023) Sustaining marine biodiversity in Singapore's heavily urbanized coast. In: Ting D, O'Brien P (eds) Progress in sustainable development: sustainable engineering practices. Elsevier, Amsterdam, pp 265–282. https://doi.org/10.1016/B978-0-323-99207-7.00006-3

Chou LM, Goh PL (1998) Singapore coral reefs: balancing development and conservation. In: Morton B (ed) The marine biology of the South China Sea. Proceedings of the Third Conference on the Marine Biology of the South China Sea, pp 355–268

Chua TE, Bonga D, Bermas-Atrigenio N (2006) Dynamics of integrated coastal management: PEMSEA's experience. Coast Manage 34(3):303–322. https://doi.org/10.1080/08920750600686695

Chun J (2004) The case for legislation in Singapore to mandate environmental impact assessment. https://ssrn.com/abstract=4887727. Accessed 31 Mar 2025

Cornelius V (2016) Singapore Infopedia: Sisters' Islands https://www.nlb.gov.sg/main/article-detail?cmsuuid=e7644831-8051-4a14-8b27-848a1bf31989#:~:text=The%20Sisters'%20Islands%20Marine%20Park,John's%20Island%20and%20Pulau%20Tekukor. Accessed 31 Mar 2025

Davison GWH, Gan JWM, Huang D, Hwang W.S, Lum SKY, Yeo D C J (eds) (2024) The Singapore Red Data book: Red Lists of Singapore biodiversity (3rd). National Parks Board, Singapore

ETH Zurich (2018) Natural capital Singapore. https://sec.ethz.ch/research/Natural-Capital.html. Accessed 31 Mar 2025

Freiss DA, Yando ES, Wong LW, Bhatia N (2020) Indicators of scientific value: an under-recognised ecosystem service of coastal and marine habitats. Ecol Indic 113:106255. https://doi.org/10.1016/j.ecolind.2020.106255

Friess DA (2017) Singapore as a long-term case study for tropical urban ecosystem services. Urban Ecosyst 20:277–291. https://doi.org/10.1007/s11252-016-0592-7

Hilton MJ, Manning SS (1995) Conversion of coastal habitats in Singapore: indications of unsustainable development. Environ Conserv 22(4):307–322. https://doi.org/10.1017/S0376892900034883

Huang D, Tun KPP, Chou LM, Todd PA (2009) An inventory of zooxanthellate scleractinian corals in Singapore, including 33 new records. Raffles Bull Zool Suppl 22:69–80

Ip YCA, Chang JJM, Oh RM, Quek ZBR, Chan YKS, Bauman AG, Huang D (2023) Seq' and ARMS shall find: DNA (meta)barcoding of Autonomous Reef Monitoring Structures across the tree of life uncovers hidden cryptobiome of tropical urban coral reefs. Mol Ecol 32:6223–6242. https://doi.org/10.1111/mec.16568

ITOPF (2020) Country and territory profile: Singapore. https://www.itopf.org/fileadmin/uploads/itopf/data/Documents/Country_Profiles/singapore.pdf. Accessed 31 Mar 2025

Jaafar Z, Huang D, Tanzil JTI, Ow YX, Yap NWL (2018) The Singapore Blue Plan 2018. Singapore Institute of Biology, Singapore, 219 pp

Lai S, Loke LHL, Hilton MJ, Bouma TJ, Todd PA (2015) The effects of urbanization on coastal habitats and the potential for ecological engineering: a Singapore case study. Ocean Coast Manage 103:78–85. https://doi.org/10.1016/j.ocecoaman.2014.11.006

Lee F, Hamid F (2009) Singapore Blue Plan 2009, 37 pp

Lim LJW, Loh JBY, Lim AJS, Tan BYX, Ip YCA, Neo ML, Tan R, Huang D (2020) Diversity and distribution of intertidal marine species in Singapore. Raffles Bull Zool 68:396–403. https://doi.org/10.26107/rbz-2020-0056

Loy RHY, Cheo PR, Lai S, Leong B, Tun K (2024) Citizen science monitoring uncovers resilience of intertidal assemblages in a tropical urban environment. Bull Mar Sci 101(1):391–408. https://doi.org/10.5343/bms.2023.0104

Malayan Nature Society (Singapore Branch) (1990) Masterplan for the conservation of nature in Singapore. http://nss.preview.orion5.iwi.com.sg/report/3634b946-cMaster%20plan%20for%20the%20conservation%20of%20nature%20in%20Singapore.pdf. Accessed 31 Mar 2025

MND (Ministry of National Development) (2024) Written answer by Ministry of National Development on measures in place to protect marine ecosystems from rising temperatures and effects of El Nino. https://www.mnd.gov.sg/newsroom/parliament-matters/q-as/view/written-answer-by-minitry-of-national-development-on-measures-in-place-to-protect-marine-ecosystems-from-rising-temperatures-and-effects-of-el-nino. Accessed 31 Mar 2025

MPA (Maritime and Port Authority of Singapore) (2019a) MPA spearheads GeoSpace-Sea initiative to harness and integrate Singapore's marine and coastal data. https://www.mpa.gov.sg/media-centre/details/mpa-spearheads-geospace-sea-initiative-to-harness-and-integrate-singapore-s-marine-and-coastal-data. Accessed 31 Mar 2025

MPA (Maritime and Port Authority of Singapore) (2019b) MPA and Our Singapore Reefs team up for marine conservation. https://www.mpa.gov.sg/media-centre/details/mpa-and-our-singapore-reefs-team-up-for-marine-conservation. Accessed 31 Mar 2025

MSE (Ministry of Sustainability and the Environment) (2022) National action strategy: addressing marine litter in Singapore. https://isomer-user-content.by.gov.sg/23/063f4f3e-aa82-4ef2-a359-76fb9bd1ad40/nasml.pdf. Accessed 31 Mar 2025

NCCS (National Climate Change Secretariat) (2025) Inter-ministerial committee on climate change. https://www.nccs.gov.sg/who-we-are/inter-ministerial-committee-on-climate-change/. Accessed 31 Mar 2025

NParks (National Parks Board) (2015) The Nature Conservation Master Plan. https://www.nparks.gov.sg/nature/nature-conservation-masterplan. Accessed 31 Mar 2025

NParks (National Parks Board) (2018) Singapore's integrated urban coastal management. https://www.nparks.gov.sg/nature/integrated-urban-coastal-management. Accessed 31 Mar 2025

NParks (National Parks Board) (2021a) Marine Climate Change Science (MCCS) Programme. https://www.nparks.gov.sg/Cuge/Programmes-Schemes/Research%20Programmes/MCCS%20Programme. Accessed 31 Mar 2025

NParks (National Parks Board) (2021b) NParks research permit application guidelines. https://www.nparks.gov.sg/-/media/nparks-real-content/biodiversity/resources-research-permit/research-permit-application-user-guide_sept2021.pdf. Accessed 31 Mar 2025

NParks (National Parks Board) (2024) Greening Singapore together: Annual report 2023–2024. https://www.nparks.gov.sg/portals/annualreport/sustainability-report.html. Accessed 31 Mar 2025

NParks (National Parks Board) (2025) Volunteer. https://www.nparks.gov.sg/get-involved/volunteer/volunteer-with-us. Accessed 31 Mar 2025

NUS (National University of Singapore) (2016) $25m boost for marine science research. https://news.nus.edu.sg/25m-boost-for-marine-science-research/#:~:text=A%20new%20$25%20million%20national,Singapore's%20marine%20science%20capabilities%20development. Accessed 31 Mar 2025

Oh RM, Neo ML, Yap NWL, Jain SS, Tan R, Chen CA, Huang D (2019) Citizen science meets integrated taxonomy to uncover the diversity and distribution of Corallimorpharia in Singapore. Raffles Bull Zool 67:306–321. https://doi.org/10.26107/RBZ-2019-0022

PSD Challenge (2025) Dealing with 400 tonnes of oil: Emergency preparedness in action. https://psdchallenge.psd.gov.sg/ideas/feature/dealing-with-400-tonnes-of-oil-emergency-preparedness-in-action. Accessed 31 Mar 2025

PUB (Public Utilities Board) (2025) Coastal protection and flood management research programme. https://www.pub.gov.sg/Public/KeyInitiatives/Flood-Resilience/Research. Accessed 31 Mar 2025

RSYC (Republic of Singapore Yacht Club), SIBiol (Singapore Institute of Biology), SUF (Singapore Underwater Federation) (1991) Singapore reef conservation and survey project-a proposal for the conservation of coral reefs in Singapore. https://coralreef.nus.edu.sg/publications/RSYC1991.pdf. Accessed 31 Mar 2025

Sea Alarm Foundation (2012) Country wildlife response profiles: Singapore. https://www.sea-alarm.org/web/wp-content/uploads/2012/05/singapore.pdf. Accessed 31 Mar 2025

SFA (Singapore Food Agency) (2024) Singapore aquaculture plan. https://www.straitstimes.com/singapore/new-plan-unveiled-to-increase-local-aquaculture-production. Accessed 31 Mar 2025

Shen Y, Ma K, Yue GH (2021) Status, challenges and trends of aquaculture in Singapore. Aquaculture 533:736210. https://doi.org/10.1016/j.aquaculture.2020.736210

Singapore Geospatial Master Plan (2024) Singapore Geospatial Masterplan: 2024–2033. https://file.go.gov.sg/singapore-geospatial-master-plan-2024.pdf. Accessed 31 Mar 2025

Singapore Green Plan (2021) The Singapore Green Plan 2030. https://www.greenplan.gov.sg/. Accessed 31 Mar 2025

SingStat (Singapore Department of Statistics) (2024) Population. https://www.singstat.gov.sg/publications/%20reference/ebook/population/population. Accessed 31 Mar 2025

Tan A (2015) First marine action plan launched to protect Singapore's underwater biodiversity. In: Straits Times. https://www.straitstimes.com/singapore/environment/first-marine-action-plan-launched-to-protect-singapores-underwater. Accessed 31 Mar 2025

Tan A (2020) Development works in Singapore to be more sensitive to wildlife under changes to EIA framework. In: Straits Times. https://www.straitstimes.com/singapore/environment/development-works-in-singapore-to-be-more-sensitive-to-wildlife-under-changes. Accessed 31 Mar 2025

Tan A (2021) New marine climate science programme in Singapore to call for proposals in November. In: Straits Times. https://www.straitstimes.com/singapore/environment/new-marine-climate-science-programme-in-spore-to-call-for-proposals-in. Accessed 31 Mar 2025

Tan KS, Chan ES, Teo S, Chou LM, Lam TJ, Taylor E (2012) A short history of TMSI@St John's Island. In: Contributions to marine science, pp vii–ix

Tan KS, Acerbi E, Lauro FM (2016a) Marine habitats and biodiversity of Singapore's coastal waters: a review. Reg Stud Mar Sci 8:340–353. https://doi.org/10.1016/j.rsma.2016.01.008

Tan KS, Koh KS, Ng JY, Goh L (2016b) The comprehensive marine biodiversity survey Singapore Strait: international workshop 2013. Raffles Bull Zool Suppl 34:1–7

Tan YHJ, Tham JKQ, Paul A, Rana U, Ang HP, Nguyen NTH, Yee ATK, Leong BPI, Drummond S, Tun KPP (2023) Remote sensing mapping of the regeneration of coastal natural habitats in Singapore: implications for marine conservation in tropical cities. Singap J Trop Geogr 44:130–148. https://doi.org/10.1111/sjtg.12454

Tan YJH, Lai S, Tun KPP (2025) "Accidental" artificial reefs—lessons from Singapore for coral and seagrass restoration along urban shorelines. Bull Mar Sci 101(1):197–209. https://doi.org/10.5343/bms.2023.0134

Tay JYL, Wong SKM, Chou LM, Todd PA (2018) Land reclamation and the consequent loss of marine habitats around the Ayer Islands, Singapore. Nat Singap 11:1–5

TCCME (Technical Committee for the Coastal and Marine Environment) (2018) Singapore's Integrated Urban Coastal Management (revised). https://www.nparks.gov.sg/docs/default-source/resources/2018/singapore-integrated-urban-coastal-management-framework-nov2018.pdf. Accessed 31 Mar 2025

Todd PA, Tanzil JTI, Yap NWL, Martin P (2021) Advancing marine science in Singapore. Marine Science Research and Development Programme, Singapore, 205 pp

URA (Urban Redevelopment Authority) (2022) Space for our dreams: Long-term plan review 2021. https://www.ura.gov.sg/-/media/Corporate/Planning/LTPR21/LTPR_publication.pdf. Accessed 31 Mar 2025

URA (Urban Redevelopment Authority) (2025a) Environmental Impact Assessment (EIA). https://www.ura.gov.sg/Corporate/Planning/Our-Planning-Process/Bringing-plans-to-Reality/Environmental-Impact-Assessment. Accessed 31 Mar 2025

URA (Urban Redevelopment Authority) (2025b) Sustaining land and space options. https://www.ura.gov.sg/Corporate/Planning/Long-Term-Plan-Review/Space-for-Our-Dreams-Exhibition/Sustain/Sustaining-Land-and-Space-Options. Accessed 31 Mar 2025

Wong TC, Yuen B, Goldblum C (eds) (2008) Spatial planning for a sustainable Singapore. Springer Netherlands. https://doi.org/10.1007/978-1-4020-6542-2

Yee ATK, Corlett RT, Liew SC, Tan HTW (2011) The vegetation of Singapore—an updated map. Gardens' Bull Singapore 63:205–212

Yeo BL (1994) The Singapore green plan. In: AMIC-ECOS seminar on urbanization, Paper no 11. Sustainable Development and the Media, Singapore

MIX
Papier aus verantwortungsvollen Quellen
Paper from responsible sources
FSC® C105338

If you have any concerns about our products,
you can contact us on
ProductSafety@springernature.com

In case Publisher is established outside the EU,
the EU authorized representative is:
Springer Nature Customer Service Center GmbH
Europaplatz 3, 69115 Heidelberg, Germany

Printed by Libri Plureos GmbH
in Hamburg, Germany